CHEMISTRY UNDER EXTREME OR NON-CLASSICAL CONDITIONS

Chemistry Under Extreme or Non-Classical Conditions

Edited by

Rudi van Eldik and Colin D. Hubbard

A JOHN WILEY & SONS, INC. and SPEKTRUM AKADEMISCHER VERLAG CO-PUBLICATION

 New York · Chichester · Brisbane · Toronto · Singapore · Weinheim

 Heidelberg · Berlin · Oxford

Copyright © 1997 jointly by John Wiley & Sons, Inc. and Spektrum Akademischer Verlag

John Wiley & Sons, Inc
605 Third Avenue
New York, NY 10158-0012
USA

Spektrum Akademischer Verlag
Vangerowstrasse 20
D-69115 Heidelberg
Germany

Telephone: (212) 850-6000

Telephone: 49 6221 91260

Address all inquiries to John Wiley & Sons, Inc.

All rights reserved. This book protected by copyright. No part of it, except for brief excerpts for review, may be reproduced, stored in a retrieval system, or transmitted in any form or by any means, electronic, mechanical, photocopying, recording, or otherwise, without permission from the publisher. Requests for permission of further information should be addressed to the Permissions Department, John Wiley & Sons, Inc., 605 Third Avenue, New York, NY 10158-0012.

While the authors, editors and publisher believe that drug selection and dosage and the specification and usage of equipment and devices, as set forth in this book, are in accord with current recommendations and practice at the time of publication, they accept no legal responsibility for any errors or omissions, and make no warranty, expressed or implied, with respect to material contained herein. In view of ongoing research, equipment modifications, changes in governmental regulations and the constant flow of information relating to drug therapy, drug reactions, and the use of equipment and devices, the reader is urged to review and evaluate the information provided in the package insert or instructions for each drug, piece of equipment, or device for, among other things, any changes in instructions or indication of dosage or usage and for added warnings and precautions.

Library of Congress Cataloging-in-Publication Data

Eldik, Rudi van.
    Chemistry under extreme or non-classical conditions / Rudi van Eldik, Colin D. Hubbard.
       p. cm.
    Includes index.
    ISBN 0-471-16561-1 (cloth : alk. paper)
    1. High pressure chemistry. 2. High temperature chemistry. 3. Sonochemistry. I. Hubbard, Colin D. II. Title.
    QD538.E43 1996
    541.3—dc20   96-27524
                                                                         CIP

Die Deutsche Bibliothek – CIP-Einheitsaufnahme

Chemistry under extreme or non-classical conditions / ed. by Rudi van Eldik/Colin D. Hubbard. – New York ; Chichester ; Brisbane ; Toronto ; Singapore ; Weinheim : Wiley ; Heidelberg ; Berlin ; Oxford : Spektrum, Akad. Verl., 1996
  ISBN 0-471-16561-1
NE: Eldik, Rudi van [Hrsg.]

The text of this book is printed on acid-free paper.

Printed in the United States of America
10 9 8 7 6 5 4 3 2 1

TABLE OF CONTENTS

	Editors' Preface	vii
	European Cooperation in the Field of Scientific and Technical Research (COST)	xi
1	**Chemical Synthesis Using High Temperature Species** J. J. Schneider	1
2	**Effect of Pressure on Inorganic Reactions** C. D. Hubbard and R. van Eldik	53
3	**Effect of Pressure on Organic Reactions** F.-G. Klärner, M. K. Diedrich and A. E. Wigger	103
4	**Organic Synthesis at High Pressure** J. Jurczak and D. T. Gryko	163
5	**Inorganic and Related Chemical Reactions in Supercritical Fluids** M. Poliakoff, M. W. George and S. M. Howdle	189
6	**Organic Chemistry in Supercritical Fluids** E. Dinjus, R. Fornika and M. Scholz	219
7	**Industrial and Environmental Applications of Supercritical Fluids** H. Schmieder, N. Dahmen, J. Schön and G. Wiegand	273
8	**Ultrasound as a New Tool for Synthetic Chemists** T. J. Mason ad J. L. Luche	317
9	**Applications of High Intensity Ultrasound in Polymer Chemistry** G. J. Price	381

10	**Chemistry Under Extreme Conditions in Water Induced Electrohydraulic Cavitation and Pulsed-Plasma Discharges** *M. R. Hoffmann, L. Hua, R. Höchemer, D. Willberg, P. Lang and A. Kratel*	**429**
11	**Microwave Dielectric Heating Effects in Chemical Synthesis** *D. M. P. Mingos and A. G. Whittaker*	**479**
12	**Biomolecules Under Extreme Conditions** *K. Heremans*	**515**
	Index	**547**

EDITORS' PREFACE

Even a casual inspection of current scientific periodicals shows there are regularly many innovations in chemical syntheses, of both organic and inorganic substances, and many creative developments in instrumental methods in the fields of spectroscopy, analytical procedures and structural determination. Specialized chemistry journals also cover these advances and in addition report the latest predictions of theoretical calculations. In the past ten to fifteen years the literature has also borne witness to a significant increase in the number of reports making use of what may be termed non-traditional or non-classical experimental approaches. In other studies, ambient or close-to-ambient conditions have been extended by orders of magnitude providing advantages for synthesis, reaction mechanism insight and in some cases novel chemistry not accessible under ambient conditions. These special conditions may be termed extreme.

This literature upsurge and interest together with two COST (European Cooperation in the Field of Scientific and Technical Research) D6 Workshops in Lausanne in 1993 and Lahnstein in 1995 gave birth to the idea of preparing a monograph on the topic of chemistry under extreme or non-classical conditions. At these meetings recent advances in various aspects covered in this monograph were reported by experts participating in COST D6 projects. Many leading investigators in these fields are authors of contributions to this monograph. The accompanying preface by Dr. Bernd Reichert of the COST Secretariat, European Commission, provides the background and general procedure for COST Actions. Further information in general, and upon current and potential future COST programmes may be obtained from Brussels using the address given.

Which methods and types of chemistry may be included within the categories of extreme or non-classical conditions, clearly could vary depending on individual perspective and semantic distinction. Indeed no attempt will be made to circumscribe those areas of chemistry which fit these categories, or define the terms of the title explicitly. Clearly in a monograph which can also serve as a whole, or in a modular sense, as a textbook for research students, only a selection of topics can be presented; a brief explanation by way of a rationale is is given below, but justification of inclusion is self-evident.

What has caused this significant growth in interest and driven many investigators to intensify their efforts in these fields? A combination of factors has contributed. Increasing environmental and public health awareness and more regulations governing use and disposal of chemicals have stimulated greater efforts toward developing "cleaner" technologies in laboratory chemical synthesis and in industrial scale production (food, pharmaceutical and agricultural industries), and methods of rendering hazardous and toxic waste materials harmless. These latter issues are being addressed in pertinent ways by microwave chemistry and sonochemistry techniques, supercritical fluid methods, and plasma applications. Understanding certain aspects of marine life and potential use in food technology are two factors leading to a widening scope for application of high pressures to biomolecules. Synthetic advantages, facile production of polymeric species, mechanistic insights not obtainable from applying other experimental variables, are all responsible for spurring the recent growth of use of high pressures in organic, inorganic and polymer chemistry. However, high pressure methods are hardly new in mechanistic and synthetic organic chemistry or in biochemistry, and uses of sound, or microwaves or supercritical fluids in chemistry applications were first reported many years ago. The emphasis here is on current trends, recent developments and future prospects.

The approach adopted in each chapter is to introduce the subject by describing the basic principles and indicating the present status, history and theoretical background, and to follow this with a description of instrumentation, methodology, and then examples, and where appropriate, industrial applications are presented. In other words the monograph is not in the mode of a collection of literature reviews *per se* of the topics, although authors have given a comprehensive set of important literature sources, *viz* texts, research publications and technical reports. Rather each topic is treated such that it can form the basis of an introduction to a student or other newcomer to the subject, but proceeds to bring the reader up to date by describing current research and development activities. Due to the obvious limitations of length, a totally comprehensive literature for each chapter is not possible; however, key references up to the middle of 1995 are included, and in some cases notes to a selection of publications up to and including early 1996 have been added.

Chapter 1 covers the field of chemistry whereby species not realizable under ambient conditions are generated at very high temperatures; subsequently these species can be subject to reaction at low temperatures to produce novel compounds and materials. The three following chapters deal respectively with high pressure inorganic chemistry, both synthetic aspects and solution reactions from a mechanistic point of view, organic chemistry where the pressure variable beside affecting synthetic efficiency can induce significant stereoselectivity, and polymer chemistry where pressure can be employed to influence the production of desired polymeric products. Chemical reactions in supercritical fluids (SCF), or exploitation of reactions in SCF for industrial processes and cleaner technology are described in Chapters 5 (Inorganic Chemistry), 6 (Organic Chemistry) and 7 (Applications of SCF). Study of the subjection of reaction solutions to sound radiation, the subsequent events and product characterization are covered in the next two chapters. Use of powerful electrical energy discharges to break chemical bonds not broken by other means is described both from a theoretical standpoint and from a practical perspective in Chapter 10. Unlike the situation pertaining to sound irradiation where the exact consequences of the effect on the medium of the radiation, and cavitation are not fully understood, there is little if any evidence to date to suggest that there is any special effect brought about by microwave radiation. The latter is the topic of Chapter 11. The dielectric heating achieved by incident microwave energy enables to advantage, different media and different reaction times to be used beyond those applicable to standard thermal heating methods. Consideration and consequences of application of temperatures and pressures well removed from those normally accepted as ambient conditions to biomolecules is the concluding topic.

The initiative to prepare the monograph was launched in May 1995. Positive responses from the contributing authors and their cooperation enabled the edited version to be delivered to the publisher in November 1995. The project would not have been possible at all without the enthusiastic support of the authors; we are indebted to them for their willingness to participate, and to do so very efficiently. We believe that our objective which is to bring knowledge and increase awareness of the impact and challenge of these non-traditional methods to a broad range of chemistry students, chemists and other scientists who have had their interest and curiosity piqued by this monograph, has been realized.

Efficient and accurate conversion of editors' notations to diskette form on a variety of word processing systems has been performed by Dr. Hans-Jörg Kremitzl, Institute for Inorganic Chemistry, University of Erlangen-Nürnberg. His knowledge and skills were of inestimable value during preparation of the edited draft.

Finally we are grateful to Dr. Bernd Reichert of the COST Secretariat for providing an introductory preface regarding COST programmes.

Erlangen, Germany RUDI VAN ELDIK
February 1996 COLIN D. HUBBARD

EUROPEAN COOPERATION IN THE FIELD OF SCIENTIFIC AND TECHNICAL RESEARCH (COST)

With its emphasis on pre-competitive research, national government financing, open participation and the initiating role of scientists, COST plays its part in a coherent structure for European research, complementing the EU Framework programmes and EUREKA.

COST cooperation takes the form of concerted Actions – essentially the coordination of national research projects. These focus on specific themes which are targeted by participating countries according to their research priorities. This coordination avoids duplication and helps build larger, more effective scientific communities in Europe. Furthermore association with COST can help projects access additional funding.

Any of COST's 25 member countries can propose concerted Actions, while participation on an Action by Action basis is open to all countries, including non-members. The research is funded nationally with the European Commission covering elements of the coordination expenses including the COST Sectetariat, workshops, evaluations, and also the mission expenses of EEA national delegates attending the Management Committee meetings.

The basis of a COST Action is the "Memorandum of Understanding" (MoU). Signed by each participating country, it lays out the Action's terms and objectives and ensures compliance with sovereignty and intellectual property rights. The document's non-legally binding status lightens the COST administrative load.

It is up to individual research institutes and universities to identify topics where mutual benefit can be drawn from collaboration. The COST framework facilitates this process. Its approach of encouraging individual investigators to initiate projects has helped to attract over 5000 scientists to participate in well over 120 Actions in fields ranging from Chemistry and Meteorology to Telecommunications and the Social Sciences.

COST ACTION D6

Chemical Processes and Reactions under Extreme or Non-Classic Conditions

The MoU of this action has been signed by 17 countries. The research is taking place within eight working groups (networks) covering the areas of chemistry at high pressure, in supercritical fluids and sonochemistry, as well as chemistry at high temperatures. The broad topics of the working groups are "Chemical Processes in Liquefied Gases and Supercritical Fluids", "Studies of Homoproteins under Extreme Conditions of Temperature and Pressure", "High Pressure Techniques directed Toward Synthesis and Discovery of New Molecules" and "Sonochemistry". Details of these projects and other aspects of COST programmes can be obtained from the COST Secretariat.

Dr. Bernd Reichert, COST Secretariat, European Commission
C.E.C., DG XII B1, B-68 5/38,
200, Rue de la Loi, B-1049, Brussels, Belgium.
Tel: 32-2-295 4517
Fax: 32-2-296 4289
E-mail: M2181@eurokom.ie

CHEMISTRY UNDER EXTREME
OR NON-CLASSICAL CONDITIONS

1

CHEMICAL SYNTHESIS USING HIGH TEMPERATURE SPECIES

Jörg J. Schneider

Institut für Anorganische Chemie, Universität/GH-Essen, Universitätsstraße 5–7, 45117 Essen, Germany

1.1 INTRODUCTION AND HISTORICAL BACKGROUND
1.2 BASIC PRINCIPLES FOR THE GENERATION OF HIGH TEMPERATURE (HT) SPECIES ON A PREPARATIVE SCALE
    1.2.1 Thermodynamic and Kinetic Viewpoints
    1.2.2 Formation of HT Species
1.3 EXPERIMENTAL TECHNIQUES FOR THE GENERATION AND REACTIONS OF HT SPECIES (ATOMS AND MOLECULES)
    1.3.1 General Techniques for the Generation of HT Species
    1.3.2 Static Reactor Systems
    1.3.3 Rotary Reactor Systems
    1.3.4 Special Developments of Reactor Designs for the Generation of HT Species
1.4 APPLICATIONS OF HT SPECIES IN CHEMISTRY
    1.4.1 Syntheses with Main Group Elements and Molecules
    1.4.2 Syntheses with Transition Elements and Molecules
    1.4.3 Syntheses Using Lanthanide Elements
    1.4.4 HT Species of the Transition Elements and Lanthanide Elements as Sources of Metal Powders, Metal Colloids, Metal Containing Polymers and Composites
1.5 REFERENCES

1.1 INTRODUCTION AND HISTORICAL BACKGROUND

Nearly all elements of the periodic table are available for high temperature (ht) synthesis in chemistry. For some of them there is no extraordinary need to use special ht techniques since they serve the chemist well with their high reactivity even below room temperature. But for most of them, especially the transition elements, the bulk state is by far too unreactive to allow an extensive chemistry of its own. Chemistry under extreme conditions using high temperatures offers the possibility to use these elements for synthesis in one of their highest reactive forms, the atomic state. In addition, high temperatures often allow the synthesis of new bi-element and tri-element molecules as well as highly reactive subvalent (coordination deficient) molecular particles. These provide an individual chemistry of their own, not available when ht methods for synthesis are not considered. Despite the possibilities which high temperatures in synthesis may offer, many experimental chemists restrict themselves to more conventional approaches and are not aware of the possibilities which ht methods in chemistry may offer them.

This chapter will deal with chemistry using extreme temperature and also energy, since formation of the atoms for example from the bulk metals needs sublimation energies in the range from 60 to 900 kJ mol^{-1} [1]. Similar inputs of energies are needed for high melting solids to produce coordination deficient molecules and to study their chemistry. However, most of the reactions of the ht species presented here are within the low temperature regime, down below $-100\,°C$.

The main emphasis of this contribution will be on the macroscale synthesis of new compounds and materials (a few hundred mg up to gram amounts) using ht synthesis techniques, for which no classical synthesis counterpart exists, and for which synthesis with ht species provides a valuable access. It is not intended to present a comprehensive review of the field, instead the topic will be highlighted in order to increase awareness of the impact of these non traditional methods and to stimulate the readers interest in the chemistry of ht species in general. The authors views will be given with an emphasis on preparative aspects of the chemistry of atoms and molecules. With respect to the high temperature species discussed, the chapter is organized as follows: chemistry of the main group and transition elements and particles as well as of subvalent molecules is described, however, species such as hydrogen, oxygen, halogen atoms, organic radicals and carbenes are not included in this article. Traditional areas such as solid state chemistry or CVD-techniques are not covered. For these topics the reader is referred to recent textbooks and monographs [2].

Ht particles derived from atoms were introduced into preparative chemistry in the mid 60s as reported by Skell and coworkers who found that carbon atoms when trapped in reactive matrices in $N_2(l)$ can form organic compounds in gram quantities [3]. As an extension of this application, in the following years, Timms introduced the usage of free transition metals into preparative chemistry [4], although reports of the use of metal vapors for the preparation of colloids in nonaqueous media date back to the year 1927 [5]. However, Timms' expansion of the method was the starting point for significant development of ht chemistry in the 70s and 80s mainly in the field of organotransition metal chemistry and in the field of ht chemistry of subvalent particles. During this time the molecular chemistry of such highly reactive particles was the focus of investigations. During the late 80s and early 90s the interest of the research community also turned to new topics. In addition to the use of free atoms and reactive molecules as synthetic building blocks in complex and organometallic chemistry, research on naked and ligated stabilized clusters and particles began to blossom and this currently gives ht species a new dimension for their use in preparative chemistry as well as in materials science [6].

1.2 BASIC PRINCIPLES FOR THE GENERATION OF HIGH TEMPERATURE (HT) SPECIES ON A PREPARATIVE SCALE

1.2.1 Thermodynamic and Kinetic Viewpoints

Due to the fact that most reactions of ht species with reactants considered in this chapter are at low temperature and reactions with other molecules usually occur during the heating period, two types of processes might occur.

1) The ht species agglomerate without reacting with a reactant molecule
2) The ht species react with reactant molecules

Kinetic control can give rise to one or another of these possibilities. A high excess of reactant makes a desired reaction of ht species with a ligand competitive with its agglomeration reaction which usually needs little or no activation energy compared to the reaction with ligand molecules (→ 1)). An equimolar mixture of ligand and ht species may yield higher nuclearity products which means that several atoms or molecules of the individual ht species are incorporated in the product molecules (→ 2)).

Thermodynamically, the energy difference between ht species and their room temperature counterparts is very high. The difference in energy for some metal atoms and solid bulk metal demonstrates this nicely (see Table 1.1) [7].

Reactions in the case of metal atoms with ligands are thermodynamically favored since the energy change in such a reaction is apparently the same as the bond energy change for the generation of vapors of all metals. This holds as long as the product remains matrix isolated or at a very low temperature. If the product is endothermic at room temperature, then dissociation of the metal complex MX occurs to form bulk metal. Only kinetic reasons may then cause the product to be stable at room temperature. The same situation holds for reactions with other ht particles, examples for which are the subvalent molecules BF, BCl, SiO, and CS. However, the energy differences are not so clearly established as in the case of metal atoms and therefore stable reaction products are observed to a much lesser extent than for metal atom species.

Table 1.1 Heats of formation ($\Delta H°$ (298 K)) of some gaseous atoms of interest in cryochemistry (values in kJ/mol of atoms)

H																	
213																	
Li	Be											B	C	N	O		
161	326											576	715	469	243		
Na	Mg											Al	Si	P	S		
108	149											326	446	314	221		
K	Ca	Sc	Ti	V	Cr	Mn	Fe	Co	Ni	Cu	Zn	Ga	Ge	As	Se		
90	177	326	473	515	397	281	416	425	430	339	125	272	372	288	210		
Rb	Sr	Y	Zr	Nb	Mo	(Tc)	Ru	Rh	Pd	Ag	Cd	In	Sn	Sb	Te		
82	164	410	611	774	659	649	669	577	381	286	111	244	301	259	200		
Cs	Ba	La	Hf	Ta	W	Re	Os	Ir	Pt	Au	Hg	Tl	Pb	Bi			
78	178	435	703	781	837	791	728	690	566	368	62	180	197	207			

Rare earths and lanthanides

Ce	Pr	Nd	(Pm)	Sm	Eu	Gd	Tb	Dy	Ho	Er	Tm	Yb	Lu
405	356	321	293	209	182	341	364	297	314	315	245	167	427

1.2. BASIC PRINCIPLES FOR THE GENERATION OF HT SPECIES

1.2.2 Formation of HT Species

Ht species can be prepared on micro- (a few mg) and macroscale (several grams) by different techniques which are described in a number of comprehensive articles and monographs on this subject [1, 8]. Atoms of the elements (see Table 1.1), or molecular species (see Table 1.2 [7]) are usually prepared from a single source, a bulk solid or liquid. For the generation of subvalent molecules, disproportionation or comproportionation reactions are the most common synthesis routes (see Scheme 1.1). Each of these types of preparations requires special experimental techniques some of which are described in Section 1.3.

Table 1.2 Selected examples of preparatively accessible subvalent ht species of main group elements.

BF	CS	
BCl	CF_2	
BC_2	CBr_2	
B_2O_2		
AlCl	SiO	P_2
AlF	SiS	PF
	SiC	PN
	SiF_2	PF_2
	$SiCl_2$	

$$2B_s + BF_3 \xrightarrow{1800\,°C} 3BF \ (90\%)$$

$$Si_s + SiCl_4 \xrightarrow{1350\,°C} 2SiCl_2 \ (95\%)$$

$$Si_s + SiF_4 \xrightarrow{1250\,°C} 2SiF_2 \ (60\%)$$

$$2B_s + 2B_2O_3 \xrightarrow{1350\,°C} 3B_2O_2$$

$$B_2Cl_4 \xrightarrow{1100\,°C} BCl + BCl_3$$

$$CS_{2g} \xrightarrow{\text{electrical discharge}} CS_g + 1/8 S_8$$

$$Si_s + SiO_{2s} \xrightarrow{1250\,°C} 2SiO$$

Scheme 1.1 Disproportionation and comproportionation reactions for the synthesis of subvalent ht compounds of group 13 and 14 elements. The pressure is less than 10^{-1} torr for all reactions.

The composition of the individual vapors of the elements or molecules of interest is well known even at different temperatures. For preparative purposes most of the elements vaporize in the monoatomic state and react with ligands as single atoms. For the main group elements, dimers, trimers, and even polymers may form a small fraction of the vapors of these elements.

A question of mutual interest in chemistry with ht species is: what makes the use of these particles, for example atoms, often superior to the use of conventional synthetic routes? Let us consider the build up of a monometallic, zerovalent organometallic complex like $[(\eta^4\text{-butadiene})_3\text{Mo}]$ (see eq. 1). The metal atoms generated to synthesize this complex are completely sterically accessible for the butadiene ligands after their ht generation. No steric requirements have to be considered since the high energy input of the naked atomic state offsets any energetics which have to be considered in a conventional synthesis such as a reduction process or a ligand dissociation from a precursor complex. Generally a huge variety of highly endothermic compounds such as $[(\eta\text{-arene})_2\text{M}]$ (M = Fe, Co, Ni) (see Section 1.4.2.) can be prepared by using different highly reactive ht species. The major problem is to isolate them under preparative conditions at ambient temperature.

$$\text{Mo}_{\text{bulk metal}} \xrightarrow{\Delta} \text{Mo}_{\text{atoms}} + 1,3\text{-butadiene}$$
$$\xrightarrow{77\,\text{K} \rightarrow \text{warm to rt}} [(\eta^4\text{-butadiene})_3\text{Mo}] \quad (1)$$

The isolated yields of the products obtained from ht processes like this are sometimes low, but a direct synthesis of a hitherto unknown compound using ht species gives ample evidence that a particular compound is stable under ambient conditions even though no classical synthesis exists, so far. The next step for the synthetic chemist might then be the search for more conventional synthesis routes which could offer access to this new compound on a larger scale. However exotic the technique in the high temperature process may be, it is worthwhile if its results inspire conventional chemists to fresh thoughts. When using this philosophy, chemistry with ht species provides a valuable method for the preparation of new prototypes of compounds. Nevertheless a major limitation of the method of using highly reactive ht species in preparative chemistry is the lack of flexibility concerning the reaction conditions. Usually cryogenic temperatures have to be used; however, there are some exceptions where reactions can be conducted at room temperature (rt) or even higher [9]. In general it is not possible to change reaction conditions in ht chemistry in the same way as it is in conventional chemistry.

1.3 EXPERIMENTAL TECHNIQUES FOR THE GENERATION AND REACTIONS OF HT SPECIES (ATOMS AND MOLECULES)

A number of different designs of apparatus as well as different experimental techniques for the macro scale generation of ht species have been reported

1.3. EXPERIMENTAL TECHNIQUES FOR HT SPECIES

[1, 8a]. A short classification of the most widely used techniques will be given here, based on general conditions used in different experimental approaches.

Syntheses with vapors of atoms or molecules can be distinguished on the basis of whether the reaction vessel used is A) static or B) rotary.

Method A): All reactant vapors are condensed at low temperatures on the walls of the reaction vessel to give a cocondensate of these vapors. Method B): The ht species are condensed into a neat liquid, a solution of ligands in inert solvents, or a suspension of solids in inert solvents. Both of the latter are more or less liquid phase techniques. By choosing either of the two ways, A) or B), the products obtained may reflect the choice of the individual synthetic technique used. Reactions in static reactor systems favor facile self association processes of ht species, whereas reactions in solutions usually favor atom coreactant interactions, mainly due to higher reaction temperatures, normally employed in solution reactions.

1.3.1 General Techniques for the Generation of HT Species

Two main techniques that are most widely used in macroscale generation of ht species are:

Resistive Heating For preparative purposes this heating technique for bulk samples is the simplest and most widely used technique for producing vapors of ht species on a macroscale. Here, the bulk sample is placed in an appropriate container and sufficient current is supplied until melting or sublimation is observed under vacuum $<10^{-2}$ Pa. Evaporation rates are then controlled by variation of the applied current. Different techniques of insulating the containers are necessary to minimize the loss of radiation and to shield the cocondensate so that no decomposition by heat radiation occurs (see Figure 1.1).

Besides a variety of molecular species (see Table 1.3), elements such as the alkali metals, the alkaline earths, or Cr, Mn, Fe, Co, Ni, Cu, Ag, Au and various others can be vaporized on a multigram scale using this simple technique of vaporization.

Electron Beam (e-Beam) Heating With e-beam heating a bulk source is bombarded with high energy electrons which are best produced from a hot filament cathode, *e.g.*, a resistively heated Mo or W wire which surrounds the bulk sample placed on a highly negatively charged copper block. The kinetic energy of the accelerated electrons heats the surface of the bulk sample up to its vaporization temperature (see Figure 1.2). This technique can be used for generation of vapors of all elements. Even the highest melting refractory metals like W, Os, and Ir can be vaporized and used as ht species in multigram quantities for synthetic purposes. Preconditions for the use of such e-beam guns are very good vacuum conditions (10^{-4}–10^{-5} Pa) during the reaction, since destructive bombardment of the cocondensate by secondary, reflected electrons must be avoided.

1. CHEMICAL SYNTHESIS USING HIGH TEMPERATURE SPECIES

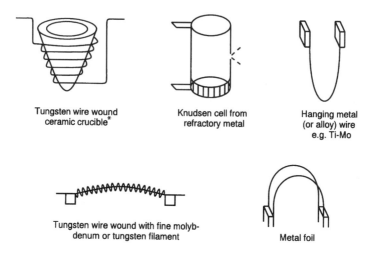

Figure 1.1 Different designs of resistively heated evaporation sources. After [8a].

Table 1.3 Temperature increase of the cocondensate surface at different distances from a metal sample. The temperature versus distance relationship has a d^2-dependence.

Distance (mm)	Temperature increase in $W \cdot mm^{-1}$	Temperature increase [K] per mm cocondensate	
		at 200 W	at 1000 W
60	0.13	26	130
80	0.07	14	70
100	0.05	10	50
125	0.03	6	30
150	0.02	4	20
200	0.01	2	10

1.3.2 Static Reactor Systems

When using a static reactor system the vapor of a ht species of interest is best produced by resistive heating from an appropriate precursor to be held in a special container (see Figure 1.1), by resistive heating, or by using a container-less method like e-beam vaporization of the precursor. This is followed by cocondensation of the vapors of the ht species together with a single or even

1.3. EXPERIMENTAL TECHNIQUES FOR HT SPECIES

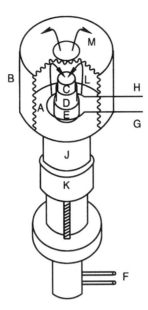

Figure 1.2 Hot filament electron gun. *A* Molybdenum hot-filament cathode; *B* cylinder to which is applied a large potential to accelerate and focus the thermionic electron beam onto the sample; *C* evaporant sample; *D* water cooled copper hearth; *E* tantalum heat shield; *F* cooling water; *G* electrical connector carrying emitter current and accelerating potential; *H* electrical connector carrying emitter current; *J* ceramic insulator bush; *K* support and focussing clamp; *L* way of accelerated electrons; *M* flux of emitted ht species. Adopted from ref. [8a].

multiple coreactants on the walls of a glass or stainless reaction vessel which is held at $N_2(l)$ temperature under a permanent vacuum $<10^{-2}$–10^{-5} Pa during the course of the reaction (see Figure 1.3). However, a multigram scale up of such a vaporization experiment in a static reactor might be sometimes limited by deleterious effects of thermal radiation from resistively heated crucibles. This drawback can be avoided by insulating the container with radiation shields or refractory wool (*e.g.*, Al_2O_3, see Figure 1.1). Another important fact to be considered when using static reactors is that typical cocondensation reactions proceed for several hours, a time in which a layer of the reactants is built up. It can be several mm thick, the thickness depending on the individual size of the reaction vessel and the amount of ligand used. Therefore, the limit of cooling capacity of the $N_2(l)$ bath might be reached after a few hours depending on the heat conduction through the layer of the cocondensate and the amount of radiant heat evolved from the vaporization source (see Figure 1.4 and Table 1.3). While being conscious of these limiting facts, multigram amounts of most of the transition metals as well as molecular species can be produced in an easy-to-use and nearly maintenance-free reactor set up as shown in Figure 1.3. When heated

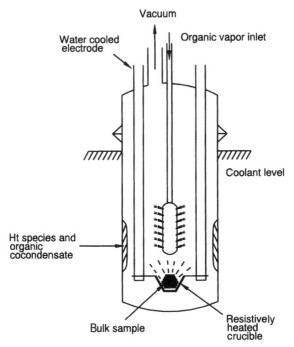

Figure 1.3 Schematic of a static metal vapor reactor used, for instance, in organometallic synthesis and in colloid preparation. Metal is evaporated from a resistively heated crucible and cocondensed with organic vapors onto the walls of a cooled evacuated vessel. After [16].

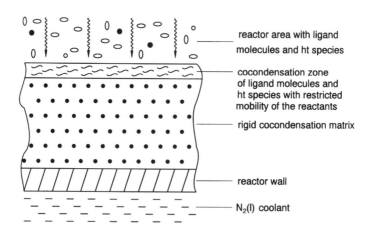

Figure 1.4 Build up of different cocondensation zones in the reaction of ht species with ligand molecules.

ligand inlet systems for such static reactors are used, even ligands with boiling points up to 300 °C at standard pressure, like 2,2'-bipyridine [10], biphenylene [11] and bimesitylene [12] can be cocondensed with the vapor of the ht species, expanding the scope of this cocondensing method from low boiling to high boiling ligand molecules as well.

1.3.3 Rotating Reactor Systems

The reaction of solutions of ligands with ht species can be performed in a rotary reactor system as shown in Figure 1.5. Vapors of the gas phase ht species are condensed directly upwards into a thin film of fluid which is spun out of a turbulently mixed liquid cooled by an external cooling bath [13]. Usually inert media were used as solvents for ligands in this technique (*e.g.*, methylcyclohexane) and the reaction temperatures are much higher than in a static reactor, often around $-100\,°C$. High speed pumping by a medium size diffusion pump, or a turbomolecular pump backed by a rotary pump, maintains a vapor pressure below 10^{-3}–10^{-4} Pa which is necessary during operation. Use of such a rotary system allows the reaction of suspensions or solutions of ligands that are nearly involatile and therefore difficult to vaporize and cocondense with vapors of ht species in a static reactor setup, as shown in Figure 1.3. In general, this method has proven to give better yields than the cocondensation procedure since the higher reaction temperatures favor reaction of the ht species with the ligands in the solution compared to the kinetically favored aggregation process which forms bulk material of the ht species. However, use of an electron beam instead of resistively heated vaporization sources in rotary solution reactors needs some special vacuum precautions. The hearth area containing the bulk precursor (see Figure 1.2) is best cooled with $N_2(l)$ instead of conventional water cooling, to obtain good vaporization results and to avoid severe degradation of the organic materials in the reaction solution, *e.g.*, due to plasma discharges, when high vacuum may eventually be lost during the reaction ($>10^{-3}$ Pa). Such

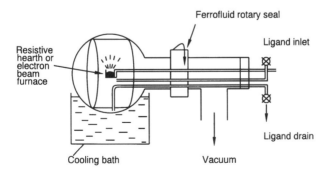

Figure 1.5 Schematic of a rotary metal vapor synthesis reactor for reacting ht species such as metal atoms with solutions of ligands. After [16].

technical problems have been solved, and commercially available rotating reactors for the generation of ht species, using single and dual e-beam vaporization techniques are available [14].

1.3.4 Special Developments of Reactors for the Generation of HT Species

Special new laboratory techniques using ht species based mainly on a static reactor design (see Figure 1.3) have been developed and allow the preparation of bimetallic catalysts, colloids, new carbon allotropic forms like fullerenes, and highly reactive subvalent molecules on a macro scale.

The so called SMAD (Solvated Metal Atom Dispersed) reactor consists of a dual electrode arrangement for allowing continuous evaporation of two bulk samples, to produce two individual ht species that can be condensed together (see Figure 1.6) [15]. Using this technique the formation of a variety of unusual heterogeneous bimetallic catalysts is possible (see Section 1.4.5).

Due to high vacuum conditions necessary when using e-beam vaporization techniques in a rotating reactor system, a new technique was introduced which allows the reaction of aerosols of ligands with metal vapors. On a rotating cold stage probe, an aerosol of a neat liquid or a polymer solution, for example, is condensed together with a flux of metal vapor generated from a sputtering gun.

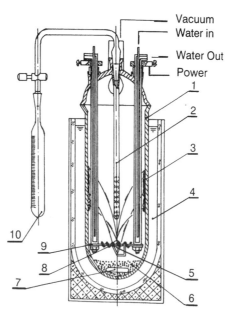

Figure 1.6 Schematic of a SMAD metal vapor reactor. (1) reactor chamber; (2) solvent vapor inlet; (3) matrix; (4) liquid N_2; (5) vaporization crucible with metal sample: (6) calcined alumina; (7) magnetic stir bar; (8) second metal sample as wire source; In (9) W rod; (10) solvent. After [96].

1.3. EXPERIMENTAL TECHNIQUES FOR HT SPECIES

In such a metal vapor/aerosol reactor metal colloids, for example, can be synthesized and isolated under anaerobic conditions (see Figure 1.7) [16].

Vaporization of metals under static pressure of an inert gas instead of high vacuum is termed the gas evaporation method and has found widespread application in the macroscale preparation of ultrafine particles [17] (see Figure 1.8). Basically a static reactor setup is used and clusters or particles of the ht species deposit on the inner walls of an evaporation chamber. The generated cluster size depends crucially on the pressure and the gas temperature. Using this method a variety of transition metal particles can be formed on a preparative scale.

For the macroscale generation of fullerenes from graphite, several static reactor systems have been developed in the last five years since the initial discovery of the synthesis of these new all carbon allotropic forms. One laboratory bench top reactor type uses a graphite rod electrode which is vertically gravity driven towards a carbon base electrode to produce a steady electrical arc between both electrodes in which a so called "carbon soot" is prepared under arcing conditions in a noble gas atmosphere (Ar, He), containing a mixture of

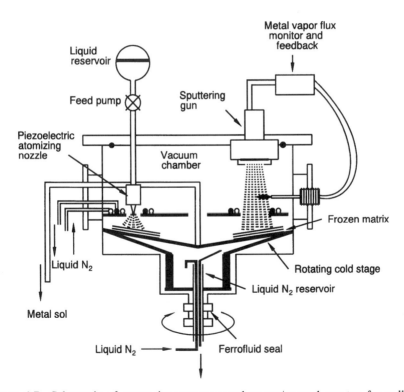

Figure 1.7 Schematic of sputtering source metal vapor/aerosol reactor for colloid preparation. After [16].

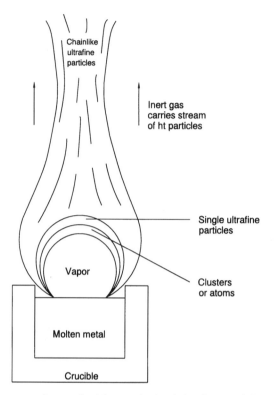

Figure 1.8 Gas evaporation method for synthesis of ultrafine particles. This method can be used to prepare ultrafine particles of any size and of any material, with little contamination except by the inert gas used in the process.

fullerenes in approximately 5–20% overall yield. The contact arc is powered by an AC generator providing up to 150 A to start the arcing process (see Figure 1.9 [18c]).

The formation of subvalent ht molecules is achieved by the use of static reactor systems which are modified to some extent in order to allow for special handling of these species. Instead of the evaporation crucibles, which contain the bulk sample necessary for the production of the individual ht particles, electrical discharge devices, for example, are necessary within the reactor, as in the synthesis of CS (see Figure 1.10, [19]) or BCl (see Figure 1.11, [20]). The latter are subvalent molecules that have been studied extensively in ht chemistry.

1.4 APPLICATIONS OF HT SPECIES IN CHEMISTRY

The organization of the material presented in the following sections is based on the periodic table of the elements. The macroscale synthesis of mononuclear,

1.4. APPLICATIONS OF HT SPECIES IN CHEMISTRY

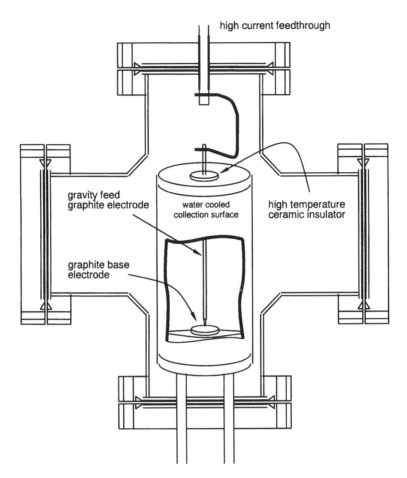

Figure 1.9 Schematic of a static reactor for fullerene synthesis using a vertically driven gravity feed device. After [18c].

and multinuclear compounds using ht species (atoms and molecules) is described in Sections 1.4.1–1.4.3 providing selected examples from recent literature. Section 1.4.4 is devoted entirely to the presentation of selected examples of the use of ht species in material science.

Since this review will deal with synthetic applications of ht species in chemistry no results of spectroscopic work will be reviewed systematically; however, some investigations will be mentioned, when necessary. This research area is of course very meaningful and represents a topic of its own, which needs to be covered separately.

1.4.1 Syntheses with Main Group Elements and Molecules

Alkali and Alkaline Earth Group Elements   By far the most widespread use of alkali and alkaline earth metals in preparative chemistry is concerned with their

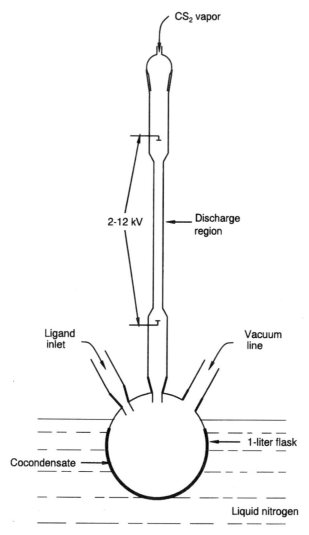

Figure 1.10 Apparatus for preparing subvalent CS in an electrical discharge. After [7].

high reducing power which reaches its peak when the atomic state of these elements is used instead of the bulk metal. Abstraction processes were studied in great detail and reactivity trends are well understood with the atoms of these elements [1]. Abstraction reactions might occur by an initial electron transfer to the LUMO of the organic compound followed by atom transfer.

Reactions of potassium vapor have found an interesting application in synthetic organometallic chemistry. Reduction of several transition metal halides in the presence of various arenes with potassium atoms produces a variety of

1.4. APPLICATIONS OF HT SPECIES IN CHEMISTRY

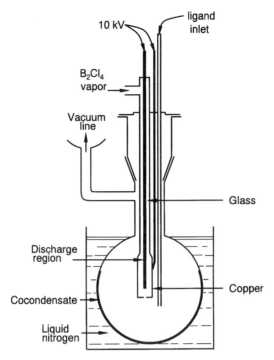

Figure 1.11 Apparatus for preparing subvalent boron trichloride, BCl, in an electrical discharge. After [7].

known, but also as found for conventional synthetic routes, still uncharacterized organometallic complexes (see Table 1.4) [21].

Especially since vaporization of alkali metals is easy to perform, this method is generally attractive in organometallic synthesis. The fact that bulk metal does not form arene metal complexes in conventional reactions might support the fact that a *one atom at a time* reduction is necessary for these reactions.

The connection between abstraction and electron transfer processes in ht chemistry of alkali and alkaline earth elements is emphasized by the selective activation of C–H bonds in benzene and toluene by Li atoms at 77 K [22a]. The thermodynamically favored product, benzyllithium, is not found. This is explained by a reaction mechanism involving single electron transfer (SET) to the LUMO of the aromatic compound as a crucial step in the reaction (see eq. 2) [22a].

$$8\,Li_{atoms} + 4\,phenyl\text{-}R + 4\,THF_{gas}$$

$$\xrightarrow[\text{cocondense}]{77\,K} [\{p\text{-}CH_3\text{-}C_6H_4\text{-}Li(THF)\}_4] + 4\,LiH(s) \qquad (2)$$

$$(R = H \text{ or } CH_3)$$

Table 1.4 Preparation of bis(arene) metal complexes using Timms' metal atom reduction technique [21].

Starting materials	Amount of K atoms	'End-point' color	Product	Yield%
[TiCl$_3$(thf)$_3$] + toluene	Excess	Green-black	[Ti(C$_6$H$_5$Me)$_2$]	19
[TiCl$_3$(thf)$_3$] + mesitylene	Excess	Black	[Ti(C$_6$H$_3$Me$_3$)$_2$]	15
[VCl$_3$(thf)$_3$] + toluene	Excess	Green-black	[V(C$_6$H$_5$Me)$_2$]	15
[VCl$_3$(thf)$_3$] + 1-methyl-naphthalene	Excess	Black	[V(C$_{10}$H$_7$Me)$_2$]	25
[CrCl$_3$(thf)$_3$] + naphthalene	Stoichiometric	Red-brown	[Cr(C$_{10}$H$_8$)$_2$]	36
[CrCl$_3$(thf)$_3$] + 1-methyl-naphthalene	Stoichiometric	Red-brown	[Cr(C$_{10}$H$_7$Me)$_2$]	40
MoCl$_5$ + toluene	Stoichiometric	Muddy-brown	[Mo(C$_6$H$_5$Me)$_2$]	10
MoCl$_5$ + 1-methyl-naphthalene	Stoichiometric	Black	[Mo(C$_{10}$H$_7$Me)$_2$]	42

In contrast to the possible Wurtz type coupling, no such reaction was observed here, and besides LiH, only ring metallated products are observed. Li atoms are reactive towards alkenes. By reacting them with ethene, 1,2-dilithioethane could be prepared and trapped as dimethylsuccinate in 8% yield after quenching with CO$_2$ and diazomethane [23].

It is not clear whether magnesium atoms might form π complexes upon cocondensation with alkenes since no discrete π complexes could be detected, but upon warming extensive isomerization takes place apparently on the surface of larger Mg particles [1, 6]. As already reported for alkali metals, insertion of Ca atoms into R–X bonds is observed and leads to organocalcium reagents from which the corresponding alkanes can be formed in excellent yields by quenching with HCl [24]. Wurtz type coupling is a side reaction which becomes the main reaction path only when benzyl halides are used.

The preparation of calcium atom/THF slurries is easy to perform and offers an attractive alternative to classical activation techniques such as amalgamation of calcium metal [25], activation with iodine [26] or use of a Grignard reagent [27].

Non-classical, solvent free Grignard reagents could be prepared in high yields by the action of Mg atoms on several organic halides. The unsolvated Grignard reagents obtained differ significantly in their reactivity when compared to the ones prepared *via* classical routes. For example acetone is enolized with formation of propane when unsolvated n-C$_3$H$_7$MgBr reacts with this ketone thus reflecting the stronger Lewis acidity of this material compared to the classical Grignard reagent [28].

Organometallic complexes of Ca, Sr and Ba using cyclic π acid ligands have been studied recently [29]. Although several [(η^5-Me$_5$-Cp)M] sandwich complexes (M = Ca, Sr, Ba) can be prepared by using conventional synthetic routes

[30], the preparation of sterically demanding complexes $[(\eta^5\text{-}Cp^R)_2M]$ (R = Si(Me)$_3$, t-butyl, M = Sr, Ba) was achieved by using the direct reaction between the metal vapors and the appropriate ligands (see eq. 3) [29].

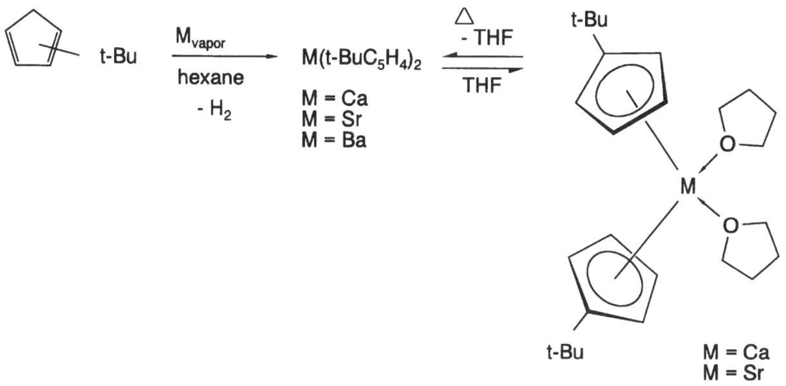

$$\text{(3)}$$

Reaction scheme: t-Bu-C$_5$H$_4$ + M$_{vapor}$ (hexane, −H$_2$) → M(t-BuC$_5$H$_4$)$_2$ (M = Ca, Sr, Ba); with THF / −THF, Δ to give the bis(t-BuCp)M(THF)$_2$ complex (M = Ca, Sr).

Recently, it has been reported that mechanically activated bulk calcium metal inserts into the Sn-Sn bond of $[(CH_3)_3Sn_2]$ to form an interesting organometallic Ca stannide in 61% yield (see eq. 4) [31].

$$Ca_{\text{bulk metal}} + [(Me_3)_3Sn - Sn(Me_3)_3] \xrightarrow{\text{THF, rt}} [\{(Me_3)Sn\}_2Ca(THF)_4] \quad (4)$$

Probably the use of metal vapor based alkaline earth atom/solvent slurries [24] might broaden the scope of this type of reaction even further since the reactivity of these slurries could be much higher using metal vapor, and might lead to new types of insertion reactions with related M-M compounds.

Group 13–16 Elements
Group 13 (B, Al, Ga, In) Preparatively useful abstraction processes (B [1, 32]) and deoxygenation reactions (B, Al [1, 32]) of these elements have been studied in detail.

Recent results concerning ht species of this group, indicate that π complex formation is especially interesting in the case of Al and is found with several π perimeters. According to ESR spectroscopic evidence, at least three different bonding modes A–C are proposed for the reaction of Al atoms with for instance alkenes in different host matrices (see Scheme 1.2) [33–35]. In the formation of the aluminacycloalkane radical a concerted mechanism *via* an Al(π-C$_2$H$_4$)$_2$ complex is assumed [33].

Phenyl-aluminiumsesquiiodide is formed on a preparative scale when Al$_{atom}$/toluene slurries are reacted with iodobenzene (see eq. 5) [36].

$$Al_x/\text{toluene} + C_6H_5I \xrightarrow{\text{toluene, 70°C}} [(C_6H_5)_3Al_2I_3] \quad (5)$$

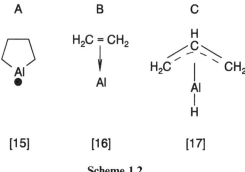

Scheme 1.2

Deep yellow colored π toluene/Al complexes are proposed as intermediates in this reaction. Similar macroscale reactions were studied for gallium [36]. Apparently such π arene stabilized complexes are also formed as intermediates when Ga and In are reacted with, for example, styrene to form colloidal metals [37].

An interesting aspect of group 13 chemistry is based on the preparation and usage of ht molecular species. BF, BCl, B_2O_2, AlCl and GaCl are subvalent molecules which can be prepared by disproportionation or comproportionation reactions at high temperatures, trapped at cryogenic temperatures, and then reacted with a variety of ligands. This type of ht chemistry was introduced by Timms in Bristol in the 70s, and many sophisticated experimental techniques have been developed in this research group. These are excellently reviewed by Timms in an important article on the ht chemistry of these species [7].

Recently the potential application of subvalent ht species, especially those of AlX and GaX (X = Cl, Br), has been subjected to extensive research [38]. Al^IX and Ga^IX can be obtained by cocondensation of Al or Ga atoms and gaseous HX in very good yields (see eq. 6).

$$E_{bulk} \xrightarrow[10^{-2} \text{ torr}]{\text{ca. } 1000\,°C} E_{atoms} \xrightarrow[10^{-2} \text{ torr } (90\%)]{\text{HX}, -196\,°C} EX + \tfrac{1}{2}H_2 \quad (6)$$

$$E = Al, Ga; X = Cl, Br$$

Crucial to the preparative breakthrough in this sector of high temperature chemistry was the observation that AlCl as well as its heavier homologue GaCl could be prepared as metastable molecules and stored in a mixture of toluene and diethylether at −78 °C for some time. A spectacular result of these studies was the synthesis of $[\{(\eta^5\text{-Me}_5\text{Cp})\text{Al}\}_4]$, the first organometallic all aluminum cluster compound [39]. The synthesis of this unique cluster has triggered high activity in this field, leading to a variety of new mixed polyhedral main group clusters [39].

Donor-stabilized $Al^{II}Br_2$ forms by comproportionation from a cocondensation solution of AlBr and anisole, in which $AlBr_3$ (10%) is also present. The actual primary product of the reaction, $AlBr_2$, is trapped at −30 °C by the donor

1.4. APPLICATIONS OF HT SPECIES IN CHEMISTRY

solvent anisole (L) in the form of a stable complex $[Al_2Br_4 \cdot L_2]$ (Figure 1.12). The presence of $AlBr_3$ in the cocondensate as an "impurity" is thus essential for the *in situ* formation of the new $Al^{II}Br_2$ species. The extended synthesis route to $[Al_2Br_4 \cdot L_2]$ *via* AlBr is necessary, since $AlBr_2$ cannot be made directly from Al atoms and HBr on a preparative scale. Interestingly, $[Al_2Br_4 \cdot L_2]$ containing Al^{II} does not disproportionate into Al^{I} and Al^{III} compounds in the solid.

Unlike this donor-stabilized $Al^{II}Br_2$ compound, $[(BrAl \leftarrow NEt_3)_4]$ is tetrameric in the crystal, and features a square Al_4 unit (Figure 1.12). $[(BrAl \leftarrow NEt_3)_4]$ already forms at around $-100\,°C$ from the cocondensate mixture of AlBr/toluene and Et_3N [38g]. Accordingly, efficient formation of $AlBr_2$ from AlBr and $AlBr_3$, as is predicted to occur during the syntheses of $[Al_2Br_4 \cdot L_2]$, is not observed under these reaction conditions, and consequently formation of an aluminum(II) system does not occur. Not until $-30\,°C$ is reached does the actual comproportionation reaction between AlBr and $AlBr_3$ occur, leading to the formation of $AlBr_2$ [38g]. The trapping of the ht species AlBr with the appropriate donor solvent in a suitable temperature range is thus of primary importance for the formation of AlX or AlX_2 molecules in these syntheses. Certainly this has to be taken into account during further studies of AlX and, possibly, of GaX [38i].

Cocondensation of AlCl with 2-butyne/pentane mixtures has lead to dimeric 1.4-dialuminacyclohexadiene $[Al_4Cl_4\{C_{16}H_{24}\}]$ (see eq. 7 and Figure 1.13) [40]. (See also note 1 added in proof at the end of this chapter.)

$$4\,AlCl + 4(H_3C)C \equiv C(CH_3) \xrightarrow[\text{2) warm up}]{\text{1) 77 K}} [Al_4Cl_4\{C_{16}H_{24}\}] \qquad (7)$$

Group 14–16 (C, Si, Ge, Sn, Pb, Bi, Te) Vapors of group 14 elements, especially that of carbon, have been studied since the pioneering days of preparative ht chemistry, mainly by the group of Skell at Penn State University. The chemistry of C_1–C_5 ht species has led to a variety of synthetic applications. However, preparatively useful reactions involving the heavier elements in this group are scarce. Free radical addition reactions of $CF_3\cdot$ and $SiF_3\cdot$ to Ge, Sn, as well as to Bi and Te atoms have led to a variety of organometallic compounds like

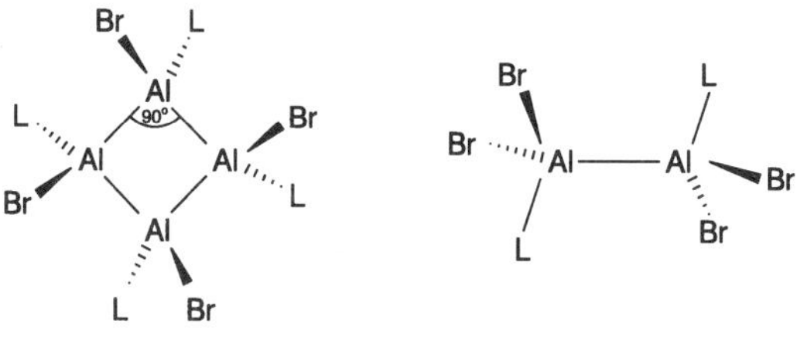

Al - Al: 2.643(3) Å   Al - Al: 2.527(6) Å

Figure 1.12

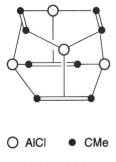

○ AlCl   ● CMe

Figure 1.13

Ge(CH$_3$)$_4$, Sn(CH$_3$)$_4$, Bi(CF$_3$)$_4$, and Te(CF$_3$)$_4$ [41]. CF$_3$ radicals codeposited with molecular PbCl$_2$ vapor lead to reasonable yields of Pb(CF$_3$)$_4$ for the first time [42]. Apparently these studies have stimulated recent work on conventional synthesis of Pb(CF$_3$)$_4$ and several other trifluoromethyl compounds of group 14 elements [43]. Bridged ansa metallocenes of Sn^{II} and Pb^{II} are accessible by direct reaction of the atoms with 6,6-dimethylfulvene on a preparative scale, albeit in low overall yields [44] (see eq. 8).

$$M(g) + 2 \text{ (fulvene)} \xrightarrow[77\,K]{\text{cocond.}} M(\text{cyclopentadienyl complex}) \quad (8)$$

M = Sn, Pb

Without any doubt the unexpected macroscale synthesis of the "fullerene family" (C$_{60}$, C$_{70}$, C$_{76}$, C$_{78}$,..) provides the most striking recent example of the successful use of ht species in preparative chemistry [45]. Under these harsh ht conditions of a carbon arc, up to a 15% overall yield of a mixture of fullerenes, such as C$_{60}$, C$_{70}$, and C$_{84}$, can be prepared by vaporizing carbon in a He or Ar atmosphere in a static reactor (see eq. 9 and Figure 1.7).

$$C_{\text{graphite}} \xrightarrow[100-200 \text{ torr He or Ar}]{\Delta, \text{ electric arc}} C_{60} + C_{70} + \text{higher fullerenes}$$
$$\text{and other carbon allotropes} \quad (9)$$

When using a vertical gravity driven electrode, the so called contact arc process, the graphite rod electrode sticks to the graphite base electrode after relatively short operation times of the electric arc, creating experimental difficulties. More sophisticated reactors use an externally controlled, driven graphite electrode set

1.4. APPLICATIONS OF HT SPECIES IN CHEMISTRY

up to generate a steady arc which allows a more continuous macroscale vaporization process. It is interesting to note, that when using a graphite filament, which is resistively heated in an inert gas atmosphere, no comparable macroscale production of fullerenes is observed. Apparently in the arc process a significant fraction of the total electrical current is carried by the arc generated plasma, but still the main fraction of the current goes through the slightly touching electrodes. Since the total contacted area of the carbon electrodes is only a small fraction of the total electrode area, most of the power input is dissipated at the contact points thus raising the temperature to that which permits instantaneous evaporation of the carbon rod. During the evaporation process the formation of a slag, which deposits on the thicker graphite base electrode, can be observed after some time. As long as this vapor deposited boundary layer remains between the two electrodes in a sufficiently thick and resistive form, the electrical power continues to be dissipated just in this small zone and carbon vaporization from the end of the graphite rod proceeeds efficiently [18d].

Obviously the ht arcing process is essential for the macroscale formation of fullerenes. Yields in the arc process are heavily dependent on a variety of parameters: a) temperature of the arc, b) diameter of the graphite electrodes, c) contact pressure of the rods, d) electrical current through the rods, e) cooling of the walls of the reaction vessel, f) type and pressure of the buffer gas used. Every single parameter has a significant influence on the yields of soluble fullerenes produced in the contact arc process and the technique requires skillful experimentalists to obtain good results.

Using essentially the same process as for empty fullerenes, filled "buckyballs", $M@C_{2n}$, are now synthetically accessible [46], when metal oxide, or better, metal carbide doped graphite rods are evaporated *via* a contact arc. For example, $La@C_{82}$ is contained in up to 1% yield in the crude carbon soot prepared by vaporization of lanthanum carbide doped graphite electrodes. Separation and purification of the endohedral fullerenes from the empty, all carbon molecules, uses special, recently developed HPLC techniques which separate "filled" and "empty" buckyballs due to their different π electron density [47].

As we complete this section it should be mentioned that beside spheroidal fullerenes, single and multiple shell nanotubes have recently become available on a macroscale by the ht contact arc process [48]. They are contained in the slag which forms on the graphite electrodes during the contact arc process and are not extractable with organic solvents, such as arenes, usually employed for the separation and isolation of the soluble fullerenes from the crude carbon soot. Due to the overwhelming response of the scientific community to the discovery of the macroscale fullerene synthesis by the ht arc process, which allowed development of the chemistry of the fullerenes on a preparative scale, the pertinent literature continues to grow at an exponential rate.

The heavier group 14 elements provide a variety of molecular ht species some of which exhibit an extensive preparative chemistry; the subvalent CS [49], SiO [7], SiS [7] and SiX_2 (X = Cl, F) [7] species, especially have been studied in

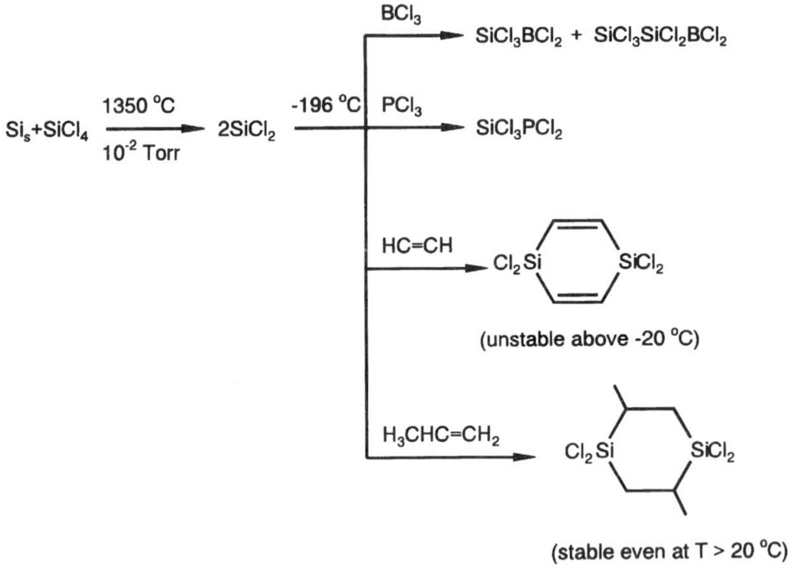

Scheme 1.3 Selected cocondensation reactions of the subvalent main group ht species $SiCl_2$ with organic and inorganic compounds at $-196\,°C$.

great detail. Scheme 1.3 shows some selected insertion and addition reactions of $SiCl_2$ which have led to preparatively useful applications of this ht molecule.

1.4.2 Syntheses with Transition Elements and Molecules

All transition elements vaporize in the monoatomic form from the bulk state. A variety of methods exists to generate these atoms. For preparative purposes, vaporization by resistance heating and e-beam are the most common ones used. However, several other generating methods such as laser [50], pulsed laser [50], arc [51] and induction heating [52] have been developed and used for many refractory elements and compounds.

Mononuclear Complexes of the Early Transition Elements Metal atom synthesis of new sandwich type organometallics using these elements has been, and still is a principal research area within ht chemistry. Several new prototypes of organometallic compounds containing early transition metals became available, for the first time, *via* direct synthesis, or from a combination of ht techniques and classical organometallic synthetic routes, during the past few years.

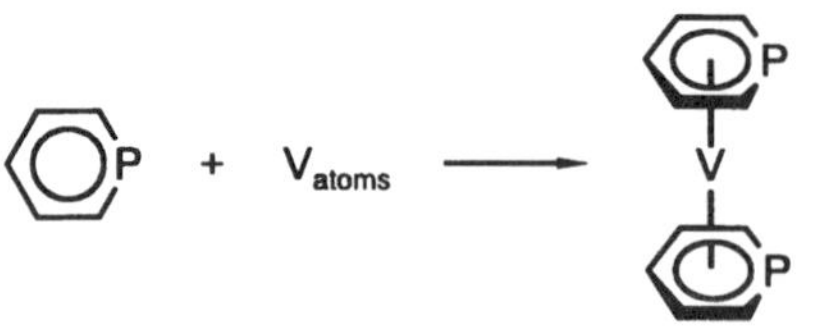

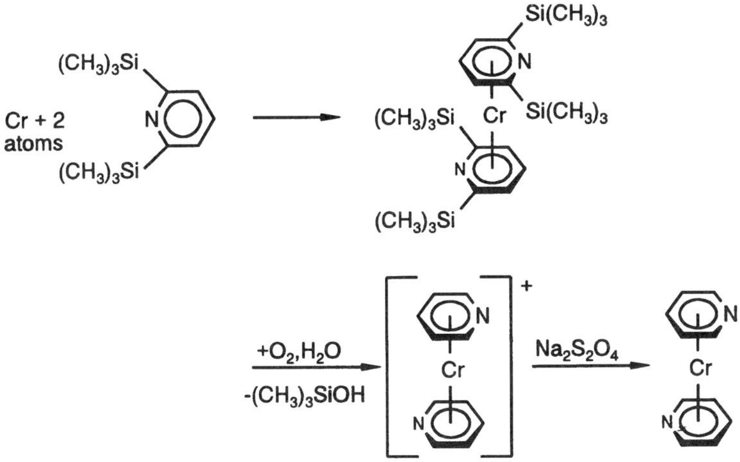

1.4. APPLICATIONS OF HT SPECIES IN CHEMISTRY 25

$$\text{Cr + 2 atoms} + \text{2,6-bis(trimethylsilyl)pyridine} \longrightarrow \text{Cr(η-2,6-(SiMe}_3)_2\text{-pyridine)}_2 \quad (11)$$

[54]

$$\text{ClMe}_2\text{Si-C}_6\text{H}_4\text{-SiMe}_2\text{Cl + Cr}_{atoms} \longrightarrow \text{bis(arene)Cr complex} \quad (12)$$

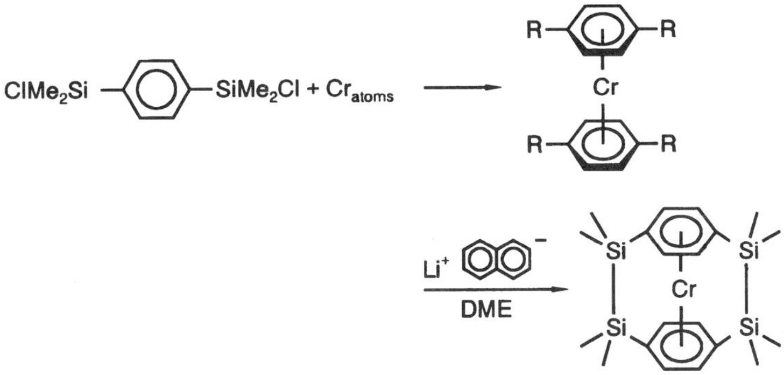

[55]

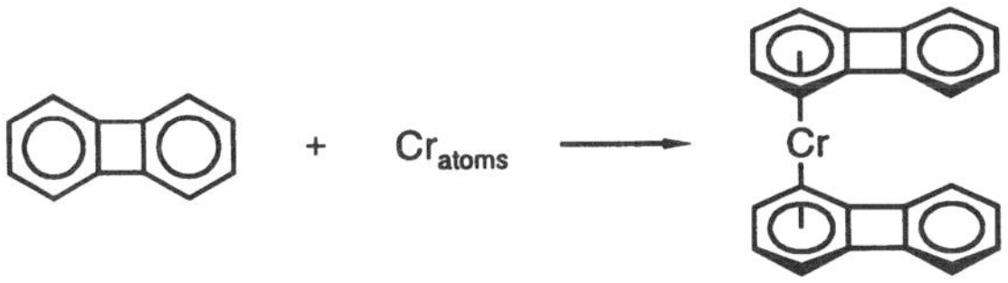

[56] (13)

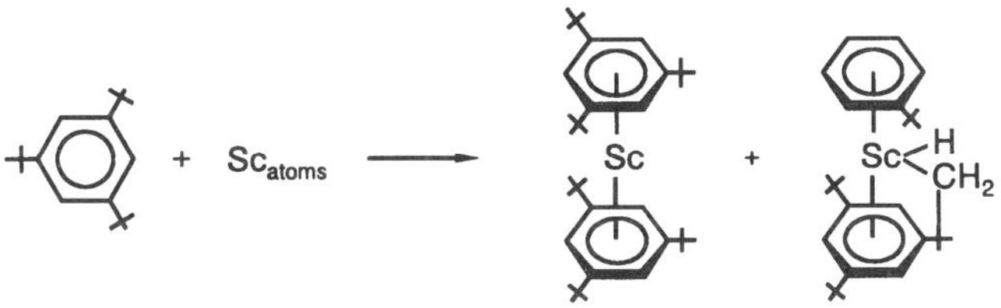

[57] (14)

$$\text{cyclopentadiene} + M_{W, Mo \text{ atoms}} \longrightarrow [Cp_2MH_2] \quad [58] \quad (15)$$

In reactions (10)–(15) a combination of zerovalent metal atoms and aromatic π ligands yield stable bis(arene)metal complexes. From a preparative point of view all of these reactions can be performed in one day giving purified reaction products after appropriate workup. Since vaporization of the metals on a multi-gram scale is possible, these complexes might also be valuable target compounds for further preparative investigations. Unreacted ligands which are usually used in excess over the metal atoms can be recovered from these reaction mixtures after separation by vacuum distillation or recrystallization and used in subsequent experiments. This technique has proven to be very valuable and allows the use of ligands which are not available in large quantities in subsequent reactions with metal atoms simply by recycling them.

In reactions (14) and (15) C–H activation processes by early transition metals, which have led to formation of hydrido sandwich complexes, are presented. In general, C–H activation processes of saturated alkanes by ht generated transition metal atoms on a microscale are well documented (*e.g.*, matrix isolation studies), but are also quite common in macroscale preparative synthesis with metal atoms [59].

Ligand Stabilized Molecular Cluster Compounds of the Early Transition Metals
Metal atom aggregation has led to stable monometallic, bimetallic, and cluster compounds of Cr, Mo and V, when those metals are vaporized into solutions of polymeric methylphenylsiloxane DC 510/50 or dimethylsiloxane-comethyl-phenylsiloxane [60] (see Figure 1.14).

These studies show that bis(phenyl)M_x sandwich molecules ($x = 1, 2, 5$) which are anchored by a polymeric chain can be stabilized in such polymers. Macroscale vanadium vapor deposition (1 mg/100 ml liquid polymer DC 510/50) yielded a clear red brown liquid that contained [η-arene)$_2$V] and [(η-arene)$_2$V$_2$], according to Raman spectroscopic studies [61]. Moreover, naked gas phase and matrix isolated homo- and heteronuclear clusters of the early transition metals comprise an important set of compounds in a field of investigation concerned with the study and interpretation of catalytic effects and the understanding of the metal clustering processes in such systems [62].

Highly Reactive Molecules of the Early Transition Elements The preparative chemistry of vapors of the refractory oxides MoO_3, WO_3, VO, M_2O_3 (M = Ti, V) has been studied in some detail recently, mainly by the groups of Timms, Klabunde and Dekock. For Mo and W oxides, polymeric complexes with 2,4-pentanedione, acetone, formic acid and methanol

1.4. APPLICATIONS OF HT SPECIES IN CHEMISTRY

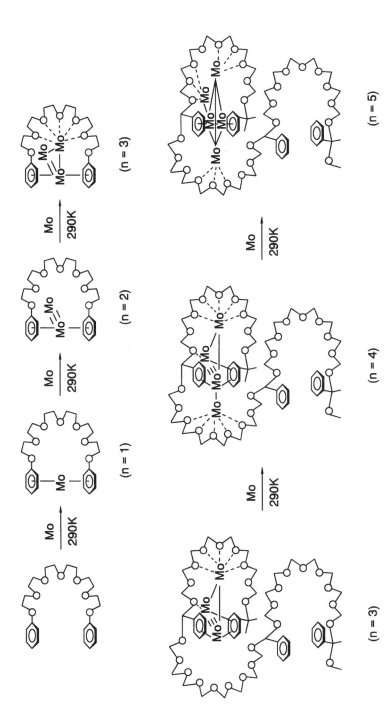

Figure 1.14 Matrix and macroscale metal vapor synthetic strategy to elucidate the role of arene groups and bis(arene) metal centres in the Mo$_{atom}$/DC510 reaction. This model accounts for the observed maximum nuclearity of n = 5 attainable in the Mo/DC510 system. After [60f].

are formed, indicating the high reactivity of such naked oxides [63]. However, synthetic utility is limited due to the formation of polymeric substances which make products difficult to handle. Mixed oxometal chlorides are accessible using the reaction of Cl_2 gas with TiO and VO vapors [64].

Further preparative studies of early transition metal compounds have not been undertaken so far. In this area there is scope for investigation since the vaporization data and composition of more than eighty halide, oxide and sulfide molecular compounds of the early transition metals are known [65], and some of them, when reacted for example with metal atoms, might well lead to interesting new coordination compounds and clusters.

Mononuclear Compounds of the Late Transition Elements The majority of the reactions performed on a preparative scale with the vapors of these elements are between the metal atoms and organic π perimeters, C_nH_n. These reactions usually allow quick access to a variety of compounds of the group 8–10 elements. A significant quantity of work has been carried out in the 70s and 80s, and a great number of complexes have been prepared during the active development phase of the metal vapor synthesis technique. Alkenes, homo- and hetero-alkynes, dienes, linear and cyclic species, and recently boranes and thiaboranes, and naturally a wide variety of arenes are all suitable ligands which allow the direct synthesis of many metal complexes by direct ht synthesis routes. Selected examples of recent work in this field will be highlighted.

Examples of bond cleavage reactions mediated by metal atoms have been reported. Nickel atoms react with the phosphaalkyne $[(CH_3)_3CC\equiv P]$ by a [2 + 2] cycloaddition and by formation of a diphosphacyclobutadiene-Ni^0 complex, and by cleavage of a P≡C triple bond of the phosphaalkyne ligand and formation of a phosphirenyl Ni^0 complex [66] (see eq. 16). Dichotomic rupture [67] of the phosphaalkyne by nickel atoms must be assumed to explain the formation of both complexes. Such a rupture was also observed recently in reactions of $[(CH_3)_3CC\equiv P]$ with Fe^0 complexes, which were derived from iron atoms [68].

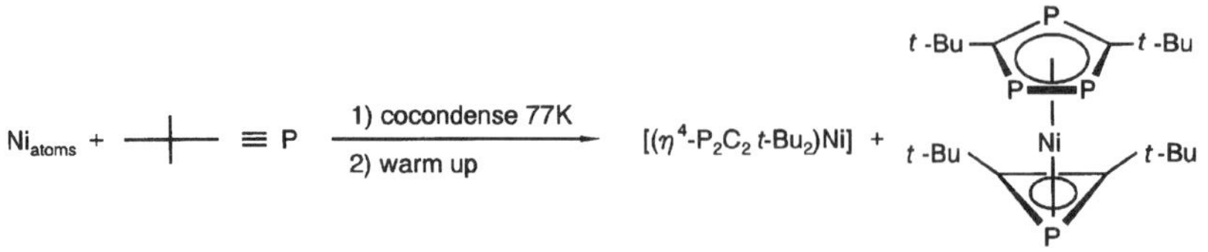

$$Ni_{atoms} + {-}\!\!\!\!{+}\!\!\!\!{-} \equiv P \xrightarrow[\text{2) warm up}]{\text{1) cocondense 77K}} [(\eta^4\text{-}P_2C_2\,t\text{-}Bu_2)Ni] + \quad (16)$$

A similar splitting of a pentamethylcyclopentadienyl (Me_5Cp) ring with extrusion of two C–CH_3 units was found in the reaction of nickel atoms with Me_5CpH [69]. A bis-ethylidyne bridged cluster $[\{(Me_5Cp)Ni\}_3\text{-}\mu_3\text{-}\{C(CH_3)\}_2]$ is one of the reaction products here. A dichotomic cleavage of an ethyne triple bond is observed in the reaction of iron atoms with substituted alkynes and cyclopentadiene to give ferrocene and substituted ferrocenes [70] (see eq. 17a). In the above reactions with the nickel atoms the mechanisms of the rupture

1.4. APPLICATIONS OF HT SPECIES IN CHEMISTRY

reactions are unknown. However, there is some evidence in the iron case that a ferracyclobutadiene complex and a Fe-carbyne complex react with each other to form the five membered alkyl substituted Cp ring of the substituted ferrocene [70a] (see eq. 17b).

$$\text{Fe}_{atoms} + \text{CpH} + \text{RC} \equiv \text{CR} \xrightarrow{77\,K} [(\eta^5\text{-Cp})_2\text{Fe}] + [(\eta^5\text{-Cp}^R)_2\text{Fe}] + H_2$$
$$+ \text{ oligomers of the alkynes} \qquad (17a)$$

R = various alkyl substituents

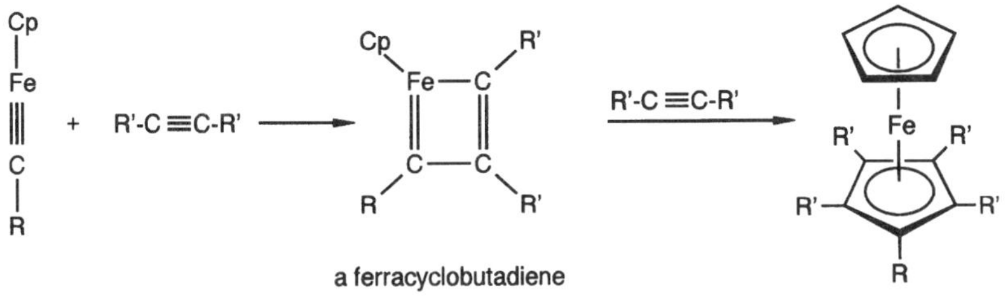

(17b)

a ferracyclobutadiene

R=H
R=Alkyl

An example of the formation of an unusual iron(IV) complex using metal vapor chemistry is the production of the highly labile intermediate, bis(η-toluene)iron(0) [71]. Oxidative addition of $HSiCl_3$ or $HSiF_3$ to this intermediate leads to the formation of $[(\eta^6\text{-toluene})Fe^{IV}(SiX_3)_2H_2]$ (X = Cl, F; see eq. 18), new members of the group of related $SiCl_3$ complexes for which there are other known examples containing Ni(II), Co(II), Cr(II) or Cr(IV) [72].

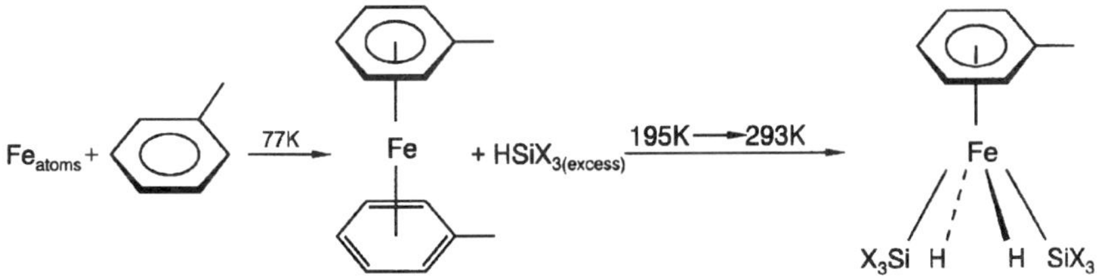

(18)

X = Cl or F

The intermediate complex in this synthesis, bis(η-toluene)iron(0), belongs to a remarkable class of compounds represented by the general formula bis(η-arene)metal(0) (arene = various alkyl substituted benzenes, metal = Fe, Co, Rh, Ni, Pd, Pt), which are exclusively accessible *via* a ht metal atom route. Particularly, the Fe, Co, and Ni complexes are stable only at temperatures up to $-50\,°C$ in excess arene solution. These complexes are described as highly labile [73]. According to a variety of spectroscopic and preparative studies one arene ligand is only weakly bound, whereas the other is η^6 bound to the metal

(Fe, Co, Ni). Despite various studies the coordination mode of the second one is still not completely clear [74], but it is assumed to be lower than η^6, probably η^4 or even η^2. This unique coordination pattern of the metal gives these complexes a high reactivity and has led to the coining of the expression "solvated metal atoms" for these compounds [73]. Especially toluene is a common solvating agent for preparative purposes and stabilizes the metal atoms in these complexes against catastrophic aggregation and formation of metal particles up to temperatures of around $-50\,°C$. Stability of the solvated metal atoms increases with more electron demanding arenes, for example, α,α,α-trifluoromethylbenzene.

In early work it was shown that solvated iron and nickel atoms can be used as a reactant to deliver the $\{(\eta^6\text{-toluene})\text{Ni}\}$ fragment, and that this could be stabilized by a variety of organic ligands to give stable organometallics not available using conventional preparation techniques [73a, 75]. Recently Zenneck used the reaction of solvated iron atoms with ethene to generate $[(\eta^6\text{-toluene})\text{iron}(\eta^2\text{-ethene})_2]$, which is highly reactive itself and has given access to many unique organometallic compounds in a variety of reactions [68a, 76]. As a result of these studies, the bis(η-toluene)iron, -nickel and -cobalt sandwich compounds can be regarded as metal atom carriers which can be used to direct the latent reactivity of the arene complexed Fe, Co or Ni atoms, or the $\{(\eta^6\text{-toluene})\text{Fe, Co or Ni}\}$ fragments, in subsequent solution phase chemistry on a preparative scale. The use of highly labile, but under low temperature conditions stable compounds of the type bis(η-toluene)M(0) (M = Fe, Co, Ni), therefore, gives the experimental chemist a possibility to "bottle" these zerovalent metal atoms in a relatively stable form and use them in bench top chemical experiments.

Very recent studies on $[(\eta\text{-toluene})_2\text{Fe}]$ have centered on the reaction of this labile molecule with a variety of N-containing ligands such as primary, secondary and tertiary amines (NH_3, NH_2CH_3, $NH_2C(CH_3)_3$, $NH(CH_3)_2$, $N(CH_3)_3$), as well as pyridine, pyrrolidine, and N,N'-tetramethylethylenediamine to form reactive, highly unstable intermediates for which definite structures have yet to be established (see Scheme 1.4). In the temperature range from -80 to $-30\,°C$, A or B, which might be such plausible intermediates, react with toluene to form the first homoleptic Fe_2/toluene complex [bis$\{(\eta^6\text{-toluene})\text{iron}\}$-$\mu_2$-$(\eta^3:\eta^3\text{-toluene})$] regardless of which N-containing ligand is used. The X-ray structure determination of the thermolabile compound (decomposition around $0\,°C$) reveals a μ-$\eta^3:\eta^3$ bridging toluene ligand capping two $\{(\eta^6\text{-toluene})\text{Fe}\}$ units which are connected by a Fe–Fe bond [2.746(1) Å]. However, in solution all three toluene ligands are η^6 bonded and are highly fluxional according to ^{1}H- and ^{13}C-NMR spectroscopy [77]. The intermediate formation of mono- or bisamine intermediates like A or B in the synthesis of [bis$\{(\eta^6\text{-toluene})\text{iron}\}$-$\mu_2$-$(\eta^3:\eta^3\text{-toluene})$] is necessary, since reaction of $[(\eta\text{-toluene})_2\text{Fe}]$ with excess toluene under similar reaction conditions, but without presence of the amines, did not lead to the formation of the synfacial di-iron complex. Further work is needed on this new Fe_{atoms}/toluene/amine system to elucidate the formation pathway for this novel di-iron complex. (See also note 2 added in proof.)

1.4. APPLICATIONS OF HT SPECIES IN CHEMISTRY

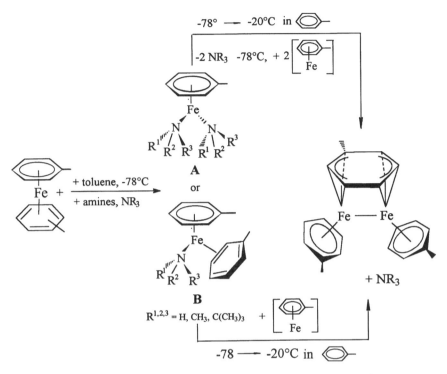

Scheme 1.4 Possible reaction pathway leading to the formation of [bis{η^6-toluene)iron-μ_2{$\eta^3:\eta^3$-toluene}] via the unstable intermediates A or B.

Ligand Stabilized Molecular Cluster Compounds of the Late Transition Elements
In general it is difficult to predict the formation of defined molecular cluster compounds using ht free metal atoms, since these species are unselective in their reactions with ligands, especially with respect to molecular cluster formation. However, there are a couple of reports on the macroscale synthesis of molecular clusters using metal atoms [78]. Nice examples of the use of gold atoms for the synthesis of larger ligand stabilized homonuclear gold clusters are given by Steggerda [78b] (see eq. 19).

$$Au_{atoms} + KSCN + PPh_3 \xrightarrow[\text{ethanol}]{173 K} Au_{11}(PPh_3)_7(SCN)_3 \qquad (19)$$

Apart from this work, it is not reported in detail whether individual experiments [78c–g] were directly aimed at studying the synthesis of metal clusters *via* a metal atom route.

In general it is more difficult to use atoms in the synthesis of cluster compounds M_xL_y than in the formation of mononuclear complexes of the type ML_y. Despite various experimental approaches *"the general problem of stabilization of clusters by ligand addition is far from being resolved"*, as Timms pointed out in 1984 [79], and this still holds.

When atoms and ligands are cocondensed in a static reactor or when atoms are condensed into a solution of ligands in an inert solvent, aggregation of atoms to form metal particles competes effectively with addition of ligands to the metal atoms to form metal complexes or even ligand stabilized metal clusters (see Scheme 1.5). High ligand to metal ratios favor metal-ligand aggregation and suppress metal-metal aggregation, thus leading to mononuclear ML_n complexes. High metal to ligand ratios drive the reaction to the right resulting in the formation of larger metal particles. This is, however, of great interest in heterogeneous catalysis (see Section 1.4.4). Using nearly equivalent amounts of metal atoms and ligand might lead to formation of molecular cluster complexes, since the reaction is now driven along the diagonal pathway (dotted line) in Scheme 1.5.

Reactions of nickel and cobalt atoms with equivalent amounts of different substituted Cp^R ligands (R = various alkyl substituents) obviously follow this hypothesis and have led to the formation of several homo-tri- and tetranuclear hydrido bridged Ni and Co cluster compounds [80]. Product selectivity and

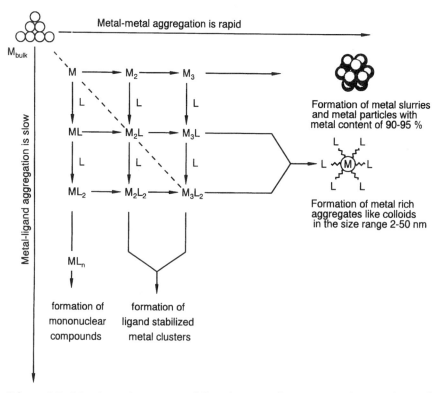

Scheme 1.5 Metal-metal versus metal-ligand aggregation as competing reaction pathways in a metal-vapor/ligand reaction system. Metal atom concentration increases from left to right; ligand concentration from top to bottom.

1.4. APPLICATIONS OF HT SPECIES IN CHEMISTRY

nuclearity depend on the steric bulk of different Cp^R ligands and mononuclear sandwich type complexes are still favored products in these reactions. Intermediate formation of $\{(\eta^5\text{-Cp})M\}$ and $\{(\eta^5\text{-Cp})MH\}$ fragments from which the clusters might be formed is assumed to occur in these reactions (see eqs. 20–26).

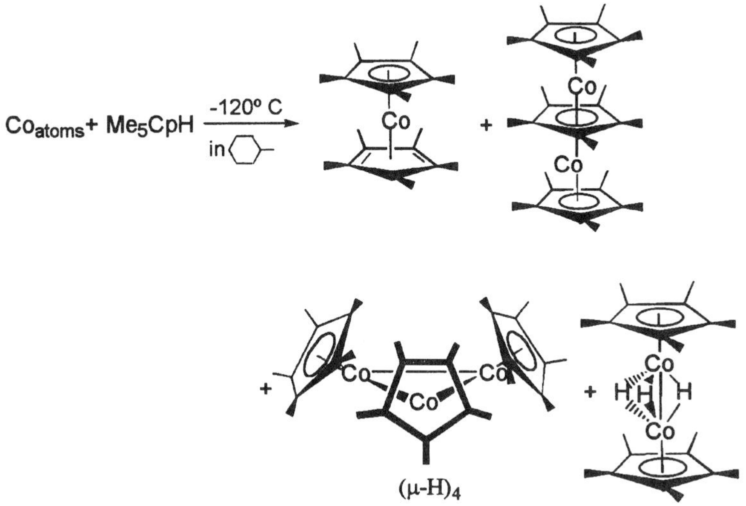

$$Co_{atoms} + Me_5CpH \xrightarrow[\text{in }\bigcirc]{-120°\,C} \quad (20)$$

$(\mu\text{-H})_4$

$$Co_{atoms} + \xrightarrow[\text{in }\bigcirc]{-120°\,C} \quad (21)$$

$(\mu\text{-H})_4$

$$Co_{atoms} + \xrightarrow[\text{in }\bigcirc]{-120°\,C} \quad (22)$$

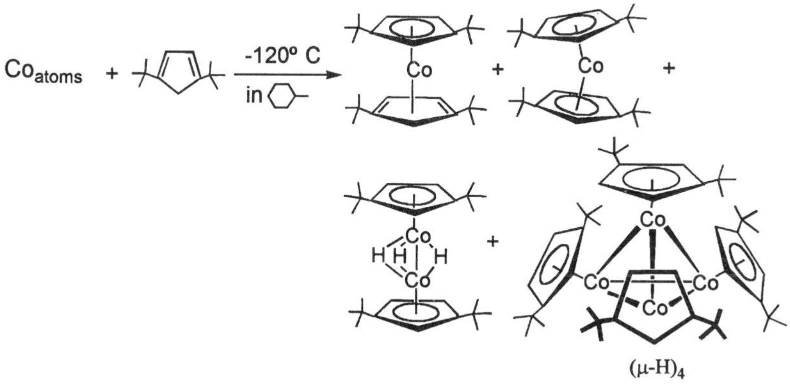

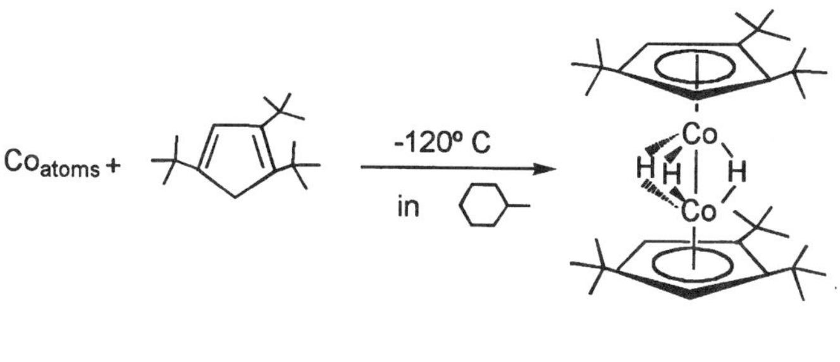

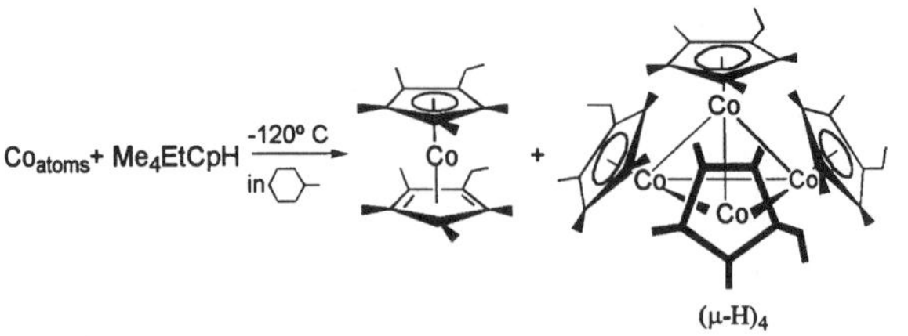

1. CHEMICAL SYNTHESIS USING HIGH TEMPERATURE SPECIES

$$\text{Co}_{atoms} + \text{Me}_4\text{EtCpH} \xrightarrow[\text{in} \bigcirc]{-120°\text{C}} \quad + \quad (\mu\text{-H})_4 \quad (23)$$

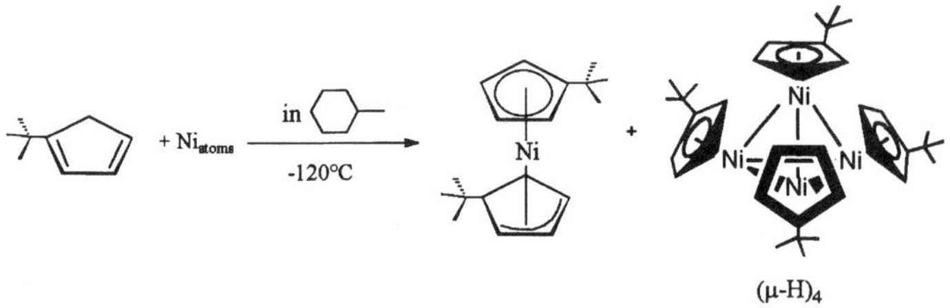

$$+ \text{Ni}_{atoms} \xrightarrow[-120°\text{C}]{\text{in} \bigcirc} \quad + \quad (\mu\text{-H})_4 \quad (24)$$

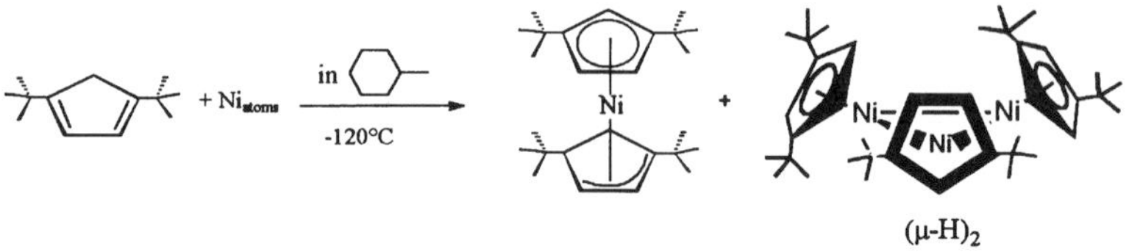

$$+ \text{Ni}_{atoms} \xrightarrow[-120°\text{C}]{\text{in} \bigcirc} \quad + \quad (\mu\text{-H})_2 \quad (25)$$

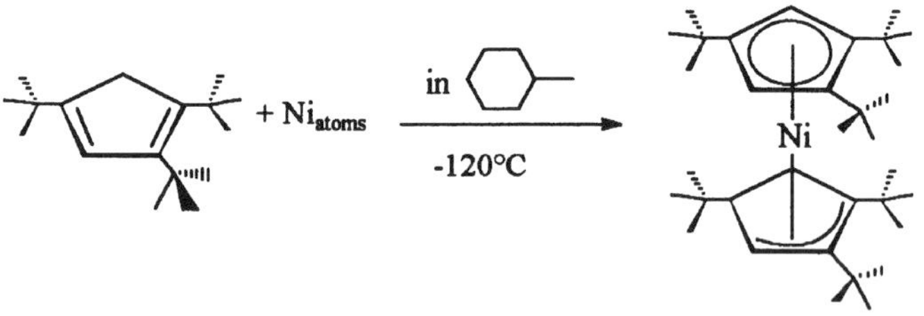

$$+ \text{Ni}_{atoms} \xrightarrow[-120°\text{C}]{\text{in} \bigcirc} \quad (26)$$

Another approach, also directly aimed at the synthesis of ligand stabilized molecular clusters, is reaction of metal atoms with suitable organometallics, either by a cocondensation procedure or by metal atom solution techniques (eq. 27).

$$\text{M}_{atoms} + \text{M}_x\text{L}_y \text{ compound} + \text{organic ligand L}_z \xrightarrow[\substack{\text{a) static reactor} \\ \text{or b) rotary solution reactor}}]{T < -100°\text{C}} \text{MM}_x\text{L}_y\text{L}_z \text{ cluster} \quad (27)$$

Early reports show that "solvated iron atoms" [$(\eta\text{-toluene})_2\text{Fe}$] react with [$\text{Mn}_2(\text{CO})_{10}$]/THF by stripping of both toluene ligands from [$(\eta\text{-toluene})_2\text{Fe}$], and formation of a heteronuclear Fe/Mn carbonylate anionic cluster [$\text{K}\{\text{Fe}_2\text{Mn}(\text{CO})_{12}\}$] [81].

Systematic investigation of the reactions of Co and Ni atoms with either [$\text{Fe}(\text{CO})_5$] or [$\text{Cp}^\text{R}\text{Co}(\text{CO})_2$] (R = Me$_5$, 1,3-$i$-propyl, 1,3-$t$-butyl) in the presence of mesitylene or Me$_5$CpH resulted in the macroscale formation of homo- and heteronuclear ligand stabilized molecular clusters (eqs. 28–31) [82].

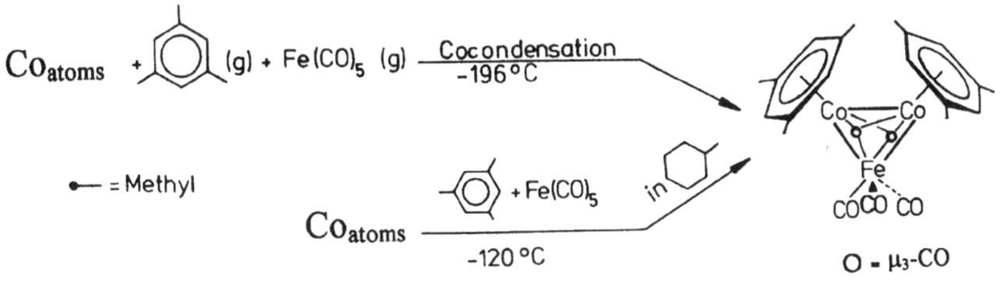

1.4. APPLICATIONS OF HT SPECIES IN CHEMISTRY

$$Co_{atoms} + \text{(mesitylene)(g)} + Fe(CO)_5 \text{(g)} \xrightarrow[-196\,°C]{\text{Cocondensation}} \text{[Co}_2\text{Fe cluster]} \quad (28)$$

• = Methyl

$$\text{(mesitylene)} + Fe(CO)_5 \xrightarrow[-120\,°C]{Co_{atoms}, \text{ in cyclohexane}} \text{[same cluster]}$$

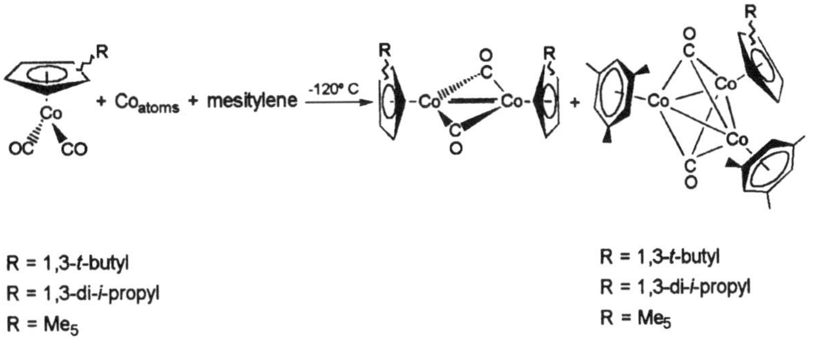

O = μ_3-CO

$$\text{(RCp)Co(CO)}_2 + Co_{atoms} + \text{mesitylene} \xrightarrow{-120°\,C} \text{[Co}_2\text{ dimer]} + \text{[Co}_4\text{ cluster]} \quad (29)$$

R = 1,3-t-butyl
R = 1,3-di-i-propyl
R = Me$_5$

R = 1,3-t-butyl
R = 1,3-di-i-propyl
R = Me$_5$

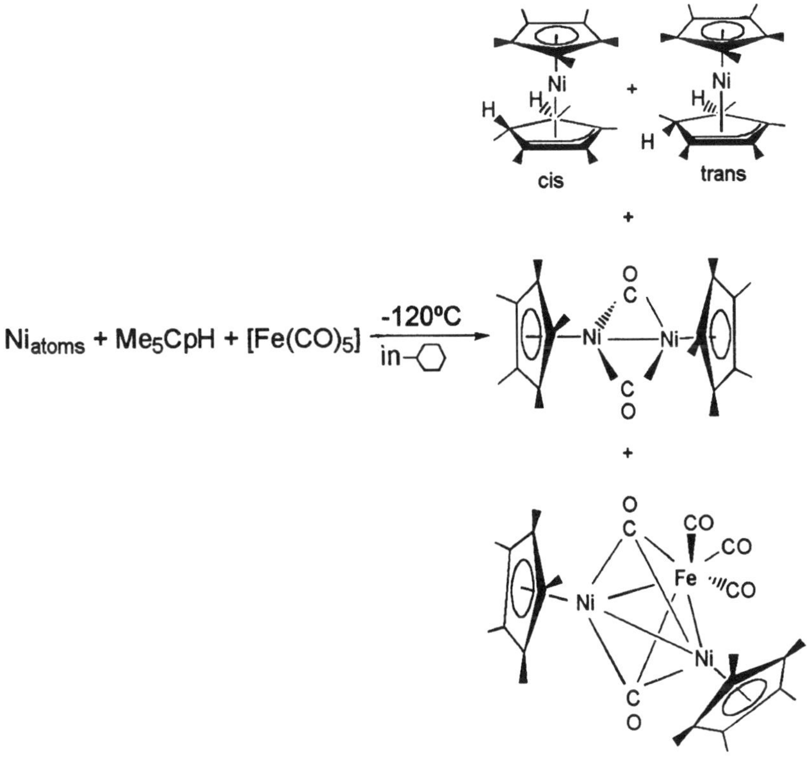

$$Ni_{atoms} + Me_5CpH + [Fe(CO)_5] \xrightarrow[\text{in cyclohexane}]{-120°C} \text{cis + trans Ni hydrides + Ni}_2(CO)_2 \text{ + Ni}_2Fe(CO)_5 \text{ cluster} \quad (30)$$

Ni_{atoms} + Me₅CpH +[(Cp)Co(CO)₂] ──→

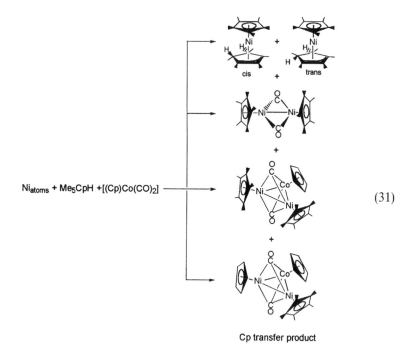

Cp transfer product

(31)

In each of these reactions metal atoms were vaporized into solutions of a mixture of the organic and organometallic ligands in methylcyclohexane. When multiple product formation occurs suitable separation techniques have to be applied. The homo- and heterotrinuclear carbonyl clusters formed obey the EAN rule, representing stable 48 VE closed shell clusters. Apparently, reactive half sandwich fragments $\{(\eta^6\text{-mesitylene})Co\}$ and $[(\eta^5\text{-Me}_5Cp)Ni\}$ are crucial in the formation of these clusters. In reaction (30) an interesting ligand transfer from the organometallic reactant complex to a naked Ni atom is observed, giving a mixture of clusters with heteronuclear Ni_2Fe cores having varying Cp ligands.

Both approaches, controlled ligand metal aggregation (see Scheme 1.5 and reactions (20)–(26)) or reaction of organometallics with metal atoms (see eqs. (27) and (28)–(31)) can lead to ligand stabilized metal clusters. However, there seems to be no universal way of favoring cluster compound formation by using single metal atoms, and other routes might be successful, too, and have to be explored in order to find additional new and unconventional approaches to the macroscale synthesis of ligand stabilized metal clusters.

1.4.3 Syntheses Using Lanthanide Elements

Mononuclear Complexes The properties and chemistry of the lanthanide metals resemble those of the alkali and alkaline earth metals in some respects, and ionic bonding is often favored by these metals. Most of the lanthanide

1.4. APPLICATIONS OF HT SPECIES IN CHEMISTRY

metals are easy to vaporize and are therefore easy to generate in their monoatomic form on a preparative scale [1]. Synthesis of the lanthanide complex bis(η^5-Me$_5$Cp)samarium(II) by the direct reaction of samarium atoms with Me$_5$CpH [83] was the starting point for an extensive organosamarium chemistry based on this complex, which has developed during the last few years and has led to a variety of unusual complexes based on this compound.

The reaction involving Y and Gd vapors resulted in the synthesis of the first zerovalent bis(η^6-arene)Y(0),Gd(0) sandwich molecules, and intensive studies followed, leading to stable complexes of this type with the metals Nd, Tb, Dy, Ho, Er and Lu. The use of the bulky arene ligand 1,3,5-tri-t-butyl benzene has proven especially useful in these investigations [84]. Some catalytic applications of ht generated lanthanide species are discussed in the following section.

1.4.4 HT Species of the Transition Elements and Lanthanide Elements as Sources of Metal Powders, Metal Colloids, Metal Containing Polymers and Composites

The growing ability of chemists to control and manipulate substances on a very small scale has opened up a new field of research on what are known as nanostructured materials. This terminology arises because the size of their building blocks is on the order of nanometers or even tens of nanometers. Apart from several classical synthetic routes to materials in this size regime known to date, non-classical ht syntheses offer new and promising ways to prepare such materials. Conventional techniques for the preparation of, for example, ultrafine particles include techniques like metal ion reduction [85] (Scheme 1.6) or decomposition of organometallics [86].

Turning to non-classical ht techniques, the gas evaporation method (see Section 1.3.4) for example, was widely used to produce clean ultrafine particles

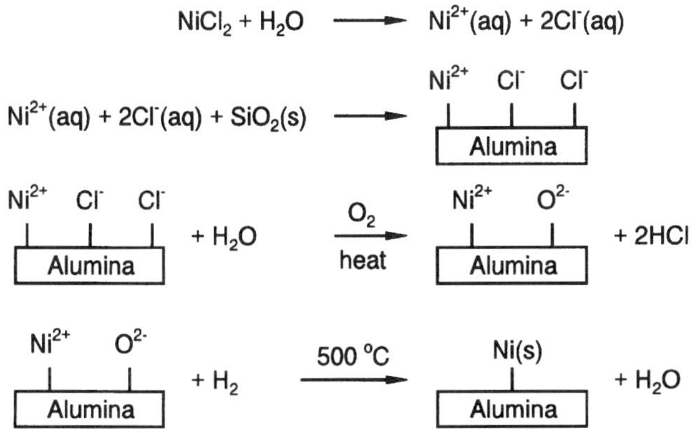

$$NiCl_2 + H_2O \longrightarrow Ni^{2+}(aq) + 2Cl^-(aq)$$

$$Ni^{2+}(aq) + 2Cl^-(aq) + SiO_2(s) \longrightarrow \underset{\text{Alumina}}{Ni^{2+}\ Cl^-\ Cl^-}$$

$$\underset{\text{Alumina}}{Ni^{2+}\ Cl^-\ Cl^-} + H_2O \xrightarrow{O_2,\ \text{heat}} \underset{\text{Alumina}}{Ni^{2+}\ O^{2-}} + 2HCl$$

$$\underset{\text{Alumina}}{Ni^{2+}\ O^{2-}} + H_2 \xrightarrow{500\ ^\circ C} \underset{\text{Alumina}}{Ni(s)} + H_2O$$

Scheme 1.6 Classical metal ion reduction procedure for the synthesis of a supported heterogeneous catalyst.

through evaporation of elements in an inert gas. Another useful technique for the generation of ht species of any material is the application of laser generated plumes which can lead to vaporization and ablation of ht species [87]. In this chapter some selected ht methods, which have led to materials with unusual catalytic and magnetic properties as well as to polymers containing metals will be highlighted.

The preparation using vapor synthesis techniques, of sub-colloidal ruthenium, rhodium, and palladium particles in non-aqueous solvents was reported recently [88]. The dimensions of these particles, which can be controlled within the 1–3 nm size regime, effectively bridge the gap between high nuclearity clusters and conventional colloidal metals. Solutions of the platinum group metal nanoparticles were prepared by metal vapor synthesis techniques using a commercial electrostatically focused electron beam rotary reactor. In a typical experiment 0.1 g platinum was evaporated during a three hour period and cocondensed at 77 K with ca. 200 ml of degassed, dried methylethylketone. After slow heating and melting, the resultant brown liquid was transferred under anaerobic conditions from the reactor flask into a Schlenk tube. In the same way rhodium nanoparticles can be prepared by cocondensation of rhodium vapor with toluene. Such procedures can be used as a general synthetic technique to prepare metal particles in the size range between large ligand stabilized molecular clusters such as $[Ni_{38}Pt_6(CO)_{48}H_{6-n}]$ [89] and a small colloid of typically 10–20 nm size [90].

Another synthetically very useful technique employing ht species which has a broad scope of applications in materials science is the use of "solvated metal atoms", already introduced in Section 1.4.3. Since these types of labile coordinated complexes are valuable reagents in organometallic chemistry, they are also good precursors to metal particles or supported metal particles after the organic ligands are completely stripped away. Particle nucleation can then take place and ultrafine particles can be prepared from these precursors. Final size and nucleation of these particles can be controlled by solvent choice, dilution effects, and warm up rates of the thermo-labile "solvated metal atoms". Particles prepared so far have crystallite sizes ranging from 3–10 nm and exhibit extremely high surface reactivities. In the presence of a solid support, nucleation and particle growth on the support surface often leads to highly dispersed heterogeneous catalysts which have been termed SMAD catalysts [91] (see Table 1.5 and Scheme 1.7).

Such powders and catalysts, supported or unsupported, often have high catalytic performance due to a very high dispersion of the active particles. During the preparation of these particles no additional chemical reduction steps are usually involved as these would cause sintering or agglomeration of the active species and lead to a decrease in catalytic activity.

Various characterization studies of SMAD catalysts have been carried out, and the following picture was developed concerning particle growth from organic media; the solvated metal atoms nucleate at surface OH sites or Lewis acid sites of the catalyst support (Al_2O_3, SiO_2, MgO or C). More surface sites

1.4. APPLICATIONS OF HT SPECIES IN CHEMISTRY

Table 1.5 Metal atom solvent combinations giving different "solvated metal atoms" species which can be used in the synthesis of SMAD catalysts [15b].

Metal M	Most appropriate solvents	Probable structure	Approximate thermal stability °C
Mn	toluene	bis(arene) sandwich	−40
Fe	toluene	bis(arene) sandwich	−30
Co	toluene	bis(arene) sandwich	−50
Ni	toluene, THF	bis(arene) sandwich tetra-THF, Ni-O-binding	−80
Zn	THF, diglyme dioxane	tetra-THF, Zn-O binding	
Cd	THF, dioxane, toluene, diglyme		
In			
Sn	THF, acetone, dioxane, toluene, ethanol, diglyme		
Pb	THF, diglyme, dioxane		
Ag	THF, toluene		
Pd	THF, toluene		
Pt	toluene	bis(arene) sandwich	
Re	toluene	arene-metal-hydride	

lead to smaller metal particles since the growth of particles proceeds until metal concentration in the solution is depleted [91a]. The particles deposited are amorphous and particle sizes are well below 25 Å. Early studies were mainly concerned with the synthesis of monometallic SMAD catalysts containing, for example, Fe, Co, and Ni metals highly dispersed on various supports. Testing these in hydrogenation or Fischer-Tropsch (FT) reactions showed that these non-classical systems lead to much higher activities than conventionally prepared catalysts [92]. Bimetallic SMAD catalysts can be prepared by simultaneous vaporization of two ht species and these were studied recently [91]. Examples are Co-Mn [92a], Fe-Mn [92b], Fe-Co [93, 94], Pt-Re [95], and Pt-Sn [96] which are in the case of Fe-Co, Fe-Mn and Co-Mn, excellent catalysts for FT reactions, hydrogenolysis (Pt-Re), and hydrocarbon reforming (Pt-Sn). The Pt-Sn system is particulary interesting since Sn is inactive itself in hydrocarbon reforming but enhances the selectivity and lifetime of the SMAD Pt-Sn catalyst. This nicely emphasizes the mutual benefit of both components in this bimetallic SMAD catalyst, by their direct combination using co-evaporation followed by *in situ* co-impregnation on the support surface.

Low valent lanthanide catalysts exhibit unusual selectivities when they are prepared from ht species since they represent highly dispersed systems with large

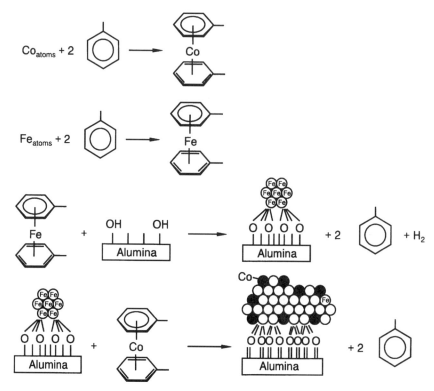

Scheme 1.7 Principle of the solvated metal atom dispersion (SMAD) technique for the generation of highly dispersed bimetallic heterogenous catalysts.

surface areas similar to the ones obtained from the transition metals. Sm and Yb particles catalyze the hydrogenation of alkenes, dienes and aromatic compounds [97], the dimerization of ethene [98] as well as the polymerization of conjugated dienes. Mechanistic studies of the selectivity patterns and H–D exchange results suggest these ht prepared rare earth particles differ significantly in their reactivity from transition metal particles.

A series of investigations led to the formation of unique colloidal Pd, Pt, and Sn particles simply by vaporization of the metals into organic solvents like acetone [99] or solutions of iso-butylaluminoxane oligomers [100]. Solvent removal can lead to formation of Pd and Sn films which have semiconductor properties. Pt_{atoms}/i-butylaluminoxane colloids react with CO and H_2O in a subsequent reaction to form the cluster anion [$\{Pt_{12}(CO)_{24}\}^{2-}$] in about 70% yield. For a compilation of selected recent work in the area of colloids by using ht derived species see [16, 101].

10 nm iron particles can be embedded into polymethylmethacrylate using "solvated iron atoms" as a metallic precursor complex [76]. High metal to ligand ratios in polymers can be obtained by direct polymerization of acetylene with Ge

1.4. APPLICATIONS OF HT SPECIES IN CHEMISTRY

or Sn atoms (up to 75% metal content). Despite this high metal content, conductivity is low due to the low ratio of unpaired electrons vs. carbon atoms in the resulting polymers [102]. Very recent reports showed that ht generated titanium and zirconium ions and atoms are able to cationically polymerize liquid monomers. This technique of metal ion and metal atom induced polymerization produces high molecular weight polymers with narrow size distribution. It even allows the incorporation of ultrafine particles into the polymers and the metal atoms themselves serve to catalyze polymerization of the host monomer [103].

Another interesting and new field for the use of ht species in materials science is macroscale preparation of nanoparticles with special magnetic properties. Several synthetic approaches using such particles have been studied [104]. One difficulty in the preparation of magnetic materials is the contamination by surface oxide layers which greatly influences the magnetic performance of ultrafine metallic particles and most often complicates the understanding of magnetic properties of such materials. Even when working under superior anaerobic conditions, surface oxidation of such particles occurs, independent of whether preparation is by conventional or non-classical ht techniques. This is a major drawback when working with high surface area particles in general. Small, mainly oxidizing contaminants have dramatic effects on such magnetic properties as susceptibility and coercivity.

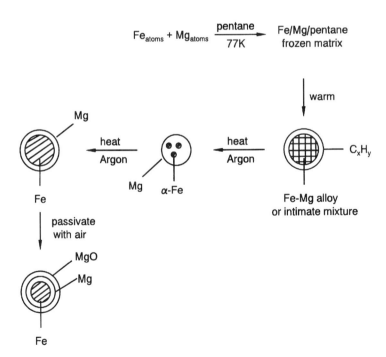

Scheme 1.8 Formation of Fe-Mg alloy nano-particles using a simultaneous co-evaporation technique followed by phase segregation.

A new promising approach in this field is the co-evaporation of Fe and Mg atoms and their co-deposition together with an organic solvent to produce novel nanoscale magnetic materials containing nanoparticles embedded into an inner protecting shell of magnesium surrounded by an outer shell of MgO (see Scheme 1.8) [104a, 105].

The relative composition of Fe and Mg can be controlled over a wide range. The preparation of such core shell structures with a *sacrificial metal* (besides Mg, Li can also be used as such a metal) has allowed for the first time magnetic studies of protected iron particles that do not possess a coating of a magnetic oxide. Comparisons of the magnetic properties of such uncoated and oxide coated particles show that the protecting shell is the key parameter for controlling magnetic properties of nanoscale particles. Another interesting technique for encapsulating nanoparticles of ferromagnetic transition metals such as Fe, Co an Ni makes use of protective graphitic shells [106]. In an electrical arc-process, metal particles and graphite encapsulated metal particles in the size

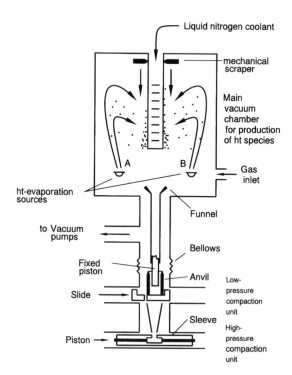

Figure 1.15 Illustration showing the classic gas phase condensation route to nanophase materials. The precursor material evaporates from sources A and/or B, condenses in a high purity inert gas and is transported *via* convection to the liquid-nitrogen-filled cold finger. The powdery material is then scraped from the cold finger, collected *via* the funnel, and consolidated, first in a low-pressure compaction device and then in a high-pressure unit, all in vacuum. After [107].

range 7–14 nm are prepared. By an acid treatment, separation of the metal and metal/carbon particles can be obtained. The complete coating of the metal particles by carbon is reflected by the intrinsic ferromagnetic properties of the interior metals. Thus, encapsulation of controlled size protected ferromagnetic nanocrystalline metals which retain their intrinsic metallic behavior can be achieved.

All of the above highlighted applications of non-classical ht techniques in materials science arise from the fact that the properties of the materials prepared depend on their nanostructure which is determined by the way the particles are prepared and then processed. From this viewpoint, non-classical ht chemistry has new synthetic routes to offer. A particularly interesting ht preparation technique for the macroscale synthesis of single phase oxide ceramics such as alumina, zirconia and titania has already gained some importance in a small scale technical process to produce these novel oxide materials. In this process a metal is evaporated in an inert gas atmosphere at reduced pressure and then condensed in the gas phase to form metal clusters or nanocrystals (this is essentially the gas evaporation method, see Section 1.3.4). These aggregate and can be oxidized in a second step to fine ceramic powder, which is then scraped from the cold finger and compacted inside the apparatus (see Figure 1.15). This compact may be sintered to yield the final polycrystalline ceramic shape [107]. To make nanocomposites, two different ht species may be vaporized and condensed at the same time. At the moment, up to 100 g per hour of such advanced nanophase material can be produced and a scale up of this process seems possible. Disks prepared from such matter have pores and grains below 8 nm, do not scatter much light, but are fairly transparent. Ceramics of this kind may find application in sensor and filtration technology in the near future [107].

ACKNOWLEDGEMENTS

The author thanks the *Deutsche Forschungsgemeinschaft* for support through a Heisenberg Fellowship and Prof. Dr. K. J. Klabunde for hospitality during a 1995 summer sabbatical at Kansas State University.

Notes added in proof:

1 As revealed from ESR spectra, cocondensation of Al atoms and HCl in Ar matrices at 4 K leads to formation of HALCl and $AlCl_2$.: Köppe, R., Kasai, P. H., J. Am. Chem. Soc., 118, 1996, 135.

Very recently new studies on the oxidative addition reaction of B atoms with NH_3 to produce HBNH, BNBH and B_2N have appeared: Thompson, C. A., Andrews, L., J. Am. Chem. Soc., 117 (1995) 10125.

Preparation and magnetic properties of matrix isolated zerovalent $Al(PF_3)_2$ have been reported: Howard, J. A., Jones, R., M. Tomietto, Inorg. Chem., 34 (1995) 3097.

2 Metal atom synthesis and reactivity studies of (η^6-cyclohepta-1,3,5-triene)(η^6-cycloocta-1,5-diene)iron(0) has been reported: Pertici, P, Breschi, C., Baretta, G. U., Marchetti, F., Vitulli, G., J. Organomet. Chem., 502 (1995) 95.

1.5 REFERENCES

[1] Klabunde, K. J., Chemistry of Free Atoms and Particles, Academic Press, New York, 1980.

[2] Selected reviews: a) Müller-Buschbaum, H., Angew. Chem., 93 (1981) 1; Angew. Chem. Int. Ed. Engl., 20 (1981) 22; b) Möhr, S., Müller-Buschbaum, H., Angew. Chem., 107 (1995) 691; Angew. Chem. Int. Ed. Engl., 34 (1995) 631; c) for a selection of preparative techniques in solid state chemistry see: Schenk, P. W., Steudel, G., Brauer, G., (Brauer, G. (ed.)), Handbuch der Präparativen Anorganischen Chemie, Teil 1, p. 40–50, F. Enke Verlag, Stuttgart, 1975; d) West, A. R., Solid State Chemistry and its Applications, John Wiley, New York, 1984; e) aspects of ht chemistry related to materials science are covered by the bimonthly journal: Margrave, J. (ed.), High Temperature and Materials Science. An International Journal, Humana Press; CVD, principles and techniques: Hitchman, M. L., Jensen, K. F., Chemical Vapor Deposition: Principles, Techniques and Applications, Academic Press, San Diego, 1993; f) Kodas, T., Hampden-Smith, M., The Chemistry of Metal CVD, VCH-Verlagsgesellschaft, Weinheim, 1994; g) Carlsson, I. O., (Rickerby, D. S., Matthews, A. (eds.)), Advance Surface Coating: a Handbook of Surface Engineering, p. 162–217, Chapman and Hall, New York, 1991.

[3] a) Skell, P. S., Wescott Jr., L. D., J. Am. Chem. Soc., 85 (1963) 1023; b) Skell, P. S., Wescott Jr., L. D., Goldstein, P., Engel, R. R., J. Am. Chem. Soc., 87 (1965) 2829.

[4] Timms, P. L., J. Chem. Soc., Chem. Commun., (1969) 1033.

[5] Roginsky, S., Schalnikoff, A., Kolloid Z., 43 (1927) 67.

[6] Klabunde, K. J., Free Atoms, Clusters, and Nanoscale Particles, Academic Press, New York, 1994.

[7] Timms, P. L., (Ozin, G. A., Moskovits, M. (eds.)), Cryochemistry, p. 61–131, John Wiley, New York, 1976.

[8] a) Blackborrow, J. R., Young, D., Metal Vapour Synthesis in Organometallic Chemistry, Springer Verlag, Berlin, 1979; b) Ozin, G. A., Moskovits, M., Cryochemistry, John Wiley, New York, 1976; c) Trost, H. K., Metallorganische Verbindungen nach Mass. Technik und Chemie der Metallatomsynthese, VCI-Verlag GmbH, Düsseldorf, 1992; d) Herberhold, M., Chem. uns. Zeit, 10 (1976) 120. e) Andrews, P. A., (Wayda, A. L., Darensbourgh, M. Y. (eds.)), Experimental Organometallic Chemistry, ACS Symposium Series, 357 (1987) 158.

[9] a) Francis, C. G., Timms, P. L., J. Chem. Soc., Chem. Commun., (1977) 466; b) Francis, C. G., Timms, P. L., J. Chem. Soc., Dalton Trans., (1980) 1401; c) Ozin, G. A., Andrews, M. P., Nazar, L. F., Huber, H. X., Francis, C. G., Coord. Chem. Rev., 48 (1983) 203; d) Hooker, P. D., Timms, P. L., J. Chem. Soc., Chem. Commun., (1988) 158; e) Hooker, P. D., Ph.D. Thesis 1990, Bristol University, England; f) Hooker, P. D., Klabunde, K. J., Chem. Mater., 5 (1993) 1089.

1.5. REFERENCES

[10] Radonovich, L. D., Eyring, M. W., Groshens, T. J., Klabunde, K. J., J. Am. Chem. Soc., 104 (1982) 2816.

[11] Elschenbroich, C., Schneider, J., Massa, W., Baum, G., Mellinghoff, H., J. Organomet. Chem., 355 (1988) 163.

[12] Elschenbroich, C., Schneider, J., Burdorf, H., J. Organomet. Chem., 391 (1990) 195.

[13] One solvent commonly used as a slush bath for cooling rotary reactors is pentane. The author never had any problems using this solvent as an external cooling bath, however a safer alternative is to use Freon-114-B2 fluorocarbon for example, as recommended by: Ittel, S. D., Tolman, C. A., Organometallics, 1 (1982) 1433.

[14] a) Ozin, G. A., Andrews, M. P., Francis, C. G., Huber, H. X., Molnar, K., Inorg. Chem., 29 (1990) 1068; b) Commercial suppliers: Torrovap Industries, Inc., 90 Nolan Court, Unit 39–40, Markham, Ontario, L3R 4L9, Canada; Planer, G. V., Ltd., Windmill Road, Sunbury-on-Thames, Middlesex, TW16 7HD, England.

[15] a) Li, Y.-X., Klabunde, K. J., J. Catal., 126 (1990) 173; b) Klabunde, K. J., Li, Y.-X., Tan, B. J., Chem. Mat., 3 (1991) 30.

[16] Bradley, J. S., (Schmid, G. (ed.)), Clusters and Colloids, chap. 6, p. 479, VCH Verlagsgesellschaft mbH, Weinheim, 1994.

[17] a) Kimoto, U., Kamiya, Y., Nonoyama, M., Ueyda, R., Jpn. J. Appl. Phys., 2 (1963) 702; b) Hayashi, C., Phys. Today, November (1987) 44; b) Niklasson, G. A., J. Appl. Phys., 62 (1987) 258.

[18] a) Koch, H. S., Khemani, K. C., Wudl, F., J. Org. Chem., 56 (1991) 4593; b) Haufler, R. E., Conclicao, J., Chibante, L. P. F., Chai, Y., Byme, N. E., Flanagan, S., Haley, M. M., OBrien, S. C., Pan, C., Xiao, Z., Billups, W. E., Ciufolini, M. A., Hauge, R. H., Margrave, J. L., Wilson, L. J., Cure, R. F., Smalley, R. E., J. Phys. Chem., 94 (1990) 8634; c) Haufler, R. E., (Kadish, U. M., Ruoff, R. S. (eds.)), Fullerenes: Recent Advances in the Chemistry and Physics of Fullerenes and Related Materials, p. 50–67, Electrochemical Society Inc. Pennington NJ, USA, 1994; d) Hirsch, A., The Chemistry of the Fullerenes, Thieme Verlag, New York, 1994.

[19] Klabunde, K. J., White, C. M., Efner, H. F., Inorg. Chem., 13 (1974) 1778.

[20] Maddren, P. S., Ph. D. Thesis, Bristol University, England, 1975.

[21] Hawker, P. N., Timms, P. L., J. Chem. Soc., Dalton Trans., (1983) 1123.

[22] a) Tacke, M., Chem. Ber., 128 (1995) 91; b) for a state of the art review of Li atom reactions with small molecules see: Manceron, L., Andrews, L., (Sapse, A. M., von Rague Schleyer, P. (eds.)), Lithium Chemistry. A theoretical and experimental overview, chap. 4, p. 89, John Wiley, New York, 1995.

[23] Van Eikema Hommes, J. R. N., Bickelhaupt, F., Klumpp, G. W., Angew. Chem., 100 (1988) 1100; Angew. Chem. Int. Ed. Engl., 27 (1988) 1083.

[24] Mochida, K., Yamanishi, T., J. Organomet. Chem., 332 (1987) 247.

[25] Bryce-Smith, D., Skinner, A. C., J. Chem. Soc., (1963) 577.

[26] Gilman, H., Schulze, F., J. Am. Chem. Soc., 48 (1926) 2463.

[27] Glacet, Z. C., Bull. Soc. Chim. Fr., 5 (1938) 895.

[28] Skell, P. S., Girard, J. E., J. Am. Chem. Soc., 94 (1972) 5518.

[29] a) Engelhardt, L. M., Junk, P. C., Raston, C. L., White, A. H., J. Chem. Soc., Chem. Commun., (1988) 1500; b) Gardiner, M. G., Raston, C. L., Kennard, C. H. L., Organometallics, 10 (1991) 3680.

[30] a) Andersen, R. A., Blom, R., Boncella, J. M., Burns, C. D., Volden, H. V., Acta Chem. Scand. Ser. A., 41 (1987) 24; b) Andersen, R. A., Boncella, J. M., Burns, C. D., Blom, R., Haaland, A., Volden, H. U., J. Organomet. Chem., 312 (1986) C49; c) Andersen, R. A., Blom, R., Burns, C. J., Volden, H. U., J. Chem. Soc., Chem. Commun., (1987) 768.

[31] Westerhausen, M., Angew. Chem., 106 (1994) 1585; Angew. Chem. Int. Ed. Engl., 33 (1994) 1493.

[32] Timms, P. L., (Ozin, G. A., Moskovits, M. (eds.)), Cryochemistry, p. 153–159, John Wiley, New York, 1976.

[33] Chenier, J. H.-B., Howard, J. A., Mile, B., J. Am. Chem. Soc., 109 (1987) 4109.

[34] a) Kasai, P. H., Jones, P. M., J. Phys. Chem., 92 (1988) 1060; b) Kasai, P. H., J. Am. Chem. Soc., 104 (1982) 1165.

[35] Histed, M., Howard, J. A., Morris, H., Mile, B., J. Am. Chem. Soc., 110 (1988) 5290.

[36] Klabunde, K. J., Murdock, T. O., J. Org. Chem., 44 (1979) 3901.

[37] Cardenas-Trivino, G., Retamal, C., Klabunde, K. J., J. Appl. Polym. Sci. Appl. Polym. Symp., 49 (1991) 15.

[38] a) Schnöckel, H., Z. Naturforsch., B 31 (1976) 1291; b) Schnöckel, H., J. Mol. Struct., 50 (1978) 267, 275; c) Tacke, M., Schnöckel, H., Inorg. Chem., 28 (1989) 2895; d) Tacke, M., Kreienkamp, H., Plaggenberg, L., Schnöckel, H., Z. Anorg. Allg. Chem., 604 (1991) 35; e) Loos, D., Schnöckel, H., Fenske, D., Angew. Chem., 105 (1993) 1124; Angew. Chem. Int. Ed. Engl., 32 (1993) 1059; f) Mocker, M., Robl, C., Schnöckel, H., Angew. Chem., 106 (1994) 946; Angew. Chem. Int. Ed. Engl., 33 (1994) 862; g) Mocker, M., Robl, C., Schnöckel, H., Angew. Chem., 106 (1994) 1860; Angew. Chem. Int. Ed. Engl., 33 (1994) 1754; h) Paetzold, P., Angew. Chem., 103 (1991) 559; Angew. Chem. Int. Ed. Engl., 30 (1991) 544; i) Schneider, J. J., Angew. Chem., 106 (1994) 1914; Angew. Chem. Int. Ed. Engl., 33 (1994) 1830.

[39] a) Dohmeier, C., Robl, C., Tacke, M., Schnöckel, H., Angew. Chem., 103 (1991) 594; Angew. Chem. Int. Ed. Engl., 30 (1991) 564; b) Schulz, S., Roesky, H. W., Koch, H. D., Sheldrick, G. M., Stalke, D., Kuhn, A., Angew. Chem., 105 (1993) 1828; Angew. Chem. Int. Ed. Engl., 32 (1993) 1729; c) Schulz, S., Hanning, L., Herbst-Irmer, R., Roesky, H. W., Sheldrick, G. M., Angew. Chem., 106 (1994) 1052; Angew. Chem. Int. Ed. Engl., 33 (1994) 969; d) Schulz, S., Schoop, T., Roesky, H. W., Hämer, L., Stiner, A., Herbst-Irmer, R., Angew. Chem., 107 (1995) 1015; Angew. Chem. Int. Ed. Engl., 34 (1995) 919.

[40] Schnöckel, H., Leimkuhler, M., Lutz, R., Mattes, R., Angew. Chem., 98 (1986) 929; Angew. Chem. Int. Ed. Engl., 25 (1986) 921.

[41] Bierschenk, T. R., Lagow, R. J., Inorg. Chem., 22 (1983) 359.

[42] Juhlke, T. J., Glanz, J. I., Lagow, R. J., Inorg. Chem., 28 (1989) 980.

[43] a) Eujen, R., Patorra, A., J. Organomet. Chem., 438 (1992) C1; b) Eujen, R., Thumpmann, U., J. Organomet. Chem., 433 (1992) 63.

[44] Tacke, M., Organometallics, 13 (1994) 4124.

1.5. REFERENCES

[45] a) Krätschmer, W., Lamb, L. D., Fostiropoulos, K., Huffmann, D. R., Nature, 347 (1990) 354; b) Haufler, R. E., Chai, Y., Chibante, L. P. F., Conceicao, J., Jin, C., Wang, L.-S., Maryyama, S., Smalley, R. E., Mater. Res. Soc. Symp. Proc., 206 (1991) 627; c) Taylor, R., Hare, J. P., Abdul-Sada, A. K., Kroto, H. W., J. Chem. Soc., Chem. Commun., (1990) 1423; d) Ajie, H., Alvarez, M. M., Anz, S. J., Beck, R. D., Diederich, F., Fostiropoulos, K., Huffman, D. R., Kratschmer, W., Rubin, Y., Schriver, K. E., Senharma, D., Whetten, R. L., J. Phys. Chem., 94 (1990) 8630; e) Scrivens, W. A., Bedworth, P. V., Tour, J. M., J. Am. Chem. Soc., 114 (1992) 7917; f) Parker, D. H., Wurz, P., Chatterjee, K., Lykke, K. R., Hunt, J. E., Pellin, M. J., Hemminger, J. C., Gruen, D. M., Stock, L. M., J. Am. Chem. Soc., 113 (1991) 7499; g) Khemani, K. C., Prato, M., Wudl, F., J. Org. Chem., 57 (1992) 3254; h) Chatterjee, K., Parker, D. H., Wurtz, P., Lykke, K. R., Gruen, D. M., Stock, L. M., J. Org. Chem., 57 (1992) 3253; i) Li, Q., Wudl, F., Thilgen, C., Whetten, R. L., Diederich, F., J. Am. Chem. Soc., 114 (1992) 3994; j) Endo, M., Kroto, H. W., J. Phys. Chem., 96 (1992) 6941; k) Drewello, T., Asmus, K.-D., Stach, J., Herzschuh, R., Kao, M., Foote, C. S., J. Phys. Chem., 95 (1991) 10554; l) Braun, T., Angew. Chem., 104 (1992) 602; Angew. Chem. Int. Ed. Engl., 31 (1992) 588; m) Bae, Y. K., Lorents, D. C., Mahotra, R., Becker, C. H., Tse, D., Jusinski, L., Mater. Res. Soc. Proc., 206 (1991) 733; n) Mittelbach, A., von Schnering, H. G., Carlsen, J., Janiak, R., Quast, H., Angew. Chem. Int. Ed. Engl., 31 (1992) 1640; o) Scrivens, W. A., Tour, J. M., J. Org. Chem., 57 (1992) 6932; p) Peters G., Jansen M., Angew. Chem., 104 (1992) 240; Angew. Chem. Int. Ed. Engl., 31 (1992) 223.

[46] a) Chai, Y., Guo, T., Jin, C., Haufler, R. E., Chibante, L. P. F., Fure, J., Wang, L., Alford, J. M., Smalley, R. E., J. Phys. Chem., 95 (1991) 7564; b) Alvarez, M. M., Gillan, E. G., Holczar, K., Kaner, R. B., Mink, S., Whetten, R. L., J. Phys. Chem., 95 (1991) 10564; c) Bandow, S., Kitagawa, H., Mitani, H., Inokuchi, H., Saito, Y., Yamaguchi, H., Hayashi, N., Sato, H., Shinohara, H., J. Phys. Chem., 96 (1992) 9609; d) Shinohara, H., Yamaguchi, H., Hayashi, N., Sato, H., Ohkohehi, M., Ando, Y., Saito, Y., J. Phys. Chem., 97 (1993) 4259; e) Kikuchi, K., Suzuki, S., Nakao, Y., Nakahara, N., Wakabayashi, T., Shiromaru, H., Saito, K., Ikemoto, I., Achiba, Y., Chem. Phys. Lett., 216 (1993) 67; f) Bandow, S., Shinohara, H., Saito, Y., Ohkohchi, M., Ando, Y., J. Phys. Chem., 97 (1993) 6101; g) Jansen, M., Peters, G., Wagner, N., Z. Anorg. Allg. Chem., 621 (1995) 689.

[47] Edelmann, F. T., Angew. Chem., 107 (1995) 1071; Angew. Chem. Int. Ed. Engl., 34 (1995) 981.

[48] a) Iijima, S., Nature, 354 (1991) 56; b) Ebbesen, T. W., Ajayan, P. M., Nature, 358 (1992) 220; c) for a recent review on the chemistry and physics of carbon nanotubes: Iijima, S., Endo, M. (guest eds.), Carbon, 33 (1995) 869.

[49] Moltzen, E. K., Klabunde, K. J., Senning, A., Chem. Rev., 88 (1988) 391.

[50] For a compilation of recent experimental developments in these fields see: Klabunde, K. J., Free Atoms, Clusters, and Nanoscale Particles, Chap. 2, p. 5–15, Academic Press, New York, 1994.

[51] Skell, P. S., Havell, J. J., McGlinchey, M. J., Acc. Chem. Res., 6 (1973) 97.

[52] a) Timms, P. L., Adv. Inorg. Chem. Radiochem., 14 (1972) 121; b) Timms, P. L., Endeavour, 27 (1968) 133.

[53] Elschenbroich, C., Nowotny, M., Metz, B., Massa, W., Graulich, J., Biehler, K., Sauer, W., Angew. Chem., 103 (1991) 601; Angew. Chem. Int. Ed. Engl., 30 (1991) 547.

[54] Elschenbroich, C., Koch, J., Kroker, J., Wünsch, M., Massa, W., Baum, G., Chem. Ber., 121 (1988) 1983.

[55] Elschenbroich, C., Hurley, J., Massa, W., Baum, G., Angew. Chem., 100 (1988) 27; Angew. Chem. Int. Ed. Engl., 27 (1988) 684.

[56] Elschenbroich, C., Schneider, J., Massa, W., Baum, G., Mellinghoff, H., J. Organomet. Chem., 355 (1988) 163.

[57] Cloke, F. G. N., Khan, K., Perutz, R. N., J. Chem. Soc., Chem. Commun., (1991) 1372.

[58] Cloke, F. G. N., Day, J. P., Green, J. C., Morley, C. P., Swain, A. C., J. Chem. Soc., Dalton Trans., (1991) 789.

[59] Selected references on microscale studies (spectroscopic work): a) Kafafi, Z. H., (Davies, J. A., Watson, P. L., Liebman, J. F., Greenberg, A. (eds.)), Selective Hydrocarbon Activation, VCH Publishers, New York, 1990; b) Klabunde, K. J., Jeong, G. H., Olsen, A. W., (Davies, J. A., Watson, P. L., Liebman, J. F., Greenberg, A. (eds.)), Selective Hydrocarbon Activation, VCH Publishers, New York, 1990; c) Armentrout, P. B., (Davies, J. A., Watson, P. L., Liebman, J. F., Greenberg, A. (eds.)), Selective Hydrocarbon Activation, VCH Publishers, New York, 1990, and references cited therein; selected references on macroscale studies (preparative work), examples for early transition metals: a) Green, M. L. H, Parkin, G., J. Chem. Soc., Chem. Commun., (1984) 1467; b) Cloke, F. G. N., Gibson, U. C., Green, M. L. H., Mtetwa, U. S. B., Prout, K., J. Chem. Soc., Dalton Trans., (1988) 2227; c) Bandy, J. A., Cloke, F. G. N., Green, M. L. H., O'Hare, D., Prout, K., J. Chem. Soc., Chem. Commun., (1984) 240; d) Green, M. L. H., O'Hare, D., J. Chem. Soc., Dalton Trans., (1987) 403; e) Green, M. L. H., O'Hare, D., J. Chem. Soc., Dalton Trans., (1986) 2409; examples for late transitions elements: f) Bandy, J. A., Green, M. L. H, O'Hare, D., Prout, K., J. Chem. Soc., Chem. Commun., (1984) 1402; g) Bandy, J. A., Green, M. L. H., O'Hare, D., J. Chem. Soc., Dalton Trans., (1986) 2477; h) Green, M. L. H., Joyner, D. S., Wallis, J. M., J. Chem. Soc., Dalton Trans., (1987) 2823; i) for a compilation of reviews dealing with C–H activation processes mediated by ht species see: Suslick, K.S. (ed.), High Energy Processes in Organometallic Chemistry, ACS Symposium Series, 333 (1987).

[60] a) Francis, C. G., Huber, H. X., Ozin, G. A., Inorg. Chem., 19 (1980) 219; b) Francis, C. G., Huber, H. X., Ozin, G. A., Angew. Chem., 92 (1980) 409; Angew. Chem. Int. Ed. Engl., 19 (1980) 402; c) Francis, C. G., Huber, H. X., Ozin, G. A., J. Am. Chem. Soc., 101 (1979) 6250; d) Ozin, G. A., Chemtech., August (1985) 448; e) Andrews, M. P., Ozin, G. A., J. Phys. Chem., 90 (1986) 3353; f) Ozin, G. A., Andrews, P. A., Nazar, L. F., Huber, H. X., Francis, C. G., Coord. Chem. Rev., 48 (1983) 203.

[61] Andrews, M. P., Ozin, G. A., Francis, C. G., Inorg. Chem., 22 (1982) 116.

[62] Klabunde, K. J., Free Atoms, Clusters, and Nanoscale Particles, chap. 4, p. 72–84, Academic Press, New York, 1994.

[63] a) Cook, N, D., Timms, P. C., J. Chem., Soc. Dalton Trans., (1983) 239; b) DeKock, C. W., McAfee, L. V., Inorg. Chem., 24 (1985) 4293.

[64] a) Groshens, T. J., Ph.D. Thesis, Kansas State University, USA, 1988; b) Groshens, T. J., Klabunde, K. J., Inorg. Chem., 29 (1990) 2979.

[65] Klabunde, K. J., Chemistry of Free Atoms and Particles, chap. 4, p. 71–75, Academic Press, New York, 1980.

1.5. REFERENCES

[66] a) Avent, G. A., Cloke, F. G. N., Flower, K. R., Hitchcock, P. B., Nixon, J. F., Vickers, D. M., Angew. Chem., 106 (1994) 2406; Angew. Chem. Int. Ed. Engl., 33 (1994) 2230; b) Cloke, F. G. N., Flower, K. R., Hitchcock, P. B., Nixon, J. F., J. Chem. Soc., Chem. Commun., (1994) 489.

[67] a) Benn, H., Wilke, G., Henneberg, D., Angew. Chem., 85 (1973) 1052; Angew. Chem. Int. Ed. Engl., 12 (1973) 1001; b) King, R. B., Ann. N.Y. Acad. Science, 295 (1977) 135.

[68] a) Zenneck, U., Angew. Chem., 102 (1990) 171; Angew. Chem. Int. Ed. Engl., 29 (1990) 126; b) Böhm, D., Knoch, F., Kummer, S., Schmidt, U., Zenneck, U., Angew. Chem., 107 (1995) 251; Angew. Chem. Int. Ed. Engl., 34 (1995) 198.

[69] Schneider, J. J., unpublished results, MPI für Kohlenforschung, Mülheim an der Ruhr, Germany, 1991.

[70] a) Cantrell, R. D., Shevlin, P. B., J. Am. Chem. Soc., 111 (1989) 2348; b) see also: Lagowski, J. J., Simons, L. H., J. Organomet. Chem., 249 (1983) 195.

[71] a) Asirvatham, V. S., Yao, Z., Klabunde, K. J., J. Am. Chem. Soc., 116 (1994) 5493; b) Yao, Z., Klabunde, K. J., Organometallics, 14 (1995), in press.

[72] a) Klabunde, K. J., Groshens, T. J., Organometallics, 1 (1982) 564; b) Lin, S. T., Groshens, T. J., Klabunde, K. J., Inorg. Chem., 23 (1984) 1; c) Choe, S. B., Kanai, H., Klabunde, K. J., J. Am. Chem. Soc., 111 (1989) 2875; d) Choe, S. B., Schneider, J. J., Klabunde, K. J., Radonovich, L. J., Ballantine, T. A., J. Organomet. Chem., 376 (1989) 419; e) Brezinski, M. M., Schneider, J. J., Radonovich, L. J., Klabunde, K. J., Inorg. Chem., 28 (1989) 2414; f) Glavee, G. N., Jagirdar, B. R., Schneider, J. J., Klabunde, K. J., Radonovich, L. J., Dodd, K., Organometallics, 11 (1992) 1043; g) Jagirdar, B. R., Palmer, R., Klabunde, K. J., Radonovich, L. J., Inorg. Chem., 34 (1995) 278; h) Jagirdar, B. R., Klabunde, K. J., J. Coord. Chem., 34 (1995) 31.

[73] a) Klabunde, K. J., Efner, H. F., Murdock, T. O., Ropple, R. J., J. Am. Chem. Soc., 98 (1976) 1021; b) for a review see: [68a]; c) Gastinger, R. G., Klabunde, K. J., Trans. Met. Chem., 4 (1979) 13.

[74] See ref. [13]–[18] as cited in [68a].

[75] a) Williams-Smith, D. K., Wolf, L. R., Skell, P. S., J. Am. Chem. Soc., 94 (1972) 4042; b) Middleton, R., Hull, J. R., Simpson, S. R., Tomlinson, C. H., Timms, P. L., J. Chem. Soc., Dalton Trans., (1973) 120; c) Ittel, S. D., Tolman, C. A., J. Organomet. Chem., 172 (1979) C47; d) Ittel, S. D., Tolman, C. A., Organometallics, 1 (1982) 1432; e) Atkins, R. M., Mackenzie, R., Timms, P. C., Turney, T. W., J. Chem. Soc., Chem. Commun., (1975) 764.

[76] Zenneck, U., Chem. uns. Zeit, 27 (1993) 208.

[77] Schneider, J. J., Specht, U., Goddard, R., Krüger, C., Ensling, J., Gütlich, P., Chem. Ber., 128 (1995) 941.

[78] a) Voellenbrock, F. A., Bouten, P. C. P., Trooster, J. M., van der Berg, J. P., Bour, J. J., Inorg. Chem., 17 (1978) 1345; b) Steggerda, J. J., Bour, J. J., van der Velden, J. W. A., Rev. Trav. Chim. Pays-Bas, 101 (1982) 165; c) Zimmermann, G. J., Wilcynski, R., Sneddon, L. G., J. Organomet. Chem., 154 (1978) C29; d) Kang, S. O., Carroll, P. J., Sneddon, L. G., Inorg. Chem., 28 (1989) 961; e) Klabunde, K. J., Groshens, T., Brezinski, M., Kennely, W., J. Am. Chem. Soc., 100 (1978) 4437; f) Schäufele, H., Pritzkow, H., Zenneck, U., Angew. Chem., 100 (1988) 1577; Angew. Chem. Int. Ed. Engl., 27 (1988) 1519; g) Vasquez, L., Pritzkow, H., Zenneck, U., Angew. Chem., 100 (1988) 705; Angew. Chem. Int. Ed. Engl., 27 (1988) 706.

[79] Timms, P. L., Proc. Roy. Soc. London A, 396 (1984) 1.
[80] a) Schneider, J. J., Goddard, R., Krüger, C., Werner, S., Metz, B., Chem. Ber., 124 (1991) 301; b) Schneider, J. J., Goddard, R., Werner, S., Krüger, C., Angew. Chem., 103 (1991) 1145; Angew. Chem. Int. Ed. Engl., 30 (1991) 1124; c) Schneider, J. J., Angew. Chem., 104 (1992) 1422; Angew. Chem. Int. Ed. Engl., 31 (1992) 1392; d) Schneider, J. J., Z. Naturforsch., 49b (1994) 689; e) Schneider, J. J., GIT, Z. f. d. Lab., 38 (1994) 1149; f) Schneider, J. J., Specht, U., Goddard, R., Krüger, C., unpublished results, MPI für Kohlenforschung, Mülheim an der Ruhr, 1993.
[81] Ozin, G. A., Coleson, K. N., Huber, H. X., Organometallics, 2 (1983) 415.
[82] a) Schneider, J. J., Goddard, R., Krüger, C., Organometallics, 10 (1991) 665; b) Schneider, J. J., Nolte, M., Krüger, C., Boese, R., J. Organomet. Chem., 468 (1994) 239; c) Schneider, J. J., Nolte, M., Krüger, C., J. Organomet. Chem., 403 (1991) C4.
[83] a) Evans, W. J., Bloom, I., Hunter, W. E., Atwood, J. L., J. Am. Chem. Soc., 103 (1981) 6507; b) Evans, W. J., (Suslick, K. S. (ed.)), High Energy Processes in Organometallic Chemistry, chap. 17, ACS Symposium Series, 333 (1987) 278.
[84] Cloke, F. G. N., Chem. Soc. Rev., 22 (1993) 17.
[85] Anderson, J. R., Structure of Metallic Catalysts, Academic Press, New York, 1975.
[86] a) Isawa, Y. (ed.), Tailored Metal Catalysts, Reidel, Dordrecht, 1985; b) Isawa, Y., J. Mol. Catal., 35 (1987) 187; c) Chaiken, J., Appl. Organomet. Chem., 7 (1993) 163.
[87] Hastie, J. W., Bonnell, D. W., Paul, A. J., Yehesekel, J., Schenk, P. K., High Temp. Mat. Science, 22 (1995) 135.
[88] Cea, F., Devenish, R. W., Goulding, T., Heaton, B. T., Kiely, C. J., Moiseev, I. I., Smith, A. K., Temple, J., Vargaftik, M., Whyman, R., Inst. Phys. Conf. Ser. No. 138: Section 10 (1993) 477.
[89] Ceriotti, A., Demartin, F., Longoni, G., Manassero, M., Marchionna, M., Piva, G., Sansoni, M., Angew. Chem., 97 (1985) 708; Angew. Chem. Int. Ed. Engl., 24 (1985) 697.
[90] Schmid, G., Chem. Rev., 92 (1992) 1709.
[91] a) Klabunde, K. J., Li, Y.-X., Tan, B.-J., Chem. Mater., 3 (1991) 30; b) Klabunde, K. J., Li, Y.-X., Khaleel, A., (Hadijpanayis, G. C., Siegel, R. W. (eds.)), Nanophase Materials, p. 757–769, Kluwer, Academic Publishers, The Netherlands, 1994; c) Klabunde, K. J. Li, Y.-X., (Davis, M. E., Suib, S. L. (eds.)), Selectivity in Catalysis, chap. 7, ACS Symposium Series, (1993) 88; d) Klabunde, K. J., Li, Y.-X., Khaleel, A., Trans Kansas Acad. Science 98 (1–2) (1995) 24.
[92] a) Meier, P. F., Permella, F., Klabunde, K. J., Imizu, Y., J. Catal., 101 (1986) 545; b) Kanai, H., Tan, B.-J., Klabunde, K. J., Langmuir, 2 (1986) 760.
[93] Klabunde, K. J., Imizu, Y., J. Am. Chem. Soc., 106 (1984) 2721.
[94] Tan, B.-J., Klabunde, K. J., Sherwood, P. M. A., Chem. Mater., 2 (1990) 186.
[95] Akhmedov, U., Klabunde, K. J., J. Mol. Catal., 45 (1988) 193.
[96] Li, Y.-X., Klabunde, K. J., J. Catal., 126 (1990) 173.
[97] Imamura, H., Ohmura, A., Haku, E., Tsuchiya, S., J. Catal., 96 (1985) 139.
[98] Imamura, H., Kitajima, K., Miyoshi, M., Tsuchiya, J. Mol. Catal., 52 (1989) L25.

1.5. REFERENCES

[99] a) Lin, S. T., Franklin, M. T., Klabunde, K. J., Langmuir, 2 (1986) 259. b) Franklin, M. T., Klabunde, K. J., (Suslick, K. S. (ed.)), High Temperature Processes in Organometallic Chemistry, chap. 15, ACS Symposium Series, 333 (1987) 246.

[100] Bradley, J. S., Millar, J., Hill, E. W., Melchior, M., ACS Symposium Series, 437 (1990) 160.

[101] a) Bradley, J. S., Millar, J. M., Hill, E. W., Behal, S., Chaudret, B., Duteil, A., Faraday Disc., 92 (1991) 255; b) Eifert, H., Metalloberfläche, 45 (1991) 2; c) Vasil'kov, A. Yu., Pribytkov, Federovskaya, E. A., Slinkin, A. A., Kogan, A. S., Sergeev, V. A., Doklady Akademii Nauk 331 (1993) 179; d) Sergeev, V. A., Udovina, L. I., Smetannikov, Yu. V., Vasil'kov, A. Yu., Pribytkov, P. V., Belavtseva, E. M., Zh. Voeoyuznogo Khim, Obsch. Im. D. I. Mendeleeva 36 (1991) 255.

[102] a) Zoellner, R. W., Klabunde, K. J., Inorg. Chem., 23 (1984) 3241; b) Zoellner, R. W., Klabunde, K. J., Chem. Rev., 84 (1984) 545.

[103] a) El-Shall, M. S., Vaun, W., J. Am. Chem. Soc., 115 (1993) 4385; b) Dagani, R., Chem. Eng. News, 22 (1993) 31.

[104] a) for selected references on the synthesis of nanoscale magnetic matter: Klabunde, K. J., Stark, J. U., Koper, O., Mohs, C., Khaleel, A., Glavee, G., Zhang, D., Sorensen, S. C., Hadjipanayis, G. C., (Hadjipanayis, G. C., Siegel, R. W. (eds.)), Nanophase Materials, p. 1–19, Kluwer Academic Publishers, Netherlands, 1994; b) Chen, J. P, Sorensen, C. M., Klabunde, K. J., Hadjipanayis, G. C., Physical Review B, 51 (1995) 11527, and references cited therein; c) Klabunde, K. J., Zhang, D., Glavee, G. N., Sorensen, C. M., Hadjipanayis, G. C., Chem. Mat., 6 (1994) 784.

[105] Zhang, D., M. S. Thesis, Kansas State University, USA, 1993.

[106] Dravid, V. P., Host, J. J., Teng, M. H., Elliot, B., Hwang, J., Johnson, D. L., Mason, T. O., Weertman, J. R., Science, 374 (1995) 602.

[107] Dagani, R., Chem. Eng. News, 21 (1992) 18.

2

EFFECT OF PRESSURE ON INORGANIC REACTIONS

Colin D. Hubbard and Rudi van Eldik

Institute for Inorganic Chemistry, University of Erlangen-Nürnberg, Egerlandstraße 1, D-91058 Erlangen, Germany

2.1 INTRODUCTION
2.2 HIGH PRESSURE CHEMISTRY: BASIC PRINCIPLES AND INSTRUMENTATION
    2.2.1 Synthesis
    2.2.2 Solution Reaction Kinetics and Chemical Mechanism
2.3 REACTIONS STUDIED AT ELEVATED PRESSURES
    2.3.1 Synthetic Studies
    2.3.2 Solvent Exchange and Ligand Substitution Reactions
    2.3.3 Addition and Elimination Reactions
    2.3.4 Electron Transfer Reactions
    2.3.5 Radiation-Induced Reactions
2.4 CONCLUDING REMARKS AND FUTURE PERSPECTIVES
2.5 REFERENCES

2.1 INTRODUCTION

Two of the first examples of the effect of pressure on inorganic reactions encountered by chemistry students are the Haber process for the synthesis of ammonia and the interconversion of allotropic forms of carbon. Further manifestations of pressure to which they are exposed are the various gas laws, pressure as a variable in the application of the Principle of Le Châtelier, osmotic pressure of solutions, hydrostatic pressure as it relates to the marine environment, and pressure as a component of many thermodynamic equations. In this chapter we aim to extend this basic knowledge and demonstrate, in a systematic way, how the application of pressure within the field of inorganic chemistry can have a significant impact on our understanding of inorganic reactions. A companion chapter (Chapter 3) covers the effect of pressure on *organic* reactions. Many of the basic principles, methods and instrumentation are common to the study of both inorganic and organic reactions, and of course there are some parallel features involved in the investigations of *biomolecules* under elevated pressure which are treated in Chapter 12.

The Haber synthesis and the production of diamond are gas and solid phase reactions respectively, but most of the chemical reactions discussed here occur in liquid solution. However, there are numerous examples of studies in solid state chemistry where pressure effects are of considerable importance. Our concern is not with studies in which pressure only brings about a physical change in a substance, but those in which chemical reactions occur. For such reactions key examples will be cited. Reactions in the gas phase at different pressures in the laboratory and in the atmosphere will not be treated here.

In general, the pressure variable can be used for the synthesis of inorganic compounds, and for the study of chemical kinetics and elucidation of reaction mechanisms. The approach to use pressure is not merely a matter of randomly choosing a reaction in either of the above contexts and subjecting it to pressure. Based on a predictive analysis of the system in question, the consequence of applying pressure should ideally result either in an improved yield of product or in an optimized yield of a specifically desired product in a preparative procedure, or it can be expected to provide mechanistic insight, not readily available from other approaches. Hence, whenever possible the experimental design is predicated on an analytical approach.

In addition to inorganic reactions themselves, we introduce examples of reactions from the field of organometallic chemistry and from bioinorganic chemistry in which reactions of metalloproteins and, more recently, metalloenzyme catalyzed reactions are yielding exciting information from investigations of the effect of pressure. Although various aspects of the effect of pressure on inorganic reactions have been the subject of books, book chapters, review articles, or compilations of data, these earlier reports were often directed at those who are already engaged in research applying high pressure, and other experts. We present the subject at an initial level so that this chapter can also be of benefit to newcomers to the field. A representative sample of earlier literature is given [1–24].

Initially the basic principles of conducting experiments at elevated pressure are presented. This is followed by a brief discussion of the types of instrumentation and methods used in high pressure chemistry, including illustrations in some cases. Some kinetic data from the literature are given and the process of calculating pertinent parameters from them is shown.

Examples of syntheses influenced by pressure are discussed first: general methodology is presented, followed by a description of progress in investigations according to the type of product, for example, metal hydrides and metal nitrides.

Mechanistic examples are then described according to the reaction classification system given in the list of contents, with emphasis on ligand substitution and ligand (solvent) exchange, electron transfer reactions and reactions induced by various forms of radiation.

Suffice to say that we do not, and could not even attempt to, cover comprehensively all classes and examples of reactions which have been investigated at elevated pressures. We recommend the reader to consult the literature to see the range of reactions which have been or may be studied, and to appreciate the tremendous growth and advances in the field of inorganic reaction high pressure chemistry, most of which have been achieved in the past fifteen years.

2.2 HIGH PRESSURE CHEMISTRY: BASIC PRINCIPLES AND INSTRUMENTATION

Pressure (force per unit area) is correctly expressed in SI units as pascals (Pa), where one pascal is one newton per square meter (Nm^{-2}). The instruments often have guages which display the pressure on a scale of units of bar, where 1 bar = 10^5 Pa. However, the pressure units employed may vary at different locations and in different contexts. In contemporary chemistry research literature, pressure is usually quoted in MPa; thus 0.1 MPa is standard atmospheric pressure and 1000 bar = 1 kbar = 100 MPa.

Which pressure ranges are under consideration in this presentation? To a large extent this depends upon the type of application we are considering. For example, in the solid state non-reacting substances may be subjected to pressures as high as 100 GPa, (and often also high temperatures). But this is not common. Pressures of about 200 MPa or less are commonly used in the study of reactions in solution for reasons outlined below. For synthetic applications in solution phase chemistry, pressures of up to 1000 and 1500 MPa are used. The type of instrumentation employed in solid state chemistry has recently been described in detail [25], and thus this topic will not be repeated here.

2.2.1 Synthesis

In chemical syntheses when elevated pressure increases the rapidity of formation of a desired product, or favors the production of one species over another (unwanted) one, clearly the higher the pressure is, the more optimized is the effect of pressure. However, in some cases at higher pressures the dependence of the rate upon pressure is non-linear, leading to a saturation of the pressure effect. Therefore, a balance must be reached between the magnitude of the pressure, the practical requirements, and the relative economics of product yield, technical sophistication and efficiency.

Instrumentation may be fairly simple; for example, an autoclave or a hydraulic press could be employed. However, if the reaction mixture needs to be periodically sampled and analyzed or monitored, additional valves and aliquot withdrawal facilities must be incorporated. The design of such apparatus and the specification of the construction of the autoclave and ancillary items will depend to some extent on the size of the reaction sample, the nature of the solvent (if any), the chemical characteristics of reactant(s) and product(s), the time frame of the reaction, the temperature, and other pertinent conditions. Commercially available equipment for high pressure synthesis is described in appropriate literature [26]. Safety aspects of conducting syntheses of inorganic substances at high pressures will be discussed at the end of this section, and in some cases, they are referred to by the practitioners of high pressure synthesis. In the specific case where the applied pressure relates to supercritical media, the synthesis of inorganic compounds using the supercritical criterion is thoroughly covered in Chapter 5. Many of the basic principles of high pressure chemistry and some aspects of the instrumentation described in the section immediately following apply equally well to solution based synthetic procedures.

2.2.2 Solution Reaction Kinetics and Chemical Mechanism

Instrumentation The instrumentation required for pursuing kinetic investigations has been described in some detail [13, 27, 28]. The general features apply equally well to inorganic and organic reactions. In the former, aqueous media are more frequently used, whereas typical organometallic or organic reactions would be conducted in non-aqueous media. The initial objective is to obtain a kinetic description in terms of a rate constant at a series of pressures. Normally the rate law would already be established from atmospheric pressure kinetic measurements. For reasons which are presented in some detail below, it is preferable to study a reaction at high pressures for which a simple rate law has been established at atmospheric pressure. General reviews of kinetic principles and practice, the initial rate law and integrated rate law methods are widespread [29-31]. Derivation of further parameters from the rate constants as a function of pressure, presented later, are the key diagnostic parameters for postulation of reaction mechanisms.

Reactions with half-lives of 10-15 minutes or longer can be studied at high pressures using reasonably simple methods. Such reactions will be termed "slow", while those reactions having shorter half-lives require special methods of reaction initiation and monitoring, and are termed "rapid", although this time distinction is not rigid and applies only to high pressure kinetics. If a slow reaction is accompanied by a change in the UV or visible absorption spectrum then simple methods are feasible. An aliquot-sampling high pressure cell, such as that depicted schematically in Figure 2.1, may be used to good effect; the reaction is initiated in a few seconds and the reaction mixture introduced into the high pressure cell. Thermal equilibrium can be attained following pressurization of the solution within about 20 minutes. Samples can be removed periodically through a standard high pressure valve, and the movable piston moves to allow for the volume reduction in the cell. The sample can be quickly

2.2. HIGH PRESSURE CHEMISTRY

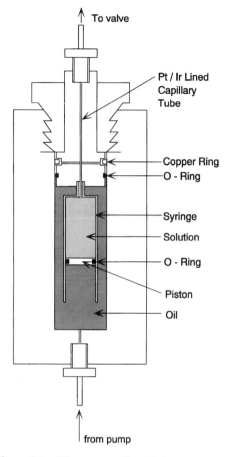

Figure 2.1 Aliquot sampling high pressure cell.

analyzed in a spectrophotometer. Pressure must be rapidly restored to the remaining solution by means of a special valve system. This arrangement has the drawback of aliquot-sampling methods, *i.e.*, the lack of continuous and automated reaction monitoring. This type of apparatus can be readily constructed for basic research or industrial laboratory purposes. This set-up is also useful for synthetic purposes; the samples can be used to monitor the conversion.

Clearly a less labor-intensive method is one in which a pressurizable cuvette or cell can be placed in the cell housing of a commercial recording spectrophotometer. Naturally a spectral change must accompany the reaction. The requirements regarding reaction time are no different from those for the aliquot-sampling method. What is different is the pressurizing equipment; the cell itself, known as a "pill box" [32], (see Figure 2.2), can contract under pressure, and the pressurizing fluid, an optically transparent one such as water or heptane, surrounds the cell and is pressurized by means of a manual hydraulic pump.

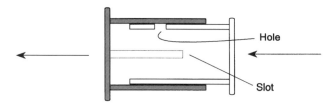

Figure 2.2 "Pill-box" for high pressure kinetic measurements in a UV-vis spectrophotometer.

Examples of reactions studied, and the information derived will be cited in later sections. Although UV/Vis spectroscopy remains the monitoring method of choice and convenience, when feasible, any other physical property of the reaction system which changes during the reaction, and is simply related to the concentration of reactant(s) and/or product(s) can be utilized in principle, for example, conductivity, optical rotation, or refractive index. However, there are fewer precedents in the literature relating to development of *in situ* analyses based on these properties.

Those reactions which are regarded as rapid can be studied over a range of pressures by a variety of techniques which have been developed into reasonably sophisticated forms over the past fifteen years in most cases. It would be beyond the scope of this section to provide the details of each of these methods. We describe and illustrate two of the most commonly used methods, and as far as is possible, provide a classification of the others as well as guidelines regarding the time span of the applicability of each method. The techniques may be classified as follows:

(1) Those in which the reactions are not so rapid that they cannot be initiated by physical mixing, albeit by a special rapid mixing device: Flow Methods.
(2) Those which are so rapid they are studied by perturbation of an equilibrium condition: Equilibrium Perturbation Methods.
(3) Those that are initiated by a very rapid burst of radiation: Radiation Induced Methods.

A summary of the applicable time ranges of these methods is given in Figure 2.3. In terms of time ranges these ᴠ ᴇ only guidelines.

According to our knowledge a high pressure adaptation has not been reported for all the methods mentioned. The reader can become familiar with the use of these methods at atmospheric pressure by referring to numerous publications [33–36]. Not all of these rapid reaction methods are available as complete units commercially. Only one instrument with added high pressure capability can be obtained commercially (a high pressure stopped-flow spectrophotometer). The design and construction of other instruments can be quite

2.2. HIGH PRESSURE CHEMISTRY

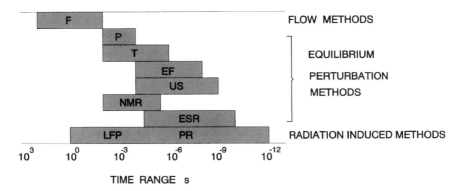

Figure 2.3 Time ranges of fast reaction methods at ambient pressure: F = flow, P = pressure jump, T = temperature jump, EF = electric field pulse, US = ultrasonic absorption, NMR = nuclear magnetic resonance, ESR = electron spin resonance, LFP = laser or flash photolysis and PR = pulse radiolysis.

challenging. Their construction has usually been carried out by university technical personnel in a few specialized laboratories.

The two most commonly used rapid reaction instruments adapted for high pressures are:

(1) The stopped-flow spectrophotometer (HPSF), and
(2) The nuclear magnetic resonance spectrometer (HPNMR), and we will focus on a general description and the performance specifications and characteristics of these two types.

(1) HPSF. The first instrument was developed by Heremans [37], and is illustrated in Figure 2.4. The overall flow path of the flow unit is vertical when the complete unit is immersed in the chamber which contains pressurizable fluid. A stepping motor was used to drive the syringes, and the dead time was in the region of 40 ms, over a pressure range up to 120 MPa. One most favorable feature of this design is that many repeat runs can be made at four or five of the selected pressures without removing the stopped-flow unit from the pressurizable vessel. The disadvantage over atmospheric pressure instruments is that the dead time is about one order of magnitude longer, thus limiting reactions that may be studied to those of $t_{1/2} > 30$ ms. Efforts have been made to rectify this and two further instruments not radically different from Heremans' version have been developed [38–40], (see Figures 2.4 and 2.5). One of them differs in the initiation method; the syringes are activated by a device which is outside the high pressure cylinder (the stepping motor is inside the cylinder) (Figure 2.5). Other HPSF instruments have been reported [41–43], and a commercial unit based on the instrument described in [38] is available. The specifications of the latter indicate a dead time approaching that of standard stopped-flow

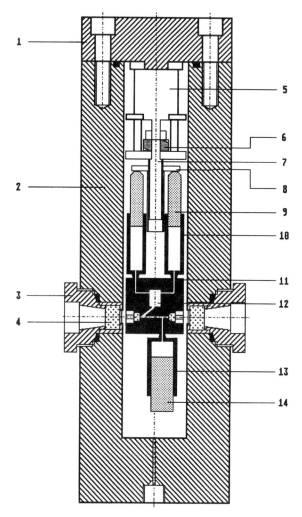

Figure 2.4 Schematic of a high pressure stopped-flow spectrophotometer, based on the design in [37]: 1 = lid to overall unit, 2 = outer vessel, 3 = window holder, 4 = quartz windows, 5 = electric stepping motor, 6 = stepping motor actuator, 7 = stopped-flow unit positioning rod, 8 = driving plate, 9 = drive syringe (inner), 10 = drive syringe (outer), 11 = block holding windows, mixer and syringe attachment points, 12 = mixing jet, 13 = stop syringe (outer), 14 = stop syringe (inner).

instruments. Many examples of reactions investigated at elevated pressures with these instruments will be described in Section 2.3.2.

(2) HPNMR. We will restrict our consideration to reactions that can be monitored by NMR methods in the liquid solution state. Design and construction

2.2. HIGH PRESSURE CHEMISTRY

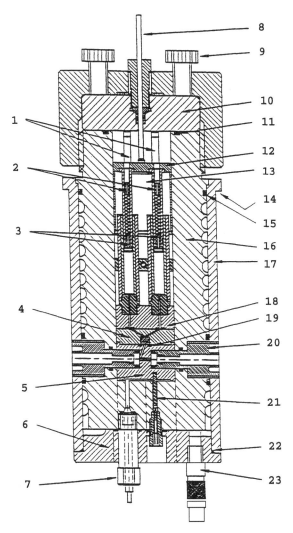

Figure 2.5 Schematic of a high pressure stopped-flow spectrophotometer, based on the design of [38] and available from Hi-Tech Scientific, Brunel Road, Salisbury, England: 1 = cell tie rods, 2 = plunger grub screws, 3 = reagent syringes, 4 = mixer mount, 5 = observation cell, 6 = lower plate, 7 = hydraulic pressure, 8 = thrust rod, 9 = disc clamping bolts, 10 = disc, 11 = top seal O-ring, 12 = anvil, 13 = syringe assembly, 14 = supporting lip, 15 = water jacket top sleeve O-ring, 16 = bomb body, 17 = outer sleeve, 18 = syringe mount O-ring, 19 = mixer O-ring, 20 = window mount, 21 = temperature sensor, 22 = bottom sleeve O-seal and 23 = coolant.

challenges are formidable in this case. A detailed account of them is outside the scope of our presentation. However, although we recommend that readers consult other authoritative sources, it is worth noting, perhaps very obviously, that the materials of the high pressure NMR units or probes as they are sometimes known, must be non-magnetic. In addition, compromises between probe type, sample size and sensitivity have to be made. Early instruments used electromagnets, and the development of early HPNMR has been reviewed [44–48]. But in the past fifteen years NMR spectrometers with superconducting magnets have been adapted for high pressure measurements up to 250 MPa [21, 49–51], and over a wide range of temperature, which varies depending on the reaction system and the solvent. Figures 2.6 and 2.7 illustrate an HPNMR probe head for use in a 400 MHz instrument at pressures up to 200 MPa. The types of reaction whose kinetics are studied by NMR are mostly those in which no net change occurs, such as those illustrated in eq. (1) for solvent exchange, and symmetrical electron transfer reactions.

$$M(H_2O)_6^{2+} + H_2O^* \rightarrow M(H_2O)_5(H_2O^*)^{2+} + H_2O \qquad (1)$$

M represents Ni or another first row transition metal ion for which NMR spectroscopy is feasible, and the asterisk on one water molecule is to show exchange has taken place. In practice the solution must contain a nucleus that can be monitored; in this case oxygen-17 can be used. The signal acquired which represents the exchange process varies with temperature, and kinetic data are obtained from these measurements. However, examination of the paper by Swift and Connick [52] on their pioneering study of the kinetics of water exchange as in eq. (1) (at atmospheric pressure) reveals that converting primary data into kinetic parameters is a much more complicated procedure than for the other methods covered. At high pressure, assuming the data quality is comparable there are no additional complications in kinetic data treatment. The construction and introduction of a high pressure probe into the spectrometer are highly specialized and costly activities, and it is unlikely that the high pressure application to NMR spectroscopy for inorganic chemistry or other branches of chemistry will become widespread in the immediate future. In fact only a few facilities are available at present. It is the technical complexity and economic factors, rather than the value of mechanistic information which are the limiting factors. Our knowledge of solvent exchange, both for water and other solvents, on solvated metal ions, and of self-exchange electron transfer, has increased tremendously due to the use of high pressure NMR spectroscopy [47].

A few practical hints for investigation of inorganic solution reactions at high pressures are worth mentioning. Some of the properties of substances which are relevant in this context are listed in Table 3.1.

Due account must be taken of solution heating upon pressurization, and cooling upon depressurization in thermal equilibration of the reacting system. Solute solubilities may be changed at elevated pressures, solvent freezing points change under pressure, solvent viscosity can be sufficiently changed by pressure

2.2. HIGH PRESSURE CHEMISTRY

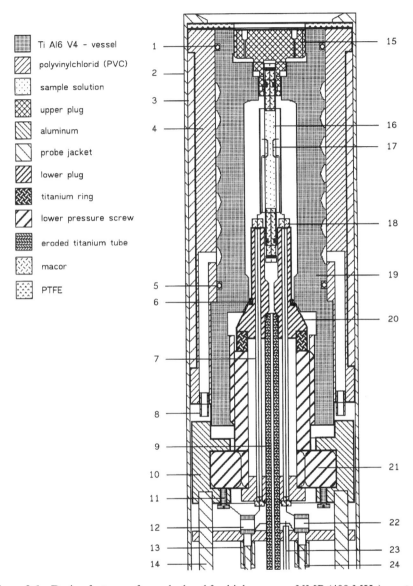

Figure 2.6 Design features of a probe head for high pressure NMR (400 MHz) spectroscopy measurements [50, 51]: 1 = O-ring, 2 = probe jacket, 3 = thermal insulating material. 4 = polyvinylchloride, 5 = O-ring, 6 = O-ring, 7 = semi-rigid coaxial cable, 8 = connection to thermostat, 9 = titanium tube, 10 = lid, 11 = screw, 12 = capacitor, 13 = capacitor holder, 14 = aluminum tube, 15 = upper plug, 16 = sample tube, 17 = saddle coil, 18 = Macor, 19 = TiAl6V4 vessel, 20 = lower plug, 21 = lower pressure screw, 22 = capacitor, 23 = coaxial cable and 24 = capacitor holder.

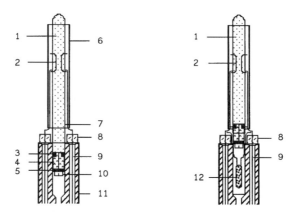

Figure 2.7 Design features of a probe head for high pressure NMR (400 MHz) spectroscopy measurements [50, 51]: (left side) 1 = sample, 2 = saddle coil, 3 = O-ring, 4 = movable piston, 5 = O-ring, 6 = glass tube, 7 = sample tube, 8 = Macor, 9 = coaxial cable, 10 = screw, 11 = lower plug, (right side) 1 = sample, 2 = saddle coil, 8 = Macor, 9 = coaxial cable, 12 = Pt-100.

so that diffusion rather than activation may control reaction rate. Consequently, awareness of these factors will improve the design and the likelihood of successful experiments.

We will conclude this section by reiterating that, while there is potential for hazards in any laboratory practice, use of high pressure *where only liquids are subjected to pressure* presents no additional danger to experimentalists, according to our knowledge and experience. If a failure in the pressurizing system occurs, either the reaction solution or the pressurizing fluid will leak and the pressure will fall immediately. Therefore, only the routine hazard of reagents themselves is extant. When compressed gases are used to pressurize the solution and the surrounding fluid, then *appropriate precautions* must be taken, restricted locations used and suitable training of personnel undertaken.

Theoretical Background The theories available to chemists to explain and predict the effects of pressure on reactions whether the objective of the investigation is a synthetic one or a mechanistic one, are thermodynamics and applications of thermodynamic principles within the absolute reaction rate (kinetic) theory [29, 30]. We are interested in how the effect of pressure manifests itself on the equilibrium constant, or ΔG°, for the reaction, or upon the free energy of activation, $\Delta G^{\neq}$. If the effect of pressure is to increase the magnitude of K, or reduce $\Delta G^{\neq}$, the reaction will have a greater yield or a faster reaction rate respectively. Qualitatively it can be said that these effects can arise from the differing pressure effects on the chemical potentials of the reactants and products (equilibrium), or reactants and transition state (kinetics). The approach favored by many is one of considering the volume per mole of initial state, the

2.2. HIGH PRESSURE CHEMISTRY

transition state and when appropriate, the product state. A reaction in which the transition state has a smaller volume than the initial state is accelerated by application of pressure. Further, if the product has a different volume from the initial state, pressure will induce a change in product yield. To be more rigorous we must first remember that the volume of a chemical species in a solution is invariably different from that of the pure substance, where the volume per mole can readily be established from the density. The quantity we are interested in is the partial molar volume, which may be thought of as the effective volume per mole of a species when its intrinsic volume is modified by the influence of solvation. It is represented by $\bar{V}$ for a mole of a substance in older literature but is now often written [29], as $V_{A,m}$ for the partial molar volume of A, (eq. 2), and

$$V_{A,m} = (\delta V/\delta n_A)_{p\,T,m_B} \quad (2)$$

where the partial derivative signifies the change of volume when the amount of substance A is increased in a binary system (A and B present), and this change is considered at constant temperature, pressure and amount of B. We shall not develop this definition further, but it parallels that of other partial molar quantities in thermodynamic theory. Determination of the partial molar volume of a substance in solution, and in the present discussion the reactant or product of a reaction under consideration is normally the solute and the second component is water as the solvent, can be accomplished by using different methods. One common method employed is to measure the density of solutions of known concentration, and obtain from these measurements the partial molar volume at infinite dilution. The density must be measured to at least six significant figures, and it is vitally important that the temperature remain constant during the measurements, preferably varying no more than $\pm 0.001\,°C$. Special instrumentation is available for these measurements [53]. Ionic substances pose the special challenge of apportioning the measured quantity into the contribution from each ion. Various conventions give rise to key single ion values, such as the partial molar volume of $H^+(aq)$, and therefore sets of values of partial molar volumes of common ions have been established [54].

The volume of activation, $\Delta V^{\neq}$, is defined as the difference between the partial molar volumes of the transition state and the initial state: i.e., $\Delta V^{\neq} = \bar{V}_{TS} - \bar{V}_{IS}$. Since $\bar{V}_{TS}$ cannot be measured, it can be obtained only if the volume of activation can be measured and the partial molar volume(s) of the reactant(s) is (are) either known or can be measured. Kinetic measurements at several pressures lead to the determination of $\Delta V^{\neq}$, since the rate constant for a reaction and $\Delta V^{\neq}$ are linked through:

$$(\delta \ln k/\delta p)_T = -\Delta V^{\neq}/RT \quad (3)$$

$$\ln k = \ln k_0 - (\Delta V^{\neq}/RT) \cdot p \quad (4)$$

where k and k_0 are the rate constant at pressure p and a pressure of 1 bar (0.1 MPa) respectively, R is the ideal gas constant and T is the absolute temperature. Eq. (3) is derived from standard thermodynamics equations, adapted to transition state theory [28, 55]. If a plot of ln k versus p is linear, then eq. (4), the integrated form of eq. (3), is valid and $\Delta V^{\neq}$ is independent of pressure. The latter is frequently the case and curvature is invariably not observed when the pressure is <150 MPa. Thus, in order to avoid the potential complications of non-linearity and subsequent fitting procedures and interpretation of the pressure dependence of $\Delta V^{\neq}$, most experiments designed to obtain this parameter use a pressure range no higher than 200 MPa. Many treatments have been proposed, however, for analyzing a non-linear dependence [55, 56]. From the form of eq. (4) it may be seen that a negative $\Delta V^{\neq}$ arises from rate acceleration with increase in P and this case is of greatest interest to chemists exploring pressure as a useable variable for improved synthesis of inorganic substances. If the rate of reaction is independent of pressure then $\Delta V^{\neq}$ is zero, and if the rate of reaction is retarded by pressure then $\Delta V^{\neq}$ is positive.

For many reactions, as we shall see in later sections, the value of $\Delta V^{\neq}$ is rarely larger than 30 cm^3 mol^{-1}. This value corresponds to a rate acceleration or retardation of about a factor of three and a half between atmospheric pressure and a pressure of about 100 MPa [57]. Therefore, variation of the temperature is often preferable in synthetic chemistry, as heating is easier in a practical sense, and the temperature has a greater effect on rate for a smaller increment of the parameter in comparison with pressure.

Concentration Scales Choice of concentration scales is important, although for kinetic parameters earlier misunderstandings have been clarified, meaning that certain corrections once thought necessary are in fact not needed [58]. Partial molar volumes are only satisfactorily defined and accounted for as pressure independent when the concentration scale is either mole fraction or molality. If the molarity scale is used a correction to partial molar volumes for the solution compressibility must be made; in dilute solution the solvent and solution compressibilities are very similar. For first order kinetic processes the rate constant dependence on pressure is identically expressed for mole fraction, molality and molarity scales. A correction term for higher order reactions is mutually canceled by a correction term derived on the basis of solution densities [58].

Calculation and Interpretation of Activation Volumes Let us look at some actual kinetic data and briefly go through the calculation of $\Delta V^{\neq}$. The reaction investigated is between an Ir(IV) species, specifically K_2IrCl_6, and 2,3-dihydroxybenzoic acid (QH_2) in acidic solution (1.0 M $HClO_4$) [59]. The detailed mechanism had been established prior to the high pressure kinetics study from analysis of the results of a kinetics investigation at atmospheric pressure [60]. The stoichiometric reaction is:

$$2\ IrCl_6^{2-} + QH_2 \rightarrow 2\ IrCl_6^{3-} + Q + 2H^+ \tag{5}$$

2.2. HIGH PRESSURE CHEMISTRY

where the product is a quinone (Q). The rate determining step is a one electron transfer, which in the presence of an excess concentration of benzenediol over the concentration of the Ir(IV) species, is pseudo first-order. The following first order rate constants, k_{obs}, were obtained at different pressures, using the same reactant solutions: 0.849, 1.04, 1.35, 1.76 and 2.31 s^{-1} (at 296.4 K), at $P = 5, 25, 50, 75$ and 100 MPa respectively. The reaction is sufficiently rapid to require the stopped-flow method.

These data may be treated by standard software, or by plotting the natural logarithm of k_{obs} versus the pressure manually, according to eq. (4). The plot is linear, indicating a pressure-independent $\Delta V^{\neq}$ over the given pressure range. The slope of the plot is equal to $-\Delta V^{\neq}/RT$. Therefore $\Delta V^{\neq}$ can be obtained; it is -26.0 ± 0.2 cm^3 mol^{-1}, for this reaction.

In synthetic applications the pressure variable is obviously of paramount value the more negative the value of $\Delta V^{\neq}$ is. Slightly negative, or zero values of $\Delta V^{\neq}$ indicate that there is no advantage in applying pressure to the reaction. In cases where pressure inhibits the progress of reaction ($\Delta V^{\neq}$ positive), synthesis prospects may be improved by the use of the temperature variable, or catalysts or other variations in reaction conditions.

Knowledge of $\Delta V^{\neq}$, particularly for an elementary reaction can be extremely valuable in terms of understanding the reaction mechanism. For many reactions the measured $\Delta V^{\neq}$ may be thought of as arising from two contributions, an intrinsic and a solvation component. Thus

$$\Delta V^{\neq} = \Delta V^{\neq}_{int} + \Delta V^{\neq}_{solv} \qquad (6)$$

The former term refers to the volume changes arising in the activated complex as a result of bond formation or breakage, (usually partially only), or other mechanistic features. In general, reactions involving bond breakage in the transition state give rise to a $\Delta V^{\neq}_{int}$ of up to about $+20$ cm^3 mol^{-1} and those involving bond formation in the transition state will exhibit a $\Delta V^{\neq}_{int}$ of up to about -20 cm^3 mol^{-1} [57]. In the absence of any contribution from the solvation component then the measured $\Delta V^{\neq}$ will indicate the intrinsic component. However, in reactions in which there is a change in charge, formation or neutralization, or increase or decrease of polar character in forming the activated complex, then $\Delta V^{\neq}_{solv}$ will contribute significantly. For example if a neutral species ionizes, solvent will be restricted (electrostriction) around the incipient ions giving rise to a negative value of $\Delta V^{\neq}_{solv}$. A reaction in which there is formation of two species from one will yield a positive intrinsic component. Thus depending on the magnitudes of the two components the measured $\Delta V^{\neq}$ may be positive or negative. In an ideal situation, when there is no solvent contribution, high pressure kinetics can provide a valuable mechanistic diagnosis. However, we should offer the elementary caution that results of an investigation employing just one variable are usually insufficient for making firm mechanistic conclusions. Another value of high pressure kinetics is the probing of solvational control of reactivity by varying the solvent for a reaction with a known mechanism. In reality many inorganic reactions contain both intrinsic and solvational contributions to $\Delta V^{\neq}$. The interpretation of $\Delta V^{\neq}$ for solvent exchange and for outer sphere electron transfer reactions has been the subject of several theoretical reports; these will be discussed below, along with the citation of appropriate literature, in the context of particular example reactions.

2.3 REACTIONS STUDIED AT ELEVATED PRESSURES

2.3.1 Synthetic Studies

The characteristic pressure dependence of chemical reactions in general and that of inorganic reactions in particular, as discussed in the following sections, create the possibility of employing the acceleration or deceleration by pressure in synthetic applications to tune the selectivity for a particular reaction product. This has been done for a number of systems in inorganic and organometallic chemistry.

The synthesis of metal nitrides and hydrides has received significant attention from researchers as a result of the direct interaction of metal surfaces with nitrogen or hydrogen at elevated pressure respectively [25]. Heating of pure Ga under nitrogen pressure resulted in the formation of GaN [61]. Such techniques were also used to study crystal growth in GaN [62]. The decomposition of GaN was also studied by allowing a sample to reach the equilibrium pressure at a specific temperature. In this way the synthesis and decomposition of GaN was able to be studied up to 1700 °C and 20 kbar, *i.e.*, 2000 MPa. Such a procedure could also be used to study the kinetics of the decomposition of GaN [63]. Thus thermodynamic as well as kinetic information can be obtained in this way as a function of temperature and pressure.

Similar techniques were employed to study the formation and decomposition of metal hydrides [64]. The interest in these compounds resulted from the potential application of metal hydrides for the storage of energy in the form of hydrogen. The most important metals for high pressure hydride synthesis include the Group VIB, VIIB and VIIIB (Groups 6–10) elements. By way of example the T-p phase diagrams for Cr–H and Mo–H are given in Figure 2.8. Similar diagrams are presented in Figure 2.9 for the formation of Mn–H and Rh–H [65, 66]. In the former case Mn–H$_n$ species with n typically 0.65 and 0.82, were identified and their lattice parameters reported [67].

In the case of synthetic reactions in solution, a $\Delta V^{\neq}$ value of $-10 \text{ cm}^3 \text{ mol}^{-1}$ corresponds to an increase in rate constant of a factor of two at a moderate pressure of 100 MPa. Typical oxidative addition reactions exhibit $\Delta V^{\neq}$ values as large as $-40 \text{ cm}^3 \text{ mol}^{-1}$, so that even application of only a few hundred atmospheres of pressure will give rise to a significant acceleration of the reaction.

In general, all addition reactions to coordinated ligands, oxidative addition reactions to low valency metal centers, and insertion reactions into M–C, M–O,

2.3. REACTIONS STUDIED AT ELEVATED PRESSURES

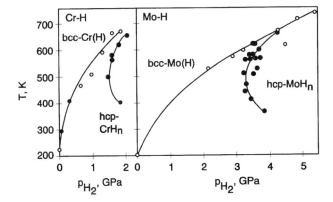

Figure 2.8 Temperature-pressure phase diagram for the formation of hydrides of (a) chromium, bcc-Cr(H) and hcp-CrH$_n$, and (b) molybdenum, bcc-Mo(H)$_n$ and hcp-MoH$_n$ [64–67]. Data points are indicated with different symbol when taken from more than one study.

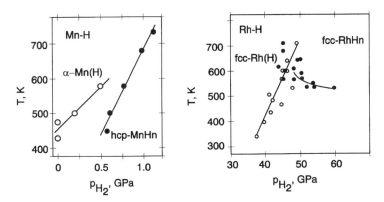

Figure 2.9 Temperature-pressure phase diagram for the formation of hydrides of manganese and rhodium [64–67]. Full circles are for hydride formation (Mn), and open circles are for hydride decomposition. The formation and decomposition of rhodium hydride are indicated with the same symbols.

and M–H bonds, are expected to exhibit a significant acceleration with increasing pressure due to the negative volumes of activation reported for such systems [20]. The yield of such reactions should therefore also exhibit a characteristic pressure dependence. Along these lines, a series of organic synthetic reactions using organometallic reagents has been investigated under pressures as high as 2500 MPa (25 kbar) [68]. In this way aldol condensation can be performed under neutral conditions for molecules that are either acid- or base-sensitive [69]. Similarly, high pressure techniques could be employed to prepare [4 + 2] cycloadducts *via* the Diels-Alder reaction between Danishefsky's diene and butyl glycolate as shown in eq. (7) [70].

$$\text{Me}_3\text{SiO} \diagup\diagdown + \text{H} \diagup\text{CO}_2\text{Bu}^n \diagdown \text{O} \diagup \text{OMe} \longrightarrow \text{Me}_3\text{SiO} \diagup\text{CO}_2\text{Bu}^n \text{ (cis)} + \text{Me}_3\text{SiO} \diagup\text{CO}_2\text{Bu}^n \text{ (trans)} \quad (7)$$

The product distribution in the [4 + 2] cycloaddition of (−)menthyl ester of (Z)-3-tributylstannylacrylic acid to 2,3-dimethyl-1,3-diene, leading to a mixture of diastereoisomers, depends on the applied pressure [71]. In the case of heat-sensitive organotin compounds, good yields of cycloaddition products, see eq. (8), could be obtained under high pressure conditions, whereas reactions at ambient pressure and high temperature resulted in low yields and reverse diastereoselectivity due to isomerization of the starting dienophiles [68].

$$\begin{array}{c} R^2 \diagup\diagdown R^1 \\ R^3 \diagup\diagdown \end{array} + \begin{array}{c} \text{SnBu}_3^n \\ \text{Ph} \end{array} \longrightarrow \begin{array}{c} R^2 \diagup\diagdown R^1 \diagdown\text{SnBu}_3^n \\ R^3 \diagup\diagdown\text{Ph} \end{array} \quad (8)$$

R^1 = H, R^2 = R^3 = Me	Z : E = 70 : 30, 0.1 MPa, 180 °C, 30 %	cis : trans = 19 : 81
	Z : E = 80 : 20, 2300 MPa, 70 °C, 80 %	cis : trans = 90 : 10
R^1 = OMe, R^2 = R^3 = H	Z : E = 75 : 25, 0.1 MPa, 180 °C, 42 %	cis : trans = 22 : 78
	Z : E = 75 : 25, 2000 MPa, 50 °C, 80 %	cis : trans = 70 : 30

Application of pressure also had a significant affect on various synthetic reactions involving substitution processes, *viz.* the silylation of tertiary alcohols [72], and rearrangement processes, *viz.* the synthesis of α-silylated esters and lactones [73]. These examples demonstrate the versatility of the application of high pressure techniques in the synthesis of organic and organometallic materials. Further details for organic syntheses at high pressures can be obtained from Chapter 4, while the topics of synthesis of inorganic and organic compounds in supercritical media are covered in Chapters 5 and 6, respectively.

2.3.2 Solvent Exchange and Ligand Substitution Reactions

Six Coordinate Species. Water and Other Solvents Most of the emphasis in this section will be on water as the solvent. In principle any inorganic ion which is stable in an aqueous medium can be of interest for exchange of coordinated water with that of the bulk solvent, as in eq. (1), which reflects the fact that

2.3. REACTIONS STUDIED AT ELEVATED PRESSURES

hydrated *cations* are almost exclusively studied. The chemical reactivity of an ion in solution depends on the stability and lability of its solvating molecules. Few mechanistic studies have been carried out on hydrated cations of the s and p block elements. In fact most investigations have focused on transition metal cations, where interest is stimulated by the rich variety of chemical reactivity exhibited by coordination compounds of these elements, many of which have several accessible oxidation states. An early example of a kinetics study of water exchange, cited above, was the determination by NMR spectroscopy of the rate constants for water exchange on the octahedral aqua ions of Mn^{2+}, Fe^{2+}, Co^{2+} and Ni^{2+} [52]. These reactions are all rapid, but vary over a few orders of magnitude, a range which could be rationalized on the basis of crystal field stabilization energies. The volumes of activation were subsequently determined for these four ions and for exchange on V^{2+} [74, 75] and have the values -4.1, -5.4, $+3.8$, $+6.1$ and $+7.2$ cm^3 mol^{-1} for V^{2+} and the other four cited ions, respectively, at 298 K.

Results for the kinetics of complex formation reactions (principally of Ni^{2+}), together with the water exchange data led to the proposal of the Eigen-Wilkins mechanism [76, 77] or the I_d mechanism in another classification system [78], whereby the formation of transition metal complexes is controlled by the rate of departure of a hydrated water molecule. Differences in the second order rate constants, k_f, for reaction of a particular metal ion with various ligands arise by virtue of the differences in magnitude of the outer sphere association constant, K_{OS}, or in the case of a charged ligand, the ion-pair formation constant, K_{IP}. Therefore, for example, if the complex formation of $Ni(H_2O)_5(NH_3)^{2+}$ is being studied, the rate constant k_f is given by eq. (9):

$$k_f = K_{OS} k_{H_2O} \quad (9)$$

It could be regarded as a potential success of high pressure kinetics that the $\Delta V^{\neq}$ values, which can nearly always be measured with greater precision than can $\Delta S^{\neq}$ values (and measured subsequent to these mechanistic proposals), allowed the distinction to be made that earlier members of the first row of transition metal cations in oxidation state (II), for example V^{2+}(aq), actually exchange water by an associatively based activation mode (I_a). The changeover point is between Mn^{2+} and Fe^{2+}, if we accept on a simple basis that an activation volume of zero would represent a pure interchange mechanism (I) (see Figure 2.10). The difference was explained in part by the different electron populations of the 3d orbitals as the first row of the transition metals is traversed. The underlying assumptions in this analysis are that electrostriction remains constant during the exchange, which in turn means that the measured volume of activation is representative only of the intrinsic contribution, and in an interchange process the bond lengths to the entering and leaving water molecules are equivalent and differ only in strength and length for the I_a and I_d mechanisms, (stronger and shorter in the former case). Although the validity of this further assumption has been examined [79] it is also implied in this approach that the

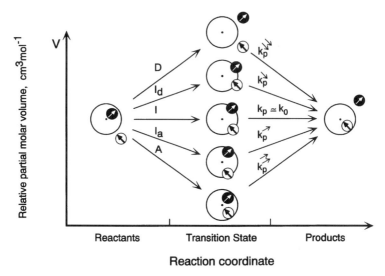

Figure 2.10 Schematic volume profiles for possible mechanisms of water exchange on $M(H_2O)_6^{2+}$ where M is a first row transition metal.

bond distances (oxygen to metal ion) of the non-exchanging water molecules remain constant during the course of the exchange.

Recently the validity of these assumptions has been challenged [80–82], and theoretical calculations purporting to describe the kinetics of exchange and assess the volume of activation have been reported. Based upon the results and a more detailed breakdown of the possible contributions to the volume of activation, it was concluded [81, 82] that all first row transition metal ions in oxidation state (II) exchange water by an I_d mechanism. We shall return to this discrepancy later.

Early work [83–87] on the kinetics of solvent exchange from hexaaqua trivalent metal ions also revealed a change in the sign of $\Delta V^{\neq}$, and the values are, in cm³ mol⁻¹, -12.1 (Ti^{3+}), -8.9 (V^{3+}), -9.6 (Cr^{3+}), -5.4 (Fe^{3+}) and $+5.0$ (Ga^{3+}). It was concluded that the first four ions exchange water associatively, Ti^{3+} strongly so, with a changeover presumably occurring between two of the ions separating Fe and Ga in the period. The exchange of water on Ga^{3+} exhibits dissociative character to a small extent. It has been implied that variation in the 3d electron populations is a principal reason behind the mechanism changeover. The volume of activation has also been measured for exchange on $Al(H_2O)_6^{3+}$ [88]. It is $+5.7$ cm³ mol⁻¹ also indicative of modest dissociative character.

It should be emphasized that the values of $\Delta S^{\neq}$ which have customarily been invoked to help explain the ordering of species and solvation effects are generally consistent with the mechanisms assigned for the water exchange on the divalent metal ions, but are not obviously compatible with the $\Delta V^{\neq}$ values for the trivalent metal ions.

2.3. REACTIONS STUDIED AT ELEVATED PRESSURES

Several trivalent ions are readily hydrolyzed and the volumes of activation for solvent exchange from the pentaaquamonohydroxy species $M(H_2O)_5(OH)^{2+}$ have been measured and reported [89]. As may be concluded from the results in Table 2.1, water exchange is considerably labilized in the hydroxy species, and $\Delta V^{\neq}$ is more positive for the latter species in the examples given, indicating a greater degree of dissociative character or lesser degree of associative mode of activation.

There are many examples of high pressure kinetics studies of water exchange when ligands other than water or the hydroxide ion occupy coordination sites in hexacoordinated metal ions. An increase in steric congestion is proposed as the basis for a shift in mechanism for water exchange on $[M(CH_3NH_2)_5(H_2O)]^{3+}$ from an I_a when M = Cr ($\Delta V^{\neq} = -3.8$ cm^3 mol^{-1}), to I_d when M = Co ($\Delta V^{\neq} = +5.7$ cm^3 mol^{-1}), while when M = Rh, $\Delta V^{\neq}$ is essentially zero, denoting only "pure" interchange [90]. Water exchange on $[Cp^*Rh(H_2O)_3]^{3+}$ and on $[Cp^*Ir(H_2O)_3]^{3+}$ studied by ^{17}O HPNMR, is governed by a dissociative interchange mechanism, based on the $\Delta V^{\neq}$ values and other results. (Cp^* is η^5-pentamethylcyclopentadienylanion) [91]. Anation reactions of related species also occur by an I_d mechanism.

Solvent Exchange (Solvents other than Water) The pioneering work of Merbach and colleagues also included high pressure kinetics (HPNMR) on exchange of solvent from $M(S)_6^{2+}$ and $M(S)_6^{3+}$ species, including some main group metals in the latter case. The solvents commonly used include methanol, acetonitrile, dimethylsulfoxide, dimethylformamide and TMP.

The following table (Table 2.2) shows the results for representative elements (3 + cations) [87, 88], transition elements (3 + cations) [6, 83–86, 92–94] and transition elements (2 + cations) [95–101]. In most cases the sign and numerical

Table 2.1 Rate constants and activation volumes for water exchange on some hexaaqua- and monohydroxypentaaquametal ions.[a]

Species[b]	k at 298 K, s^{-1}	$\Delta V^{\neq}$, cm^3 mol^{-1}
Ga^{3+}	4.0×10^2	+ 5.0
Ga(OH)$^{2+}$	1.1×10^5	+ 6.2
Fe^{3+}	1.6×10^2	− 5.4
Fe(OH)$^{2+}$	1.2×10^5	+ 7.0
Cr^{3+}	2.4×10^{-6}	− 9.6
Cr(OH)$^{2+}$	1.8×10^{-4}	+ 2.7
Ru^{3+}	3.5×10^{-5}	− 8.3
Ru(OH)$^{2+}$	5.9×10^{-4}	+ 0.9
Rh^{3+}	2.2×10^{-9}	− 4.2
Rh(OH)$^{2+}$	4.2×10^{-5}	+ 1.3

[a] Values obtained from [89].
[b] The aqua ligands are omitted.

Table 2.2 Volumes of activation for solvent exchange on trivalent or divalent cations.[a]

S	3 + Ions							3 + Ions					2 + Ions				
	Al	Ga	In	Sc	Ti	V	Cr	Fe	V	Mn	Fe	Co	Ni	Cu
H_2O	+5.7	+5.0			−12.1	−8.9	−9.6	−5.4	−4.1	−5.4	+3.8	+6.1	+7.2	
DMSO	+15.6	+13.1					−11.3	−3.1						
DMF	+13.7	+7.9					−6.3	−0.9			+8.5	+6.7	+9.1	
TMP	+22.5	+20.7	−21.4	−21.3										
CH_3OH										−5.0	+0.4	+8.9	+11.4	+8.3
CH_3CN										−7.0	+3.0	+9.9	+9.6	
												+7.7	+7.3	
												+6.7		
NH_3													+5.9	

[a] All values in cm³ mol⁻¹ and are taken from [6, 83–88, 92–94, 95–101].

2.3. REACTIONS STUDIED AT ELEVATED PRESSURES

values of $\Delta V^{\neq}$, and therefore the mechanistic assignment, parallel the situation for water, where comparison is possible. However, as with exchange of water, it is difficult to estimate the limiting value for solvent exchange that permits a distinction between an A mechanism and an I_a mechanism with substantial associative character, and between a D and an I_d mechanism with significant dissociative character.

Recent studies on solvents exchanging on M^{2+} ions, where M = Mn, Fe, Co and Ni, have employed the bidentate ligand ethylenediamine (en) as a solvent [102, 103]. The kinetics were monitored by HPNMR and values of $\Delta V^{\neq}$ were -0.6 (278.4 K), -1.2 (324.5 K), $+0.9$ (332.4 K) and $+11.4$ cm^3 mol^{-1} (383 K) respectively, for the four metal ions. These values taken together with other structural information led to the conclusion that en exchanges by a dissociative interchange mechanism and there is no changeover in the mechanism as promulgated for monodentate solvents. The $\Delta S^{\neq}$ values of $-43, -1, +16$ and $+11$ J mol^{-1} K^{-1}, for the four metal ions respectively, are not readily accommodated within this explanation. However, the kinetic chelate effect and the chelate strain effect [102, 103] provide scope for resolving an apparent contradiction.

We will now turn to ligand substitution reactions, first by considering formation reactions in which a hexaaqua metal ion, typically divalent, has one or more water molecules substituted by an incoming ligand. The number of water molecules substituted depends on the dentate number of the ligand. The I_d mechanism remains operative for bidentate and higher dentate ligands, except in some unusual cases where sterically controlled substitution prevails [104, 105], since the rate determining step is still loss of the first water molecule. In principle, interpretation of $\Delta V^{\neq}$ is not straightforward because, in the case of complex formation occurring by an I_d mechanism, the measured value is a composite:

$$\Delta V^{\neq} = \Delta \bar{V}_{OS} + \Delta V^{\neq}_{H_2O} \qquad (10)$$

This arises because the second order rate constant is a product (eq. 9). However, it has been shown [106] that a reasonable estimation of the volume change for formation of the outer sphere complex can be made. Its value is small, and therefore a correction to obtain the volume change for the kinetic step can be simply calculated.

A selection of available data is assembled in Table 2.3 for complex formation from $M(H_2O)_6^{2+}$ [106–108], where $\Delta V^{\neq}$ has been corrected to be the value for the interchange step in some cases.

The values are all finitely positive indicating (as may be predicted from the water exchange results) that the mechanism for these three ions is I_d. The reactions are rapid and the high pressure kinetics were studied by the T-jump method using laser heating or with a HPSF instrument. Negative values of $\Delta V^{\neq}$ were found for the formation of $V(SCN)^+$ [109] studied by HPSF and for formation of $Mn(bpy)^{2+}$ [110], supporting an I_a mechanism for the earlier

Table 2.3 Activation volumes for formation of some transition metal complexes.

MM	L[a]	$\Delta V^{\neq}$, cm^3 mol^{-1}
Co	PADA	+ 7.2
Co	bpy	+ 5.9
Co	(gly)$^-$	+ 5
Ni	PADA	+ 7.7
Ni	bpy	+ 5.9
Ni	(gly)$^-$	+ 7
Ni	iso	+ 7.4
Zn	bpy	+ 7.1
Zn	(gly)$^-$	+ 4

[a] PADA is pyridine-2-azo-p-dimethylaniline, bpy is 2,2'-bipyridine, (gly)$^-$ is the anion of glycine, and iso is isoquinoline.

members of the first row transition metal ions, as would be predicted on the basis of experimental values for water exchange (*vide supra*). Formation of the V(SCN)$^+$ ion affords an excellent illustration of a relative volume profile [109] from which the compact nature of the transition state can be shown.

Other more recent examples reveal interesting characteristics. For example, the labilization introduced in the hydrolyzed form of aqua iron(III), Fe(H$_2$O)$_5$(OH)$^{2+}$, gives rise to an I$_d$ mechanism upon complexation with desferrioxamine B or acetohydroxamic acid, whereas the less reactive hexaaqua Fe^{3+} species complexes these ligands *via* an I$_a$ mechanism [111].

Substitution at Octahedral Cr(III) and Co(III) High pressure kinetic measurements were vital in resolving the mechanisms of aquation of pentaammine complexes of Cr(III) and Co(III). The complexes used, Cr(NH$_3$)$_5$X^{3+} and Co(NH$_3$)$_5$X^{3+}, contain a neutral X which is essential to avoid variable and unquantifiable changes in elecrostriction if X is charged. For a series of neutral X, $\Delta V^{\neq}$ was found to be negative for Cr(III) and positive for Co(III), denoting from the sign and magnitudes of the values (see Table 2.4) I$_a$ and I$_d$ mechanisms respectively, a distinction not possible from analysis of a variety of kinetic and other information [112]. Hence this resolution represents a prominent success in the application of high pressure kinetics.

Five Coordinate Species Very early work [76, 113] established that for six coordinate species the lability of remaining coordinating water molecules could be increased considerably by the presence of other ligands in the coordination sphere of the metal ion. However, when certain polydentate ligands coordinate to some of the later members of the M^{2+} ions of the first row transition metal series, the coordination number is changed from six to five. Four sites are

2.3. REACTIONS STUDIED AT ELEVATED PRESSURES

Table 2.4 Activation volumes for aquation of chromium(III) and cobalt(III) pentaammine complexes with neutral leaving groups.[a]

Species	L	$\Delta V^{\neq}$, cm^3 mol^{-1}
$Cr(NH_3)_5L^{3+}$	OH_2	-5.8
	$OS(CH_3)_2$	-3.2
	$OCHNH_2$	-4.8
	$OC(NH_2)_2$	-8.2
	$OC(NHCH_3)_2$	-3.8
	$OCHN(CH_3)_2$	-7.4
	$OC(CH_3)N(CH_3)_2$	-6.2
	$OP(OCH_3)_2$	-8.7
$Co(NH_3)_5L^{3+}$	OH_2	$+1.2$
	$OS(CH_3)_2$	$+2.0$
	$OCHNH_2$	$+1.1$
	$OC(NH_2)_2$	$+1.3$
	$OC(NHCH_3)_2$	$+1.5$
	$OCHN(CH_3)_2$	$+2.6$
	$OHCH_3$	$+2.2$
	$OHCH_2CH_3$	$+2.9$
	$OHCH(CH_3)_2$	$+3.8$
	$OC(NH_2)(NHCH_3)$	$+0.3$
	$OCH(NHCH_3)$	$+1.7$

[a] Taken from [100].

occupied by the ligand and the remaining coordinated water molecule is much less labile than water molecules of the fully aquated ions. The contrasting behavior of $Cu(tren)^{2+}$ (five coordinate) and $Cu(Me_6tren)^{2+}$ (five coordinate) with respect to both solvent exchange and ligand substitution [114–117] has been investigated recently. Me_6tren is 2,2′,2″-tris(dimethylamino)triethylamine and tren is the parent unmethylated ligand. Lincoln, and Merbach et al. [114–116] studied the exchange of DMF on the 2 + cations of Mn, Co and Cu and the exchange of DEF (diethylformamide) on Co^{2+} and Cu^{2+}, all coordinated by Me_6tren, using proton HPNMR. The volume of activation is modestly negative for exchange of DMF on the Mn complex ion (-6 cm^3 mol^{-1}), and on the Co complex ion (-2.7 cm^3 mol^{-1}), and for DEF exchange on the corresponding cobalt ion (-1.3 cm^3 mol^{-1}). For $Cu(Me_6tren)S^{2+}$, DMF and DEF are exchanged with positive $\Delta V^{\neq}$ values, $+6.5$ and $+5.3$ cm^3 mol^{-1} respectively.

These results suggest different activation modes, associative for Mn and Co, and dissociative for Cu, but the character is not well defined as being different from exact interchange in any species. Consideration of the d orbital occupancies and stereochemical differences in the trigonal bipyramidal geometry led to an explanation for the mechanistic difference. The original literature may be consulted for further details [114–116].

HPNMR and the high pressure temperature-jump method, with joule heating were used respectively to obtain $\Delta V^{\neq}$ for the water exchange on, and for ligand substitution by pyridine and substituted pyridines in Cu(tren)H$_2$O^{2+} [117]. The findings illustrate interesting differences from the reactivity of Cu(H$_2$O)$_6^{2+}$, and demonstrate the special nature of Cu(II)tren complexes. The hexaaqua species exchanges water three orders of magnitude faster than does the tren complex, and probably by an I$_d$ mechanism [118]. The volumes of activation for water exchange on Cu(tren)H$_2$O^{2+} are -4.7 cm^3 mol^{-1} and in the range of -7.5 to -10 cm^3 mol^{-1} for ligand substitution, denoting an I$_a$ mechanism. Since a relaxation method could be used, $\Delta V^{\neq}$ for the reverse of the substitution reactions could also be obtained. This permitted the construction of a relative volume profile, and it is illustrated in Figure 2.11. Volume reduction also accompanies formation of the activated complex for replacement of a ligand by water, and the formation is therefore also an I$_a$ process.

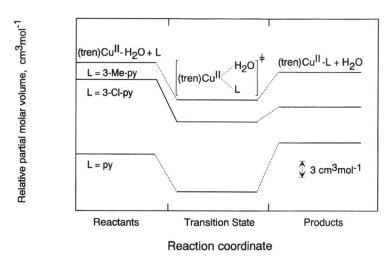

Figure 2.11 Volume profile for the reaction between Cu(tren)(H$_2$O)$^{2+}$ and pyridine and substituted pyridines at 298 K [117].

Four Coordinate Complexes Water exchange on square planar Pd(H$_2$O)$_4^{2+}$ and Pt(H$_2$O)$_4^{2+}$ is characterized by $\Delta V^{\neq}$ values of -2.2 and -4.4 cm^3 mol^{-1} respectively [119–121]. Analysis of complex formation kinetics at atmospheric pressure, and the negative values of $\Delta S^{\neq}$ for water exchange had already pointed toward an I$_a$ mechanism. The volume of activation values support this proposal, but other explanations may be possible; for example, an entering water molecule associates equatorially and there is bond extension (metal to oxygen) in the two axial positions of a trigonal bipyramidal configuration which will offset the volume collapse due to bond formation.

The activation parameters for dimethylsulfide exchange on trans-$Pd(Me_2S)_2Cl_2$ in different apolar solvents are in the -60 to -80 J mol^{-1} K^{-1} range ($\Delta S^{\neq}$) and the -5 to -8 cm^3 mol^{-1} range ($\Delta V^{\neq}$) [122]. The small variation in $\Delta V^{\neq}$ with the change in solvent was taken as evidence of a small electrostriction component, and hence a negative value for the intrinsic contribution, suggesting an associative process.

Fewer studies of solvent exchange have been conducted on tetrahedral complexes. The kinetics of triphenyl phosphine (TPP) exchange on $CoBr_2(TPP)_2$ in an appropriately deuterated chloroform solvent, studied by proton HPNMR yielded a value of $\Delta V^{\neq}$ of -12 cm^3 mol^{-1} [123]. Other facts, such as the form of the rate law (second order) and a negative $\Delta S^{\neq}$, all argue in favor of an associative process, but it is not obvious whether a limiting A mechanism is followed.

Ligand Substitution in Square Planar Complexes A classical example of this type of reaction is the solvolysis reaction of $Pd(L)Cl^+$ where L is a terdentate ligand such as diethylenetriamine (dien), and the metal is in oxidation state (II). Introduction of methyl and ethyl substituents causes rate retardation of up to five orders of magnitude, but $\Delta V^{\neq}$ is virtually invariant at about -12 cm^3 mol^{-1} [124], which is much more negative than for water exchange on the tetraaqua species, $Pd(H_2O)_4^{2+}$ ($\Delta V^{\neq} = -2.2$ cm^3 mol^{-1}) [119]. This difference can be ascribed to the change in electrostriction when the leaving group is anionic during the formation of this five coordinate intermediate. Volume profiles can be constructed as values of $\Delta V^{\neq}$ for the reverse anation reaction are available in some cases.

Solvent Exchange on Ions of the Lanthanide Elements Early attempts to measure the kinetics of water exchange on the first row inner transition elements (f block elements) in oxidation state three were bedeviled by the extreme rapidity of the exchange. The measurements are still rendered difficult by unfavorable magnetic properties in some cases. As a result of doubt about coordination numbers and geometry of coordination of solvents in contrast to the relative certainty regarding the transition metal ions and ions of main group elements, the kinetics of solvent exchange were less amenable to study. Very recently molecular dynamics simulation methods have been used to ascertain the coordination numbers of water throughout the series [125]. The exchange rate constants and activation parameters for the exchange of DMF on several $Ln(DMF)_8^{3+}$ ions (Ln = a lanthanide element) were obtained [126, 127]. Initially the values shown in Table 2.5 presented an interpretative challenge. That is $\Delta S^{\neq}$ and $\Delta V^{\neq}$ are not obviously reconcilable. However, it was proposed that the negative $\Delta V^{\neq}$ values arose from a decrease in the degrees of freedom accompanying formation of a nine coordinate transition state, whereas the positive volume of activation was considered to arise from a combination of a negative component due to penetration of a ninth ligand into the first coordination sphere and a larger positive contribution from the bond

Table 2.5 Activation parameters for DMF exchange on $Ln(DMF)_8^{3+}$ in DMF.[a]

Ln	$\Delta H^{\neq}$, kJ mol^{-1}	$\Delta S^{\neq}$, J mol^{-1} K^{-1}	$\Delta V^{\neq\,b}$, cm^3 mol^{-1}
Tb	14.1	−58.3	+5.2
Dy	13.8	−68.5	+6.1
Ho	15.3	−68.1	+5.2
Er	23.6	−29.6	+5.4
Tm	33.2	+9.9	+7.4
Yb	39.3	+40.0	+11.8

[a] Ln = lanthanide element.
[b] For Tm the temperature was 255 K, and for all other ions $\Delta V^{\neq}$ was obtained within the range 235–239 K.

lengthening to the leaving ligand and non-exchanging ligands. However, mechanistic assignment is more difficult than with exchange on species of lower coordination number.

Solvent Exchange and Ligand Substitution in Organometallic Systems We conclude the section on solvent exchange and ligand substitution kinetics by describing the role of high pressure kinetics in providing insight into the mechanism of organometallic reactions, which are frequently reactions involving metal-carbonyl species.

Replacement of CO by $P(OMe)_3$ in $Cr(CO)_4$(phen) (where phen is 1,10-phenanthroline) gives a $\Delta V^{\neq}$ value of $+14$ cm^3 mol^{-1}, whereas for the corresponding Mo compound $\Delta V^{\neq}$ is -21 cm^3 mol^{-1}. This is diagnostic of a mechanistic changeover from D to A, which is a reflection of the different size of the metal center [128].

As may be anticipated, most of these reactions are studied in aprotic solvents [129]. The species $M(CO)_5THF$, where M = Cr, Mo or W and THF is tetrahydrofuran, can be prepared in solution and the solvent molecule can be subjected to substitution by ligands such as piperidine, $P(C_6H_5)_3$ and $P(OC_2H_5)_3$, in THF. Using normal stopped-flow and HPSF methods kinetic and activation parameters can be obtained. The volumes of activation are negative in all nine reactions, but progressively more so in the order Cr < Mo < W and for larger entering nucleophiles. The observed rate constant is a composite of other constants but the results can be rationalized as indicating a changeover to a more associative character from Cr to W. A further pertinent discussion of this type of system is found in the section on photo-induced substitution reactions.

2.3.3 Addition and Elimination Reactions

A large variety of inorganic reactions may be included in this category, some of which could easily be classified also as substitution reactions. It is not practicable to classify all of the possible reactions. We would rather draw attention to

2.3. REACTIONS STUDIED AT ELEVATED PRESSURES

a few reactions in which use of the pressure variable has had a significant impact on mechanistic understanding.

One type of reaction is oxidative addition where a four coordinate metal complex undergoes conversion to an oxidized six coordinate species. Such reactions are frequently accompanied by a large negative $\Delta V^{\neq}$ value which is a consequence of a transition state in which there is bond formation, and charge development causing increasing solvent electrostriction.

When H_2, CH_3I or HCl are added to Vaska's compound ([trans-IrCl(CO)(PPh$_3$)$_2$]) [130] or related compounds, negative values of $\Delta V^{\neq}$ were found and they exhibited a solvent dependence. The variation of $\Delta V^{\neq}$ could be correlated with a polarizability function, q_p, of the solvent permitting an extrapolation to yield a $\Delta V^{\neq}_{int}$ of -17 to -18 cm^3 mol^{-1} [131]. Simultaneous formation of two Ir–H bonds to produce a *cis*-dihydrido complex upon H_2 addition was the explanation proposed, while addition of CH_3I resulted in a linear transition state, [I–CH$_3$–IrL$_4$].

A more recent example from organometallic chemistry is the addition reaction of *p*-cyanoaniline to an α, β-acetylenic pentacarbonyl chromium carbene complex [132]. The reaction, carried out in acetonitrile, was monitored in a pill-box cell in a conventional spectrophotometer. However, for less strongly electron-withdrawing groups than the cyano substituent the stopped-flow method was required. The $\Delta V^{\neq}$ values are more negative for the slower reacting anilines, which can be attributed to a later transition state and a greater degree of bond formation. Subsequent reaction occurs rapidly.

The kinetics of oxidative addition of CH_3I to the Pd(II) species, PdMe$_2$(bpy), to form PdIMe$_3$(bpy) in acetone have been investigated at ambient and at higher pressures [133]. A $\Delta V^{\neq}$ value of -11.9 cm^3 mol^{-1} tends to confirm the S_N2 mechanism proposed initially. The complementary reaction, reductive elimination of C_2H_6 from the Pd(IV) species, PdIMe$_3$(bpy), leaving PdIMe$_2$(bpy) likewise in acetone, was also investigated over a range of pressure. This allowed the construction of a volume profile. The latter reaction slowed down upon pressure application yielding $\Delta V^{\neq}$ of $+17 \pm 1$ cm^3 mol^{-1}, a value consistent with species production in the overall reaction, a change in the oxidation state of Pd from (IV) to (II) and probably considerable bond breakage in the transition state.

2.3.4 Electron Transfer Reactions

A succinct account of the background and early results of the study of electron transfer reactions at elevated pressures is available [10]. More recent developments of applying the Marcus-Hush theory [134, 135] to prediction and calculation of $\Delta V^{\neq}$ values in comparison with experimental values for some outer sphere electron transfer (OSET) reactions were also reported therein. Several other accounts of high pressure effects on electron transfer reactions, as well as more advanced treatises on the subject in general, may be consulted [1, 136–141]. Classification of redox reactions as inner sphere or outer sphere was made over 40 years ago, and indeed initially $\Delta V^{\neq}$ was thought [142] to be a criterion which could be used to distinguish between those reactions in which transfer between two metal centers occurs through a shared (bridging) first coordination sphere ligand (inner sphere), or between the metal complexes with their coordination spheres intact (outer sphere, OSET).

Much more attention has been given to OSET reactions, and perhaps particularly to symmetrical self-exchange reactions because, in principle, they are more amenable to a theoretical approach and have the simplification of a zero reaction volume. An example is eq. (11),

$$MnO_4^- + MnO_4^{2-} \rightarrow MnO_4^{2-} + MnO_4^- \qquad (11)$$

which was investigated by HPNMR [143], and exhibits a $\Delta V^{\neq}$ value of $-21 \text{ cm}^3 \text{ mol}^{-1}$. This reaction is catalyzed by some counter ions, notably Na^+ and K^+. This catalysis manifests itself in very significantly different $\Delta V^{\neq}$ values, $+4$ and $-1 \text{ cm}^3 \text{ mol}^{-1}$ respectively. Metal ion catalysis of other types of anion-anion OSET reactions has been previously observed [144], and earlier similar findings discussed [145]. However, in variable pressure investigations such catalysis can cause complications in the separation of the components of the measured $\Delta V^{\neq}$.

Other methods of monitoring self-exchange reactions include isotopic labeling of the metal of a self exchanging complex species (^{59}Fe [146] and ^{60}Co [147], for example), and the stopped-flow technique using circular dichroism as the physical property monitored [148]. Frequently, stopped-flow or conventional spectrophotometry is used for following the kinetics of other redox reactions at various pressures.

Efforts begun over 20 years ago by Stranks [1] and continued by Swaddle [10], Wherland [149–151] and others [152–155] in recent years, have been aimed at adapting the Marcus-Hush theory to rate constants obtained as a function of pressure to permit calculation of $\Delta V^{\neq}$ and to compare those values with experimental data, particularly for these symmetrical self-exchange reactions, and in addition, more recently for some bioinorganic reactions. A typical volume profile for a non-symmetrical OSET reaction between a ruthenium complex and cytochrome c is presented in Figure 2.12 [155]. Although not exclusively so, many OSET reactions occur in aqueous solution and this applies to the examples cited thus far. However, some interest in investigating the effect of solvents on these processes [150] is beginning to emerge.

A full description of the Marcus-Hush theory and the various adaptations to it, both at atmospheric and at higher pressures is not possible here. We describe the generally accepted way of calculating $\Delta V^{\neq}$ [10, 139], and then apply both this calculation method and the procedure for correlating (theoretical/experimental) rate constants themselves for a specific reaction.

For self-exchange reactions, $\Delta V^{\neq}$ can be expressed initially as a sum of four terms:

$$\Delta V^{\neq} = \Delta V^{\neq}_{IR} + \Delta V^{\neq}_{SR} + \Delta V^{\neq}_{COUL} + \Delta V^{\neq}_{DH} \qquad (12)$$

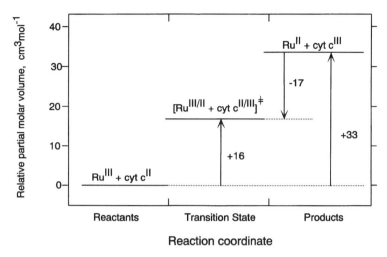

Figure 2.12 Volume profile for the reaction of the isonicotinamide pentaammine ruthenium complex with cytochrome c at 298 K [155].

In eq. (12) $\Delta V_{IR}^{\neq}$ is an internal reorganization term, $\Delta V_{SR}^{\neq}$ is a solvent reorganization term, $\Delta V_{COUL}^{\neq}$ is a term for coulombic or electrostatic work change and $\Delta V_{DH}^{\neq}$ is a Debye-Hückel activity coefficient term, each of which contributes to the overall change. This expression assumes that the electron transfer process is adiabatic, that is one in which the electron transfer follows almost every encounter of reactants in which the reorganizational conditions are satisfied.

A detailed analysis of self-exchange reactions, the results of which are collected from a variety of sources, indicates that the correlation (experimental/theoretical) is "good" for $Fe(phen)_3^{2+/3+}$, $Cu(dmp)_2^{2+/3+}$, $Ru(hfac)^{0/1-}$, $Fe(C_5H_5)_2^{1+/0}$, $Mn(CN\text{-}t\text{-}Bu)_6^{2+/1+}$ and $Fe(H_2O)_6^{2+/3+}$ [156–159] in aqueous solution, but is progressively less satisfactory in polar organic solvents, and inappropriate for reactions in solvents with low dielectric constants: dmp is 2,9-dimethyl-1,10-phenanthroline, hfac is the hexafluoroacetylacetonato ligand, and $Fe(C_5H_5)_2^{1+/0}$ is the ferrocinium ion or ferrocene. When a discrepancy occurs the results are rationalized within the adaptation to the Marcus theory, but with non-adiabaticity being invoked to explain the differences; this applies in the case of the $Co(en)_3^{2+/3+}$ self exchange reaction and to a lesser extent for eq. (10). However, in the cases of $Co(en)_3^{2+/3+}$ and $Co(phen)_2^{2+/3+}$ an adiabatic pathway with attendant factors concerned with Co(II)/(III) spin states has been discussed in some detail, and has not been completely ruled out. Figure 2.13 illustrates the agreement (based on adiabatic electron transfer) and non-agreement of a selection of self-exchange reactions.

For the final part of the section on electron transfer, it is worth devoting closer attention to a reaction which has been thoroughly investigated and is understood, rather than to continue providing further information in the manner of

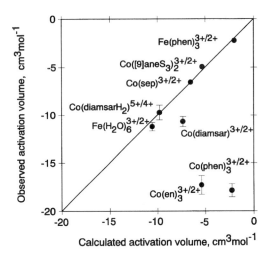

Figure 2.13 Electron transfer self-exchange reactions: plot of $\Delta V^{\neq}$ (calculated) versus $\Delta V^{\neq}$ (experimental). This figure has been redrawn from Figure 3 [185], an assembly of results from earlier publications: sep = sepulchrate = 1,3,6,8,10,13,16,19-octaazabicyclo[6,6,6]eicosane, [9]aneS$_3$ = 1,4,7-trithiacyclononane, diamsar = diaminosarcophagine = 1,8-diamino-1,3,10,13,16,19-hexaazabicyclo[6,6,6]eicosane. Other abbreviations are given in the footnote of Table 2.3.

a catalogue. The example chosen for this case study is the reaction between cytochrome c and the metal complex Co(terpy)$_2^{2+/3+}$ [160]. The effect of pressure on proteins and reactions of proteins, including cytochrome c is presented in Chapter 12. In addition to the interest from a bioinorganic chemical point of view, this reaction system illustrates how a prudent choice of reaction, based on equilibrium and redox properties as well as potential kinetic properties, can allow a thorough examination of the actual kinetic properties and their concurrence, or otherwise, with the relevant theoretical arguments. This then optimizes the opportunity for significant advancement in this field of endeavor. The reaction also provides the additional challenge of establishing a volume profile, since there is a non-zero reaction volume as distinct from the relatively straightforward self-exchange reactions considered above.

The reaction studied may be written, as in eq. (13),

$$\text{Cyt c}^{\text{II}} + \text{Co(terpy)}_2^{3+} \rightleftarrows \text{Cyt c}^{\text{III}} + \text{Co(terpy)}_2^{2+} \tag{13}$$

where terpy is 2,2′,2″-terpyridine, and the II and III superscripts on Cyt c signify the reduced and oxidized forms of the redox protein cytochrome c, an essential biological electron transporter which is vital in cellular oxidations in both plants and animals. The reaction is not only reversible, but has a low driving force (an equilibrium constant close to 1), and is therefore conveniently studied in both

2.3. REACTIONS STUDIED AT ELEVATED PRESSURES

directions. This enables a relative volume profile to be established, and a reaction volume to be obtained, without necessarily resorting to density measurements and calculation of the partial molar volumes for reactants and products. The empirical second order rate constants at 298 K are $1.43 \pm 0.06 \times 10^3$ M^{-1} s^{-1} and $1.70 \pm 0.05 \times 10^3$ M^{-1} s^{-1} at pH 7.2 and ionic strength 0.1 M, for reaction in the forward (k_f) and reverse (k_d) directions. These rate constants are products, in each case, of the precursor complex equilibrium formation constant, K_{OS}, and the actual electron transfer rate constant, k_{ET}, which characterizes the formation of the successor complex from the precursor complex. Thus in the forward direction:

$$\text{Cyt c}^{II} + \text{Co(terpy)}_2^{3+} \overset{K_{os}}{\rightleftharpoons} \{\text{Cyt c}^{II} \cdot \text{Co(terpy)}_2^{3+}\} \tag{14}$$

$$\{\text{Cyt c}^{II} \cdot \text{Co(terpy)}_2^{3+}\} \xrightarrow{K_{ET}} \{\text{Cyt c}^{III} \cdot \text{Co(terpy)}_2^{2+}\} \tag{15}$$

By using the Marcus-Hush theory it is possible to determine the reaction rate constant, providing the overall equilibrium constant for the reaction, expressed as K_{12}, is known, and the self exchange rate constants for each of the redox partners can be calculated. In the expression: $k_{12} = (k_{11} k_{22} K_{12})^{1/2}$, k_{11} and k_{22} are the self-exchange rate constants for Cyt c$^{II/III}$ and Co(terpy)$_2^{3+/2+}$ respectively. It is assumed that other terms in the theoretical treatment, namely work terms and the factor f_{12} can justifiably be neglected in this case. Calculation gave rise to a k_{11} value of 350 M^{-1}s^{-1} and a k_{22} value in the range of 1.9×10^3 to 3.4×10^3 M^{-1} s^{-1}. With $K_{12} = 0.9$, k_{12} and k_{21} values of 7.8×10^2 to 1.03×10^3 M^{-1} s^{-1} and 8.7×10^2 to 1.15×10^3 M^{-1} s^{-1} are obtained respectively. These are in reasonable agreement with the experimental values. The forward reaction has an activation volume of $+18.4 \pm 1.2$ cm^3 mol^{-1} and the reverse reaction is accelerated by pressure with a value of -18.0 ± 1.4 cm^3 mol^{-1}.

Clearly it is important to determine whether the theory can be adapted to provide estimated values of $\Delta V^{\neq}$ to compare with the experimental values. Eq. (12) must be modified for non-symmetrical reactions which have an overall volume change (as distinct from self-exchange reactions) by adding the term $\lambda^{\neq} \Delta \bar{V}$ and by adding βRT. The latter is a contribution from the pre-exponential work terms, related to solvent compressibility. The reaction volume is $\Delta \bar{V}$ and $\lambda^{\neq}$ is a parameter which gives the location of the transition state relative to products and reactants. In this case calculation [160] using a formula for $\lambda^{\neq}$ from the Marcus theory gave a value of 0.5, which yields a $\lambda^{\neq} \Delta \bar{V}$ value of $+17.5$ cm^3 mol^{-1}, ($\Delta \bar{V}$ is $+35$ cm^3 mol^{-1}). The other contributions, in cm^3 mol^{-1}, have been calculated to be, $\Delta V^{\neq}_{COUL} = -5.1$, -3.9, $\Delta V^{\neq}_{DH} = +6.3$, $+4.9$, $\Delta V^{\neq}_{SR} = -4.7$, -4.6 for the forward and reverse reactions respectively, with the internal rearrangement contribution being regarded as negligible and the βRT term contributing $+1.3$ cm^3 mol^{-1}. Taken together these contributions yield a $\Delta V^{\neq}_{12}$ value of $+15.3$ cm^3 mol^{-1} and a $\Delta V^{\neq}_{21}$ value of

$-19.8 \text{ cm}^3 \text{ mol}^{-1}$. When a correction is made to account for the solvent rearrangement around cytochrome c being less than the calculated values, the $\Delta V^{\neq}$ terms become $+18.9$ and $-16.3 \text{ cm}^3 \text{ mol}^{-1}$ respectively, which are in reasonably good agreement with the experimental values. The volume profile is presented as Figure 2.14. In addition to the result that the transition state is, in a volume sense, halfway along the reaction coordinate of the reaction volume change, it is concluded that the volume changes occur mostly on the cobalt complex and not on the protein. Reactions of cyt $c^{II/III}$ with $Co(bpy)_3^{2+/3+}$ and

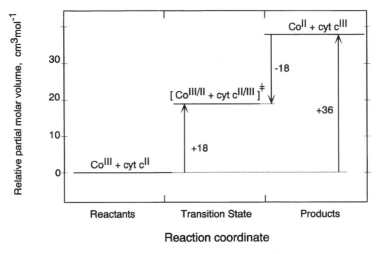

Figure 2.14 Volume profile for the reaction of $Co(terpy)_2^{2+/3+}$ with cytochrome $c^{II/III}$ at 298 K [160].

with $Co(phen)_3^{2+/3+}$ are also amenable to study at high pressure [161], and in addition to volume of activation determinations for the forward and reverse redox reactions in each case, reaction volumes measured spectrophotometrically, from density measurements and from pressure dependent electrochemistry measurements all combine to give self-consistent volume profiles.

There are many other examples of good correlations between experiment and theory for outer-sphere electron transfer reactions, and details of those studies can be consulted [154]. In some cases, however, it is necessary to invoke non-adiabaticity of the electron transfer in order to achieve a correlation. But in general there is acceptance of the applicability of electron transfer theory as applied to atmospheric pressure kinetics, and now there is increasing evidence that the adaptations of the theory to high pressure kinetics provide successful correlations.

2.3.5 Radiation-Induced Processes

Reactions Initiated by Photolysis or Pulse Radiolysis In this section we will focus on examples where the combination of high pressure kinetic and

2.3. REACTIONS STUDIED AT ELEVATED PRESSURES

photolysis or radiolysis techniques has been employed in an effort to improve the understanding of the intimate mechanism of the induced process. Work from our group in this area has benefited from intensive collaborations with many other groups included in the cited references. Examples of various types of photo- and radiation-induced reactions are presented.

Of the three possible reaction pathways for thermal ligand substitution reactions, *viz.* associative (A), dissociative (D) or interchange (I), that exhibit characteristic pressure dependences (see Section 2.3.2), the question that arises is whether these possibilities also exist for photo- and radiation-induced processes. Earlier reports from our laboratories and others have shown that chemical and physical processes that occur in the electronic excited state of an inorganic or organometallic molecule exhibit characteristic pressure dependencies [162, 163]. The pressure dependence of the observed quantum yield for a photosubstitution reaction, along with the pressure dependence of the excited state lifetime, enable the elucidation of the intimate nature of the substitution mechanism in terms of the known mechanisms referred to above [20]. This was successfully accomplished for a series of Rh(III) and Cr(III) ammine complexes [164]. Photosubstitution reactions of $M(CO)_6$ complexes shown in eq. (16), are all accompanied by significantly positive $\Delta V^{\neq}$ values that indicate a dissociative mechanism [165].

$$M(CO)_6 + L \xrightarrow{h\nu\,(313\,nm)} M(CO)_5L + CO \qquad (16)$$

M = Cr, Mo, W; L = piperidine, pyridine, acetonitrile. The photosubstitution mechanism of CO in $M(CO)_4$(phen) (M = Cr, Mo, W) is controlled by the nature of the excited state, *viz.* dissociative for LF excitation versus associative for MLCT excitation. The apparent discrepancy in the literature regarding these mechanisms could be resolved by studying the pressure dependence of the photosubstitution reactions as a function of irradiation wavelength, since the results clearly produced positive volumes of activation for LF excitation compared to negative values for MLCT excitation [166]. Similar investigations were employed to resolve the photochemical reaction mechanisms of $(CO)_5ReMn(CO)_3(\alpha$-diimine$)$ and $CpFe(CO)_2(COCH_3)$ complexes [167, 168].

Flash photolysis techniques were used with great success to study the substitution behavior of reactive solvento intermediates of the type $M(CO)_5S$ [169, 170]. The effect of pressure on such reactions has clearly demonstrated the crucial role played by the size of the metal center M, the bulkiness of L and the binding properties of the solvent S in controlling the nature of the substitution mechanism. It was also possible to study the displacement of a coordinated solvent molecule *via* ring-closure of a potential bidentate ligand such as a P-olefin in cis-$(CO)_4W(S)(PPh_2(CH_2)_nCH=CH_2)$ ($n = 1$ to 4) [171]. The $\Delta V^{\neq}$ values increased from $+7.7$ to $+10.5$ cm^3 mol^{-1} on increasing n from 1 to 4, which corresponds to a dissociative interchange mechanism for the ring-closure reaction. When the entering nucleophile in the case of ligand substitution of M(CO)$_5$S is a bidentate ligand, then the rapid displacement of S is followed by a slower displacement of CO during the ring-closure reaction. The series of data in Table 2.6 clearly demonstrates a changeover from an associative interchange to a dissociative interchange mechanism with increasing steric hindrance on the chelate ligand L [172].

The binding of small molecules such as O$_2$, CO and NO to ferrous hemes and hemoproteins is of fundamental importance to the transport of such molecules in biological systems. Such processes can be studied *via* the application of flash-photolysis techniques. The pressure dependence of such reactions can assist the mechanistic clarification of the binding process following the flash-induced dissociation step. Typical $\Delta V^{\neq}$ data for the addition of neutral ligands to model heme systems, *viz.* protoheme dimethyl ester (PHDME) and monochelated protoheme (MCPH), are summarized in Table 2.7 [173]. For the slower reactions, recombination characterized by a negative $\Delta V^{\neq}$ value is rate-determining, whereas for the faster reactions the binding becomes diffusion

Table 2.6 Kinetic data for ring-closure of Mo(CO)$_5$L complexes.

Solvent	L[a]	k (298 K) s^{-1}	$\Delta H^{\neq}$ kJ mol^{-1}	$\Delta S^{\neq}$ J mol^{-1} K^{-1}	$\Delta V^{\neq}$ cm^3 mol^{-1}
Toluene	en	3.0×10^{-5}	72 ± 7	−92 ± 22	−5.4 ± 0.8
Toluene	dabR$_2$	1.1×10^{-3}	78 ± 5	−40 ± 17	−9.5 ± 0.4
Toluene	dpbpy	1.4	69 ± 2	−4 ± 6	+5.4 ± 0.5
Toluene	dmbpy	2.6	65 ± 1	−20 ± 3	−5.6 ± 0.4
Toluene	bpy	3.1	62 ± 1	−26 ± 3	−3.9 ± 0.6
Fluorobenzene	phen	1.1×10^4	47 ± 2	−9 ± 7	−2.9 ± 0.2

[a] dabR$_2$ = 1,4-diisopropyl-1,4-diazabutadiene; dpbpy and dmbpy are respectively the 4,4′-diphenyl and 4,4′-dimethyl derivatives of 2,2′-bipyridine (bpy). Other abbreviations are given earlier in the text.

Table 2.7 Bimolecular addition of neutral ligands to five-coordinate ferrous model heme complexes in toluene: volume of activation data.

Heme complex[a]	L	k_{on} (298 K), M^{-1} s^{-1}	$\Delta V^{\neq}$, cm^3 mol^{-1}
MCPH	CO	1.1×10^7	−19.3 ± 0.4
MCPH	O$_2$	1.0×10^8	−11.3 ± 1.0
(MeNC)PHDME	MeNC	3.9×10^8	+11.6 ± 0.8
(t-BuNC)PHDME	t-BuNC	2.5×10^8	+9.9 ± 1.0
(1-MeIm)PHDME	1-MeIm	1.5×10^8	+10.9 ± 3.1

[a] MNPH in monochelated protoheme, PHDME is the protoheme dimethyl ester and Im is imidazole

2.3. REACTIONS STUDIED AT ELEVATED PRESSURES

controlled in toluene as solvent and is slowed down by increasing pressure due to a significant increase in solvent viscosity [173].

The reversible binding of dioxygen to Co(II) complexes can also be studied using flash-photolysis techniques. A typical volume profile for such a system is presented in Figure 2.15, from which it can clearly be concluded that ligand substitution on Co(II) controls the binding of dioxygen [174]. This step is then followed by a rapid electron-transfer reaction during which Co(II)-O_2 is converted to Co(III)-O_2^-, which is accompanied by a large volume collapse due to intrinsic and solvational changes. The effect of pressure on the overall equilibrium constant resulted in a reaction volume of -22.4 ± 0.4 cm^3 mol^{-1}, which is in excellent agreement with the reaction volume calculated from the difference in the activation volumes for the forward and reverse reactions, viz. -22.6 cm^3 mol^{-1}.

Similar techniques were used to study the association of sperm whale myoglobin with a series of neutral ligands in aqueous medium. The results in Table 2.8 [173, 175, 176] demonstrate that only the binding of CO is characterized by a negative $\Delta V^{\neq}$ value in line with a bond formation process. The positive $\Delta V^{\neq}$ values for the other ligands are ascribed to rate-determining entering of the ligand into the protein, which may be accompanied by significant desolvation and presumably conformational changes on the protein chain. A combination of these data for the binding of O_2 and CO with those obtained for the reverse reaction with the aid of stopped-flow techniques, resulted in the volume profiles presented in Figure 2.16 [176]. The significantly different volume profiles for O_2 and CO can be accounted for in terms of different rate-determining steps for the "on" reaction and the nature of the binding of the ligands in the protein pocket.

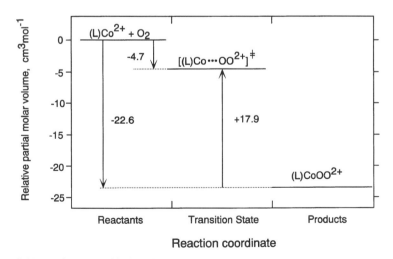

Figure 2.15 Volume profile for the reaction of O_2 with the Co(II)L complex, (L is hexamethyl cyclam) at 298 K [174].

Table 2.8 Rate constants and $\Delta V^{\neq}$ values for the bimolecular addition of various ligands to deoxymyoglobin in aqueous buffer solution.

L	k_{on} (298 K), $M^{-1} s^{-1}$	$\Delta V^{\neq}$, $cm^3 mol^{-1}$
CO	5.2×10^5	-10.0 ± 0.8
O_2	2.5×10^7	$+5.2 \pm 0.5$[a]
	1.3×10^7	$+7.8 \pm 1.3$
CH_3NC	1.4×10^5	$+8.8 \pm 1.0$
t-BuNC	2.1×10^3	$+9.3 \pm 0.3$

[a] Data obtained using the temperature-jump method.

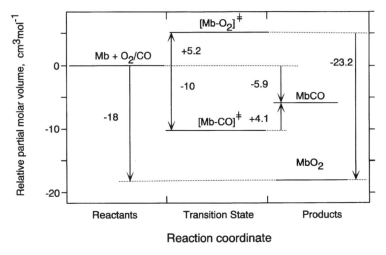

Figure 2.16 Volume profile for the reactions of CO and O_2 with myoglobin at 298 K [176].

High pressure pulse-radiolysis techniques can also be used to study other reactions of molecular oxygen. For instance the oxidation of $Cu^I(phen)_2$ by dioxygen proceeds *via* a Cu^I-O_2 transient in which a copper–oxygen bond is formed followed by the rapid formation of $Cu^{II}(phen)_2$ and O_2^-. This process is characterized by a $\Delta V^{\neq}$ value of -22 cm^3 mol^{-1}, which is indeed close to the reaction volume expected for the binding of dioxygen [177]. This technique has also been successfully applied in the study of the formation and cleavage of metal–carbon σ bonds. For instance, the reaction of the methyl radical with a Co(II) complex shown in reaction (17) is characterized by a reaction volume of -16 cm^3 mol^{-1} [178].

$$Co^{II}(nta)(H_2O)_2^- + \cdot CH_3 \rightarrow Co^{III}(nta)(CH_3)(H_2O)^- + H_2O \qquad (17)$$

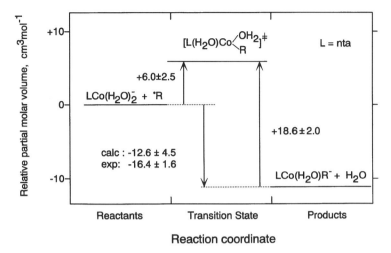

Figure 2.17 Volume profile for the reaction of methyl radicals with the nitrilotriacetate complex of Co(II) at 298 K [178].

The volume profile in Figure 2.17 clearly indicates a significantly higher partial molar volume for the transition state than for either the reactant or product states. This is interpreted in terms of an I_d mechanism in which the forward reaction is controlled by solvent exchange on $Co(nta)(H_2O)_2^-$. The large volume ollapse following the transition state is ascribed to metal–carbon bond formation which is accompanied by oxidation of Co(II) to Co(III).

A similar result was found for the reaction of aquated Cr(II) with 10 different aliphatic radicals, for which $\Delta V^{\neq}$ varied between $+3.4$ and $+6.3\,cm^3\,mol^{-1}$ [179]. These data were interpreted in terms of a water-exchange-controlled formation of the Cr–R bond, from which it followed that water exchange on aquated Cr(II) occurs according to an I_d mechanism. The typical volume profile in Figure 2.18 closely resembles that for the $Co(nta)(H_2O)_2^-$ system. Once again the large volume collapse following the transition state must be due to Cr–R bond formation and the conversion of Cr^{II}–R to Cr^{III}–R^-. These and more recent studies on the interaction of metal complexes with free radicals suggest that for non-diffusion controlled processes, the radicals can be treated as normal nucleophiles in ligand substitution processes which are often controlled by solvent exchange on the metal complex [180].

High pressure flash-photolysis and pulse-radiolysis can be used to induce intermolecular and intramolecular electron-transfer processes. The first work involving the application of high pressure techniques in such studies focused on a number of long-distance electron-transfer reactions in ruthenated cytochrome c. The intramolecular electron-transfer reactions in horse heart $(NH_3)_5Ru^{II}$–His 33 and *candida krusei* $(NH_3)_5Ru^{II}$–His 39 are accompanied by a significant pressure-induced acceleration with corresponding $\Delta V^{\neq}$ values of -17.7 ± 0.9 and $-18.3 \pm 0.7\,cm^3\,mol^{-1}$, respectively [181]. The intermolecular process

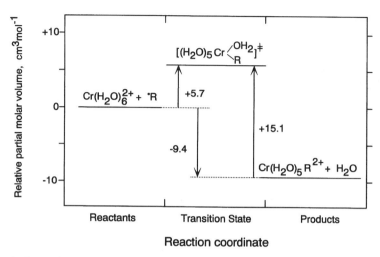

Figure 2.18 Volume profile for the reaction of an aliphatic radical with aqua Cr(II) species at 298 K [179].

between $Ru(NH_3)_6^{2+}$ and horse heart cytochrome c exhibits a similar pressure acceleration with a $\Delta V^{\neq}$ value of -15.6 cm^3 mol^{-1}. Electrochemical measurements at elevated pressure performed recently indicated that the overall reaction volume for the cited redox reactions is approximately twice the values quoted for the activation volumes [182]. This means that the transition state for such an electron-transfer process lies approximately halfway between the reactant and product states along the reaction coordinate on a partial molar volume scale. In addition, the observed volume changes mainly arise from changes in electrostriction on the ruthenium ammine centers, since it could be shown very convincingly that oxidation or reduction of cytochrome c itself exhibits almost no pressure dependence [182] (for further information on the effect of pressure on electron-transfer reactions see Section 2.3.4). Finally, a meaningful pressure-induced acceleration was reported for the Fe(II) to Ru(III) electron transfer in (bpy)$_2$(im)Ru(His 72)-cyt c compared to almost no effect for the His 33 analogue [183]. This was interpreted in terms of the effect of a slight compression of cyt c under pressure that in turn shrinks the through-space gaps that are essential for the electron-tunneling pathways between the heme and His 72. Such a compression is not expected to affect covalent and hydrogen bond lengths, which accounts for the absence of a pressure acceleration in the case of the His 33 modified protein.

2.4 CONCLUDING REMARKS AND FUTURE PERSPECTIVES

This chapter has by necessity presented only a selection of inorganic reactions which have been subjected to pressure. A comprehensive tabulation of the data

2.4. CONCLUDING REMARKS AND FUTURE PERSPECTIVES

available at the time [20] and more recent frequent updates of results from solution reaction kinetics studies [18, 19] may be consulted to ascertain the wide range of reactions investigated. This contribution has chronicled the main features of instrumental developments and some of the more important reports in the history of the field of inorganic high pressure kinetics, as well as providing examples from the recent literature.

A decade ago the limitations of transition state theory (TST) with respect to interpretation of volumes of activation were pointed out [184], and other theoretical models and considerations were discussed. The use of TST remains widespread, and the simplicity of its application is no doubt partly responsible. In many experiments the range of pressure applied is less than 150 MPa, in which case the pressure dependence of the volume of activation is not detectable. The adaptations to the Marcus-Hush theory to account for high pressure kinetics of electron transfer reactions predict a pressure dependence of the volume of activation for self-exchange redox reactions [139, 143], although this is not expected to be detected when reaction takes place in solvents with high dielectric constants.

These adaptations have matured and considerable progress has been made in successful correlations of theoretical values of volumes of activation with those experimentally acquired for self-exchange reactions [139, 143], a success which is now being enjoyed in studies of OSET non-symmetrical and bioinorganic reactions [139, 143, 154, 160, 161]. A recent report [185] presents an analysis of self-exchange in Co(II)/Co(III) complexes, and explains the lack of accordance with the Marcus-Hush theory of some self-exchanging cobalt couples as attributable to structural changes associated with spin equilibria of Co(II) species prior to electron transfer, rather than to non-adiabaticity which had not previously been ruled out. This conclusion emerged from studies of self-exchange in cobalt-cage complexes in which the structural changes are suppressed and the value of $\Delta V^{\neq}$ found comports closely with the predicted value.

Theoretical calculations of the volumes of activation for solvent exchange on trivalent and divalent cations are beginning to emerge [80–82]. The initial understanding and interpretation of a pure interchange mechanism being characterized by a volume of activation of zero has been assailed. It has been recognized that assumptions inherent to the interpretation of volumes of activation are just that [79], and efforts to make appropriate corrections and adjustments that would reduce the number or eliminate these assumptions are needed. Nevertheless the enthalpy calculations which form the basis of the results and the proposed step by step mechanism, and the proposal that all first row transition metal dications exchange water *via* an I_d mechanism also contain assumptions [81, 82]. Therefore, it is not yet transparently clear that conclusions regarding the mechanism of water exchange based upon experimental findings, with the implied assumptions, are erroneous. Further calculations using different models are being undertaken with the view to resolving this problem [186].

A changeover from an interchange mechanism to a limiting dissociative mechanism has been proposed for water exchange on the series of octa-substituted aqua ions of Nd(III), Eu(III), Tb(III), Dy(III) and Ho(III) [187]; the first has a near zero volume of activation while the latter four have $\Delta V^{\neq}$ values between $+7$ and $+10 \, cm^3 \, mol^{-1}$ for water exchange. Unexpected rate differences for water exchange on substituted lanthanide ions [187] and the discovery of a change in mechanism for water exchange on the aqua Gd^{3+} ion (associative) in contrast to dissociative for heavily substituted Gd(III) ions [188-190] demonstrate that current and future activities in the field of high pressure inorganic chemistry have a significant contribution to make to chemistry enlightenment.

In the future, in addition to further investigations of the inorganic and organometallic reactions of the classes described in this chapter, the use of the high pressure variable will be extended to provide information on more complex systems. For example, recently [191] the first *complete* volume profile for an enzyme catalyzed reaction (the hydration of CO_2 and the dehydration of HCO_3^- catalyzed by human carbonic anhydrase II) was established. The coordination chemistry of zinc in the metalloenzyme and its role in the catalytic cycles can be explored through the volume profile and by comparing it with that for the spontaneous reaction [192]. The general mechanistic features were confirmed, and new insights into the mechanism of the dehydration reaction were obtained. In another sub-area, photochemical reactions at different pressures are providing interesting details of various features of the ligand to metal bonding and structural characteristics of photosensitive organometallic compounds and the nature of excited state species [193]. Developments in HPNMR spectroscopy for higher field spectrometers also herald greater opportunities [194].

Considering the variety of chemistry currently studied and likely to be studied in the future, the next decade promises to generate new mechanistic challenges to rival the exciting progress over the past ten years.

ACKNOWLEDGEMENTS

The authors gratefully acknowledge financial support from the Deutsche Forschungsgemeinschaft, Fonds der Chemischen Industrie and the Volkswagen Stiftung. They greatly appreciate the stimulating contacts with research students, postdoctoral associates and visiting scientists over a period of many years. This contribution would not have been possible without their efforts; much of the research reported in this chapter is a consequence of their endeavors. We are indebted to them, and to Dr. Anton Neubrand and Dr. Achim Zahl for their skillful assistance in the preparation of the figures.

Note Added in Proof:

Water exchanges from $Ln(PDTA)(H_2O)_2^-$, where Ln = Tb, Dy, Er, Tm, Yb and PDTA = propylenediaminetetraacetate, with a dramatic decrease in rate constant with decreasing ionic radius across the lanthanide series [195]. On the basis of $\Delta V^{\neq}$ values this decrease in lability is accompanied by a change in mechanism from associatively to dissociatively activated.

2.5. REFERENCES

Water and acetonitrile exchange on trans-[Os(en)$_2$(η^2-H$_2$)S]$_2$, (S = H$_2$O or CH$_3$CN) is characterized by almost zero $\Delta V^{\neq}$ values [196]. Substitution reactions proceed by an I$_d$ or by a D mechanism [196].

A molecular dynamics simulation has been employed to study water exchange on the tripositive aqua ions of Nd, Sm and Yb [197]. The coordination number changes from 9 to 8 across the series and this is coupled to a changeover in exchange mechanism. A $\Delta V^{\neq}$ of $+4.5$ cm^3 mol^{-1} is calculated for [Nd(H$_2$O)$_9$] (I$_d$ mechanism), but exchange on [Sm(H$_2$O)$_9$]$^{3+}$ cannot be classified as D, I or A. Exchange on [Yb(H$_2$O)$_8$]$^{3+}$ is by an I$_a$ mechanism.

How water exchanges on aqua Cu(II) ions has long been an intriguing question. Resolution of this question has been elusive; complications in interpretation arise because of the consequences of Jahn-Teller (J-T) distortion. A recent paper [198] describes the kinetics of water and DMF exchange on Cu^{2+}. For DMF a dissociative activation mechanism is proposed. $\Delta V^{\neq}$ for water exchange is $+2.0$ cm^3 mol^{-1}; and this increase corresponds to dissociatively activated loss of water from an axial position. The small value of $\Delta V^{\neq}$ arises because the exchanging molecule is already more distant from the metal than in the absence of J-T distortion.

Substitution in labile chromium complexes, specifically [Cr(HEDTA)OH$_2$] and [Cr(EDTA)]$^-$ by SCN$^-$ has been investigated [199]. The anomalous substitutional lability of Cr^{III}EDTA complexes is due to activation by transient chelation of the pendant arm of quinquedentate EDTA.

$\Delta V^{\neq}$ values for nucleophile addition to (μ_2-H)$_2$Os$_3$(CO)$_{10}$ in chlorobenzene show a clear correlation with the cone angle for P-donor nucleophiles [200]. Rate constant/activation volume correlations have also been described for the substitution of ligands in pentacyanoferrate complex ions by CN$^-$ [201].

An assessment of $\Delta V^{\neq}$ data for acid and base hydrolysis for a wide variety of iron(II)diimine complexes in aqueous and aqueous solvent mixtures has been made [202]. The interpretation is complicated by large electrostriction effects.

The intermolecular electron transfer kinetics between pentaammineruthenium complexes and cytochrome c have been investigated [203]. One objective was to ascertain the effect of substituents on pyridine, the sixth ligand of the complex. Whether pyridine interacts in a specific way with the heme-edge groove is not conclusively answered. A comparative pressure dependence of the formation and deoxygenation kinetics of oxyhemerythrin and oxyhemocyanin has been reported [204].

Solvent exchange on cobalt(II) ions in 1,3-propanediamine (tn) and n-propylamine has been studied, and compared with corresponding reactions of ethylenediamine [205]. Exchange of tn has a $\Delta V^{\neq}$ value of $+6.6$ cm^3 mol^{-1}.

Thus the prophesy that a multitude of fascinating chemistry would be revealed using high pressure and other techniques in the next decade is already being realized within a few months.

2.5 REFERENCES

[1] Stranks, D. R., Pure Appl. Chem., 38 (1974) 303.
[2] Kelm, H. (ed.), High Pressure Chemistry, Reidel, Dordrecht, 1978.
[3] Lawrence, G. A., Stranks, D. R., Acc. Chem. Res., 12 (1979) 403.
[4] Isaacs, N. S., Liquid Phase High Pressure Chemistry, John Wiley, Chichester, 1981.
[5] Palmer, D. A., Kelm, H., Coord. Chem. Rev., 36 (1981) 89.
[6] Merbach, A. E., Pure Appl. Chem., 54 (1982) 1479.
[7] Blandamer, M. J., Burgess, J., Pure Appl. Chem., 55 (1983) 55.
[8] Moore, P., Pure Appl. Chem., 57 (1985) 347.
[9] van Eldik, R. (ed.), Inorganic High Pressure Chemistry, Kinetics and Mechanism, Elsevier, Amsterdam, 1986.
[10] Swaddle, T. W., (van Eldik, R. (ed.)), Inorganic High Pressure Chemistry, Kinetics and Mechanism, Chapter 5, page 273, Elsevier, Amsterdam, 1986.
[11] Merbach, A. E., Pure Appl. Chem., 59 (1987) 161.
[12] van Eldik, R., Jonas, J. (eds.), High Pressure Chemistry and Biochemistry, Reidel, Dordrecht, 1987.
[13] Kotowski, M., van Eldik, R., Coord. Chem. Rev., 93 (1989) 19.
[14] van Eldik, R., (Twigg, M. V. (ed.)), Mechanisms of Inorganic and Organometallic Reactions, Volume 3, Chapter 15, page 399, Plenum, New York, 1985.
[15] van Eldik, R., (Twigg, M. V. (ed.)), Mechanisms of Inorganic and Organometallic Reactions, Volume 4, Chapter 15, page 433, Plenum, New York, 1986.
[16] van Eldik, R., (Twigg, M. V. (ed.)), Mechanisms of Inorganic and Organometallic Reactions, Volume 5, Chapter 15, page 377, Plenum, New York, 1988.
[17] van Eldik, R., Schneider, K., (Twigg, M. V. (ed.)), Mechanisms of Inorganic and Organometallic Reactions, Volume 6, Chapter 15, page 437, Plenum, New York, 1989.
[18] van Eldik, R., (Twigg, M. V. (ed.)), Mechanisms of Inorganic and Organometallic Reactions, Volume 7, Chapter 15, page 371, Plenum, New York, 1991.
[19] Neubrand, A., van Eldik, R., (Twigg, M.V. (ed.)), Mechanisms of Inorganic and Organometallic Reactions, Volume 8, Chapter 15, page 399, Plenum, New York, 1994.
[20] van Eldik, R., Asano, T., le Noble, W. J., Chem. Rev., 89 (1989) 549.
[21] Akitt, J. W., Merbach, A. E., NMR Basic Principles and Progress, 24 (1990) 189.
[22] van Eldik, R., Merbach, A. E., Comments Inorg. Chem., 12 (1992) 341.
[23] van Eldik, R., (Winter, R., Jonas, J. (eds.)), High Pressure Chemistry, Biochemistry and Materials Science, pages 309, 329, Kluwer, Dordrecht, 1993.
[24] van Eldik, R., Pure Appl. Chem., 65 (1993) 2603.
[25] Jurczak, J., Baranowski, B. (eds.), High Pressure Chemical Synthesis, Elsevier, Amsterdam, 1989.
[26] See literature cited in Chapters 3 and 4.
[27] Magde, D., van Eldik, R., (Isaacs, N., Holzapfel, W. (eds.)), High Pressure Techniques in Chemistry and Physics: A Practical Approach, Oxford University Press, Oxford, in press.

2.5. REFERENCES

[28] Hubbard, C. D., van Eldik, R., Instrumentation Sci. Tech., 23 (1995) 1.
[29] For example, Atkins, P. W., Physical Chemistry, 5th Edition, Oxford University Press, Oxford, 1994.
[30] Moore, J. W., Pearson, R. G., Kinetics and Mechanism, 3rd Edition, John Wiley, New York, 1981.
[31] Wilkins, R. G., Kinetics and Mechanism of Reactions of Transition Metal Complexes, 2nd Edition, VCH, Weinheim, 1991.
[32] le Noble, W. J., Schlott, R., Rev. Sci. Instrum., 47 (1976) 770.
[33] Caldin, E. F., Fast Reactions in Solution, Blackwell Scientific, Oxford, 1964.
[34] Gibson, Q. H., Milnes, L., Biochem. J., 91 (1964) 161.
[35] Weissburger, A., Techniques of Organic Chemistry, Volume VIII, Part II, Investigations of Rates and Mechanisms of Reactions, Interscience, New York, 1967.
[36] Strehlow, H., Rapid Reactions in Solutions, VCH, Weinheim, 1992.
[37] Heremans, K., Snauwaert, J., Rijkenberg, J., Rev. Sci. Instrum., 51 (1980) 806.
[38] Nichols, P. J., Ducommun, Y., Merbach, A. E., Inorg. Chem., 22 (1983) 3993.
[39] van Eldik, R., Palmer, D. A., Schmidt, R., Kelm, H., Inorg. Chim. Acta, 50 (1981) 131.
[40] van Eldik, R., Gaede, W., Wieland, S., Kraft, J., Spitzer, M., Palmer, D. A., Rev. Sci. Instrum., 64 (1993) 1355.
[41] Balny, C., Saldana, J. L., Dahan, N., Anal. Biochem., 139 (1984) 178.
[42] Takisawa, N., Sasaki, M., Amita, F., Osugi, J., Chem. Lett., (1979) 671; Sasaki, M., Amita, F., Osugi, J., Rev. Sci. Instrum., 50 (1979) 1073.
[43] Funahashi, S., Ishihara, K., Tanaka, M., Inorg. Chem., 20 (1981) 51; Ishihari, K., Funahashi, S., Tanaka, M., Rev. Sci. Instrum., 53 (1982) 1231.
[44] Jonas, J., Adv. Magn. Reson., 6 (1973) 73.
[45] Jonas, J., (Kelm, H. (ed.)), High Pressure Chemistry, page 65, Reidel, Dordrecht, 1978.
[46] Jonas, J., (van Eldik, R., Jonas, J. (eds.)), High Pressure Chemistry and Biochemistry, page 193, Reidel, Dordrecht, 1987.
[47] Merbach, A. E., (van Eldik, R., Jonas, J. (eds.)), High Pressure Chemistry and Biochemistry, page 311, Reidel, Dordrecht, 1987.
[48] Morishima, I., (Jannasch, H. W., Marquis, R. E., Zimmermann, A. M. (eds.)), Current Perspectives in High Pressure Biology, page 315, Academic Press, New York, 1987.
[49] Jonas, J., (Winter, R., Jonas, J. (eds.)), High Pressure Chemistry, Biochemistry and Materials Science, page 393, Kluwer, Dordrecht, 1993.
[50] Zahl, A., Neubrand, A., Aygen, S., van Eldik, R., Rev. Sci. Instrum., 65 (1994) 882.
[51] Zahl, A., Ph.D. Thesis, Universität Witten/Herdecke, Witten, Germany, 1994.
[52] Swift, T. J., Connick, R. E., J. Chem. Phys., 37 (1962) 307.
[53] Anton Parr, Graz, Austria.
[54] Millero, F. J., (Horne, R. A. (ed.)), Water and Aqueous Solutions: Structure, Thermodynamic and Transport Processes, Wiley-Interscience, London, 1972.
[55] van Eldik, R., (van Eldik, R. (ed.)), Inorganic High Pressure Chemistry: Kinetics and Mechanism, Chapter 1, page 1, Elsevier, Amsterdam, 1986.

[56] Eckert, C. A., Annu. Rev. Phys. Chem., 23 (1972) 239.
[57] Asano, T., le Noble, W. J., Chem. Rev., 78 (1978) 407.
[58] Hamann, S. D., le Noble, W. J., J. Chem. Ed., 61 (1984) 658.
[59] Mentasti, E., Pelizzetti, E., Biaocchi, C., J. Chem. Soc., Dalton Trans., (1977) 132.
[60] Hubbard, C. D., Gerhard, A., van Eldik, R., Inorg. Chem., 30 (1991) 5023.
[61] Madar, R., Jacob, G., Hallais, J., Fruchart, R. J., Cryst. Growth, 31 (1975) 197.
[62] Karpinski, J., Porowski, S., Jun, J. J., Cryst. Growth, 66 (1984) 1.
[63] Baranowski, B., Filipek, S. M., Chapter 4 in [25].
[64] Ponyatovskii, E. G., Belash, I. T., Dokl. Akad. Nauk SSSR, 229 (1976) 1171.
[65] Antonov, V. E., Belash, I. T., Ponyatovskii, E. G., Dokl. Akad. Nauk SSSR, 248 (1979) 635.
[66] Ponyatovskii, E. G., Antonov, V. E., Belash, I. T., (Prokhorov, A. M., Prokhorov, A. S. (eds.)), Problems in Solid State Physics 109, Mir Publ., Moscow, 1984.
[67] Antonov, V. E., Belash, I. T., Malyshev, V. Y., Ponyatovskii, E. G., Platinum Metals Rev., 28 (1984) 158.
[68] Jurczak, J., Rahm, A., Chapter 11 in [25].
[69] Yamamoto, Y., Maruyama, K., Matsumoto, K., J. Am. Chem. Soc., 105 (1983) 6963.
[70] Jurczak, J., Golekiowski, A., Rahm, A., Tetrahedron Lett., 27 (1986) 853.
[71] Rahm, A., Ferkous, F., Jurczak, J., Golekiowski, A., Synth. React. Inorg. Met. Org. Chem., 17 (1987) 937.
[72] Dauben, W. G., Gerdes, J. M., Look, G. C., Synthesis, (1986) 532.
[73] Yamamoto, Y., Maruyama, K., Matsumoto, K., Organometallics, 3 (1984) 1583.
[74] Ducommun, Y., Newman, K. E., Merbach, A. E., Inorg. Chem., 19 (1980) 3696.
[75] Ducommun, Y., Zbinden, D., Merbach, A. E., Helv. Chim. Acta, 65 (1982) 1385.
[76] Eigen, M., Wilkins R. G., Adv. Chem. Series, ACS, 1965.
[77] Holyer, R. H., Hubbard, C. D., Kettle, S. F. A., Wilkins, R. G., Inorg. Chem., 4 (1965) 929.
[78] Langford, C. H., Gray, H. B., Ligand Substitution Processes, W. A. Benjamin, New York, 1965.
[79] Merbach, A. E., (van Eldik, R. (ed.)), Inorganic High Pressure Chemistry: Kinetics and Mechanism, Chapter 2, page 68, Elsevier, Amsterdam, 1986.
[80] Kang, S. K., Lam, B., Albright, T. A., O'Brian, J. F., New. J. Chem., 15 (1991) 757.
[81] Akesson, R., Pettersson, L. G. M., Sandstrom, M., Siegbahn, P. E. M., Wahlgren, U., J. Phys. Chem., 97 (1993) 3765.
[82] Akesson, R., Pettersson, L. G. M., Sandstrom, M., Wahlgren, U., J. Am. Chem. Soc., 116 (1994) 8705.
[83] Hugi, A. D., Ph.D. Thesis, University of Lausanne, 1984.
[84] Hugi, A. D., Helm, L., Merbach, A. E., Helv. Chim. Acta, 68 (1985) 508.
[85] Xu, F-C., Krouse, H. R., Swaddle, T. W., Inorg. Chem., 24 (1985) 267.
[86] Swaddle, T. W., Merbach, A. E., Inorg. Chem., 20 (1981) 4212.
[87] Hugi-Cleary, D., Ph.D. Thesis, University of Lausanne, 1984.
[88] Hugi-Cleary, D., Helm, L., Merbach, A. E., Helv. Chim. Acta, 68 (1985) 545.

2.5. REFERENCES

[89] Laurenczy, G., Rapaport, I., Zbinden, D., Merbach, A. E., Mag. Res. Chem., 29 (1991) 545.

[90] González, G., Moullert, B., Martinez, M., Merbach, A. E., Inorg. Chem., 33 (1994) 2330.

[91] Dadci, L., Elias, H., Frey, U., Hörnig, A., Koelle, U., Merbach, A. E., Paulus, H., Schneider, J. F., Inorg. Chem., 34 (1995) 306.

[92] Carle, D. L., Swaddle, T. W., Can. J. Chem., 51 (1973) 3795.

[93] Meyer, F. K., Monnerat, A. R., Newman, K. E., Merbach, A. E., Inorg. Chem., 21 (1982) 774.

[94] Lo, S. T. D., Swaddle, T. W., Inorg. Chem., 14 (1975) 1878.

[95] Meyer, F. K., Newman, K. E., Merbach, A. E., J. Am. Chem. Soc., 101 (1979) 5588.

[96] Meyer, F. K., Newman, K. E., Merbach, A. E., Inorg. Chem., 18 (1979) 2142.

[97] Helm, L., Lincoln, S. F., Merbach, A. E., Zbinden, D., Inorg. Chem., 25 (1986) 2550.

[98] Sisley, M. J., Yano, Y., Swaddle, T. W., Inorg. Chem., 21 (1982) 1141.

[99] Yano, Y., Fairhurst, M. T., Swaddle, T. W., Inorg. Chem., 19 (1980) 3267.

[100] Monnerat, A. R., Moore, P., Newman, K. E., Merbach, A. E., Inorg. Chim. Acta, 47 (1981) 139.

[101] Cossy, C., Ph.D. Thesis, University of Lausanne, 1986.

[102] Soyama, S., Ishii, M., Funahashi, S., Tanaka, M., Inorg. Chem., 31 (1992) 536.

[103] Aizawa, S., Matsuda, K., Tajima, T., Maeda, M., Sugata, T., Funahashi, S., Inorg. Chem., 34 (1995) 2042.

[104] Kustin, K., Pasternack, R. F., Weinstock, E. M., J. Am. Chem. Soc., 88 (1966) 4639.

[105] Kustin, K., Kowalak, A., Pasternack, R. F., Pettrucci, S., J. Am. Chem. Soc., 89 (1967) 3126.

[106] Caldin, E. F., Grant, M. W., Hasinoff, B. B., J. Chem. Soc., Faraday Trans., I 68 (1972) 2247; Grant, M. W., J. Chem. Soc., Faraday Trans. I, 69 (1973) 560.

[107] van Eldik, R., Mohr, R., Inorg. Chem., 24 (1985) 3396.

[108] Ishihara, K., Funahashi, S., Tanaka, M., Inorg. Chem., 22 (1983) 2564.

[109] Nichols, P. J., Ducommun, Y., Merbach, A. E., Inorg. Chem., 22 (1983) 3993.

[110] Doss, R., van Eldik, R., Inorg. Chem., 21 (1982) 3993.

[111] Birus, M., van Eldik, R., Inorg. Chem., 30 (1991) 4559.

[112] Curtis, N. J., Lawrance, G. A., van Eldik, R., Inorg. Chem., 28 (1989) 329, and references cited therein.

[113] Hammes, G. G., Steinfeld, J. I., J. Am. Chem. Soc., 84 (1962) 4639.

[114] Lincoln, S. F., Hounslow, A. M., Pisaniello, D. L., Doddridge, B. G., Coates, J. H., Merbach, A. E., Zbinden, D., Inorg. Chem., 23 (1984) 1090.

[115] Lincoln, S. F., Hounslow, A. M., Doddridge, B. G., Coates, J. H., Merbach, A. E., Zbinden, D., Inorg. Chim. Acta, 100 (1985) 207.

[116] Helm, L., Meier, P., Merbach, A. E., Tregloan, P. A., Inorg. Chim. Acta, 73 (1983) 1.

[117] Powell, D. H., Merbach, A. E., Fabian, I., Schindler, S., van Eldik, R., Inorg. Chem., 33 (1994) 4468.

[118] Powell, D. H., Helm, L., Merbach, A. E., J. Chem. Phys., 95 (1991) 9258.
[119] Helm, L., Elding, L. I., Merbach, A. E., Helv. Chim. Acta, 67 (1984) 1453.
[120] Ducommun, Y., Nichols, P. J., Helm, L., Elding, L. I., Merbach, A. E., Journal de Physique, 45 (1984) C8-221.
[121] Helm, L., Elding, L. I., Merbach, A. E., Inorg. Chem., 24 (1985) 1719.
[122] Tubino, M., Merbach, A. E., Inorg. Chim. Acta, 71 (1983) 149.
[123] Meyer, F. K., Earl, W. L., Merbach, A. E., Inorg. Chem., 18 (1979) 88.
[124] Breet, E. L. J., van Eldik, R., Inorg. Chem., 23 (1984) 1865.
[125] Kowall, T., Foglia, F., Helm, L., Merbach, A. E., J. Am. Chem. Soc., 117 (1995) 3790.
[126] Pinsaniello, D. L., Nichols, P. J., Ducommun, Y., Merbach, A. E., Helv. Chim. Acta, 65 (1982) 1025.
[127] Pisaniello, D. L., Helm, L., Meier, P., Merbach, A. E., J. Am. Chem. Soc., 105 (1983) 4528.
[128] Schneider, K. J., van Eldik, R., Organometallics, 9 (1990) 1235.
[129] Wieland, S., van Eldik, R., Organometallics, 10 (1991) 5865.
[130] Walper, M., Kelm, H., Z. Phys. Chem. N. F., 113 (1978) 207; Schmidt, R., Geis, M., Kelm, H., Z. Phys. Chem. N. F., 92 (1974) 223.
[131] Kelm, H., Stieger, H., J. Phys. Chem., 77 (1973) 290.
[132] Pipoh, R., van Eldik, R., Organometallics, 12 (1993) 2668.
[133] Dücker-Benfer, C., van Eldik, R., Canty, A. J., Organometallics, 13 (1994) 2412.
[134] Marcus, R. A., J. Chem. Phys., 24 (1956) 966, 979.
[135] Hush, N. S., Trans. Faraday Soc., 57 (1961) 557.
[136] Cannon, R. D., Electron Transfer Reactions, Butterworths, London, 1980.
[137] Sutin, N., Prog. Inorg. Chem., 30 (1983) 441.
[138] Marcus, R. A., Sutin, N., Biochim. Biophys. Acta, 811 (1985) 265.
[139] Swaddle, T. W., Inorg. Chem., 29 (1990) 5017.
[140] Winkler, J. R., Gray, H. B., Chem. Rev., 92 (1992) 369.
[141] Isied, S. S., Ogawa, M. Y., Wishart, J. F., Chem. Rev., 92 (1992) 381.
[142] Discussed within [10].
[143] Spiccia, L., Swaddle, T. W., Inorg. Chem., 26 (1987) 2265.
[144] Bruhn, H., Nigam, S., Holzwarth, J. F., Faraday Discuss. Chem. Soc., 74 (1982) 129; Rampi Scandola, A., Scandola, F., Indelli, A., J. Chem. Soc., Faraday Trans. I, 81 (1985) 2967.
[145] Pethybridge, A. D., Prue, J. E., Prog. Inorg. Chem., 17 (1972) 327.
[146] Jolley, W. H., Stranks, D. R., Swaddle, T. W., Inorg. Chem., 29 (1990) 1948.
[147] Jolley, W. H., Stranks, D. R., Swaddle, T. W., Inorg. Chem., 29 (1990) 385.
[148] Grace, M. R., Swaddle, T. W., Inorg. Chem., 32 (1993) 5597.
[149] Wherland, S., Inorg. Chem., 22 (1983) 2349.
[150] Murguia, M. A., Wherland, S., Inorg. Chem., 30 (1991) 139.
[151] Anderson, K. A., Wherland, S., Inorg. Chem., 30 (1991) 624.
[152] Burgess, J., Hubbard, C. D., Inorg. Chim. Acta, 64 (1982) L71.

2.5. REFERENCES

[153] Hubbard, C. D., Bajaj, H. C., van Eldik, R., Burgess, J., Blundell, N. J., Inorg. Chim. Acta, 183 (1991) 1.
[154] Bänsch, B., Martinez, P., van Eldik, R., J. Phys. Chem., 96 (1992) 234.
[155] Bänsch, B., Meier, M., Martinez, P., van Eldik, R., Su, C., Sun, J., Isied, S. S., Wishart, J. F., Inorg. Chem., 33 (1994) 4744.
[156] Doine, H., Swaddle, T. W., Inorg. Chem., 27 (1988) 665.
[157] Doine, H., Yano, Y., Swaddle, T. W., Inorg. Chem., 28 (1989) 2319.
[158] Neilson, R. M., Hunt, J. P., Dodgen, H. W., Wherland, S., Inorg. Chem., 25 (1986) 1964; Stebler, M., Neilson, R. M., Siems, W. F., Hunt, J. P., Dodgen, H., Wherland, S., Inorg. Chem., 66 (1988) 2763.
[159] Kirchner, K., Dang, S.-Q., Stebler, M., Dodgen, H., Wherland, S., Hunt, J. P., Inorg. Chem., 28 (1989) 3604.
[160] Meier, M., van Eldik, R., Inorg. Chim. Acta, 225 (1994) 95.
[161] Meier, M., van Eldik, R., Inorg. Chim. Acta, 242 (1996) 185.
[162] Wieland, S., van Eldik, R., Coord. Chem. Rev., 97 (1990) 155.
[163] Ford, P. C., Crane, D. R., Coord. Chem. Rev., 100 (1991) 153.
[164] van Eldik, R., (van Eldik, R., Jonas, J. (eds.)), High Pressure Chemistry and Biochemistry, pages 198 and 357, Reidel, Dordrecht, 1987.
[165] Wieland, S., van Eldik, R., J. Phys. Chem., 94 (1990) 5865.
[166] Wieland, S., Bal Reddy, K., van Eldik, R., Organometallics, 9 (1990) 1802.
[167] Rossenaar, B. D., van der Graaf, T., van Eldik, R., Langford, C. H., Stufkens, D. J., Vleck, A., Inorg. Chem., 33 (1994) 2865.
[168] Ryba, D. W., van Eldik, R., Ford, P. C., Organometallics, 12 (1993) 104.
[169] Zhang, S., Dobson, G. R., Bajaj, H. C., Zang, V., van Eldik, R., Inorg. Chem., 29 (1991) 3477.
[170] Zhang, S., Zang, V. M., Bajaj, H. C., Dobson, G. R., van Eldik, R., J. Organomet. Chem., 397 (1990) 279.
[171] Zang, V., Zhang, S., Dobson, C. B., Dobson, G. R., van Eldik, R., Organometallics, 11 (1992) 1154.
[172] Bal Reddy, K., Hoffmann, R., Konya, G., van Eldik, R., Eyring, E. M., Organometallics, 11 (1992) 2319.
[173] Taube, D. J., Projahn, H.-D., van Eldik, R., Magde, D., Traylor, T. G., J. Am. Chem. Soc., 112 (1990) 6880.
[174] Zhang, M., van Eldik, R., Espenson, J. H., Bakac, A., Inorg. Chem., 33 (1994) 130.
[175] Projahn, H.-D., Dreher, C., van Eldik, R., J. Am. Chem. Soc., 112 (1990) 17.
[176] Projahn, H.-D., van Eldik, R., Inorg. Chem., 30 (1992) 3288.
[177] Goldstein, S., Czapski, G., van Eldik, R., Cohen, H., Meyerstein, D., J. Phys. Chem., 95 (1991) 1282.
[178] van Eldik, R., Cohen, H., Meyerstein, D., Angew. Chem., Int. Ed. Engl., 30 (1991) 1158.
[179] van Eldik, R., Gaede, W., Cohen, H., Meyerstein, D., Inorg. Chem., 31 (1992) 3695.
[180] van Eldik, R., Cohen, H., Meyerstein, D., Inorg. Chem., 33 (1994) 1566.
[181] Wishart, J. F., van Eldik, R., Sun, J., Su, S., Isied, S. S., Inorg. Chem., 31 (1992) 3986.

[182] Sun, J., Wishart, J. F., van Eldik, R., Shalders, R. D., Swaddle, T. W., J. Am. Chem. Soc., 117 (1995) 2600.
[183] Meier, M., van Eldik, R., Chang, I.-J., Mines, G. A., Wutke, D. S., Winkler, J. R., Gray, H. B., J. Am. Chem. Soc., 116 (1994) 1577.
[184] van Eldik, R., (van Eldik, R. (ed.)), Inorganic High Pressure Chemistry, Kinetics and Mechanism, Chapter 8, Elsevier, Amsterdam, 1986.
[185] Shalders, R., Swaddle, T. W., Inorg. Chem., 34 (1995) 4815.
[186] Hartman, M., Clark, T., van Eldik, R., submitted for publication.
[187] Pubanz, D., González, G., Powell, D. H., Merbach, A. E., Inorg. Chem., 34 (1995) 4447.
[188] Micskei, K., Powell, D. H., Helm, L., Brücher, E., Merbach, A. E., Magn. Reson. Chem., 31 (1993) 1011.
[189] Micskei, K., Helm, L., Brücher, E., Merbach, A. E., Inorg. Chem., 32 (1993) 3844.
[190] González, G., Powell, D. H., Tissières, V., Merbach, A. E., J. Phys. Chem., 98 (1994) 53.
[191] Zhang, X., van Eldik, R., Hubbard, C. D., J. Phys. Chem., 100 (1996) 9161.
[192] Palmer, D. A., van Eldik, R., J. Solution Chem., 11 (1982) 339.
[193] Wenfu, F., van Eldik, R., Inorg. Chim. Acta, in press.
[194] Zahl, A., van Eldik, R., work in progress.
[195] Graeppi, N., Powell, D. H., Laurenczy, G., Zékány, L., Merbach, A. E., Inorg. Chim. Acta 235 (1995) 311.
[196] Frey, U., Li, Z.-W., Matras, A., Inorg. Chem., 35 (1996) 981.
[197] Kowall, T., Foglia, F., Helm, L., Merbach, A. E., Chem. Eur. J., 2 (1996) 285.
[198] Powell, D. H., Furrer, P., Pittet, P. -T., Merbach, A. E., J. Phys. Chem., 99 (1995) 16622.
[199] Beswick, C. L., Shalders, R. D., Swaddle, T. W., Inorg. Chem., 35 (1996) 991.
[200] Neubrand, A., Pöe, A. J., van Eldik, R., Organometallics 14 (1995) 3249.
[201] Alshehri, S., Burgess, J., van Eldik, R., Hubbard, C. D., Inorg. Chim. Acta, 240 (1995) 305.
[202] Burgess, J., Hubbard, C. D., Comments Inorg. Chem., 17 (1995) 283.
[203] Meier, M., Sun, J., Wishart, J. F., van Eldik, R., Inorg. Chem., 35 (1996) 1564.
[204] Projahn, H.-D., Schindler, S., van Eldik, R., Fortier, D. G., Andrew, C. R., Sykes, A. G., Inorg. Chem., 34 (1995) 5935.
[205] Aizawa, S., Iida, S., Matsuda, K., Funahashi, S., Inorg. Chem., 35 (1996) 1338.

3

EFFECT OF PRESSURE ON ORGANIC REACTIONS

Frank-Gerrit Klärner*, Matthias K. Diedrich and Arne E. Wigger
Institut für Organische Chemie, Universität GH Essen, D-45117 Essen, Germany

3.1 INTRODUCTION
3.2 APPARATUS, OPERATION, AND BASIC PRINCIPLES
3.3 CYCLOADDITION REACTIONS
    3.3.1 Diels-Alder [4 + 2] Cycloadditions
    3.3.2 [2 + 2] Cycloadditions
    3.3.3 1,3 Dipolar Cycloadditions
    3.3.4 [6 + 4] and [8 + 2] Cycloadditions
    3.3.5 Cheleotropic Reactions
3.4 THE EFFECT OF PRESSURE ON COMPETITIVE REACTIONS
3.5 MOLECULAR REARRANGEMENTS
3.6 ENE REACTIONS
3.7 THE RELATIONSHIP BETWEEN ACTIVATION OR REACTION VOLUME AND RING-SIZE
3.8 CONCLUDING REMARKS
3.9 REFERENCES

3.1 INTRODUCTION

Pressure in the range of 1–20 kbar (units of pressure: 1 kbar = 100 MPa = 0.1 GPa = 1013.25 atm) strongly influences the rate and equilibrium position of many chemical reactions. Processes accompanied by a decrease in volume are accelerated by pressure and the equilibria are shifted toward the side of products, while those accompanied by an increase of volume are retarded and the equlibria are shifted toward the side of reactants. Therefore, the application of high pressure may be particularly useful in controlling the course of competitive and consecutive reactions as well as those proceeding to equilibrium, and can lead to an improvement of chemo-, regio-, and stereoselectivity. It is not the purpose of this chapter to provide a complete survey of organic reactions which have been investigated at elevated pressures. There are many excellent monographs (*e.g.*, [1–3]) and reviews (*e.g.*, [4–15]) on this topic which cover the literature up to early 1990. After an introduction into the basic concepts necessary to understand pressure effects on chemical processes (also see Chapter 2) our major objective is to review literature of the past five to ten years.

3.2 APPARATUS, OPERATION, AND BASIC PRINCIPLES

Static high pressure in the range of 1–20 kbar frequently used for the investigation of organic reactions in compressed fluids or solids can be generated with relatively simple devices, *e.g.*, one consisting of a hydraulic pump operating in the range up to several hundred bars, a pressure intensifier having a transmission ratio between 1:10 and 1:25, a high-pressure vessel, a gauge to measure the pressure and several valves to separate the reaction vessel from the pressure generating system and to release the pressure after the reaction has been completed. The reaction mixture is usually sealed in a flexible PTFE tube and, hence, is well protected from the hydraulic oil frequently used to transmit the pressure from the intensifier to the vessel. A schematic setup of equipment operating at the University of Essen is shown in Figure 3.1. The setup can be simplified even further if the high-pressure part of the intensifier is directly used as the reaction vessel.

For kinetic analyses it is necessary to determine the rate of reaction, and eventually the position of equilibrium, from its time-dependence at various pressures. Nowadays, this can be done by spectroscopic on-line measurements using high-pressure cells especially designed for different commercially available spectrometers, such as FT-IR [15], UV-Vis [15], or NMR [16]. Another method is to connect one exit of the high-pressure vessel containing the reaction mixture with a valve having a micro spindle which allows small samples (ca. 100–200 µl) to be removed from the reaction mixture without releasing the total pressure and to be analyzed by GC or HPLC. The pressure can be kept constant during the release of sample by the operation of an electric pump controlled by the pressure measuring transducer *via* a computer. High-pressure equipment and operation has been described in detail [1, 3], and a list of some suppliers delivering commercially available high-pressure equipment is cited [17].

Pressure influences the physical properties of matter such as boiling and melting point, density, viscosity, solubility, dielectric constant, and conductivity. Before carrying out high-pressure experiments it is important to have some knowledge of these effects. The melting points of most liquids used as common

3.2. APPARATUS, OPERATION, AND BASIC PRINCIPLES

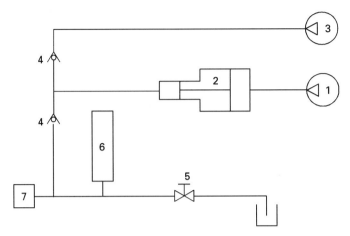

Figure 3.1 Schematic setup of a 14 kbar apparatus (t_{max} = 150 °C); 1 L.P. Drive pump Enerpac BPM 14453 electric (to 700 bar); 2 14 kbar Intensifier, Transmission 1:23; 3 Hand operated pump to 700 bar; 4 Ball type non-return valve; 5 Exhaust valve; 6 Pressure vessel with external band heater; 7 Pressure transducer for 14 kbar with Fylde digital indicator

organic solvents are raised by increasing the pressure (Table 3.1). To perform a reaction in compressed solution it is necessary to use a solvent which does not solidify under the chosen conditions. The pressure-induced increase of the melting points (ca. 15–20 °C per 1 kbar), however, makes it possible to run reactions in thermally relatively stable matrices. The solubility of gases in liquids is increased and that of solids is usually decreased by raising the pressure. Therefore, the solid solute of a saturated solution may precipitate during the generation of pressure and no longer be accessible for the reaction. The viscosity of liquids increases approximately two-fold for each kilobar increase. It is

Table 3.1 Pressure induced increase of the melting points (m.p.) of common organic solvents [1].

	m.p. (°C/1 bar)	m.p. (°C)	at (p/kbar)
Acetone	− 94.8	+ 20	(8.0)
Benzene	+ 5.5	+ 114	(5.5)
Cyclohexane	+ 6.5	+ 97	(4.0)
Ethanol	− 117.3	+ 25	(20.0)
Diethylether	− 116.3	+ 35	(12.0)
Dichloromethane	− 96.7	+ 25	(13.0)
Pentane	− 130.0	0	(12.0)
Toluene	− 95.1	+ 30	(9.6)
Water	0.0	− 9	(1.0)

particularly important to recognize this effect for reactions containing diffusion-controlled steps. Finally, the compressibility of liquids is usually small compared to that of gases. At 1 kbar it varies between 4 (water) and 18 percent (pentane) of the original volume at 1 bar and rapidly approaches an upper limit at higher pressure. For that reason experiments with compressed liquids and solids are far less potentially hazardous than those with compressed gases. A detailed discussion of the effects of pressure on physical properties of matter is available in the literature [1].

The effect of pressure on chemical equilibria and rates of reactions can be described by the well known eqs. (1) and (2):

$$\left(\frac{\partial \ln K}{\partial p}\right)_T = -\frac{\Delta V}{RT} \tag{1}$$

$$\left(\frac{\partial \ln k}{\partial p}\right)_T = -\frac{\Delta V^{\neq}}{RT} \tag{2}$$

K: equilibrium constant; k: rate constant; p [bar]: pressure, ΔV, $\Delta V^{\neq}$ [cm^3 mol^{-1}]: volume of reaction and activation, respectively; $R = 83.14$ [cm^3 bar K^{-1} mol^{-1}]; T [K]: absolute temperature.

The volume of reaction corresponds to the difference between partial molar volumes of reactants and products, eqs. (3), (4). Within the scope of transition state theory the volume of activation can be considered to be a measure of the partial molar volume of the transition state (TS) with respect to the partial molar of the reactant state, eq. (5).

$$n_1 R_1 + n_2 R_2 + \cdots \rightleftharpoons m_1 P_1 + m_2 P_2 + \cdots \tag{3}$$

$$\Delta V = \sum_{i=1}^{i} m_i V_i (\text{products}) - \sum_{i=1}^{i} n_i V_i (\text{reactants}) \tag{4}$$

$$\Delta V^{\neq} = V^{\neq} (\text{TS}) - \sum_{i=1}^{i} n_i V_i (\text{reactants}) \tag{5}$$

Volumes of reaction can be determined in three ways: (a) from the pressure-dependence of the equilibrium constant (from the plot of ln K vs. p employing eq. (1)). Since volumes of reaction are pressure-dependent themselves and there is up to now no theory explaining their pressure-dependence, empirical equations are used to calculate the volumes of reaction from the fit of the pressure-dependence of equilibrium constants in a comparable way to the determination of activation volumes (*vide infra*); (b) from the measurement of partial molar volumes of all reactants and products derived from the densities, d, of the solution of each individual component measured at various concentrations, c, by using eq. (6) and extrapolation of the apparent molar volume Φ vs. $c \to 0$; or (c) from the direct measurement of the difference between the volumes of

3.2. APPARATUS, OPERATION, AND BASIC PRINCIPLES

reactants and products employing dilatometry.

$$\Phi = \frac{M}{d_0} - \frac{1000}{c} \cdot \frac{d - d_0}{d_0} \qquad (6)$$

M: molecular weight of the solute; d: density of the solution; d_0: density of the pure solvent

To a first approximation the molar volumes of many neat compounds ($V_{neat} = M/d$) and, hence, the reaction volumes can be calculated with additive group increments which were derived by Exner [18] empirically for many groups such as CH_3, CH_2, or CH from the molar volumes known for many different types of compounds. This method is comparable to that of the calculation of enthalpies of formation by the use of Franklin [19] or Benson [20] group increments. In all cases where the volume of reaction could be determined by at least two independent methods the data were in good agreement [21]. Most of the volumes of reaction are determined by method (b) from partial molar volumes of reactants and products.

Volumes of activation can be unambiguously determined only from the pressure-dependences of rate constants. Attempts to obtain volumes of activation from the correlation of the rate constants with the solubility parameter δ [22], or the cohesive energy density parameter (ced) [23], which are related to the internal pressure of solvents, have not led to clear-cut results. Volumes of activation derived from pressure-dependence and the solvent-dependence, using the ced parameter for the same reaction, differ from each other quite substantially [23].

Volumes of activation as well as volumes of reaction are pressure-dependent. There is no theory explaining this pressure-dependence which would allow the volumes of activation to be determined over a larger pressure range. Therefore, several empirical equations are employed to fit the pressure-dependencies of rate constants; volumes of activation are calculated from these fits. Of the many different equations used in the literature [24] only four eqs. (7)–(10) are discussed here.

$$\ln k(p) = a + b \cdot p \qquad (7)$$

$$\ln k(p = 0) = a; \quad \Delta V^{\neq} = - b \cdot R \cdot T$$

$$\ln k(p) = a + b \cdot p + c \cdot p^2 \qquad (8)$$

$$\ln k(p = 0) = a; \quad \Delta V^{\neq} = - b \cdot R \cdot T; \quad \Delta \beta^{\neq} = c \cdot 2 \cdot R \cdot T$$

$$\Delta \beta^{\neq} = (\partial \Delta V / \partial p)_T \text{ (compressibility coefficient of activation)}$$

$$\ln [k(p)/k(p = 1)] = a \cdot p + b \cdot p/(1 + c \cdot p) \qquad (9)$$

$$\Delta V^{\neq} = - (a + b) \cdot R \cdot T$$

$$\ln [k(p)/k(p = 1)] = a \cdot p + b \cdot \ln(1 + c \cdot p) \qquad (10)$$

$$\Delta V^{\neq} = - (a + b \cdot c) \cdot R \cdot T$$

Since $\Delta V^{\neq}$ and ΔV are pressure-dependent we need to select a pressure to which volumes of activation and reaction refer so that the values can be compared with one another. The choice has universally been that of zero pressure ($p = 0$). The values at $p = 0$ differ only by immeasurably small amounts from those at atmospheric pressure ($p \approx 1$ bar), that is, from the volumes of reaction calculated from the partial molar volumes of the reactants and products determined at atmospheric pressure. Eq. (7) is only applicable in the low-pressure range (< 2000 bar) where the dependence of $\ln k(p)$ on pressure p is usually linear. This method, which is the simplest and in many cases also the most reliable, requires a very precise measurement of the rate constants at relatively low pressures (1–2000 bar) where the effect of pressure on the rate constants is relatively small. If data over a larger pressure range are to be used, where the pressure-dependence of $\ln k(p)$ is nonlinear (Figure 3.2), the quadratic equation (8) is frequently applied. In eq. (8) the constant b corresponds to the volume of activation at zero pressure and constant c has been assigned to the compressibility coefficient of activation $\Delta \beta^{\neq}$ which has a meaning similar to the compressibility coefficient of liquids. The advantage of eq. (8) is its simplicity (quadratic least squares fit); the disadvantage, however, is that its shape with a maximum or minimum is not realistic and the errors of the measured k values are strongly reflected in constant b and hence in the desired volume of activation. Therefore, Asano and Okada [24] have suggested the new three-parameter equations (9) and (10) which avoid the disadvantage of eq. (8). Eqs. (9) and (10), however, cannot be solved by simple least squares analyses and require an iterative optimization which can be performed by using a simplex or Marquardt routine [25]. The calculation of activation volumes by different methods has been critically reviewed [24, 26].

In two reviews [4, 5], le Noble, Asano, and van Eldik have compiled the volumes of activation and reaction known for almost all types of organic, organometallic, and inorganic reactions up to early 1987. A useful summary derived from these data is shown in Table 3.2. In applying it, one should be

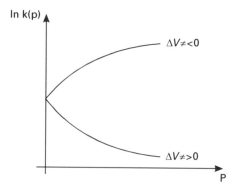

Figure 3.2 Non-linear profile of the relationship between rate of reaction $\ln k(p)$ and pressure p.

Table 3.2 Contributions to account for activation volumes of various elementary processes [4].

Mechanistic feature	Contribution in cm^3 mol^{-1}
Homolytic bond cleavage	ca. + 10
Homolytic association	ca. − 10
Bond deformation	ca. 0
Ionization	ca − 20[a]
Neutralization	ca. + 20[a]
Cncentration	ca. − 5[a]
Charge dispersal	ca. + 5[a]
Displacement	ca. − 5
Steric hindrance	(−)
Diffusion control	ca. > + 20[a]

[a] Dependent on the solvent polarity.

aware that contributions to the activation volumes listed there for several elementary processes are no more than estimates from the averages and do not consider special structural features or the solvent-dependence of activation volumes in ionic reactions (see Chapter 2).

Processes accompanied by a decrease in volume, such as C–C bond formation, in which the bond distance between two carbon atoms decreases from the van der Waals distance of ca. 3.6 Å to the bonding distance of ca. 1.5 Å are accelerated by raising the pressure and equilibria are shifted toward the side of products (volume of activation and reaction $\Delta V^{\neq}$, $\Delta V < 0$). The reverse reaction, a homolytic bond cleavage, leads to an increase in volume ($\Delta V^{\neq}$, $\Delta V > 0$). Pressure induces a deceleration of such a process and a shift in equilibrium toward the side of reactants. However, in an ionization, such as an ionic dissociation, the attractive interaction between the ions produced and the solvent molecules leads to a contraction of volume that is generally much stronger than the expansion of volume resulting from the bond dissociation. Thus, the overall dominant effect, *electrostriction*, leads to negative volumes of activation and reaction ($\Delta V^{\neq}$, $\Delta V < 0$). Neutralization of charges releases the solvent cage leading to positive volumes of activation and reaction ($\Delta V^{\neq}$, $\Delta V > 0$). A similar but less pronounced trend due to the effect of *electrostriction* is observed for charge concentration (increase) and charge dispersal (decrease), respectively. Volumes of activation and reaction in polar reactions are usually strongly dependent on the solvent polarity. Thus, it is customary to divide the activation and reaction volume into an intrinsic part ($\Delta V_i^{\neq}$, ΔV_i) and a solvent-dependent part ($\Delta V_s^{\neq}$, ΔV_s) according to eqs. (11) and (12). From the solvent-dependence of ΔV and $\Delta V^{\neq}$, the ΔV_s term or $\Delta V_s^{\neq}$ term can sometimes be estimated by using the Drude-Nernst equation, eq. (13), which describes the relation between volume of electrostriction (ΔV_s, $\Delta V_s^{\neq}$) the newly generated or neutralized charges (q) in products or transition state, radius (r), the dielectric constant (ε) of the solvent and its pressure-dependence [1]

$$\Delta V^{\neq} = \Delta V_i^{\neq} + \Delta V_s^{\neq} \tag{11}$$

$$\Delta V = \Delta V_i + \Delta V_s \tag{12}$$

$$\Delta V_s, \Delta V_s^{\neq} = -\frac{N_0 \cdot q^2}{2r \cdot \varepsilon^2} \cdot \left(\frac{\partial \varepsilon}{\partial p}\right)_T \tag{13}$$

N_0: Avogadro number

An increase in steric crowding in the transition or product states results in a volume contraction ($\Delta V^{\neq}, \Delta V < 0$). In the case of diffusion control the rate of reaction depends on the viscosity of the medium. As already pointed out, pressure induces an increase in the dynamic viscosity and, hence, a deceleration of diffusion-controlled processes.

Table 3.3 shows for different values of the activation volume how large the maximum effect of pressure on the rate of reaction can be. At 1 kbar (1000 bar) the effect is small; even a reaction having a highly negative activation volume such as $\Delta V^{\neq} = -30$ cm^3 mol^{-1} is only 3.4 times faster than it is at 1 bar. At 7 kbar this factor is calculated to be 5,000 and at 10 kbar as much as 180,000. Such large effects, however, are usually not observed since the plot describing the relation between ln $k(p)$ and p is, generally, non-linear and begins to level off at pressures higher than 2 to 5 kbar due to the pressure-dependence of the activation volumes. In favorable cases factors of several thousands can be observed experimentally by raising the pressure from 1 bar to 10 or 14 kbar. The values of the pressure-induced rate acceleration observed for the Diels-Alder reaction of isoprene with acrylonitrile listed in Table 3.4 provide an illustration of the magnitude of the experimental effects of pressure. From these

Table 3.3 The pressure-dependence of rate constants; the maximum values of rate retardation and acceleration are calculated for four different values of the activation volume $\Delta V^{\neq}$ at 25 °C.

	$k(p)/k$ (1 bar) = exp[$-\Delta V^{\neq}/RT \cdot (p - 1$ bar)] $\Delta V^{\neq}$ (cm^3 mol^{-1})			
p (kbar)	+10	−10	−20	−30
1	0.67	1.5	2.2	3.4
3	0.30	3.4	11	38
5	0.13	7.5	56	420
7	0.06	17	280	4800
10	0.02	56	3200	180000

Table 3.4 **Pressure-induced rate acceleration of the Diels Alder reaction of isoprene with acrylonitrile at 21 °C** ($\Delta V^{\neq} = -35.4 \text{ cm}^3 \text{mol}^{-1}$, $\Delta V = -37.0 \text{ cm}^3 \text{mol}^{-1}$).

p [bar]	1000	1500	2000	3000	5000	8000	10000
$k(p)/k$ (1 bar)	3.4	7.0	10.5	24.4	74.4	561	1650

values the activation volume, in parentheses, was calculated by graphical extrapolation [27].

In the following sections we discuss the effect of pressure on cycloadditions and molecular rearrangements as representative examples. A major aspect will be the effect of pressure on selectivity.

3.3 CYCLOADDITION REACTIONS

3.3.1 Diels-Alder [4+2] Cycloadditions

Many Diels-Alder [4 + 2] cycloadditions show a powerful pressure-induced acceleration, which is exploited for synthetic purposes [28]. The activation volumes resulting from the pressure-dependence of the rate constants are usually highly negative ($\Delta V^{\neq} \approx -25$ to $-50 \text{ cm}^3 \text{mol}^{-1}$) as shown in Tables 3.5 and 3.6. For a comparison of activation volumes with the corresponding reaction volumes it is necessary to determine both data at the same temperature which is, however, not feasible in most cases. The measurement of the pressure-dependence of the rate constants frequently requires a temperature different from that used for the determination of partial molar volumes of reactants and products (in general room temperature). Therefore, the activation volumes have to be extrapolated to room temperature or the reaction volumes, correspondingly, to the temperature of reaction. The measurement of the temperature-dependence of activation volumes requires a large collection of experimental data. To the best of our knowledge only one case, the Diels-Alder dimerization of isoprene, has been reported in the literature [29]. With modern thermostated densimeters it is much easier to determine the temperature-dependence of partial molar volumes and, hence, of reaction volumes. From these data El'yanov extrapolated a generally applicable equation shown in Table 3.5 (footnote a) to describe the temperature-dependence of activation and reaction volumes. The temperature-dependence obtained experimentally for the isoprene dimerization is in accord with the El'yanov equation.

In Table 3.5 Diels-Alder reactions are compiled showing ratios of activation volume to reaction volume that are smaller than or close to unity ($\Theta = \Delta V^{\neq}/\Delta V \leq 1$) and in Table 3.6 those that are even larger than unity ($\Theta > 1$). Within the scope of the transition state theory, activation volumes can be considered to be an indication of the relative partial molar volumes of

Table 3.5 Volume data for selected Diels-Alder reactions showing ($\Delta V^{\neq}/\Delta V$) ratios smaller than unity ($\Theta < 1$)

	Reaction	Solvent	T (°C)	$\Delta V_T^{\neq}$ [a)]	$\Delta V_{25}^{\neq}$ [a]	ΔV_{25} [a]	Θ [b]	Ref.
(1)		n-BuCl	40	−23.7	−22.2	−33.0	0.67	[30]
(2)	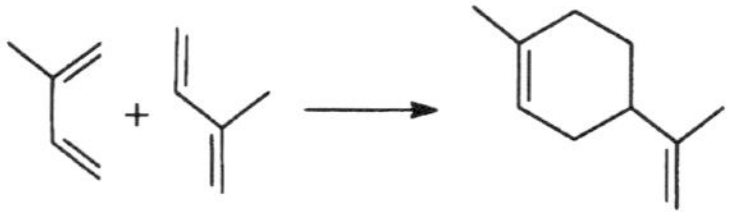		70	−42.0	−35.0	−47.8	0.73	[29]
(3)	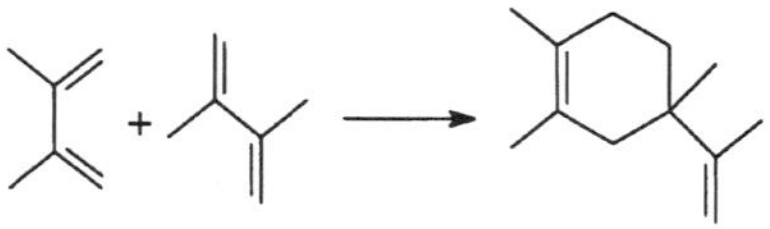	n-BuBr	70	−33.0	−27.5	−44.9	0.61	[31]
(4)	 E = CO2CH3		30	−32.7	−32	−36.7	0.87	[30]
(5)	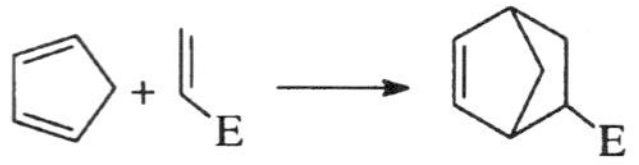	n-BuCl	40	−30.1	−28.3	−33.5	0.84	[30]

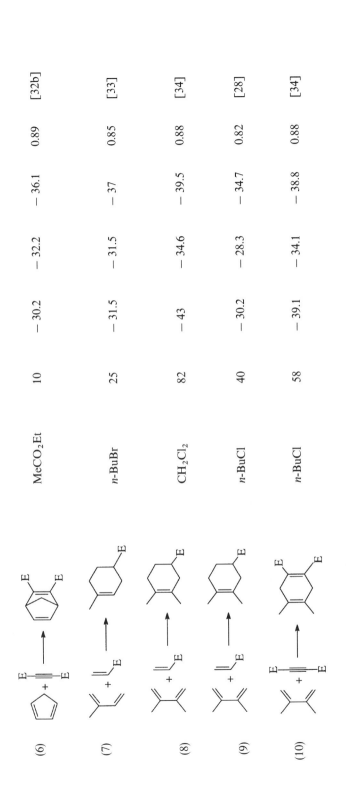

[a] in cm³ mol⁻¹; $\Delta V_T^{\neq}$ determined from the pressure-dependence of the rate constant at temperature T; $\Delta V_{25}^{\neq}$ determined from the temperature-dependence of the activation volumes or extrapolated by using the El'yanov equation $\Delta V_{25}^{\neq} = \Delta V_T^{\neq}/[1 + 4.43 \cdot 10^{-3} \text{ K}^{-1} \cdot (T - 25, °C)]$ [35b, c].
[b] $\Theta = \Delta V_{25}^{\neq}/\Delta V_{25}$
[c] At 20 °C.

Table 3.6 Volume data for selected Diels-Alder reactions showing ($\Delta V^{\neq}/\Delta V$) ratios larger than unity ($\Theta > 1$).

Reaction	Solvent	T (°C)	$\Delta V_T^{\neq}$ [a]	$\Delta V_{25}^{\neq}$ [a]	ΔV_{25} [a]	Θ [b]	Ref.
(1) 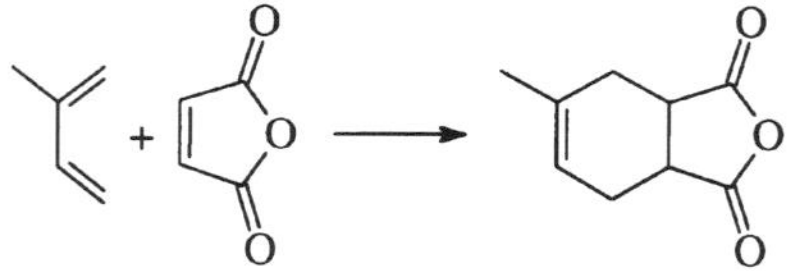	(i-Pr)$_2$O		−38.5	−36.9	−38.3	0.96	
	Me$_2$CO		−39.0	−37.3	−35.9	1.04	
	MeCN		−37.5	−35.9	−34.5	1.04	[32a]
	n-BuCl		−38.0	−36.4			
	AcOEt		−37.4	−35.8	−36.8	0.97	
	ClCH$_2$CH$_2$Cl		−37.0	−35.4	−35.5	0.99	
	CH$_2$Cl$_2$	35	−39.8	−38.1	−33.4	1.14	
(2)	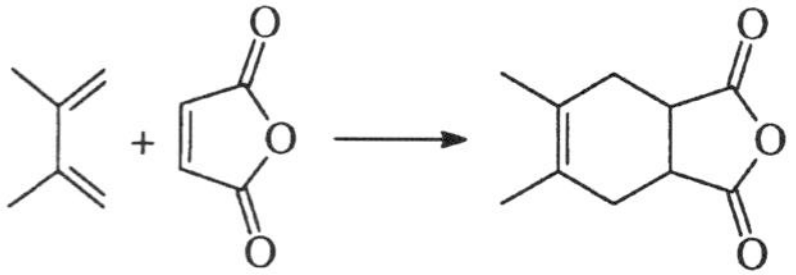 n-BuCl	30	−41.3	−40.4	−35.5	1.14	[30]
(3)	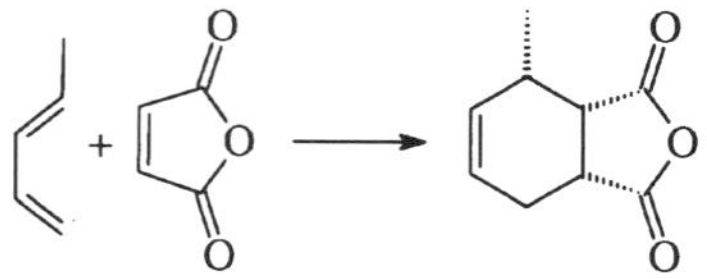 n-BuCl	35	−44.7	−42.8	−31.9	1.22	
(4)	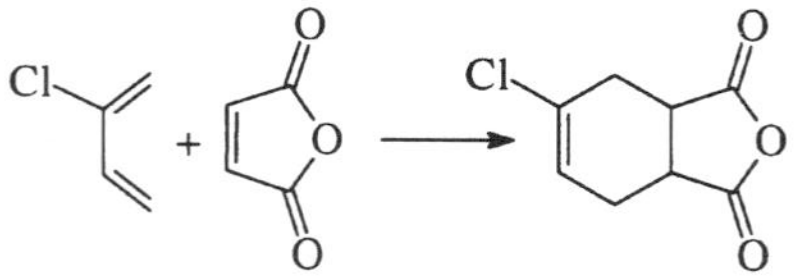 MeCN	65	−41.6	−35.3	−36.9	0.96	[37]

(5)	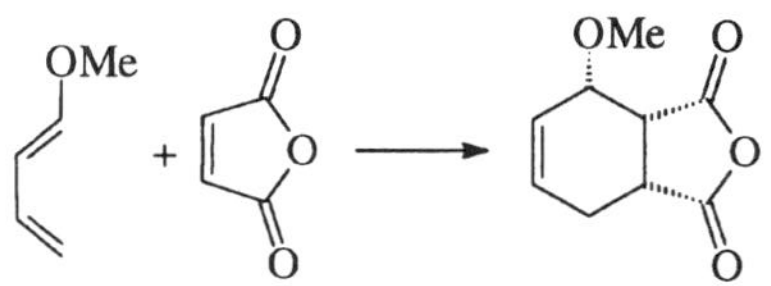 OMe + [maleic anhydride] → [OMe-substituted cyclohexene anhydride]	MeCN	−32.0	−30.6	−32.4	0.94	[36]
		(MeO)$_2$CH$_2$	−53.6	−51.3	−32.2	1.6	
		n-BuCl	−45.4	−43.5	−35.5	1.23	
		MeNO$_2$	−43.7	−41.2	−28.2	1.46	
		ClCH$_2$CH$_2$Cl	−43.7	−41.8	−30.4	1.37	
(6)	[cyclohexadiene] + [maleic anhydride] → [bicyclic anhydride]	35					[37]
		CH$_2$Cl$_2$	−37.2	−35.6			
(7)	2 [acrolein] → [dihydropyran aldehyde]	35					[38]
		n-C$_7$H$_{16}$	−37.0	−30.8	−28.6	1.08	
(8)	2 [methyl vinyl ketone] → [dihydropyran methyl ketone]	70					[38]
		n-C$_7$H$_{16}$	−41.0	−34.2	−32.3	1.06	
(9)	[acrolein] + [methyl vinyl ketone] → [dihydropyran methyl ketone]	70					[38]
		n-C$_7$H$_{16}$	−36.5	−30.8			
(10)	[methyl vinyl ketone] + [acrolein] → [methyl-dihydropyran aldehyde]	70					[38]
		n-C$_7$H$_{16}$	−35.0	−29.2	−29.7	0.98	

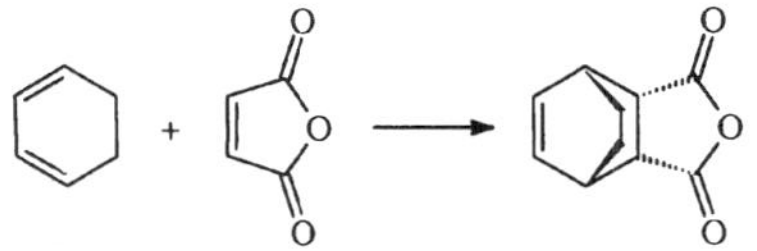

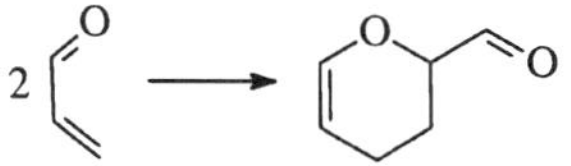

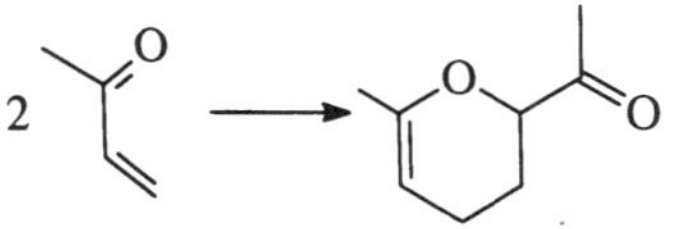

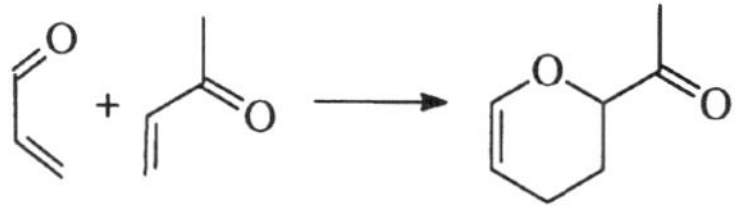

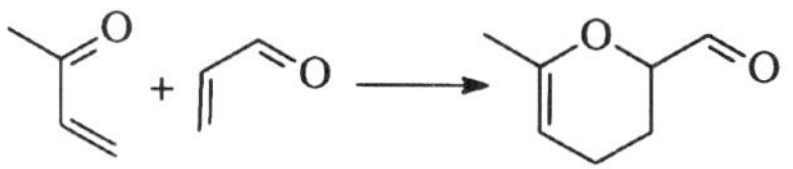

[a] In cm^3 mol^{-1}; $\Delta V_T^{\neq}$ determined from the pressure-dependence of the rate constant at temperature T; $\Delta V_{25}^{\neq}$ determined from the temperature-dependence of the activation volumes or extrapolated by using the El'yanov equation $\Delta V_{25}^{\neq} = \Delta V_T^{\neq}/[1 + 4.43 \cdot 10^{-3} \text{ K}^{-1} \cdot (T - 25, °\text{C})]$ [b] $\Theta = \Delta V_{25}^{\neq}/\Delta \bar{V}_{25}$.

transition states. Accordingly, the relative transition state volumes of the reactions listed in Table 3.5 are close to the corresponding product volumes while those listed in Table 3.6 are even smaller. These surprising results could be confirmed by two independent studies of the reactions shown in Scheme 3.1. In both Diels-Alder reactions (that of furan with acrylonitrile [39] and that of N-benzoylpyrrole with N-phenylmaleic imide [40]) the ratio ($\Delta V^{\neq} : \Delta V$) turned out to be larger than unity ($\Theta > 1$). The adducts isolated from both reactions undergo smooth retro-Diels-Alder reactions and show a pressure-induced acceleration corresponding to negative volumes of activation in agreement with the values of $\Theta > 1$ determined for the forward reactions.

The result that the relative transition state volume in several Diels-Alder reactions (Table 3.6) is smaller than the product volume, is surprising and not well understood. This finding seems to be contradictory to the generally accepted relation between molecular structure and its volume. In the transition state the new bonds between diene and dienophile are only partially formed (according to quantum mechanical calculations the distance of the newly forming σ bonds in the transition state is in the range between 2.1 and 2.3 Å) [41]. As we will discuss in more detail later, the intrinsic volumes of the individual structures, the so called van der Waals volumes V_w, can be calculated by using the structural parameters obtained from X-ray analyses, force-field calculations, or quantum mechanical calculations and the van der Waals radii of the different types of atoms derived from X-ray data. The van der Waals volumes of the transition states of Diels-Alder reactions calculated by this method are generally larger than those of the products (vide infra). Grieger and Eckert [32] gave two

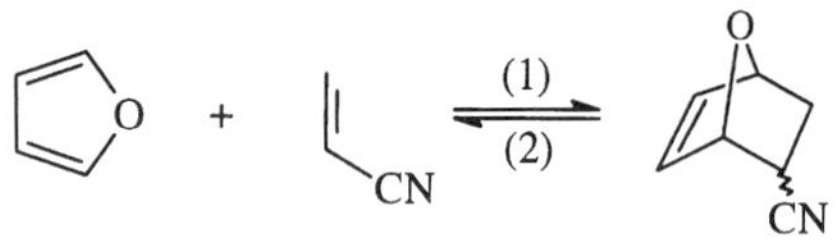

(1): 37 °C, cyclohexane: $\Delta V^{\neq} = -30.3$ cm^{-3} mol^{-1}; $\Delta V = -28.7$ cm^3 mol^{-1}; $\Theta = 1.06$
(2): 80 °C, cyclohexane: $\Delta V^{\neq} = -2.0$ cm^3 mol^{-1} [39].

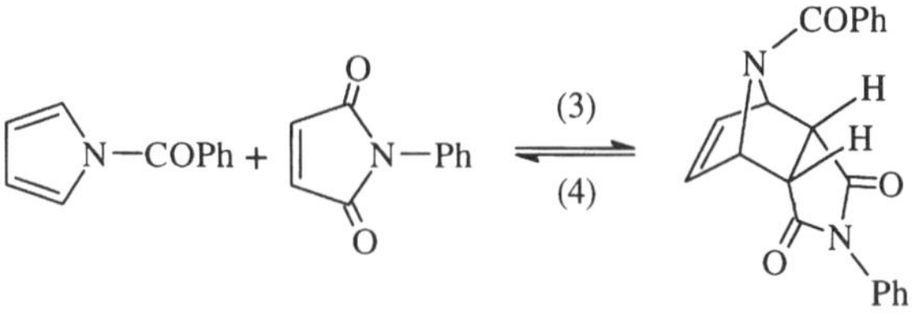

(3): $\Delta V^{\neq}_{\text{calc.}} = -30.7$ cm^3 mol^{-1}; $\Delta V = -22.4$ cm^3 mol^{-1}; $\Theta = 1.37$
(4): 38 °C, CHCl$_3$: $\Delta V^{\neq} = -8.3$ cm^{-3} mol^{-1}; $\Delta V = 22.4$ cm^3 mol^{-1} [40]

Scheme 3.1

3.3. CYCLOADDITION REACTIONS

explanations of the ratio $(\Delta V^{\neq}:\Delta V) > 1$ in the Diels-Alder reactions with maleic anhydride as dienophile: *viz.* a larger dipole moment and secondary orbital interactions [42] in the transition state from which secondary orbital interactions can only occur in the transition states of *endo*-Diels-Alder reactions. By plotting partial molar volumes of the reactants, the transition state, and the product, respectively, in various solvents against $[3/(2\varepsilon + 1)^2] \cdot (d\varepsilon/dp)$ using the Kirkwood equation the authors estimated the dipole moment of the transition state in the reaction of isoprene with maleic anhydride to be 5.3×10^{-30} C m (1.6 Debye) larger than that of maleic anhydride. The polarity difference between the transition state and the product, however, was minimal. Thus, the effect of electrostriction should be operative to a similar extent in the transition state as well as in the product and may not explain the observed difference between $\Delta V^{\neq}$ and ΔV. Therefore, the authors concluded that the secondary orbital interactions must be the primary reason for the more negative activation volume. But Seguchi, Sera, and Maruyama [43] observed a very small difference between the activation volumes of the *endo*- and *exo*-Diels-Alder reaction of dimethyl maleate with 1,3-cyclopentadiene ($\delta\Delta V^{\neq} = \Delta V^{\neq}$ (*endo*) $- \Delta V^{\neq}$ (*exo*) $= -0.8$ cm^3 mol^{-1}). This finding seems to rule out that secondary orbital interactions are important and induce a large contraction of the transition state volume in the *endo* reaction. Therefore, this issue remains unresolved. As we shall discuss later, the molecular packing of the entire ensemble consisting of solute and solvent molecules and its reorganization during the course of reaction are most important for the magnitude of activation and reaction volume. An effective packing around the globular transition state (which may be also due to the restricted vibrations and rotations) may contribute to the observed difference between $\Delta V^{\neq}$ and ΔV of the Diels-Alder reactions listed in Table 3.6.

In a noteworthy systematic study, Jenner, Papadopoulos, and le Noble [44] determined the activation and reaction volumes as well as the activation enthalpies and activation entropies for the Diels-Alder reactions of hexachlorocyclopentadiene with various cycloalkenes as dienophiles (Table 3.7) which can be classified as Diels-Alder reactions with reverse electron demand [45]. The activation volumes increase (become less negative) with increasing ring-size of the dienophile whereas the reaction volumes remain almost constant for all reactions. Thus, the ratios $\Theta = (\Delta V^{\neq}:\Delta V)$ are diminished and the transition-state volumes become less product-like. The activation entropies show a trend comparable to the activation volumes becoming less negative with increasing ring size of the dienophiles indicating a lessening of the transition state ordering (exceptions are the reactions with cyclopentene and cyclodecene). The authors considered these findings to be an indication of a possible change in mechanism from a concerted process to a stepwise process involving the formation of diradical intermediates

In so called *homo*-Diels-Alder reactions, in which double bonds are replaced by three-membered rings or two π bonds of the reactive 1,3-diene are bridged by a sp^3-hybridized group, a powerful pressure-induced acceleration and hence

Table 3.7 Kinetic data for the Diels-Alder reaction of hexachlorocyclopentadiene with *cis*-cycloalkenes in *n*-decane extrapolated to 25 °C [44].

n	Cycloalkane	$\Delta V^{\neq\,a}$	ΔV^a	Θ^b	$\Delta S^{\neq\,c}$	$\Delta H^{\neq\,d}$
1	Cyclopentene	− 34.9	− 33.2	1.05	− 49	12.3
2	Cyclohexene	− 33.4	− 34.3	0.97	− 51	14.8
3	Cycloheptene	− 27.4	− 35.6	0.76	− 38	15.8
4	Cyclooctene	− 25.8	− 34.5	0.75	− 31	18.5
6	Cyclodecene	− 22.6	− 35	0.65	− 33	20.3
	Norbornene	− 28.6	− 33.3	0.86	− 42	14.8

a In cm^3 mol^{-1}; b $\Theta = \Delta V^{\neq} : \Delta V$; c In cal mol^{-1} K^{-1}; d In kcal mol^{-1} at 363 K.

highly negative activation volumes, comparable to those of the ordinary Diels-Alder reactions, have been observed. Examples of the pressure effect on *homo*-Diels-Alder reactions are shown in Scheme 3.2. The reaction of homofuran with fumaronitrile (Scheme 3.2, entry (8)) seems to be an exception in that the activation volume is significantly less negative than those in other cases. With the use of fumaronitrile and maleonitrile ((*E*)- and (*Z*)–NC–CH=CH–CN) as dienophiles it was established that the reaction occurs stereospecifically with retention of configuration in the dienophiles and non-stereospecifically with respect to the ratio (*endo:exo*). The relatively small pressure effect was explained in terms of a "late" transition state (with respect to the bond reorganization in the homofuran) in which the central cyclopropane bond has already been extensively cleaved and the bonding to the dienophile is still weak.

3.3.2 [2+2] Cycloadditions

[2 + 2] cycloadditions involving ketene derivatives as one or both reaction partners are assumed to be rare examples of concerted [π^2s + π^2a] cycloadditions [49]. The activation volumes determined for the [2 + 2] cyclodimerization of dimethylketene [50] and the [2 + 2] cycloadditions of diphenylketene to various enolethers [50–52] turned out to be highly negative. Kelm, Huisgen and coworkers studied the mechanism of the reaction of diphenylketene with *n*-butylvinylether (Scheme 3.3) in great detail [53]. Although the rate constants at

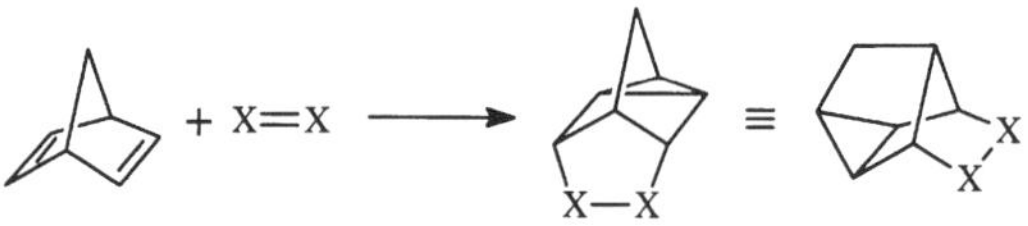

3.3. CYCLOADDITION REACTIONS

	X=X	Solvent	T (°C)	$\Delta V_T^{\neq\,a}$	$\Delta V_{25}^{\neq\,a}$	ΔV_{25}^{a}	Θ^a	Ref.
(1)	E–C≡C–E[b]	Ph–H	90	−40.5	−32.0	−35.5	0.89	[46]
(2)	(CH)$_2$C=C(CN)$_2$	Ph–Me	40	−30.0	−28.1	−32.9	0.85	[46]
		MeCN		−32.8	−30.7	−36.2	0.94	
(3)	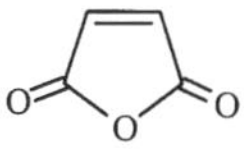 (maleic anhydride)	EtOAc	113	−45.0	−32.4	−29.5	1.10	[34]
(4)	CH$_2$=CH–CN	CHCl$_3$	90	−32.0	−24.9	−29.1	0.86	[47]
(5)	HC≡C–E[b]	CHCl$_3$	90	−43.5	−33.8	−38.8	0.87	[47]
(6)	E–C≡C–E[b]	CHCl$_3$	40.5	−37.3	−34.9	−39.4	0.89	[34]
(7)	E–N=N–E[b]	CHCl$_3$	50	−28.2	−25.4	−29.6	0.90	[47]
(8)	(E)-CN–CH=CH–CN	(CD$_3$)$_2$CO	70	−19				[48]

[a] See footnotes (a) and (b) in Table 3.5;   [b] E = CO$_2$Me;   [c] E = CO$_2$Et.

Scheme 3.2

atmospheric pressure could be successfully correlated with the term containing the dielectric constants of the solvents used $[(\varepsilon - 1)/(2\varepsilon + 1)]$, indicating an increase of polarity during the reaction, the very large solvent dependence of $\Delta V^{\neq}$ (Scheme 3.3) was erratic and not understandable. The authors found a fairly good correlation between the partial molar volumes of the reactants and the solvent cohesion energy densities (ced), but the correlation failed for those of the transition state and the product. Thus the pressure effect does not provide further insight into the mechanism of this reaction.

From detailed mechanistic studies, Huisgen and coworkers [54] concluded that the [2 + 2] cycloaddition of tetracyanoethene (TCNE) to vinylethers is a stepwise reaction passing through an interceptable dipolar intermediate. The

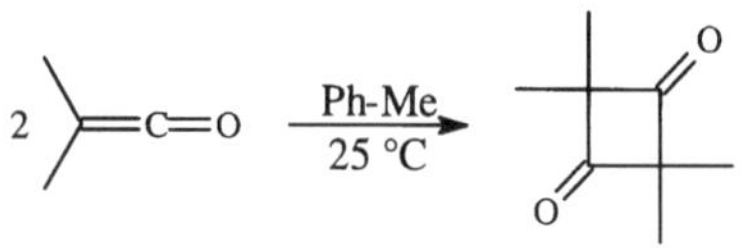

Scheme 3.3

$= -30 \text{ cm}^3 \text{ mol}^{-1}$ [50]

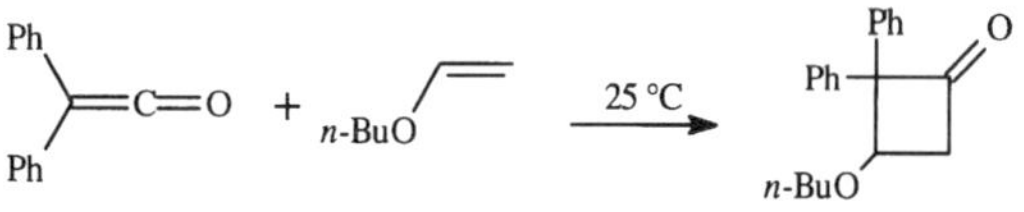

[53]

Solvent	$\Delta V^{\neq\ a}$	$\Delta V^{\ a}$
Ph–CN	−22	−10.4
cy-C$_6$H$_{12}$	−26	−31.3
CH$_2$Cl$_2$	−29	−26.2
Ph–Cl	−30	−26.4
Ph–H	−44	−23.0
Ph–Me	−52	−29.0

[a] In cm^3 mol^{-1}.

Scheme 3.3

polar [2 + 2] cycloadditions show a powerful pressure-induced acceleration [50–53]. The observation, that the activation volumes of these reactions are generally more negative than the corresponding reaction volumes ($\Theta = \Delta V^{\neq} : \Delta V > 1$), has been confirmed by a study of the cycloreversion of a TCNE-vinylether adduct where the dipolar intermediate could be trapped by methanol (Scheme 3. This finding can be rationalized by means of the effect of electrostriction mentioned above. Due to this effect the volumes of the dipolar intermediate and the polar transition state for its formation are smaller than those of the reactants and product. A study of the pressure-dependence of the reaction between TCNE and n-butylvinylether allowed the activation volumes (observed in different solvents, Scheme 3.4, entry 1) to be divided into the $\Delta V_i^{\neq}$ and $\Delta V_S^{\neq}$ according to eq. (11).

3.3.3 1,3-Dipolar Cycloadditions

1,3-Dipolar cycloadditions usually occur stereospecifically with retention of configuration in the dipolarophiles, and their rates are relatively insensitive to solvent polarity [58]. These results have been taken as evidence for a concerted process. Only a few exceptions are known where the non-stereospecific course of reaction suggests a stepwise mechanism [59]. Activation and reaction volumes were measured for a few reactions (Scheme 3.5). They are generally in the range

3.3. CYCLOADDITION REACTIONS

(1) RO-CH=CH₂ + (NC)₂C=C(CN)₂ → [RO⁺=CH–CH₂–C⁻(CN)–C(CN)₂] → cyclobutane with RO, (CN)₂, (CN)₂

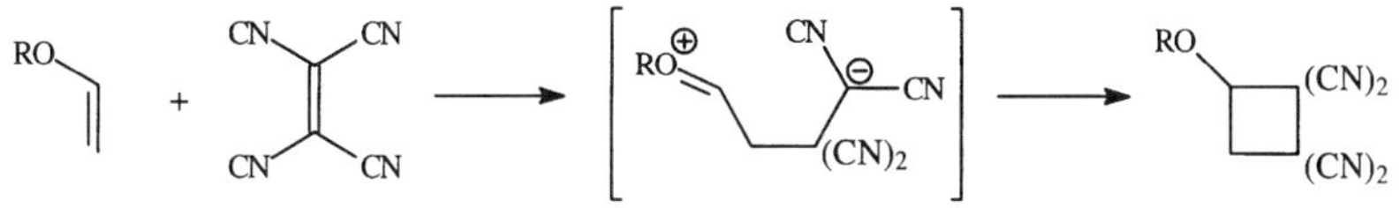

R	Solvent	T (°C)	$\Delta V^{\neq}$ [a]	ΔV [a]	Θ [b]	Ref.
Et	CH$_2$Cl$_2$	25	−55	−31	1.77	[55]
n-Bu	CCl$_4$	30	−50			[56]
n-Bu	MeCN	30	−29			[56]

(2) cyclobutane (EtO, Me, (CN)₂, (CN)₂) ⇌ [EtO⁺=CH–CH(Me)–C⁻(CN)–C(CN)₂] → open-chain product (EtO, OMe, H, Me, CN, (CN)₂)

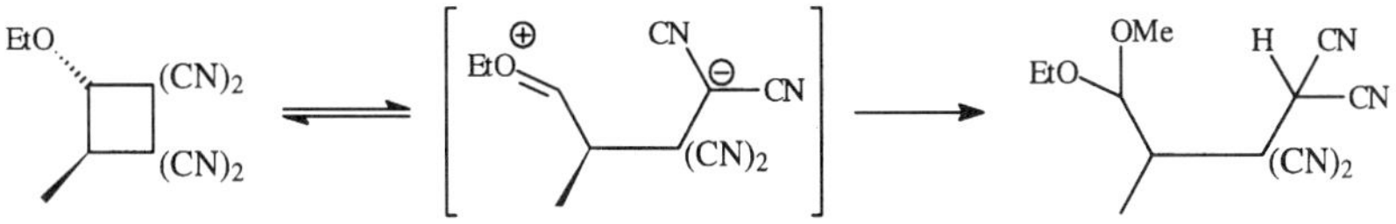

in MeOH at 25 °C: $\Delta V^{\neq}$ [a] = −16.7   [57]

[a] In cm^3 mol^{-1};   [b] $\Theta = \Delta V^{\neq}/\Delta V$.

Scheme 3.4

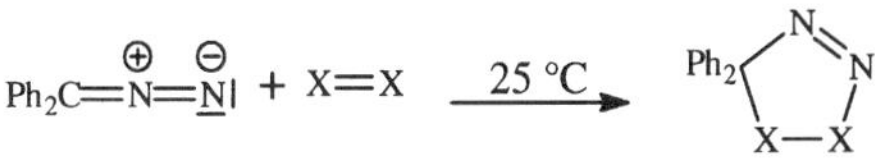

Ph$_2$C=N⁺=N⁻ + X=X $\xrightarrow{25\ °C}$ Ph$_2$-pyrazoline (N=N, X–X)

X=X	Solvent	$\Delta V^{\neq}$ [a]	ΔV [a]	Θ [b]	Ref.
E–C≡C–E [c]	Ph–Me	−23.2	−26.8	0.87	[25]
	MeCN	−15.3	−27.8	0.55	
(E)-E–CH=CH–E	Ph–Cl	−20.9	−27.0	0.77	[61]
	Ph–Me	−22.3			[62]
(Z)-E–CH=CH–E	Ph–Cl	−23.6	−25.1	0.94	[61]
	Ph–Me	−23.4	−24.5	0.96	[62]

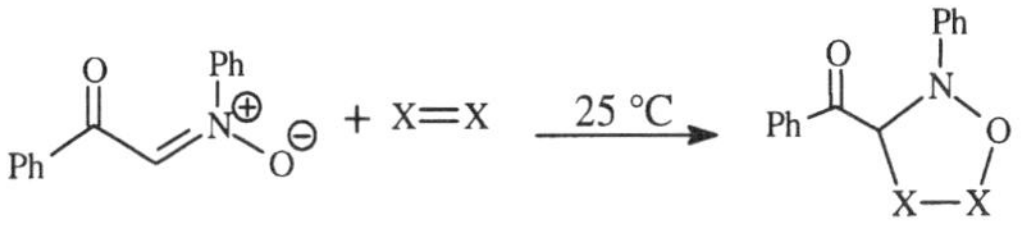

Ph-C(=O)-CH=N⁺(Ph)-O⁻ + X=X $\xrightarrow{25\ °C}$ isoxazoline (Ph-C(=O), Ph-N, O, X–X)

		$\Delta V^{\neq}$ [a]	ΔV [a]	Θ [b]	Ref.
(E)-E–CH=CH–E	Ph–Me	−21.7	−22.7	0.96	[62]

[a] In cm^3 mol^{-1};   [b] $\Theta = \Delta V^{\neq} : \Delta V$;   [c] E = CO$_2$Me.

Scheme 3.5

of $\Delta V^{\neq} = -(21 \pm 3) \text{ cm}^3 \text{ mol}^{-1}$ and $\Delta V = -(25 \pm 2) \text{ cm}^3 \text{ mol}^{-1}$ [60]. The relatively large ratios ($\Delta V^{\neq} : \Delta V$) have been regarded as an indicator for a pericyclic mechanism.

The absolute values of ΔV and $\Delta V^{\neq}$ for 1,3-dipolar cycloadditions, however, are about 5 to 10 cm³ mol⁻¹ smaller than those of Diels-Alder reactions. No clear explanation has been given for this difference. But one can assume, that one reason might be the effect of pressure on cyclization. According to information provided later in Scheme 3.25, the effect seems to be larger for the formation of a six-membered ring rather than of a five-membered ring.

3.3.4 [6+4] and [8+2] Cycloadditions

The pressure-dependence of the orbital symmetry allowed [6 + 4] cycloaddition of tropone with 1,3-dienes was studied first by le Noble and Ojosipe [63] who reported extremely small absolute values of $\Delta V^{\neq}$ and ΔV. A reinvestigation by Takeshita and his coworkers [64] showed, however, that the activation and reaction volumes of these cycloadditions are in the same order of magnitude as those of Diels-Alder reactions. Dogan [65] confirmed this finding in a study of the reaction between 1,3-butadiene and tropone in which a [6 + 4] cycloaddition competes with a [4 + 2] Diels-Alder reaction. The activation volume of the overall reaction was again found to be highly negative (Scheme 3.6). But the ratio between the [6 + 4] and [4 + 2] cycloadduct turned out to be almost pressure independent, which means that the difference between the activation volumes ($\delta \Delta V^{\neq}$) is almost zero and hence the activation volumes for both reactions are of the same value.

Tropone can also react as a tetraene component in [8 + 2] cycloadditions including the C=O double bond. Tropone reacts *e.g.*, with 1,1-diethoxyethene (at 120 °C, 10 h, 1 bar) to give the corresponding [4 + 2], [8 + 2], and [6 + 4] cycloadducts in yields of 1.1, 9.1, and 3.1%, respectively (conversion of tropone: 16%). At 3 kbar, 120 °C, the [4 + 2] and [8 + 2] cycloadducts were formed in yields of 13 and 17%, respectively (conversion of tropone: 30%) [67]. Tropone reacts with 2,3-dihydrofuran in a similar fashion leading to the corresponding [8 + 2] and [4 + 2] cycloadducts. The product ratio is again pressure-dependent [68]. The heptafulvene derivative shown in Scheme 3.7 can undergo a [8 + 2] cycloaddition leading to methyl azulene-1-carboxylate obviously after elimination of CO_2 and ethanol from the undetected primary cycloadduct. The [8 + 2] cycloaddition competes with [4 + 2] cycloadditions. Study of the pressure effect on the competitive reactions showed that the formation of the [4 + 2] cycloadduct is reversible even at 10 kbar, and that the [4 + 2] cycloadduct is not directly converted to methyl azulene-1-carboxylate. Thus, the azulene formation can only occur *via* the intermediate [8 + 2] cycloadduct [69].

3.3.5 Cheleotropic Reactions

Cheleotropic reactions were defined by Woodward and Hoffmann [70] to be processes in which two σ bonds directed to the same atom are formed or cleaved

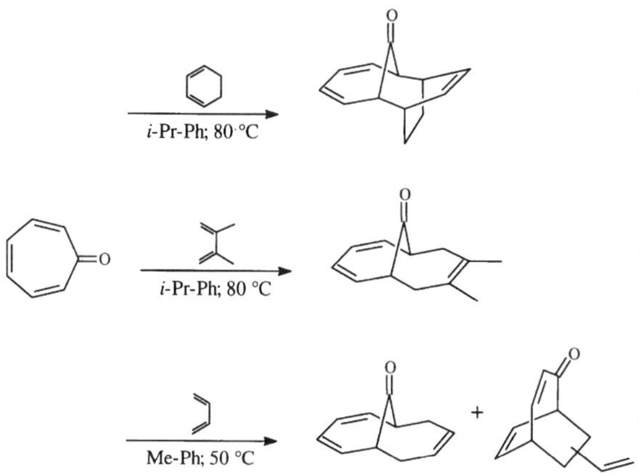

3.3. CYCLOADDITION REACTIONS

	$\Delta V^{\neq}$ [a]	ΔV	Ref.
(cycloheptatrienone + cyclohexadiene, i-Pr-Ph; 80 °C)	-37.6[b]	-36.1[c]	[64]
(cycloheptatrienone + 2,3-dimethylbutadiene, i-Pr-Ph; 80 °C)	-33.1[b]	-34.9[c]	[64]
(cycloheptatrienone + propene, Me-Ph; 50 °C) [6+4] + [4+2]	-31.0[d]	-33.6[c]	[65]

Product ratio   ($p = 0.9$ kbar)   10.0 : 1
                ($p = 6.9$ kbar)   10.8 : 1

$$\delta\Delta V^{\neq a} = \Delta V^{\neq}[6+4] - \Delta V^{\neq}[4+2] = -0.3$$

[a] In cm³ mol⁻¹;   [b] At 80 °C;   [c] At 60 °C [66];   [d] At 50 °C.

Scheme 3.6

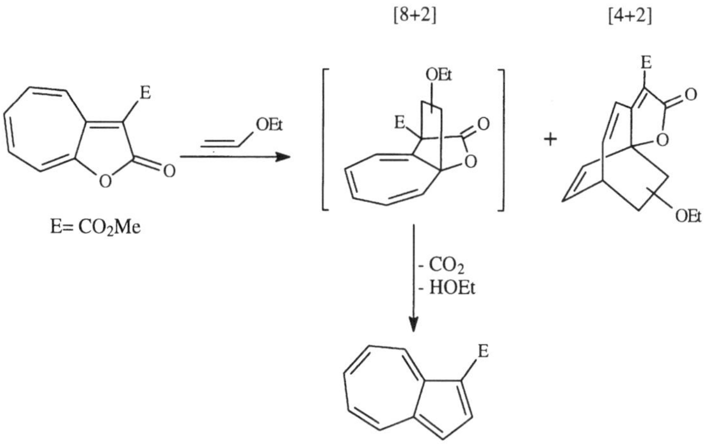

E = CO₂Me

Scheme 3.7 [69].

in one step. The addition of a singlet carbene to an alkene is an example of a non-linear cheleotropic reaction. Turro, Moss, and coworkers [71] generated phenylhalogenocarbenes (Ph–C–X: X = F, Cl, Br) by flash photolysis of the corresponding azirines and studied the pressure-dependence of their addition to tetramethylethene and (E)-2-pentene at room temperature up to 2 kbar (Scheme 3.8). The activation volumes obtained from these measurements were less negative (by about 20 to 30 cm^3 mol^{-1}) than those obtained for Diels-Alder reactions. This result can be explained again by the interdependence between the effect of pressure and ring size. According to Scheme 3.25 the smallest volume contraction is expected for the formation of three-membered rings.

The addition of SO$_2$ to 1,3-dienes (Scheme 3.8) is an example of a linear cheleotropic reaction. The activation volume for the reaction between SO$_2$ and 2,3-dimethyl-1,3-butadiene was found by Isaacs and Laila [72] to be more negative than the reaction volume ($\Theta = \Delta V^{\neq} : \Delta V = 1.06$). Comparable to many Diels-Alder reactions the relative transition state volume is product-like. Due to the large Θ value one might speculate that in the rate determining step the Diels-Alder adduct (the six-membered ring sulfinic ester) [73] is formed followed by a rearrangement to the observed five-membered ring sulfone.

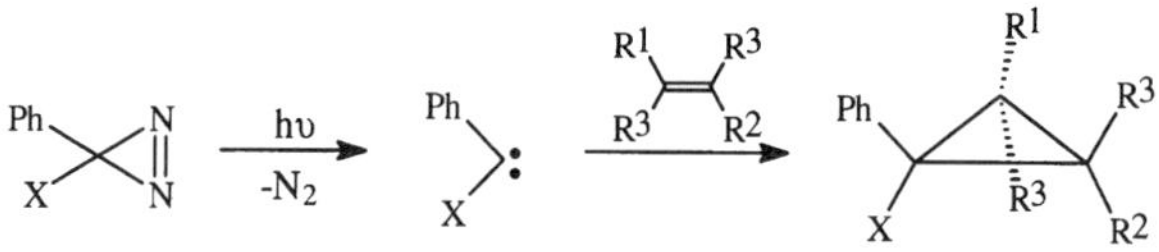

X	R^1	R^2	R^3	$\Delta V^{\neq}$ a
F	Me	Me	Me	-17
	Me	Et	H	-18
Cl	Me	Me	Me	-14
	Me	Et	H	-15
Br	Me	Me	Me	-10
	Me	Et	H	-12

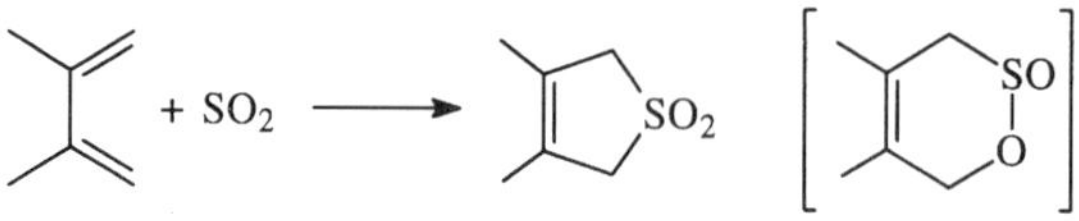

$\Delta V^{\neq a} = -35; \Delta V = -33; \Theta = 1.06$

a In cm^3 mol^{-1}.

Scheme 3.8

3.4 THE EFFECT OF PRESSURE ON COMPETITIVE REACTIONS

In competitive processes pressure can have a strong influence on the product ratio and, hence, on the selectivity provided that the activation volumes of the competing reactions are different. For example, if the difference between the activation volumes $\delta \Delta V^{\neq} = 5$ or $10 \text{ cm}^3 \text{ mol}^{-1}$, an improvement in selectivity by a factor of 4 and 16, respectively, at 7 kbar can be expected provided that the pressure-dependence of the product ratio fits eq. (14).

Selectivity in competitive reactions: $A \begin{array}{c} \xrightarrow{k_1} B \\ \xrightarrow{k_2} C \end{array}$

Pressure-dependence of the product ratio:

$$\left(\frac{[B]}{[C]}\right)_p = \left(\frac{[B]}{[C]}\right)_{1 \text{ bar}} \cdot \exp\left(-\delta\Delta V^{\neq} \cdot \frac{(p-1)}{RT}\right) \quad (14)$$

$$\delta\Delta V^{\neq} = \Delta V^{\neq}(A+B) - \Delta V^{\neq}(A+C)$$

Example: $T = 298 \text{ K } (25\,°\text{C})$, $R = 83.14 \text{ cm}^3 \text{ bar K}^{-1} \text{ mol}^{-1}$

p [bar]	[B]/[C]	B [%]	
1	1.0	50	
7000	4.1	80	($\delta\Delta V^{\neq} = -5 \text{ cm}^3 \text{ mol}^{-1}$)
7000	16.1	94	($\delta\Delta V^{\neq} = -10 \text{ cm}^3 \text{ mol}^{-1}$)

In the previous section we have already discussed that the activation volumes of many Diels-Alder [4+2] cycloadditions are usually highly negative, sometimes even more negative than the corresponding reaction volumes ($\Delta V^{\neq}/\Delta V \geq 1$). This observation has been regarded as an indication of a concerted mechanism. In order to test this hypothesis and to gain further insight into the often more complex mechanism of Diels-Alder reactions, the effect of pressure on competing [4+2] and [2+2] or [4+2] and [4+4] cycloadditions has been investigated. In competitive reactions the difference between the activation volumes and hence the transition-state volumes is derived directly from the pressure-dependence of the product ratio. All [2+2] or [4+4] cycloadditions listed in Tables 3.8 and 3.9 doubtlessly occur in two steps *via* diradical intermediates and can, therefore, be used as internal standards of the activation volumes expected

Table 3.8 Activation volumes $\Delta V^{\neq}$ (cm^3 mol^{-1}), given in parentheses, and differences in activation volumes $\delta \Delta V^{\neq}$ (cm^3 mol^{-1}) for competing [4+2] and [2+2] or [4+4] cyclodimerizations.

	Reaction	[4+2]-cycloadducts ($\Delta V^{\neq}$)		[2+2] or [4+4] cycloadducts ($\Delta V^{\neq}$)		$\delta V^{\neq}$	Ref.
(1)	2 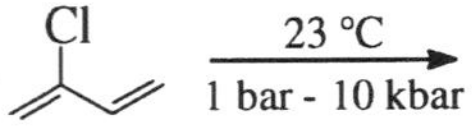 ⟶ 23 °C, 1 bar - 10 kbar	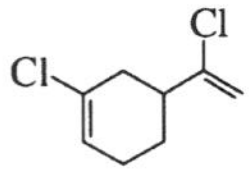 **2** (−31)	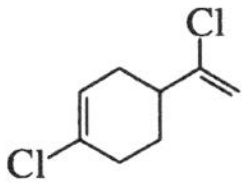 **3** (−29)	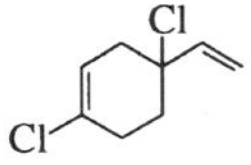 **4** (−22)	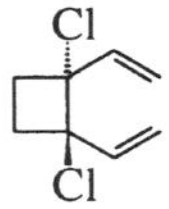 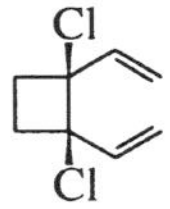 **5** (−22) **6** (−22)	−9 −7 0	[74]
(2)	2 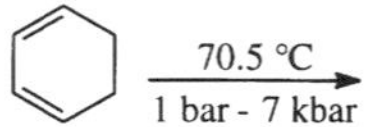 ⟶ 70.5 °C, 1 bar - 7 kbar	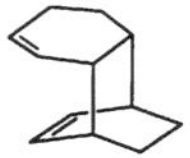 **8** (−28)	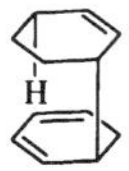 **9**[a] (−32)	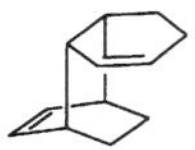 **10** (−22)	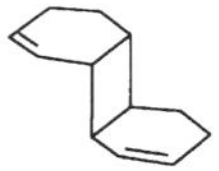 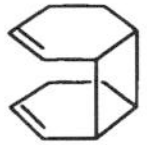 **11** (−18) **12** (−22)	−10[b] −14[c] 0[d]	[75]
(3)	2 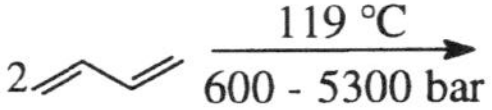 ⟶ 119 °C, 600 - 5300 bar	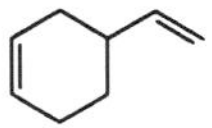 **14** (−38.4)			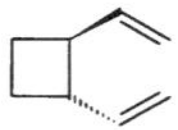 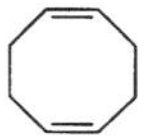 **15** (−20.4) **16** (−34.0)	−17.5[e] −4.4[f]	[76]

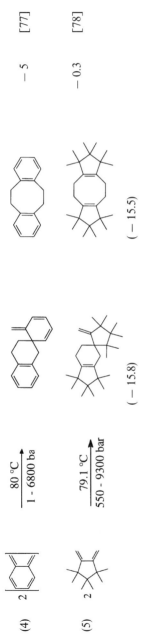

(−15.8) (−15.5)

−5 [77]

−0.3 [78]

---

a [6 + 4]-ene reaction;
b $\Delta V^{\neq}$ (**8**) − $\Delta V^{\neq}$ (**11**);
c $\Delta V^{\neq}$ (**9**) − $\Delta V^{\neq}$ (**11**);
d $\Delta V^{\neq}$ (**10**) − $\Delta V^{\neq}$ (**12**);
e $\Delta V^{\neq}$ (**14**) − $\Delta V^{\neq}$ (**15**);
f $\Delta V^{\neq}$ (**14**) − $\Delta V^{\neq}$ (**16**).

Table 3.9 Activation volumes $\Delta V^{\neq}$ (cm^3 mol^{-1}), given in parentheses, and differences in activation volumes $\delta\Delta V^{\neq}$ (cm^3 mol^{-1}) for competing [4 + 2] and [2 + 2] or [4 + 4] cyclodimerizations.

	Reaction	[4+2] cycloadducts	[2+2] or [4+4] cycloadducts	$\delta\Delta V^{\neq}$	Ref.
(1)	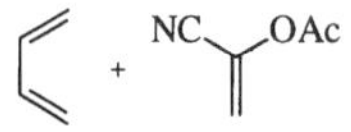	80 °C, 1 bar - 9 kbar		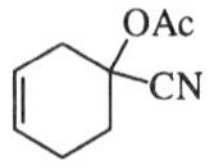 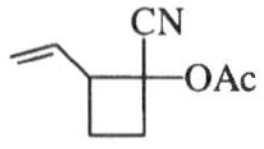 −11.5	[80], [81]
(2)	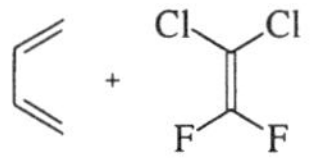	80 °C, 1 bar - 10 kbar		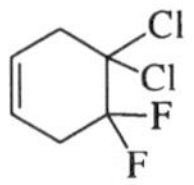 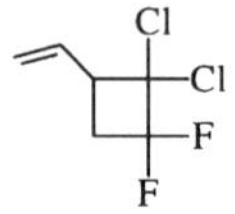 0	[81], [82], [83]
(3)	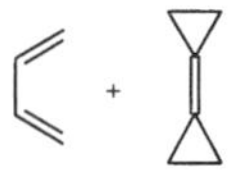	100 °C, 1 bar - 7 kbar		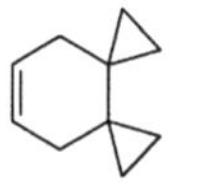 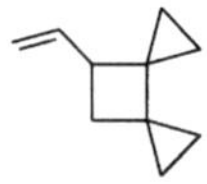 2	[84], [65]
(4)	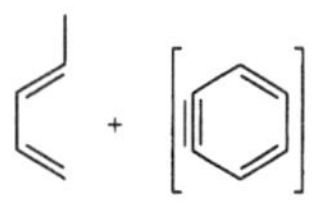	40 °C, 1 bar - 4 kbar		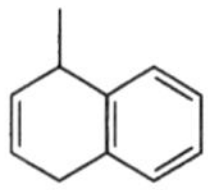 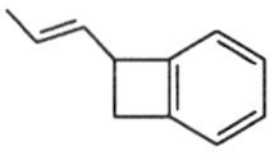 0	[85], [65]
(5)	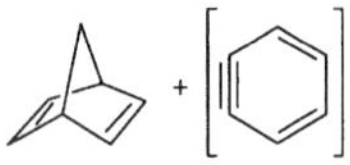	40 °C, 1 bar - 5.8 kbar		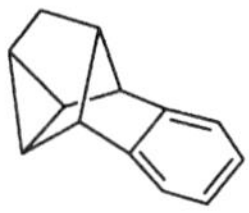 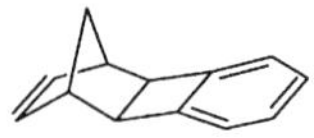 0	[86], [65]
(6)	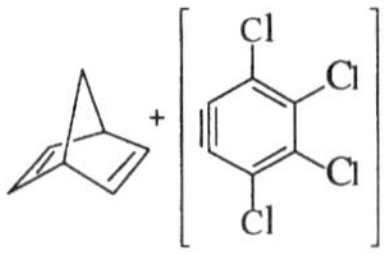	25 °C, 1 bar - 5 kbar		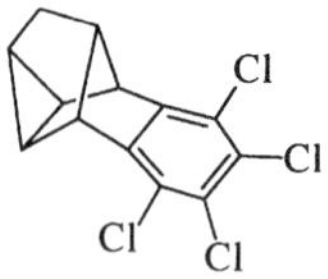 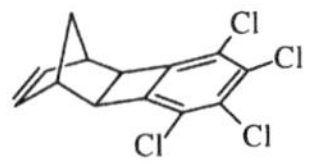 −0.7	[87]

for stepwise processes. Thus, a relatively simple measurement of the pressure-dependence of the product ratio should give important information about the mechanism of Diels-Alder reactions.

The results of the investigation of the pressure effect on competing [4 + 2] and [2 + 2] or [4 + 4] cycloadditions are summarized in Tables 3.8 and 3.9. In the thermal dimerization of chloroprene (**1**), the activation volumes for two [4 + 2] cycloadditions leading to **2** and **3** were found to be smaller than those of the third [4 + 2] and the [2 + 2] cycloadditions leading to **4**, **5**, and **6**, respectively.

3.4. THE EFFECT OF PRESSURE ON COMPETITIVE REACTIONS

Stewart [74] explained these results in terms of concerted Diels-Alder reactions competing with stepwise [2 + 2] cycloadditions. According to its larger (less negative) activation volume, the third Diels-Alder adduct **4** should also be formed in a non-concerted process. Similarly it can be concluded from the pressure-dependence of the dimerization of 1,3-cyclohexadiene (**7**) that the *endo*-Diels-Alder dimer **8** and the [6 + 4]-ene product **9** are formed concertedly while the *exo*-Diels-Alder adduct **10** and the [2 + 2] cyclodimers **11** and **12** arise *via* diradical intermediates. According to the activation volume data the Diels-Alder dimerization of 1,3-butadiene [79] and *o*-quinodimethane (Table 3.8, entries (3) and (4), respectively) fall into the same class of concerted processes as those discussed for **1** and **7**, while the Diels-Alder dimerization of hexamethyl-bis(methylene)cyclopentane seems to occur in stepwise fashion. According to the activation volume data summarized in Table 3.9 only the Diels-Alder reaction of 1,3-butadiene with α-acetoxyacrylonitrile seems to proceed concertedly while all other Diels-Alder and *homo*-Diels-Alder adducts are probably formed in stepwise processes comparable to the corresponding competitive [2 + 2] cycloadditions.

Stereochemical investigations of the chloroprene and 1,3-butadiene dimerizations with specifically deuterated derivatives confirm the conclusions drawn from activation volume data. In the dimerization of (*E*)-1-deuteriochloroprene the diastereomeric Diels-Alder adducts **4a**-D_2 and **4b**-D_2 (Scheme 3.9) are formed in a (59:41) ratio. The non-stereospecific course provides clear cut evidence that this Diels-Alder dimerization proceeds in a stepwise fashion, as suggested by the activation volumes, passing through, most likely, a diradical intermediate where rotation about C–C single bonds can substantially compete with a ring-closure reaction. The [2 + 2] cyclodimerization leading to a mixture of **5a**-D_2, **5b**-D_2, and **5c**-D_2 also occurs non-stereospecifically as expected.

The Diels-Alder dimerization of *Z,Z*-1,4-dideuterio-1,3-butadiene occurs with 97% *cis*-stereoselectivity at atmospheric pressure (1 bar) and > 99% stereospecificity at high pressure (6.8–8 kbar). These findings provide good evidence for the stereospecific Diels-Alder mechanism in competition with a small amount of the non-stereospecific reaction, which can be almost completely suppressed by high pressure, and confirm the conclusion drawn from the different activation volumes found for the [4 + 2] and [2 + 2] cyclodimerization of 1,3-butadiene. From the pressure-dependence of the product ratio (*endo*:*exo*) extrapolation indicates that the *endo*-Diels-Alder reaction shows a slightly more negative activation volume than the corresponding *exo*-reaction ($\delta \Delta V^{\neq} \approx -2.5$ cm^3 mol^{-1}) although both reactions are evidently pericyclic.

One question that needs to be addressed is: why are the activation volumes of pericyclic reactions smaller (more negative) than those of the corresponding stepwise reactions? In the past it was assumed that the simultaneous formation of two new bonds in a pericyclic [4 + 2] cycloaddition leads to a larger contraction of volumes than the formation of one bond in the rate determining transition state of the stepwise reaction. The interpretation presented is limited by the scope of the Eyring transition state theory where the activation volume, as mentioned above, is considered to be the reaction volume for the formation

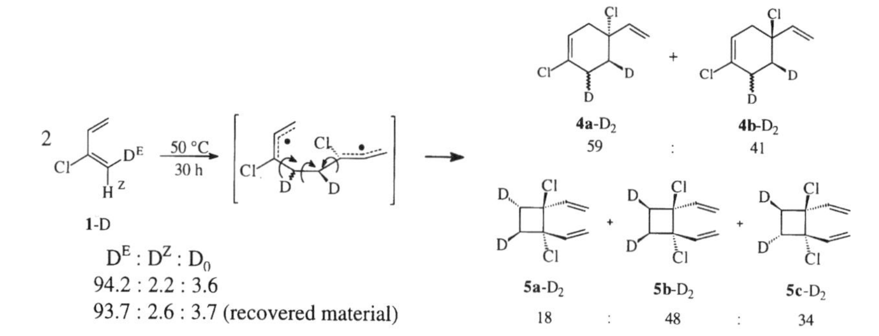

1-D

$D^E : D^Z : D_0$
94.2 : 2.2 : 3.6
93.7 : 2.6 : 3.7 (recovered material)

4a-D_2 59 : **4b-D_2** 41

5a-D_2 18 : **5b-D_2** 48 : **5c-D_2** 34

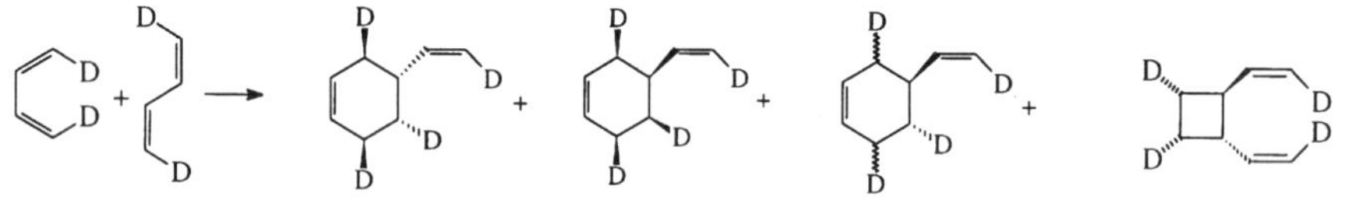

exo-cis   *endo-cis*   *endo/exo-trans*   *cis:trans* 58:42

p	T [°C]	[cis : trans]	[endo : exo]
1	138	97 : 3	56 : 44
1 bar	120	97 : 3	60 : 40
6.8 kbar	120	> 99 : <1	73 : 27
8.0 kbar	120	> 99 : <1	71 : 29

Scheme 3.9 [13], [76].

of a transition state and does not incorporate dynamic effects related to pressure-induced changes in viscosity [88]. In order to uncover the effect of differences in bonding on the volumes of transition states, the intrinsic volumes (the so-called van der Waals volumes, V_W) of ground and transition state structures for the pericyclic and stepwise cycloadditions of ethene to 1,3-butadiene (the prototype of Diels-Alder reactions) were calculated following the method of Nakahara et al. [61] (Table 3.10). The necessary structural parameters, bond lengths and bond angles, required for such a calculation have been determined by force-field or quantum mechanical methods, and van der Waals radii were derived from X-ray data [89].

The ratio V_W/V is defined as the packing coefficient η. Table 3.10 contains η values for cyclohexene and the three isomeric hexadienes as examples. According to these values, the molar van der Waals volume V_W occupies only approximately 50 to 60% of the macroscopic molar volume V. The empty space between the single molecules can be attributed to the so-called void volume and expansion volume required for the thermal motions and collisions of the molecules in the liquid state [90]. Generally η is found to be larger for cyclic

3.4. THE EFFECT OF PRESSURE ON COMPETITIVE REACTIONS

Table 3.10 Comparison between molar volumes V, van der Waals volumes V_W (cm^3 mol^{-1}), and packing coefficients η for the pericyclic and stepwise cycloaddition of ethene to 1,3-butadiene.

Compound	d	$V = M/d$ [a]	V_W [a,b]	$\eta = V_W/V$
CH$_2$=CH$_2$		59.9[c]	25.5	0.426
CH$_2$=CH–CH=CH$_2$		83.2[c]	44.8	0.539
CH$_2$=CH–CH$_2$–CH$_2$–CH=CH$_2$	0.6880	119.4	63.9	0.535
CH$_2$=CH–CH$_2$–CH=CH–CH$_3$	0.7000	117.7	63.9	0.544
CH$_2$=CH-CH=CH-CH$_2$-CH$_3$	0.7050	116.5	63.8	0.548
cyclohexene	0.8102	101.4	59.1	0.583
[18]$^{\neq}$ (2.24 Å)		109.1[d]	63.8	0.583
19 (1.54 Å)		118.7[e]	64.4	0.542
[20]$^{\neq}$ (1.84 Å)		120.4[e]	65.3	0.542

[a] In cm^3 mol^{-1}.
[b] For the calculation of van der Waals volumes the following van der Waals radii were used: $R_W(H) = 1.17$ Å; $R_W(C) = 1.80$ Å.
[c] Calculated values with volume increments [18].
[d] Calculated molar volumes of the transition structures [41b] with the packing coefficient of cyclohexene ($\eta = 0.583$).
[e] Calculated values with the average of the packing coefficients determined for the three isomeric hexadienes ($\eta = 0.542$).

compounds than for the corresponding acyclic ones. The van der Waals volume of the Diels-Alder reaction [61, 91] ($\Delta V_W = 59.1 - (44.8 + 25.5) = -11.2$ cm^3 mol^{-1}) can be calculated to be only roughly a quarter of the experimentally accessible volume of reaction ($\Delta V = 101.4 - (83.2 + 59.9) = -41.7$ cm^3 mol^{-1}). Consequently a significant part of the observed ΔV results from the higher packing of the cyclic product (compared to the acyclic starting materials) rather than from the changes in bonding. The difference between the van der Waals volumes of activation calculated for the pericyclic reaction ($\Delta V_W^{\neq} = 63.6 - (44.8 + 25.5) = -6.7$ cm³ mol⁻¹) and the stepwise reaction ($\Delta V_W^{\neq} = 65.3 - (44.8 + 25.5) = -5.0$ cm³ mol⁻¹) is small ($\delta \Delta V_W^{\neq} -1.7$ cm³ mol⁻¹) and is inconsistent with the experimental data listed in Tables 3.8 and 3.9. The comparison between the van der Waals volumes of the hexenediyl **19** and the transition structure [**20**]$^{\neq}$ for its formation shows that the effect on volumes caused by the change of bond lengths is rather small. In order to explain the finding that the activation volume of a pericyclic reaction is significantly more negative than that of the corresponding stepwise process, it has been assumed that the packing coefficient of the pericyclic transition state [41b] is similar to that of the cyclic product, whereas the transition state of the stepwise process [41b] is acyclic and has, therefore, a smaller packing coefficient. The difference between the activation volumes calculated by using the packing coefficients of cyclohexene and the average of the three hexadienes for the transition state structures of the pericyclic and stepwise Diels-Alder reactions ($\Delta V^{\neq} = 109.1 - (59.9 + 83.2) = -34.0$ cm³ mol⁻¹ and $\Delta V^{\neq} = 120.4 - (59.9 + 83.12) = -22.7$ cm³ mol⁻¹) respectively, is $\delta \Delta V^{\neq} = -11$ cm³ mol⁻¹. This value is well in accord with the experimental findings (Tables 3.8 and 3.9). Therefore, the analysis of activation volumes seems to provide important information regarding whether the geometry of a transition state is cyclic or acyclic. The conclusions drawn from this simple analysis are strongly supported by Monte Carlo simulations resulting in activation and reaction volumes for the Diels-Alder reaction of ethene and 1,3-butadiene and the dimerization of 1,3-butadiene [76].

The acivation and reaction volumes simulated with the Monte Carlo method using the model of hard spheres for the pericyclic Diels-Alder reaction of ethene with 1,3-butadiene are of the same order of magnitude ($\Delta V^{\neq} = -27.0$ cm³ mol⁻¹, $\Delta V = -30.7$ cm³ mol⁻¹, $\Delta V^{\neq} : \Delta V = 0.88$), consistent with the experimental findings for many Diels-Alder reactions discussed above. The results for the simulated activation and reaction volumes of the butadiene dimerization are summarized in Table 3.11. The *endo*-Diels-Alder dimerization passing through the pericyclic *endo*-transition state has accordingly the most negative activation volume, which is only slightly more negative than that of the corresponding *exo*-reaction passing through the corresponding *exo*-transition state ($\delta \Delta V_{calc.}^{\neq} = \Delta V_{endo}^{\neq} - \Delta V_{exo}^{\neq} = -1.8$ cm³ mol⁻¹). The activation volume calculated for the stepwise [2+2] cyclodimerization is substantially less negative than those calculated for the pericyclic = Diels-Alder reactions mentioned above ($\delta \Delta V_{calc.}^{\neq} = \Delta V_{endo}^{\neq} - \Delta V^{\neq} [2+2] = -18$ cm³ mol⁻¹). The calculated values are indeed in surprisingly good agreement with the experimental data. The difference between the activation volumes found experimentally for the *endo*- and *exo*-Diels-Alder dimerizations is $\delta \Delta V_{exp.}^{\neq} = -2.5$ cm³ mol⁻¹ and that for the competing [2+2] and [4+2] cyclodimerizations is ($\delta \Delta V_{exp.}^{\neq} = \Delta V_{endo}^{\neq} - \Delta V^{\neq} [2+2] = -13.3$ cm³ mol⁻¹. A comparably good agreement between calculated and experimental data was also found for reaction volumes (Table 3.11). It must be noted that the van der Waals volumes of

3.4. THE EFFECT OF PRESSURE ON COMPETITIVE REACTIONS

Table 3.11 Monte-Carlo (MC) simulation using mixtures of 24 transition structures [41b] or product structures + 208 s-*trans*-butadiene structures. Van der Waals radii R_W (Å): C 1.80, H 1.17; van der Waals volumes (V_W) are given in parentheses. Experimental values at 21 °C. All volume data in cm^3 mol^{-1} [76].

		$\Delta V^{\neq}$		ΔV	
		exp.	calc.	exp.	calc.
[endo TS] (81.2)	cyclohexene-vinyl (77.5)	−27.9	−31.5	−33.5	−35.4
[exo TS] (81.1)	cyclohexene-vinyl	−25.4	−29.7		
2 propene → [biradical 1.89] (83.2)		−14.6	−13.5		
↓ [biradical 1.55] (82.6)	cyclobutane			−24.2	−29.8
	cyclooctadiene (76.9)$^{a)}$ (76.6)$^{b)}$			−43.5	−34.4^a / −40.5^b

^{a} Chair conformation;
^{b} Twist-boat conformation

the transition structures listed in Table 3.11 do not differ significantly from one another. The preference of Diels-Alder reactions over [2 + 2] cyclodimerizations at high pressure is therefore due to a configurational effect which can be explained by the different packing of cyclic and acyclic states.

From a similar calculation of van der Waals volumes and packing coefficients, Nakahara and coworkers [62] predicted the activation and reaction volume for the hypothetical prototype of the pericyclic 1,3-dipolar cycloaddition shown in Table 3.12. With the assumption, that the packing coefficient of the pericyclic transition state is product-like, the ratio of activation volume to reaction volumes was calculated to be $\Theta = \Delta V^{\neq} : \Delta V = 0.82$ consistent with

Table 3.12 Activation and reaction volumes predicted for the hypothetical prototype of a pericyclic 1,3-dipolar cycloaddition by calculation of van der Waals volumes and packing coefficients η derived from the partial molar volumes of reactants and product [62a].

$$H-C\equiv\overset{\oplus}{N}-\overset{\ominus}{O} \;+\; H-C\equiv C-H \;\longrightarrow\; \left[\begin{array}{c}\overset{N}{\diagup}\diagdown O \\ 2.18\,\text{Å}\,\underset{\text{---}}{}\,2.21\,\text{Å}\end{array}\right] \;\longrightarrow\; \overset{N}{\underset{O}{\diagup\diagdown}}$$

$V_W{}^a$	24.1	23.1	43.4	39.0
V^a	48	54	71	64
η^b	ca. 0.50[c]	0.43	0.61[d]	0.61

$\Delta V_W{}^a = 39.0 - (24.1 + 23.1) = -8.2 \qquad \Theta = 0.46$
$\Delta V_W^{\neq\,a} = 43.4 - (24.1 + 23.1) = -3.8$
$\Delta V^a = 64 - (48 + 54) = -38$
$\Delta V^{\neq\,a} = 71 - (48 + 54) = -31 \qquad \Theta = 0.82$

[a] In cm³ mol⁻¹;
[b] $\eta = V_W/V$;
[c] Taken from $\eta(\text{H–}\bar{\text{N}}\text{–}\overset{+}{\text{N}}\equiv\text{N})$;
[d] Assumed that $\eta^{\neq}$ of pericyclic transition state is identical with η of the cyclic product.

experimental results shown in Scheme 3.5. The ratio between the corresponding van der Waals volumes was calculated to be only $\Theta_W = \Delta V_W^{\neq} : \Delta V_W = 0.49$, inconsistent with the experimental findings.

The effect of pressure on the *endo/exo*-diastereoselectivity and regioselectivity of cycloadditions is less well understood than that on the competitive [4 + 2] and [2 + 2] cycloadditions discussed in the previous section. In Scheme 3.10 the effect of pressure on the ratio (*endo:exo*) is summarized for various Diels-Alder reactions. In some cases such as the reactions of 1,3-cyclopentadiene with several acrylic acid derivatives, the dimerization of (*E*)-1-phenyl-1,3-butadiene or the addition of furan to the substituted maleic anhydride (Scheme 3.10, entry (1), (5), and (6), respectively) the effect is negligibly small. In other cases, however, such as the dimerization of 1,3-cyclohexadiene, entry (3), or the hetero-Diels-Alder reaction with reverse electron demand, entry (7), the effect is quite substantial. In combination with the pressure-induced acceleration of the overall reaction this effect can be very useful for synthetic purposes. A good example of this is the hetero-Diels-Alder reaction, entry (7). By using high pressure the reaction can be performed at a lower temperature leading to a further increase in diastereoselectivity. Another prominent example is the Diels-Alder reaction of furan with the substituted maleic anhydride studied first by Dauben et al. [93] and later by Grieco et al. [97] which occurs only at high pressure, or catalyzed by $LiClO_4$ at atmospheric pressure, leading to a precursor of cantharidin. High

3.4. THE EFFECT OF PRESSURE ON COMPETITIVE REACTIONS

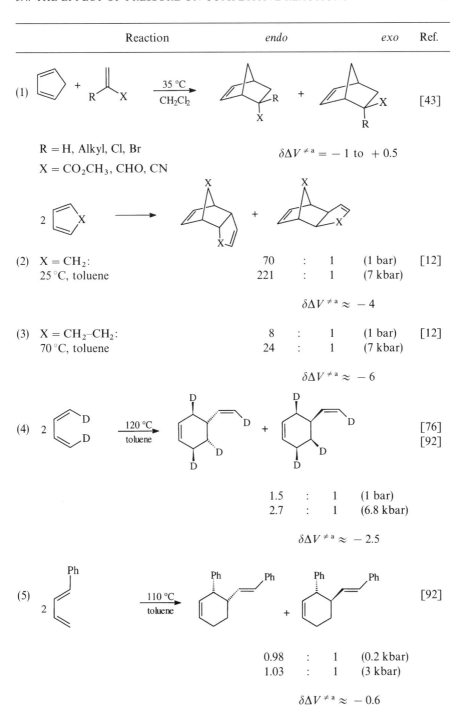

Scheme 3.10

Reaction	endo	exo	Ref.

(6) 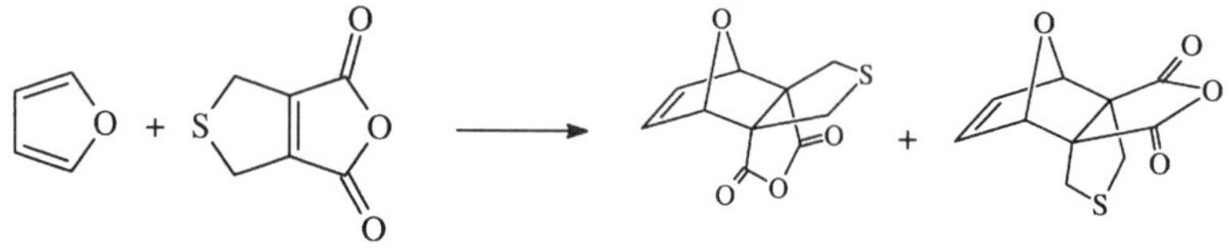 furan + S-containing anhydride

20 °C, 4–15 kbar   1 : 4   [93]
(pressure independent)

5.0 M LiClO$_4$, Et$_2$O   1 : 5.7   [99]

(7) 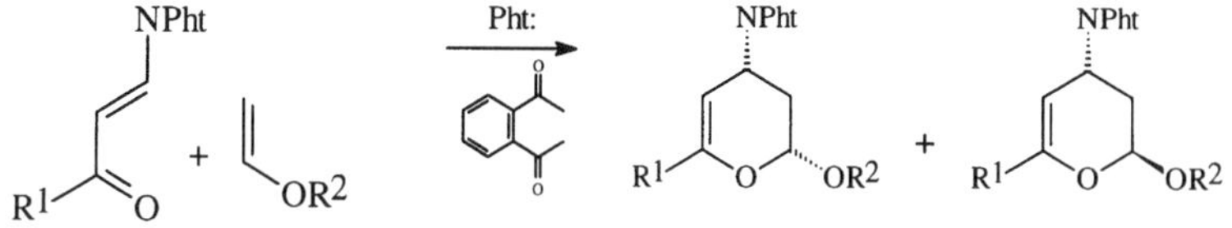 NPht-enone + vinyl ether → dihydropyran products [94]

$R^1 = CCl_3$, $R^2 = Et$: 60 °C, CH$_2$Cl$_2$

			endo : exo	$\delta\Delta V^{\neq a}$
	(1 bar)		2 : 1	-5.9
	(6.5 kbar)		6.5 : 1	
0.5 °C	(6 kbar)		14 : 1	

$R^1 = CF_3$, CO_2Me, $R^2 = Et$:

$R^1 = CCl_3$, $R^2 = i\text{-}Bu$, $i\text{-}Pr$, $t\text{-}Bu$, $PhCH_2$, $p\text{-}MeOC_6H_4$ $\delta\Delta V^{\neq a} = -(4.5 \text{ to } 6.9)$

(8) 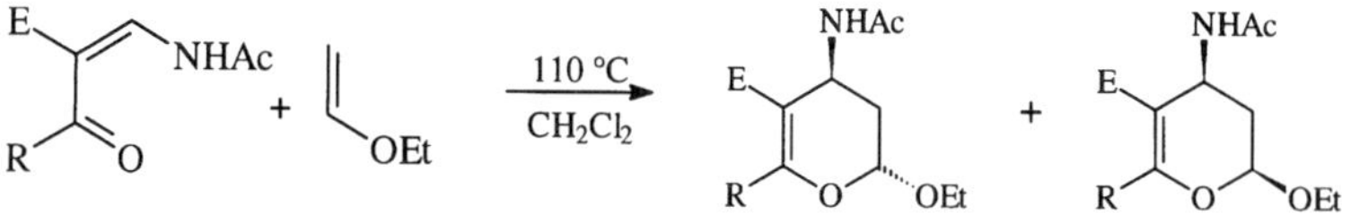 enone + ethyl vinyl ether → trans / cis dihydropyrans

110 °C, CH$_2$Cl$_2$ [26] [95]

$R = H, Et, i\text{-}Pr$; $E = CO_2Me$ $\delta\Delta V^{\neq a} = \sim 0 \text{ to } -4.6$

(9) 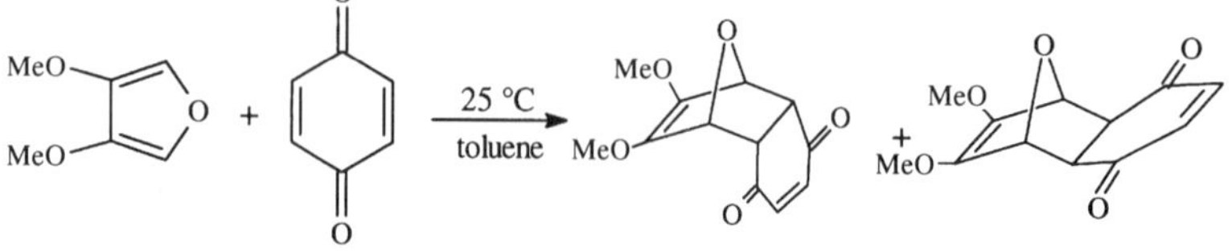 dimethoxyfuran + benzoquinone → endo/exo adducts

25 °C, toluene   [96]

$R = MeO$

	endo : exo	
	89 : 11	(7 kbar)
	83 : 17	(11 kbar)
	54 : 46	(19 kbar)

$\delta\Delta V^{\neq a} = +4$

---

[a] $\delta\Delta V^{\neq} = \Delta V^{\neq}(endo) - \Delta V^{\neq}(exo)$ in cm^3 mol^{-1}.

Scheme 3.10 Continued.

3.4. THE EFFECT OF PRESSURE ON COMPETITIVE REACTIONS

pressure or the catalyst allows the reaction to proceed at room temperature while the high temperature, required at atmospheric pressure and without any catalyst, causes a shift of the equilibrium toward the starting materials due to the unfavorable activation entropy of bimolecular cycloadditions.

Breslow [101], Grieco [99] and others [102] have found that the rates of many reactions such as cycloadditions can be strongly enhanced by conducting them in water or in a saturated $LiClO_4$-ether solution comparable to the rates of the same reactions at high pressure in conventional organic solvents. Suggested origins of these effects are high internal solvent pressure, hydrophobic association, micellar catalysis, solvent polarity, and hydrogen bonding. An effect of internal solvent pressure comparable to that of external pressure can certainly be excluded considering the ratio (*endo*:*exo*) in the Diels-Alder reaction of cyclopentadiene with methyl acrylate (Scheme 3.11), which was found to be almost pressure independent but strongly solvent dependent comparable to the Lewis-acid catalyzed reactions. Blake and Jorgensen [103] found in a Monte-Carlo simulation of the solvent effect on the Diels-Alder reaction between cyclopentadiene (CP) and methylvinylketone (MVK) that the interaction between water and the transition state leads to a substantial stabilization whereas

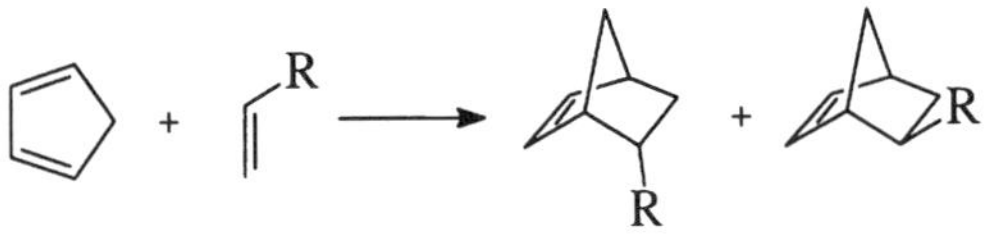

R	Solvent	T (°C)	p	endo : exo	$\delta \Delta V^{\neq \, a}$	Ref.
(1) CO_2CH_3	CH_2Cl_2	35	1 bar	3.85 : 1	− 0.52	[43]
			3 kbar	3.98 : 1		
	cyc-C_6H_{12}	25	1 bar	2.9 : 1		[98]
	EtOH	25	1 bar	5.2 : 1		
	H_2O	25	1 bar	9.3 : 1		
(2) $CO_2C_2H_5$	H_2O	25	1 bar	4.0 : 1		[97]
	5 M $LiClO_4/Et_2O$	25	1 bar	8.0 : 1		
(3) CO_2CH_3	CH_2Cl_2	0	—	4.6 : 1		[100]
			$BF_3 \cdot Et_2O$	32 : 1		
			$AlCl_3 \cdot Et_2O$	49 : 1		
			$SnCl_4$	19 : 1		
			$TiCl_4$	24 : 1		
			$AlCl_3$ in $MeNO_2$	24 : 1		

[a] In $cm^3 \, mol^{-1}$.

Scheme 3.11

the interaction between water and the reactants or adduct is small. Propane as solvent has, accordingly, no significant influence on transition state, reactants, or adduct. The authors concluded that the aqueous acceleration of the reaction between CP and MVK contains besides the hydrophobic association a non-hydrophobic component stemming from enhanced polarization of the transition state that leads *inter alia* to stronger hydrogen bonds at the carbonyl oxygen. The various methods for acceleration and selectivity enhancement of Diels-Alder reactions were recently reviewed by Pindur and coworkers [104].

A study of the reaction between furan or 1-methylfuran and acrylic acid derivatives by Jenner [109] showed that in these cases the Diels-Alder addition is less sensitive to pressure in aqueous solution than in an organic solvent such as CH_2Cl_2. Isaacs *et al.* found, however, that the pressure effect on the Lewis acid or $LiClO_4$ catalyzed Diels-Alder reaction of isoprene with N-phenylmaleimide is larger (more negative activation volume) than that on the corresponding uncatalyzed reaction [110]. Similar results were also obtained for the Diels-Alder reaction between 9-anthracenemethanol and N-ethylmaleimide in 1-butanol and water. The reaction of 2-pyrone with enol ethers reported by Posner and Ishihara [105] illustrates that the combination of high pressure and Lewis-acid catalyst can have a synergic effect. This reaction occurs only at high pressure in the presence of the catalyst. The application of a chiral catalyst, however, does not induce any enantioselectivity; only the racemic adduct is formed. A pressure-induced increase of enantioselectivity from 4.5% ee at 1 bar to 20.4% ee at 5 kbar was observed by Tietze, Buback and coworkers [106] in the intramolecular Diels-Alder reaction catalyzed by a chiral titanium complex, shown in Scheme 3.12, entry (2). The pressure effect on enantio- and diastereoselectivity has been studied for many uncatalyzed Diels-Alder reactions and has been recently reviewed by van Eldik, Asano, and le Noble [5] and by Ibata in reference [3]. Scheme 3.12, entries (3) and (4), shows two representative examples.

Scheme 3.13 shows an example of competitive "domino" and "pincer" cycloadditions with *syn-o,o'*-dibenzene derivatives [111]. The selectivity strongly depends on the nature of the acetylenic dienophile as well as of the *syn-o,o'*-dibenzene derivative. Preferential formation of the "pincer" adduct and of the "domino" adduct occurs respectively with dicyanoacetylene (DCA), (Scheme 3.13, entry (1)), and with dimethyl acetylenedicarboxylate (DMAD), entry (2). This finding seems to indicate that the different steric demand of the two dienophiles may be responsible for the selectivity. The approach of the linear DCA to the center of the two cyclohexadiene rings may be supported by non-covalent bonding interactions between the orthogonal π bonds of DCA and the inner faces of the electron-rich cyclohexadiene units while the sterically larger ester groups in DMAD may prevent this orientation. Thus, DMAD approaches preferentially from the outer face of one diene unit. In the reaction of DCA with the dibenzo-substituted bis-diene, entry (3), one of the benzene rings can successfully compete with one cyclohexadiene ring to complex DCA, so that the formation of the "domino" adduct is favored. High pressure induces a large

3.4. THE EFFECT OF PRESSURE ON COMPETITIVE REACTIONS

	Reaction			Ref.
(1) 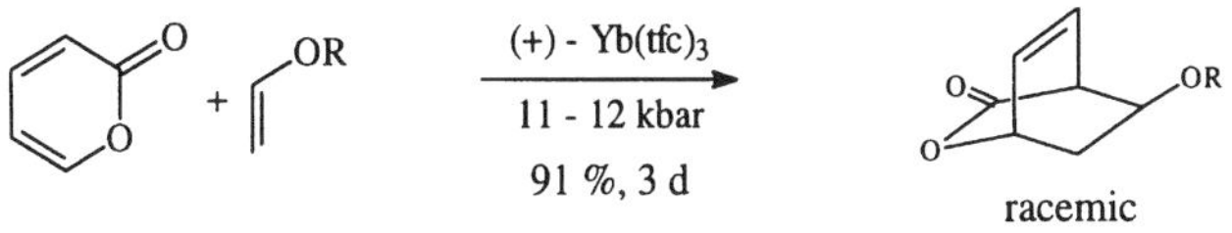	pyranone + vinyl ether OR	(+)-Yb(tfc)$_3$ 11-12 kbar 91 %, 3 d	bicyclic adduct OR racemic	[105]
(2) 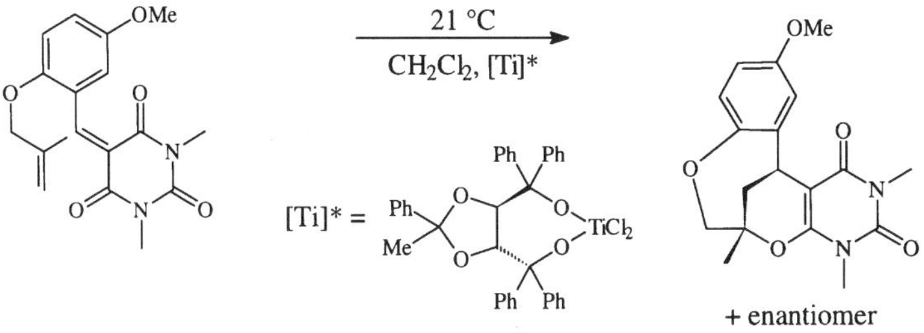	chromene-barbiturate OMe	21 °C CH$_2$Cl$_2$, [Ti]* [Ti]* = Ph,Me-dioxolane-TiCl$_2$ with Ph,Ph groups	tricyclic product OMe + enantiomer 1 bar: 4.5 % ee 5 kbar: 20.4 % ee	[106]
(3)	cyclopentadiene + β CH$_2$OBz / BzO lactone α	CH$_2$Cl$_2$	α-endo + α-exo + β-endo + β-exo	[107]

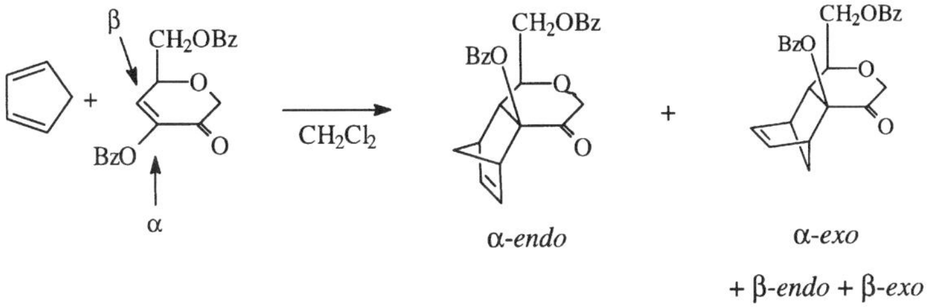

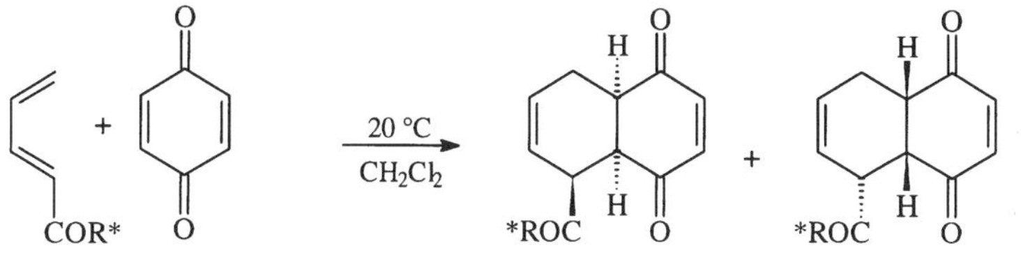

	α:β	α endo : exo	β endo : exo
200 °C, 1 bar:	3:1	8 : 1	5 : 1
TiCl$_4$, −78 °C:	4:1	1.2 : 1	1 : 1.3
15 kbar, 25 °C:	3:1	5 : 1	4.3 : 1

(4)	diene-COR* + benzoquinone	20 °C CH$_2$Cl$_2$	decalindione *ROC + decalindione *ROC 75 : 25 (15 kbar)		[108]

R*: Ph-substituted (−)-menthyloxy (23 h, yield: 62 %)

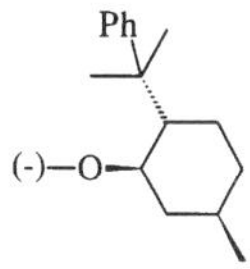

Scheme 3.12

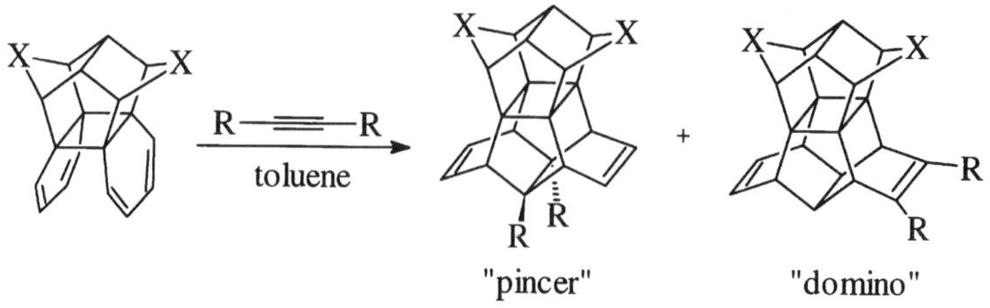

Scheme 3.13 [111].

		T (°C)	Time (h)	"pincer"	:	"domino"
(1)	X = CH$_2$, R = CN					
	1 bar	100	4	74	:	26
	12 kbar	25		85	:	15
(2)	X = CH$_2$, R = CO$_2$CH$_3$					
	1 bar	100	80	3	:	97
	14 kbar	25		3	:	97
(3)	X = o-C$_6$H$_4$, R = CN					
	1 bar	25		0	:	100
	1 bar	80		4	:	96
	1 bar	100		12	:	88
	14 bar	25		0	:	100

rate enhancement but no significant change of selectivity. This finding supports the conclusion that each reaction, the "pincer" as well as the "domino" cycloaddition, consists of two consecutive Diels-Alder reactions.

Dolbier and Weaver [112] investigated the effect of pressure on the stereo- and regioselectivity in a stepwise [2 + 2] cycloaddition of 1,1-difluoroallene with (Z)-β-deuteriostyrene (Scheme 3.14). In order to explain the pressure-induced increase in stereoselectivity the authors concluded that in the diradical intermediate at high pressure, the ring-closure reactions leading to (Z)-configurated methylenecyclobutane derivatives are favored over bond rotation which is a prerequisite for the formation of (E)-configurated derivatives.

Since many different effects on the transition states of the competitive reactions such as attractive or repulsive interactions (*e.g.*, hydrogen bonding, π–π interactions, or steric hindrance) or polarization may be responsible for the observed selectivity, it is usually difficult to explain or even predict the influence of pressure on the selectivity of the reactions discussed in this section. An exception, where the effect of pressure can be better understood today, is apparently the competition between pericyclic and stepwise cycloadditions.

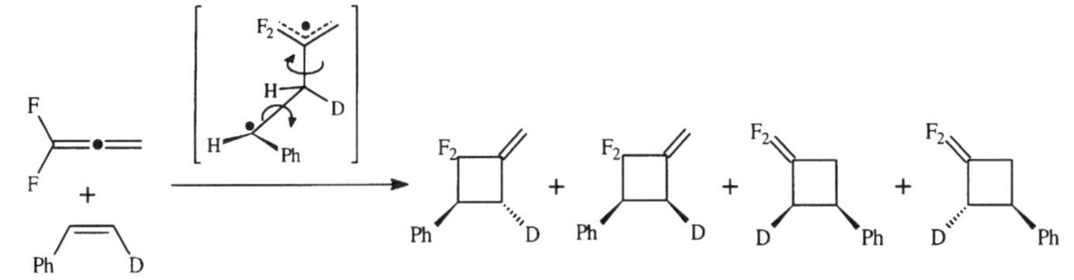

3.5. MOLECULAR REARRANGEMENTS

Scheme 3.14 [112].

	Z		E		Z		E
1.8 kbar:	66.3	:	33.7		88.0	:	12.0
		86.1 %				13.9 %	
13.0 kbar:	85.5	:	14.5		95.2	:	4.8
		69.6%				30.4 %	

3.5 MOLECULAR REARRANGEMENTS

Many pericyclic rearrangements show a pressure-induced acceleration which is characterized by a negative volume of activation [113]. The effect, which is usually smaller than that of intermolecular cycloadditions, may be explained by invoking different packing coefficients of cyclic and acyclic states as already discussed for the pericyclic and stepwise cycloadditions.

On the basis of stereochemical and kinetic investigations, most Cope rearrangements are regarded as being pericyclic processes [114]. The van der Waals volumes calculated for the parent 1,5-hexadiene and the pericyclic transition state are approximately the same (Scheme 3.15) [115, 41]. This is understandable since in the symmetrical transition state the bond breaking and making have proceeded to the same extent so that the effects of the two processes on the van der Waals volume compensate and no great overall effect of pressure on the Cope rearrangement is to be expected. If it is assumed that, by analogy with the pericyclic and stepwise cycloadditions already mentioned, the transition state here also exhibits a larger packing coefficient because of its cyclic geometry, the activation volume ought to be negative. The activation volume can be estimated at approximately $-10 \text{ cm}^3 \text{ mol}^{-1}$ if the packing coefficient determined for cyclohexene is used for the unknown packing coefficient of the transition state. In fact, negative activation volumes of the expected magnitude were found for the Cope rearrangements and related Claisen rearrangements shown in Scheme 3.16. However, the reacting compounds are highly polar, so

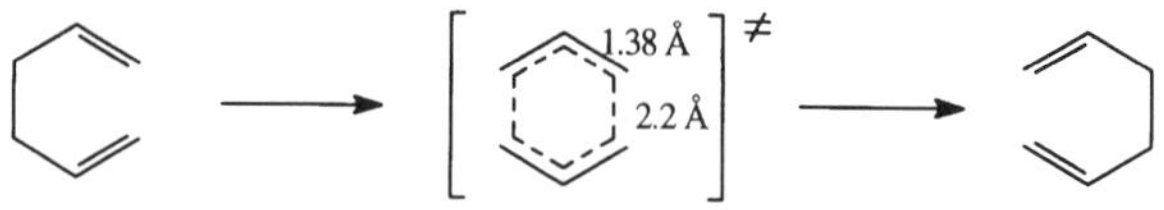

142   3. EFFECT OF PRESSURE ON ORGANIC REACTIONS

V_W	63.9	63.6	$\Delta V^{\neq} = -0.3$
V	119.4	109.1	$\Delta V^{\neq} = -10.3$

All volumes are given in $cm^3\,mol^{-1}$. The structural parameters necessary for the calculation of the van der Waals volume for the transition state (TS) were taken from *ab initio* calculations [41a, 115]. The partial molar volume for the TS was calculated from the equation:

$V\,(TS) = V_W(TS)/(\eta(\text{cyclohexene});\ \eta \equiv V_W/V = 0.5829\ (\text{cyclohexene}).$

Scheme 3.15

		T (°C)	Solvent	$\Delta V_T^{\neq\,a}$	Ref.
(1)	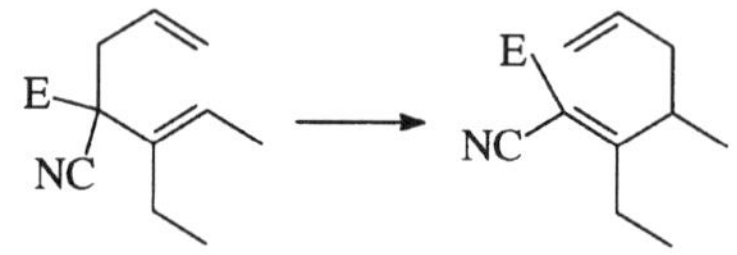	119	Decalene	−6.7	[116]
(2)	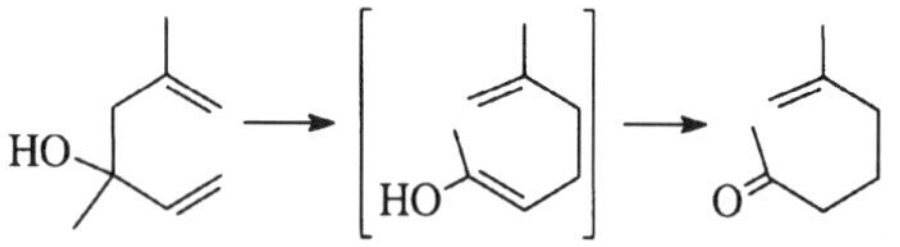	180	N-methyl-pyrrolidinone	−9.7	[117]
(3)	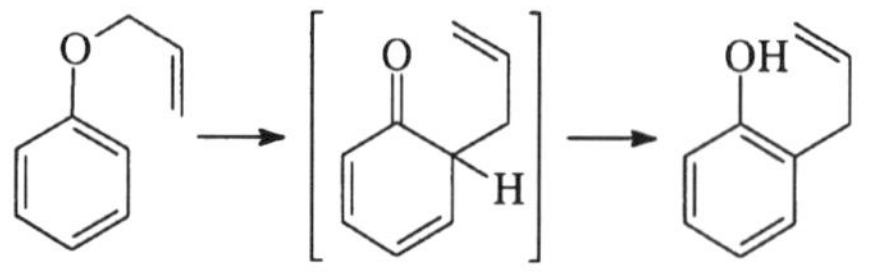	160	Decalene	−7.7	[116]
(4)		130.4	Neat	−18	[116]

[a] In $cm^3\,mol^{-1}$.

Scheme 3.16

3.5. MOLECULAR REARRANGEMENTS

the negative activation volumes could also be due to electrostriction effects rather than be a consequence of the cyclic transition states.

The activation volumes obtained from the pressure-dependence of the Cope rearrangements in pure hydrocarbons (Scheme 3.17), in which electrostriction effects caused by polar substituents should be negligible, were in good agreement with that predicted for the parent system. This explains why the degenerate Cope rearrangement in bullvalene, investigated by Merbach, le Noble, and coworkers [118] employing pressure- and temperature-dependent NMR spectroscopic measurements, shows no significant pressure effect ($\Delta V^{\neq} = -0.5$ cm^3 mol^{-1}). As a result of the fixed stereochemistry due to the rigid bullvalene skeleton no new cyclic interaction, in the sense discussed here, appears in the transition state.

In Scheme 3.18 the activation volume data for some potential sigmatropic [1,n] carbon or hydrogen shifts ($n = 3$–9) are summarized. Analogously to the Cope rearrangement (sigmatropic [3,3] carbon shift) the activation volumes turned out to be negative in cases of a pericyclic mechanism while the activation volumes are positive in cases of a dissociative mechanism. The [1,4] shift of a benzyl or benzhydryl group in 1-alkoxypyridine-N-oxides, Scheme 3.18, entry (3), is particularly instructive. From the completely different pressure-dependences of the two reactions, le Noble and Daka [123] concluded that the shift of the benzyl group occurs *via* a pericyclic mechanism while that of the benzhydryl group proceeds *via* a dissociative radical-pair mechanism. The conclusion drawn from the different activation volumes is in full accord with the stereochemical finding of retention of configuration in the PhCHD migration and the observation of a CIDNP (Chemically Induced Dynamic Nuclear Polarization) effect in the Ph$_2$CH migration [124].

In the transition state of the electrocyclization of (Z)-1,3,5-hexatriene to 1,3-cyclohexadiene (Scheme 3.19, entry (1)) a new six-membered ring develops analogously to that of the Cope rearrangement. The electrocyclization is accelerated by an increase in pressure. The activation volume determined at different temperatures listed in Scheme 3.19 is about -10 cm^3 mol^{-1} and corresponds to those of the Cope rearrangements (Scheme 3.17). In the temperature range of about 20 °C investigated, the activation volume does not show any significant temperature-dependence within the experimental error limits of ± 1 cm^3 mol^{-1}. From the volume data shown in Scheme 3.19 the packing coefficient of the transition state is calculated to equal approximately that of the cyclic product and differs substantially from that of the acyclic reactant. This result provides good evidence for the assumption used in the explanation of the pressure effect on pericyclic reactions. From the complete volume data set of the (Z)-1,3,5-hexatriene → 1,3-cyclohexadiene isomerization the activation volume of the reverse reaction, the electrocyclic ring-opening, can be extrapolated to be slightly positive ($\Delta V^{\neq} = +4$ cm^3 mol^{-1} (1,3-cyclohexadiene → (Z)-1,3-hexatriene)). The electrocyclic ring-opening of heavily substituted cyclobutene derivatives, however, shows negative activation volumes of different size depending on the substitution pattern (Scheme 3.19, entry (2)). This result indicates that

	Reaction	$\Delta V^{\neq}$	Ref.
(1)	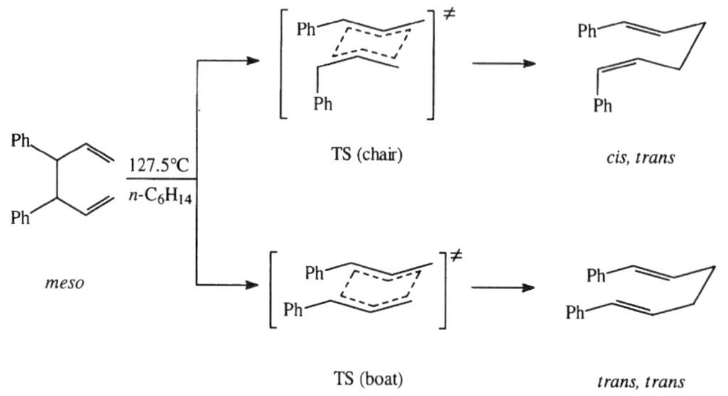 Ph Ph meso 127.5°C n-C6H14 TS (chair) cis, trans TS (boat) trans, trans	$-(13.3 \pm 0.6)$ $-(8.8 \pm 0.7)$	[119]
(2)	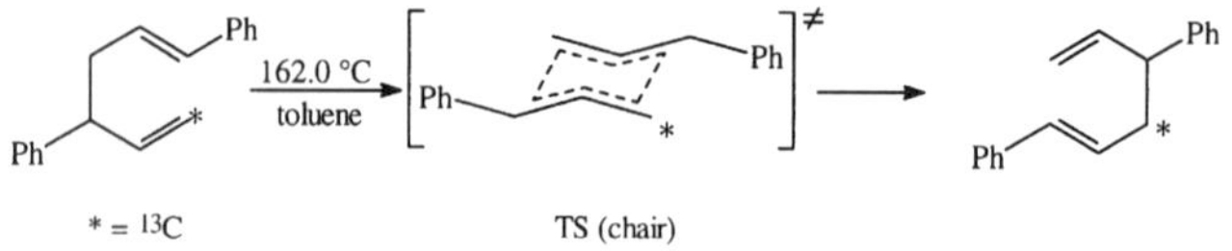 Ph 162.0 °C toluene * = 13C TS (chair)	$-(9.1 \pm 0.5)$	[120]
(3)	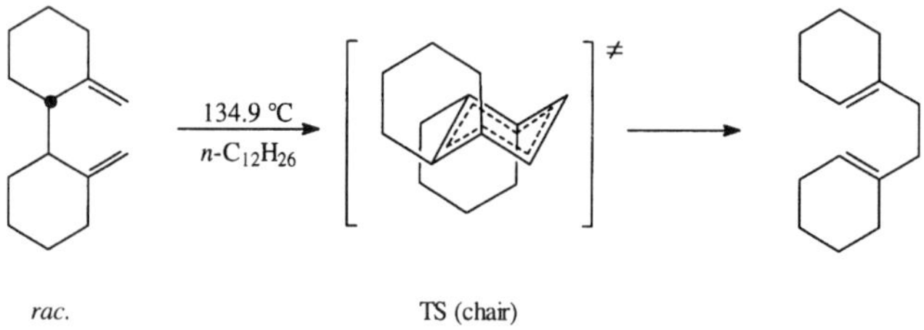 134.9 °C n-C12H26 rac. TS (chair)	$-(12.8 \pm 0.8)$	[121]
(4)	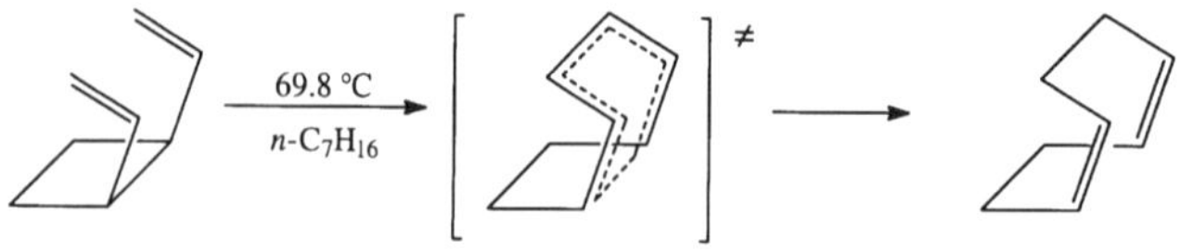 69.8 °C n-C7H16	$-(13.4 \pm 0.6)$	[120]
(5)	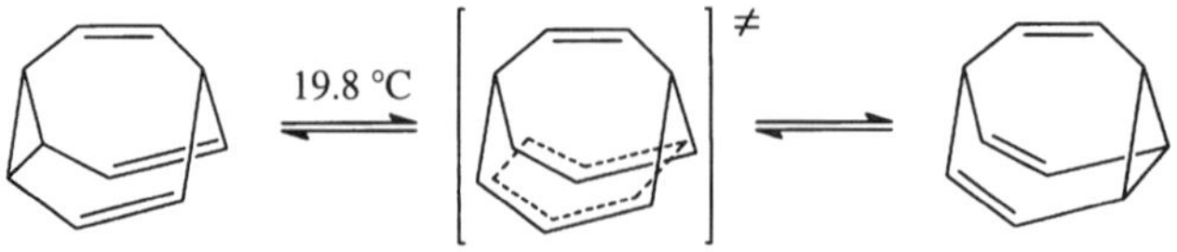 19.8 °C	-0.5	[122]

Scheme 3.17

3.5. MOLECULAR REARRANGEMENTS

		T (°C)	Solvent	$\Delta V_T^{\neq\,a}$	Ref.
(1)	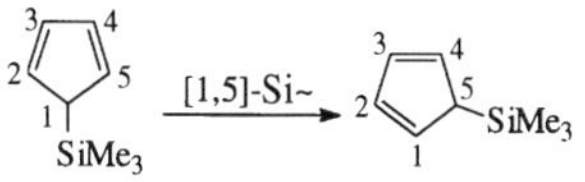 cyclopentadiene-SiMe₃ $\xrightarrow{[1,5]\text{-Si}\sim}$ cyclopentadiene-SiMe₃	68	Benzene-Freon	-12.5	[122]
(2)	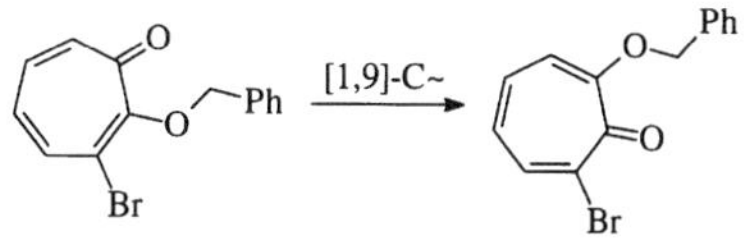 tropone-OCH₂Ph, Br $\xrightarrow{[1,9]\text{-C}\sim}$ tropone rearranged	< 130	i-Pr–Ph	-11.1	[125]
(3)	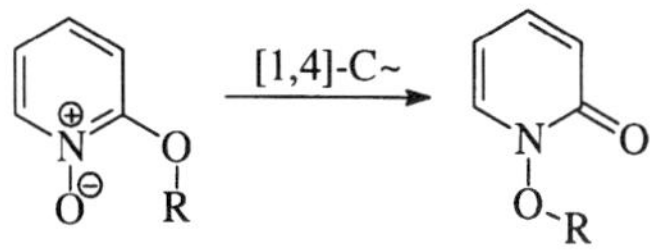 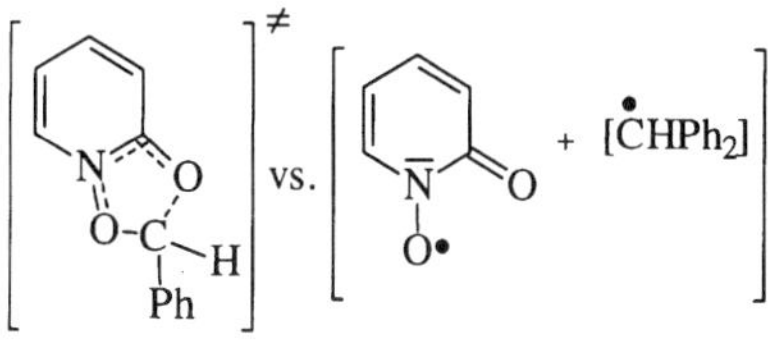 pyridine N-oxide-OR $\xrightarrow{[1,4]\text{-C}\sim}$ N-OR pyridone; R = PhCH₂; R = Ph₂CH; [concerted TS] vs. [radical pair + ĊHPh₂]	100	Diglyme	-30 / $+10$	[123]
(4)	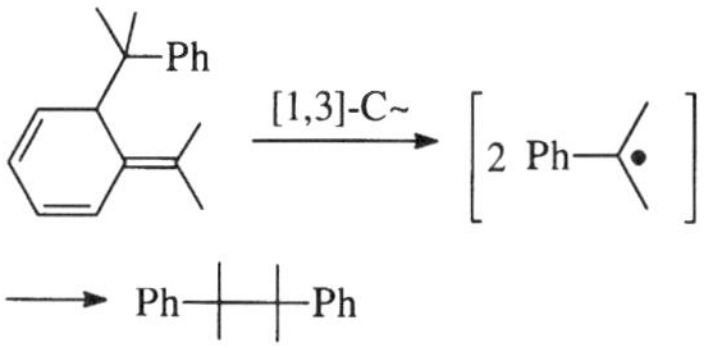 cyclohexadiene with CPh(iPr) $\xrightarrow{[1,3]\text{-C}\sim}$ [2 Ph–C•] $\longrightarrow$ Ph–C–C–Ph	20	Cyclohexane	$+6$	[22d]
(5)	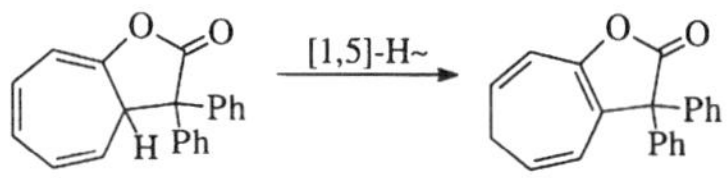 cycloheptatriene-lactone HPh,Ph $\xrightarrow{[1,5]\text{-H}\sim}$ rearranged Ph,Ph	130	n-Bu₂O	-2.2	[126]
(6)	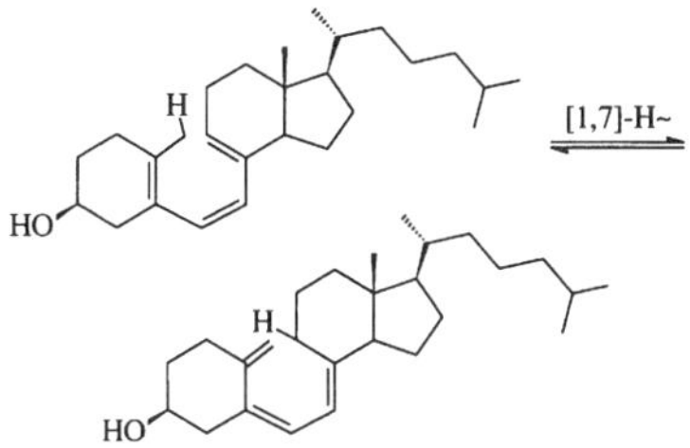 cholesterol-type structure $\xrightleftharpoons{[1,7]\text{-H}\sim}$ rearranged structure	20		<-5	[127]

[a] In cm³ mol⁻¹.

Scheme 3.18

146   3. EFFECT OF PRESSURE ON ORGANIC REACTIONS

	T [°C]	$\Delta V_T^{\neq\,a}$	$\Delta V_T^{\,a}$	Θ	Ref.
(1) Benzene dimerization transition state (2.24 Å) → cyclohexadiene 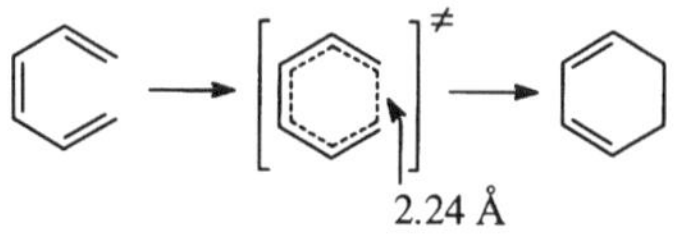	101.2	−9.8	−14.4	0.68	[136]
	108.1	−10.8	−14.8	0.73	
	117.5	−10.9	−15.2	0.72	
	122.4	−10.3	−15.4	0.67	

$V_W^{\,a}$	61.2	58.6	57.0	[131]
$V^{\,b}$	118.7	107.9	103.9	
η	0.5156	0.5431	0.5486	

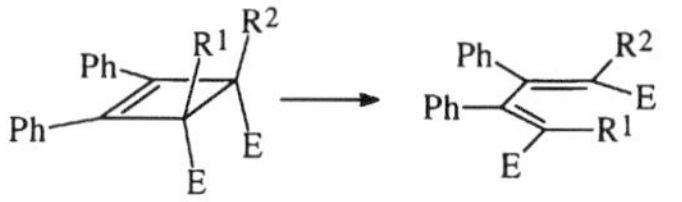

(2) Ph-substituted cyclopropene rearrangement [132], [133]

$E = CO_2Me$

	T [°C]	$\Delta V_T^{\neq}$	ΔV_T	Θ	
$R^1 = H, R^2 = H$		−1 to −2			
$R^1 = Me, R^2 = Me$		−7			
$R^1 = H, R^2 = Me$	70	−12.7		−23[c]	

(3) Dewar benzene → benzene; hexamethyl Dewar benzene → hexamethylbenzene; tetracyanobenzvalene → tetracyanobenzene

	T [°C]	$\Delta V_T^{\neq}$	ΔV_T	Ref.
 Dewar benzene → benzene	20	+5		[128]
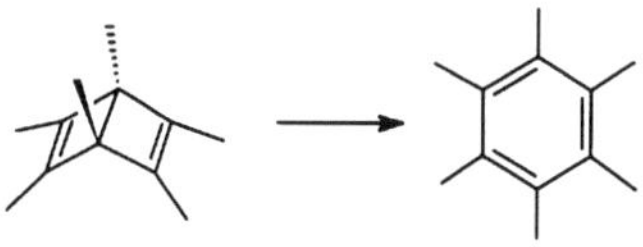 Hexamethyl Dewar benzene → hexamethylbenzene	140	−12	−22	[128], [129]
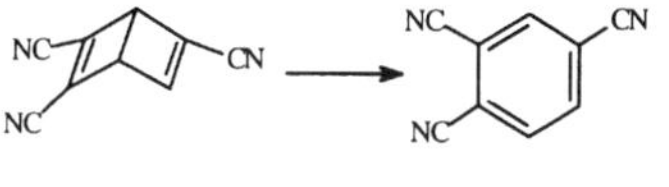 Tetracyanobenzvalene → tetracyanobenzene	51.3	≈ 0		[134]

[a] In $cm^3\,mol^{-1}$; the reaction volume ΔV_T was calculated from the partial molar volumes V_T determined by the temperature dependence of the densities of reactant or product according eq. (6);
[b] 108.1 °C in toluene;
[c] In toluene.

Scheme 3.19

other effects such as an increase in steric crowding contribute to the activation volume, over compensating the effect of ring-opening. A clear cut example is the ring-opening of Dewar benzene to benzene. The isomerization of the parent Dewar benzene is retarded by pressure ($\Delta V^{\neq} = +5\,cm^3\,mol^{-1}$) [128] whereas the isomerization of the hexamethyl derivative is accelerated by pressure ($\Delta V^{\neq} = -12\,cm^3\,mol^{-1}$) [129]. The negative volume of activation of the latter isomerization can again be explained by steric crowding of the six methyl groups which is larger in the planar hexamethylbenzene than in the non-planar precursor thus over compensating the volume-increasing effect of ring-opening.

In intramolecular Diels-Alder reactions, two new rings are formed. There are examples of relatively large pressure-induced accelerations which can be exploited for preparative purposes (Scheme 3.20, entries (1) to (3)). These compounds, without exception, contain polar groups and are not, therefore, very suitable for the analysis of the relation between pressure effect and ring formation. The strong solvent dependence of the activation volume of the intramolecular Diels-Alder reaction shown in Scheme 3.20, entry (2), was considered to be largely the result of the strongly solvent dependent partial molar volume of the reactant, $V_{(reactant)}$, whereas the partial molar volume of the transition state ($V^{\neq} = \Delta V^{\neq} + V_{(reactant)}$) appears to be almost unaffected by the solvent. The activation volumes for the intramolecular Diels-Alder reactions in the pure hydrocarbon systems (Scheme 3.20, entries (4) and (5)) were both found to be $-24.8 \text{ cm}^3 \text{mol}^{-1}$ or -37.6 and $-35.0 \text{ cm}^3 \text{mol}^{-1}$, respectively. The absolute values here are approximately twice as large as or even larger than those observed for the Cope rearrangements or the electrocyclization of 1,3,5-hexatriene to 1,3-cyclohexadiene. From this it was estimated that each additional five- or six-membered ring formed in the rate-determining step of reactions contributes about -10 to $-15 \text{ cm}^3 \text{mol}^{-1}$ to the activation volume.

A particularly instructive example is the thermolysis of (Z)-1,3,8-nonatriene in which an intramolecular Diels-Alder reaction competes with a sigmatropic [1,5] hydrogen shift (Scheme 3.21). The use of high pressure reveals a reversal of the selectivity. At 150 °C and 1 bar the [1,5] hydrogen shift is preferred possessing a monocyclic transition state. At 7.7 kbar, on the other hand, the intramolecular Diels-Alder reaction is preferred due to its bicyclic transition state.

3.6 ENE REACTIONS

Using the concept introduced in the previous section the pressure effect on Alder ene reactions, associated with the already discussed sigmatropic [1,5] H shift, can be explained. The reactions show a powerful pressure-induced acceleration. In many cases (Scheme 3.22, entries (1) and (2)) the absolute values of activation volumes were found to be larger than those of the corresponding reaction volumes and, hence, the ratio ($\Delta V^{\neq} : \Delta V$) to be larger than unity ($\Theta > 1$) comparable to the Diels-Alder reactions listed in Table 3.6. This result was taken as an evidence of the pericyclic nature of the ene reactions.

The van der Waals volumes of reactants, transition state structure, and product were calculated for the prototype of the ene reaction between propene and ethene (Scheme 3.23). The structural parameters necessary for this calculation were taken in the case of reactants and products from force-field calculations (MM2) and in the case of the transition state structure from *ab initio*

3. EFFECT OF PRESSURE ON ORGANIC REACTIONS

	Solvent	$\Delta V_T^{\neq\,a}$	$\Delta V_{25}^{\neq}$	ΔV_{20}^{a}	Ref.
(1) 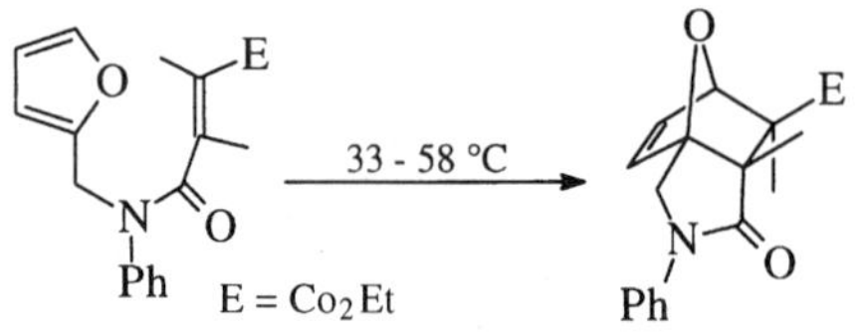 furfuryl N-Ph acrylamide (E = CO$_2$Et), 33–58 °C → bicyclic adduct		−25		−23	[135]
(2) 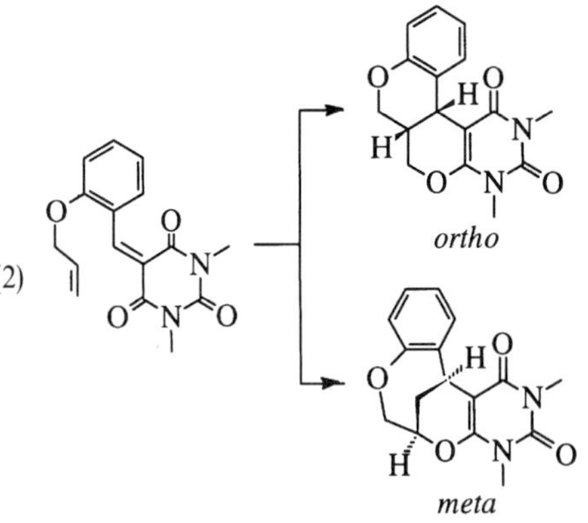 ortho product	CH$_2$Cl$_2$	−33.1	−23.6	−30	
	THF	−34.2	−24.4	−35	
	Ph–Me	−17.0	−9.5	−18	
	MeCN	−13.4	−12.5	−15	[136]
meta product	CH$_2$Cl$_2$	−32.1			
	THF	−32.7			
	Ph–Me	−15.2			
	MeCN	−12.1			

T = 110 °C; *ortho* : *meta* = (3.8 - 6.8) : 1

	Solvent	$\Delta V_T^{\neq\,a}$			Ref.
(3) 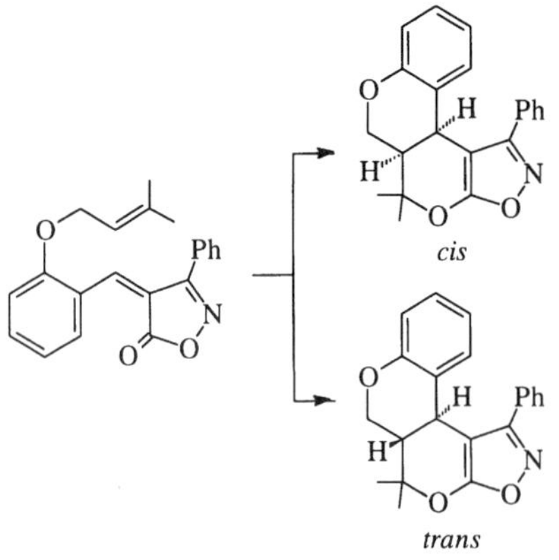 cis	CH$_2$Cl$_2$	−19.4			[137]
trans	CH$_2$Cl$_2$	−17.9			

T = 70 °C; *cis* : *trans* = (5.5 - 6.8) : 1

Scheme 3.20

calculations [41]. In this case the van der Waals volume of the transition state structure was calculated to be smaller than that of the product [142] and, hence, the ratio ($V_W^{\neq} : V_W$) is larger than unity. With the reasonable assumption, that the packing coefficient of the transition structure is, at least, equal to but probably larger than that of the product, one can predict that the

3.6. ENE REACTIONS

	Solvent	$\Delta V_T^{\neq\,a}$	$\Delta V_{25}^{\neq}$	ΔV_{20}^{a}	Ref.
(4) cis	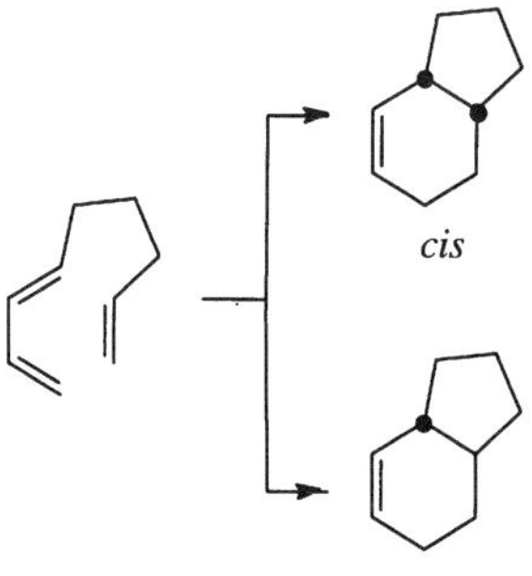 n-C$_6$H$_{14}$	-24.8			[138]
(4) trans	n-C$_6$H$_{14}$	-24.8			

T = 153.2 °C; cis : trans = 3.0 : 1

	Solvent	$\Delta V_T^{\neq\,a}$	$\Delta V_{25}^{\neq}$	ΔV_{20}^{a}	Ref.
(5) cis	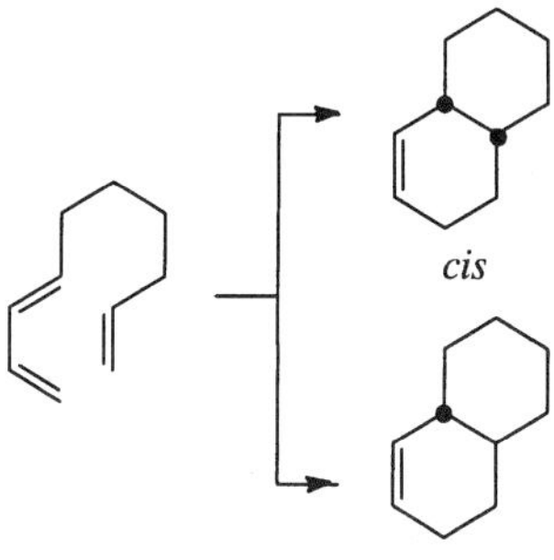 n-C$_7$H$_{16}$	-37.6			[130]
(5) trans	n-C$_7$H$_{16}$	-35.0			

T = 172.5 °C; cis : trans = 1.2 : 1

[a] In cm^3 mol^{-1}; the $\Delta V_{20}^{\neq}$ values were extrapolated from the $\Delta V_T^{\neq}$ values by using the El'yanov equation; see footnote a of Table 3.5.

Scheme 3.20 Continued.

ratio between activation and reaction volume ($\Delta V^{\neq} : \Delta V$) is larger than unity, too.

In the case of ene reactions between cycloalkenes or alkenes and diethyl azodicarboxylate (DEAD) (e.g., Scheme 3.22, entry (3)), the ratio was found to be smaller than unity ($\Theta < 1$). This result can be considered as an indication of a stepwise process in which a pericyclic transition state is not involved. Stephenson and Mattern, however, observed that the ratios S/R and k_H/k_D in the ene reaction with DEAD as enophile shown in Scheme 3.22, entry (4), were roughly equal, which was explained in terms of a concerted ene reaction. To explain this discrepancy Jenner and coworkers proposed a mechanism comparable to the formal ene reactions of alkenes with singlet oxygen or triazolindiones as

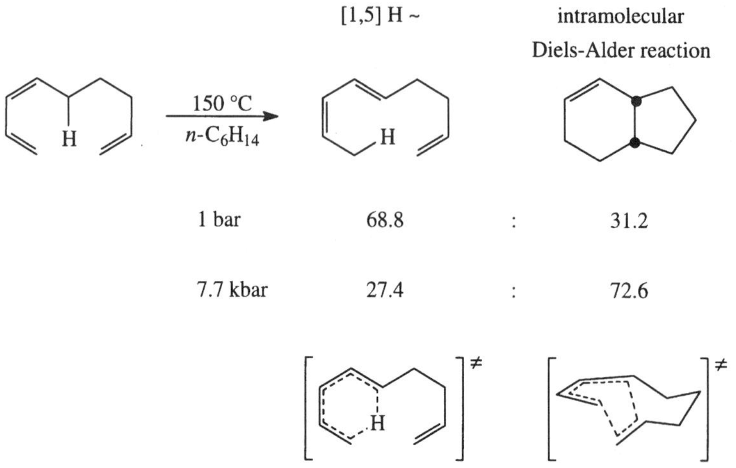

Scheme 3.21 [138].

enophiles, in which the first step is the formation of three-membered rings between the alkenes and one center of the enophile prior to the hydrogen transfer. To classify the mechanism of the ene reaction with DEAD it woud be desirable to obtain the pressure dependence of the rate of the reaction studied by Stephenson and Mattern [141].

3.7 THE RELATIONSHIP BETWEEN ACTIVATION OR REACTION VOLUME AND RING-SIZE

The investigation of the pressure effect on the rearrangement and cleavage of *trans*-1,2-divinyl cyclobutane showed that the volume of reaction depends not only on the number but also on the size of the newly forming rings. In contrast to the Cope rearrangement of *cis*-1,2-divinylcyclobutane (Scheme 3.17, entry (4)) the competitive reactions of *trans*-1,2-divinylcyclobutane leading to 4-vinylcyclohexene, 1,5-cyclooctadiene and 1,3-butadiene are slowed by pressure and the volumes of activation become positive, consistent with the hypothesis of the closure of the cyclobutane ring leading to an acyclic diradical intermediate (Scheme 3.24). Because the product ratio shows no significant pressure dependence, the activation volumes of the individual reactions are essentially equal. It was concluded here that in the diradical intermediate neither ring closure reactions nor cleavage are product-determining in contrast to the [2 + 2] cycloaddition shown in Scheme 3.14. Probably pressure-independent rotations about C–C bonds in the diradical determine the distributions among the three products

The volumes of reaction determined for the isomerization of *trans*-1,2-divinylcyclobutane to 4-vinylcyclohexene or 1,5-cyclooctadiene, in which six- and

3.7. ACTIVATION OR REACTION VOLUME AND RING-SIZE RELATIONSHIP

Scheme 3.22

	$\Delta V_T^{\neq\,a}$	$\Delta V_{25}^{\neq\,a}$	ΔV_{25}^{a}	Θ^a	Ref.
(1)	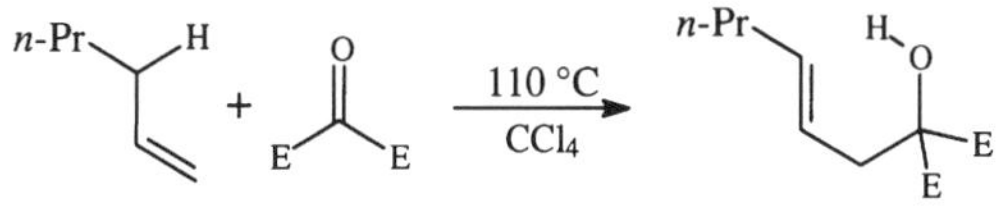 −39.0	−28.4	−27.0	1.05	[139]
(2)	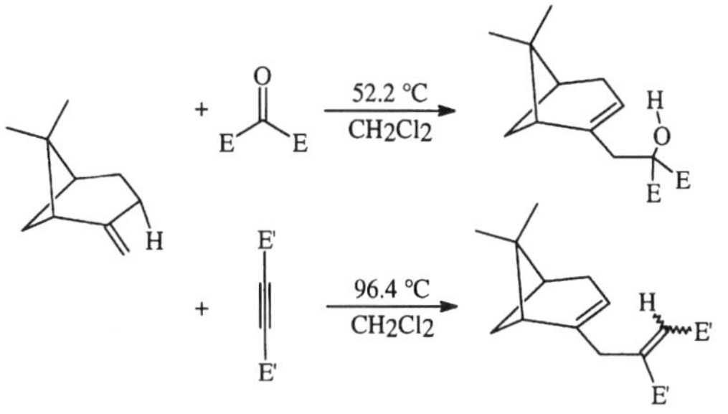 −35.0	−31.3	−29.4	1.06	[140]
	−52.0	−39.6	−35.4	1.12	
(3)	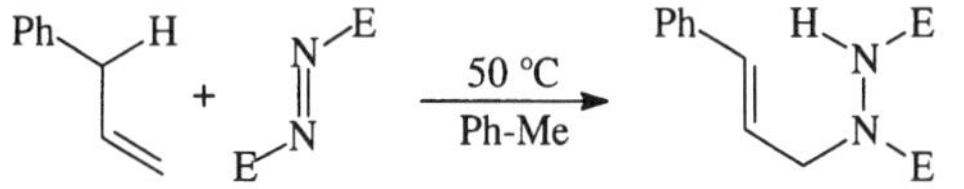	−27.0	−35.8	0.75	[140]
(4)					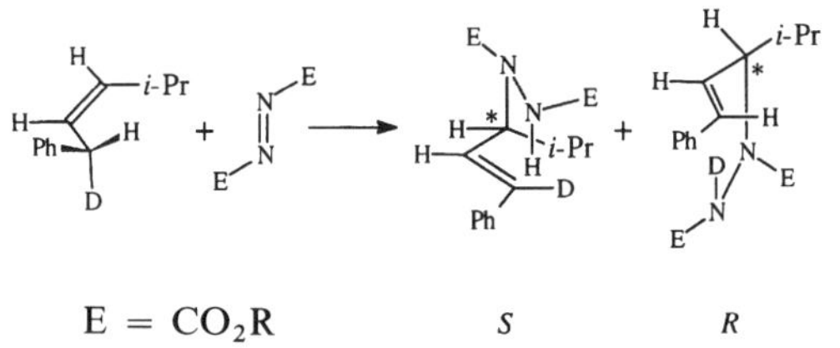 [141]

$E = CO_2R$

$S/R \approx k_H/k_D \approx 3$

[a] See footnotes a and b of Table 3.5.

eight-membered rings are formed, respectively, at the expense of a four-membered ring, were found to be substantially negative. This observation of the decrease in volume from the four- to the six- or eight-membered ring indicates that the activation volumes of cyclizations also depend on the size of the newly forming ring. The van der Waals volumes of the ground and transition state structures do not differ from each other appreciably and therefore do not explain the observed activation and reaction volumes.

The volumes of reaction calculated for the hypothetical cyclizations of n-alkenes to the corresponding cycloalkanes by the use of experimentally observed partial molar volumes (Scheme 3.25) [130] confirm the trend derived from the

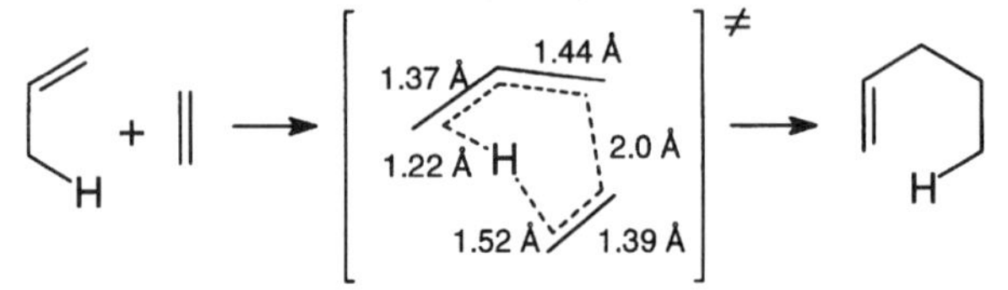

Scheme 3.23

				$\Delta V^{\neq a}$	ΔV^a	Θ^b
V_w^a	35.1	25.6	54.6	55.8	−6.1	−5.3	1.15
V^a	76.2	59.9	94–105	108.6	−(42−31)	−27.5	1.5−1.13
η^c	0.4606	0.4273	0.58−0.52	0.5193

a In cm³ mol⁻¹;
b $\Theta = \Delta V^{\neq}/\Delta V$;
c $\eta = V_w/V$, for the calculation of the partial molar volume of the transition state ($V^{\neq} = V_w^{\neq}/\eta^{\neq}$) the unknown packing coefficient $\eta^{\neq}$ was assumed to be within the range of the ene product and cyclohexane.

Scheme 3.23 [41].

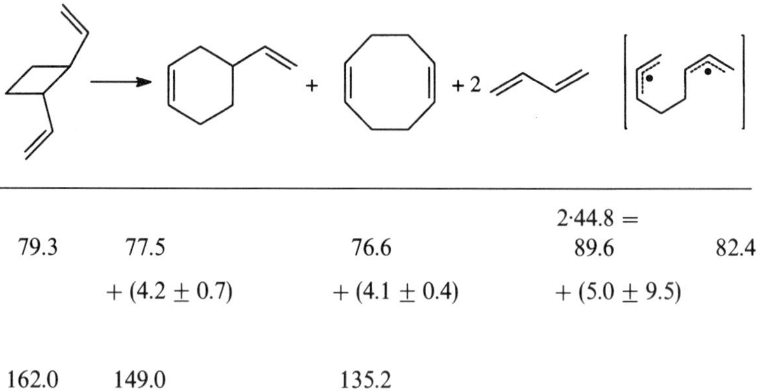

Scheme 3.24

V_w^a:	79.3	77.5	76.6	2·44.8 = 89.6	82.4
$\Delta V^{\neq a}$ (159.6 °C):		+(4.2 ± 0.7)	+(4.1 ± 0.4)	+(5.0 ± 9.5)	
V^a	162.0	149.0	135.2		
$\eta = V_w/V$	0.4895	0.5201	0.5666	0.4958	
V^a (20.0 °C)	140.0	130.4	122.6	2·83.2 = 166.4	
ΔV^a:		130.4 − 140.0 = −9.6	122.6 − 140.0 = −17.4	166.4 − 140.0 = +26.4	

a In cm³ mol⁻¹.

Scheme 3.24 [143].

3.7. ACTIVATION OR REACTION VOLUME AND RING-SIZE RELATIONSHIP

	$\Delta V_W{}^a$	$\Delta V{}^a$	ΔH^b	ΔS^c	ΔG^b
 alkene → triangle	−1.72	−5.5	7.86	−7.0	9.95
 alkene → square	−2.50	−6.56	6.43	−10.3	9.50
 alkene → pentagon	−3.77	−14.72	−13.46	−13.1	−9.56
 alkene → hexagon	−4.38	−16.45	−19.47	−21.0	−13.21
 alkene → heptagon	−4.71	−21.2	−13.41	−19.6	−7.57
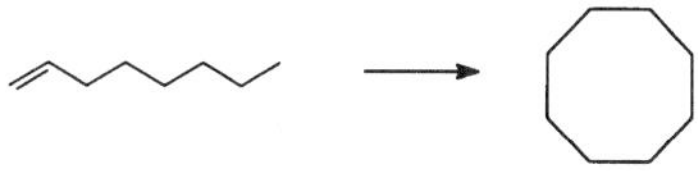 alkene → octagon	−4.89	−25.63	−9.88	−18.78	−4.28
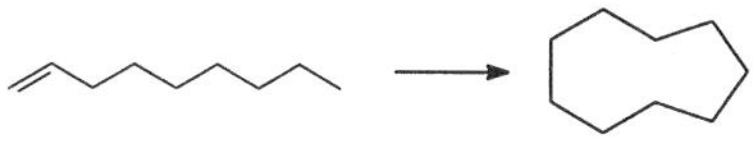 alkene → 9-ring	−4.65	−30.93			
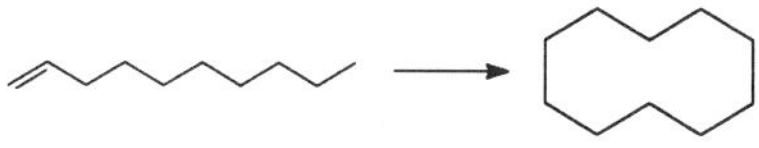 alkene → 10-ring	−4.58	−32.28			
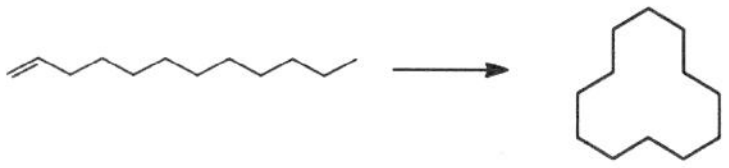 alkene → 11-ring	−4.70	−32.81			
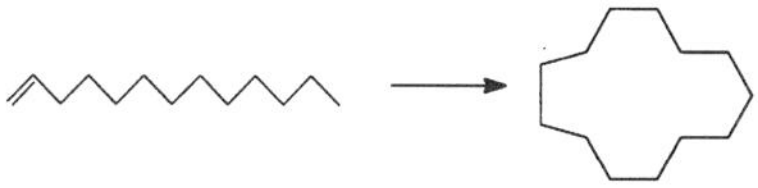 alkene → 12-ring	−4.74	−32.31			
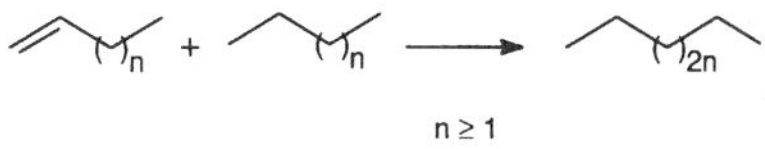 alkene + alkene → 2n ($n \geq 1$)	−4.6	−27.56			

[a] $cm^3\,mol^{-1}$. V (n-alkene) calculated by the use of Exner increments. V (cycloalkane) determined from density measurements in n-hexane.
[b] $kcal\,mol^{-1}$.
[c] $cal\,mol^{-1}\,K^{-1}$.

Scheme 3.25 [130].

ring enlargements shown in Scheme 3.24. The volumes of reaction decrease continuously up to the formation of the ten-membered ring and, then, seem to be constant for the larger rings, whereas the van der Waals volumes of reaction are approximately equal, with the exceptions of the cyclopropane, cyclobutane, and cyclopentane formations and, therefore, independent of the ring-size. It is interesting to note that the ΔV values do not correlate with any of the other thermodynamic parameters such as enthalpy, entropy, or Gibbs enthalpy of reaction. Provided that the activation volumes depend similarly on the ring-size, the formation of larger rings should be dramatically accelerated by pressure. The intramolecular Diels-Alder reactions of (E)-1,3,8-nonatriene and (E)-1,3,9-decatriene, in which either new five- and six-membered rings or two new six-membered rings are formed, seem to be the first examples of the validity of this assumption (Scheme 3.25, entries (4) and (5)). Furthermore, this ring size effect explains why the activation volumes for the formation of the three-membered rings in the cheleotropic reactions shown in Scheme 3.8 and for the five-membered rings in the 1,3-dipolar cycloadditions (Scheme 3.5) are substantially less negative than those for the formation of six-membered rings in the Diels-Alder reactions (Tables 3.5 and 3.6). The comparison between the cyclization of a n-alkene and the intermolecular addition of n-alkenes to a n-alkane shows that in the case of the larger rings the intramolecular cyclization should even be

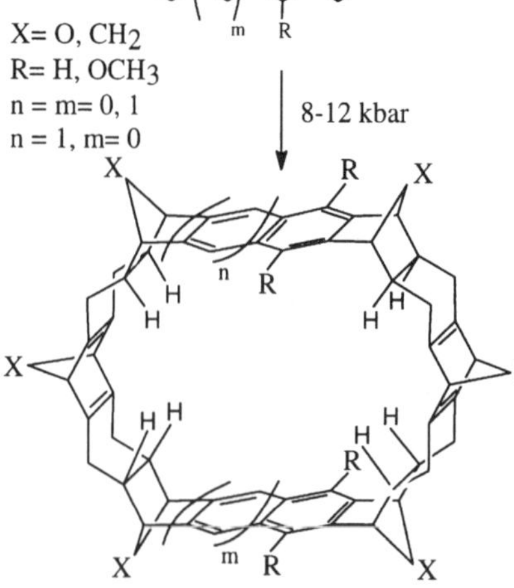

Scheme 3.26

favored over the intermolecular addition. Thus, the application of high pressure should be a favorable method for the synthesis of medium- or large-sized rings. The high-pressure synthesis of macrocycles reported by Jurczak and coworkers [144] and discussed in the following chapter may benefit *inter alia* from this effect. Another example is the synthesis of macrocycles by the use of repetitive Diels-Alder reactions which only succeed at high pressure, as shown in Scheme 3.26 [145, 146]. Obviously, both consecutive reactions, the inter- and intramolecular Diels-Alder reactions, are accelerated by pressure, so that only the desired macrocycles are observed as products.

3.8 CONCLUDING REMARKS

The packing coefficient, $\eta = V_W/V$, has been demonstrated to be a valuable tool to explain the effect of pressure on many pericyclic reactions. The finding that the η value of cyclic structures is larger than that of the corresponding acyclic structures explains the preference of pericyclic reactions over stepwise reactions at high pressure and the negative activation volumes of many pericyclic rearrangements. The magnitude of η depends on the number and size of the newly forming rings. This explains for example why intramolecular Diels-Alder reactions involving a bicyclic transition state are favored over rearrangements involving monocyclic transition states.

ACKNOWLEDGEMENTS

We are grateful to the Deutsche Forschungsgemeinschaft, Ministerium für Wissenschaft und Forschung des Landes Nordrhein-Westfalen, and Fonds der Chemischen Industrie for financially support of our work. F.-G. K. thanks the coworkers mentioned in the references for their committed and skillful collaboration and Mrs. I. Reiter for her skilled assistance with the preparation of this manuscript.

3.9 REFERENCES

[1] Isaacs, N. S., Liquid Phase High Pressure Chemistry, John Wiley, Chichester, 1981.
[2] le Noble, W. J. (ed.), Organic High Pressure Chemistry, Elsevier, Amsterdam, 1988.
[3] Matsumoto, K., Morrin Acheson, R. (eds.), Organic Synthesis at High Pressure, John Wiley, New York, 1991.
[4] Asano, T., le Noble, W. J., Chem. Rev., 78 (1978) 407.
[5] van Eldik, R., Asano, T., le Noble, W. J., Chem. Rev., 89 (1989) 549.
[6] le Noble, W. J., Kelm, H., Angew. Chem., 92 (1980) 887; Angew. Chem. Int. Ed. Engl., 19 (1980) 841.

[7] le Noble, W. J., Chem. uns. Zeit, 17 (1983) 152.
[8] Jenner, G., J. Chem. Soc., Faraday Trans. I, 81 (1985) 2437.
[9] Matsumoto, K., Sera, A., Uchida, T., Synthesis, (1985) 1.
[10] Matsumoto, K., Sera, A., Synthesis, (1985) 999.
[11] le Noble, W. J., (van Eldik, R., Jonas, J. (eds.)), High Pressure Chemistry and Biochemistry, pp 279–293 and 295–310, Reidel, Dordrecht, 1987.
[12] Klärner, F.-G., Chem. uns. Zeit, 23 (1989) 53.
[13] Klärner, F.-G., Ruster, V., Zimny, B., Hochstrate, D., High Pressure Res., 7 (1991) 133.
[14] Isaacs, N. S., Tetrahedron, 47 (1991) 8463.
[15] Buback, M., Angew. Chem., 103 (1991) 658; Angew. Chem. Int. Ed. Engl., 30 (1991) 641.
[16] Frey, U., Helm, L., Merbach, A. E., High Pressure Res., (1990) 237; Frey, U., Elmroth, S., Moullet, B., Elding, L. I., Merbach, A. E., Inorg. Chem., 30 (1991) 5033.
[17] a) Nova Swiss, Nova Werke AG, Vogelsangstraße 24, CH-8307 Effretikon, Switzerland; b) Bernd Dieckers GmbH, Meß- und Prüfanlagen-Technik, Hagwinkel 46, D-47877 Willich-Neersen, Germany; c) Andreas Hofer Hochdrucktechnik GmbH, Friedrich-Freye-Straße 56-61, D-45481 Mülheim an der Ruhr, Germany; d) Unipress Equipment, Division High Pressure Research Center, Polish Academy of Science, ul. Sokolowska 29/37, 04-142 Warszawa, Poland.
[18] Exner, O., Empirical calculations of molar volumes, chapter 2 in ref. [2].
[19] Franklin, J. L., Ind. Eng. Chem., 41 (1949) 1070.
[20] Benson, S. W., Thermochemical Kinetics, 2nd ed., John Wiley, New York, 1976.
[21] le Noble, W. J., (van Eldik, R., Jonas, J. (eds.)), High Pressure Chemistry and Biochemistry, Reidel, Dordrecht, 1987.
[22] a) Olsen, H., Snyder, J. P., J. Am. Chem. Soc., 100 (1978) 285. The authors derived the volumes of activation for the N_2 elimination of various bicyclic and tricyclic azo compounds from the solvent dependence by using the solubility parameter δ. A direct comparison with activation volumes determined from the pressure-dependence is impossible here since activation volumes have been determined by the use of the latter method only for the N_2 elimination of acyclic azo compounds. b) Neumann Jr., R. C., Binegar, G. A., J. Am. Chem. Soc., 105 (1983) 134; c) Neumann Jr., R. C., Lockyer Jr., G. D., J. Am. Chem. Soc., 105 (1983) 3982; d) Neuman Jr., R. C., Amrich Jr., M. J., J. Org. Chem., 45 (1980) 4629.
[23] Swieton, G., von Jouanne, J., Kelm, H., Huisgen, R., J. Org. Chem., 48 (1983) 1035.
[24] Asano, T., Okada, T., J. Phys. Chem., 88 (1984) 238.
[25] The residual sum of squares and, hence, the volumes of activation in the least squares calculation according to eqs. (10) or (11) can be performed *e.g.* by the use of Mathcad 5.0, MathSoft, Inc.
[26] Tietze, L. F., Hübsch, T., Ott, C., Kuchta, G., Buback, M., Liebigs Ann., (1995) 1.
[27] Jenner, G., Angew. Chem., 87 (1975) 186; Angew. Chem. Int. Ed. Engl., 14 (1975) 137.
[28] Reviews on the synthetic application of pressure-induced Diels-Alder reactions are found in ref. [2] chapter 11 (Jurczak, J.: Synthesis); ref. [3]: chapter 9 and 10 (Ibata, T.: Diels-Alder rections of acyclic and alicyclic dienes and Diels-Alder reactions of heterocyclic dienes, respectively); ref. [8]; ref. [10].

3.9. REFERENCES

[29] Rimmelin, J., Jenner, G., Tetrahedron, 30 (1974) 3081. A recent measurement of the pressure and temperature-dependence of the electrocyclic ring-closure of Z-1,3,5-hexatriene to 1,3-cyclohexadiene in the range of 200 to 2500 bar and 100 to 125 °C does not show a significant temperature-dependence of the activation volume (see Scheme 3.19). Diedrich, M. K., Klärner, F.-G., unpublished results.

[30] Seguchi, K., Sera, A., Maruyama, K., Bull. Chem. Soc. Jpn., 47 (1974) 2252.

[31] Jenner, G., Rimmelin, J., Tetrahedron Lett., 21 (1980) 3039.

[32] a) Grieger, R. A., Eckert, C. A., J. Am. Chem. Soc., 92 (1970) 2918, 7149; Trans. Faraday Soc., 66 (1970) 2579; b) McCabe, J. R., Eckert, C. A., Ind. Eng. Chem. Fundamentals, 13 (1973) 168.

[33] Brun, C., Jenner, G., Tetrahedron, 28 (1972) 3113.

[34] Jenner, G., New. J. Chem., 15 (1991) 897.

[35] a) Beck, K., Hünig, S., Klärner, F.-G., Kraft, P., Artschwager-Perl, U., Chem. Ber., 120 (1987) 2041; b) El'yanov, B. S., Gonikberg, E. M., J. Chem. Soc., Faraday Trans. 1, 75 (1979) 172; c) El'yanov, B. S., Vasylvitskaya, E. M., Rev. Phys. Chem. Jpn., 50 (1980) 169.

[36] Grieger, R. A., Eckert, C. A., Ind. Eng. Chem. Fundamentals, 10 (1971) 369.

[37] McCabe, J. R., Eckert, C. A., Ind. Eng. Chem. Fundamentals, 13 (1974) 168.

[38] Rimmelin, J., Jenner, G., Abdi-Oskoui, H., Bull. Soc. Chim. Fr., (1977) 341.

[39] Jenner, G., Papadopoulos, M., Rimmelin, J., J. Org. Chem., 48 (1983) 748.

[40] George, A. V., Isaacs, N. S., J. Chem. Soc., Perkin Trans. II, (1985) 1845.

[41] a) Houk, K. N., Li, Y., Evanseck, J. D., Angew. Chem., 104 (1992) 711; Angew. Chem. Int. Ed. Engl., 31 (1992) 682; b) Li, Y., Houk, K. N., J. Am. Chem. Soc., 115 (1993) 5414.

[42] According to recent quantum mechanical calculations the importance of secondary orbital interactions which have also been frequently used to explain the *endo* diastereoselectivity of Diels-Alder reactions seems to be questionable and to be reserved for special cases such as the addition of cyclopropene to various dienes. Karcher, T., Sicking, W., Sauer, J., Sustmann, R., Tetrahedron Lett., 33 (1992) 8027; Sustmann, R., Sicking, W., Tetrahedron, 48 (1992) 10293; Apeloig, Y., Matzner, E., J. Am. Chem. Soc., 117 (1995) 5375.

[43] Seguchi, K., Sera, A., Maruyama, K., Tetrahedron Lett., 14 (1973) 1585.

[44] Jenner, G., Papadopoulos, M., le Noble, W. J., Nouveau J. de Chimie, 7 (1983) 687.

[45] Sauer, J., Sustmann, R., Angew. Chem., 92 (1980) 773; Angew. Chem. Int. Ed. Engl., 19 (1980) 779.

[46] Jenner, G., Papadopoulos, M., Tetrahedron Lett., 23 (1982) 4333.

[47] Jenner, G., Papadopoulos, M., Nouveau J. de Chimie, 7 (1983) 463.

[48] Klärner, F.-G., Schröer, D., Chem. Ber., 122 (1989) 179.

[49] Ghosez, L., O'Donnell, M. J., Pericyclic Reactions of Cumulenes, (Marchand, A. P., Lehr, R. E. (eds.)), Pericyclic Reactions, Vol. II, Academic Press, New York, 1977.

[50] Isaacs, N. S., Rannala, E., J. Chem. Soc., Perkin Trans. II (1975) 1555.

[51] Swieton, G., von Jouanne, J., Kelm, H., Proc. 4th Int. Conf. High Pressure 1974, (1975) 652.

[52] Fleischmann, F., Kelm, H., Tetrahedron Lett., 14 (1973) 3773.

[53] Swieton, G., von Jouanne, J., Kelm, H., Huisgen, R., J. Chem. Soc., Perkin Trans. II (1983) 37.

[54] Huisgen, R., Acc. Chem. Res., 10 (1977) 117, 199.

[55] von Jouanne, J., Kelm, H., Huisgen, R., J. Am. Chem. Soc., 101 (1979) 151.

[56] Fleischmann, F., Kelm, H., Tetrahedron Lett., 14 (1973) 3773.

[57] le Noble, W. J., Mukhtar, R., J. Am. Chem. Soc., 97 (1975) 5938.

[58] Sustmann, R., Heterocycles, 40 (1995) 1; Padwa, A. (ed.), 1,3-Dipolar Cycloaddition Chemistry, Volumes 1 and 2, Wiley Interscience, New York, 1984.

[59] Huisgen, R., Mloston, G., Langhals, E., J. Am. Chem. Soc., 108 (1986) 6401.

[60] a) Zhulin, V. M., Makarov, Z. G., Krayushkin, M. M., Zhulavleva, E. B., Beskopyl'nyi, A. M., Dokl. Akad. Nauk SSSR, 280 (1985) 917; b) Zhulin, V. M., Zhuravleva, E. B., Dokl. Akad. Nauk SSSR, 290 (1986) 383.

[61] Yoshimura, Y., Osugi, J., Nakahara, M., Bull. Chem. Soc. Jpn., 56 (1983) 680.

[62] a) Yoshimura, Y., Osugi, J., Nakahara, M., J. Am. Chem. Soc., 105 (1983) 5414; b) Komornicki, A., Goddard, J. D., Schaefer III, H. F., J. Am. Chem. Soc., 102 (1980) 1763. Predicted 4-31 G SCF transition state structure for acetylene plus fulminic acid reaction.

[63] le Noble, W. J., Ojosipe, B. A., J. Am. Chem. Soc., 97 (1975) 5939.

[64] Takeshita, H., Sugiyama, S., Hatsui, T., J. Chem. Soc., Perkin Trans. II (1986) 1491; Sugiyama, S., Takeshita, H., Bull. Chem. Soc. Jpn., 60 (1987) 977.

[65] Dogan, B. M. J., Organische Reaktionen unter hohem Druck; der Druckeffekt auf Konkurrenzreaktionen, Ph.D. Thesis, Ruhr-Universität Bochum, 1984.

[66] The partial molar volumes of tropone, 1,3-butadiene, and the [6 + 4] cycloadduct were measured by Dogan [65] to be V = 88.8, 83.1 and 153.2 cm^3 mol^{-1}, respectively, at 21 °C and extrapolated to be 99.8, 87.9 and 154.1 cm^3 mol^{-1}, respectively, at 60 °C using the values measured by Takeshita et al. [64] for the reaction of tropone with 2,3-dimethylbutadiene.

[67] Takeshita, H., Nakashima, H., Sugiyama, S., Mori, A., Bull. Chem. Soc. Jpn., 61 (1988) 573.

[68] Li, Z., Mori, A., Takeshita, H., Nagano, Y., Chem. Express, 7 (1992) 213.

[69] Mori, A., Nukii, Y., Takeshita, H., Nozoe, T., Heterocycles, 35 (1993) 863.

[70] Woodward, R. B., Hoffmann, R., Angew. Chem., 81 (1969) 797; Angew. Chem. Int. Ed. Engl., 8 (1969) 781.

[71] Turro, N. J., Okamoto, M., Gould, I. R., Moss, R. A., Lawrynowicz, W., Hadel, L. M., J. Am. Chem. Soc., 109 (1987) 4973.

[72] Isaacs, N. S., Laila, A., J. Phys. Org. Chem., 7 (1994) 178.

[73] It is known that cyclic sulfinic esters undergo a mutual interconversion to the corresponding 1,3-dienes and SO_2. Jung, F., Molin, M., van den Elzen, R., Durst, T., J. Am. Chem. Soc., 96 (1974) 935; Heldeweg, R. F., Hogeveen, H., J. Am. Chem. Soc., 98 (1976) 2341.

[74] Stewart Jr., C. A., J. Am. Chem. Soc., 94 (1972) 635.

[75] Klärner, F.-G., Dogan, B. M. J., Ermer, O., Doering, W. v. E., Cohen, M. P., Angew. Chem., 98 (1986) 109; Angew. Chem. Int. Ed. Engl., 25 (1986) 108.

[76] Deiters, U., Klärner, F.-G., Krawczyk, B., Ruster, V., J. Am. Chem. Soc., 116 (1994) 7646.

[77] a) Roth, W. R., Scholz, B. P., Chem. Ber., 114 (1981) 3741; b) Bartmann, M., Zur Chemie des o-Chinodimethans und seiner Homologen, Ph. D. Thesis, Ruhr-Universität Bochum, 1980; c) ref. [65].

[78] Baran, J., Mayr, H., Ruster, V., Klärner, F.-G., J. Org. Chem., 54 (1989) 5016.

[79] The surprisingly small activation volume found for the formation of the very minor [4 + 4] cyclodimer **16** in the 1,3-butadiene dimerization seems to be inconsistent with a stepwise mechanism. The small ratio $\Theta = \Delta V^{\neq} : \Delta V = 0.54$ which is substantially smaller than that of the formation of **14**, $\Theta = 0.79$, and almost equal to that of the formation of **15**, $Q = 0.59$, may indicate that **13** is formed *via* a stepwise [4 + 4] cycloaddition passing through a (Z,Z)-configuration diradical intermediate.

[80] Little, J. C., J. Am. Chem. Soc., 87 (1965) 4020.

[81] Ruster, V., Organische Reaktionen unter hohem Druck; Konkurrenz von [2 + 2]- und [4 + 2]-Cycloadditionen, Diplomarbeit 1987; Konkurrierende Cycloadditionen unter hohem Druck., Ph. D. Thesis, Ruhr-Universität Bochum, 1991.

[82] Bartlett, P. D., Schmeller, K. E., J. Am. Chem. Soc., 90 (1968) 6077.

[83] Swenton, J. S., Bartlett, P. D., J. Am. Chem. Soc., 90 (1968) 2056.

[84] Kaufmann, D., de Meijere, A., Angew. Chem., 85 (1973) 151; Angew. Chem. Int. Ed. Engl., 12 (1973) 159.

[85] de Camp, M. R., Levin, R. H., Jones, M. J., Tetrahedron Lett., 15 (1974) 3575.

[86] Simmons, H. E., J. Am. Chem. Soc., 83 (1961) 1657; Martin, H. D., Kagabu, S., Schiwek, H. J., Tetrahedron Lett., 41 (1975) 3311.

[87] le Noble, W. J., Mukhtar, R., J. Am. Chem. Soc., 96 (1974) 6191.

[88] a) Kowa, L. N., Schwarzer, D., Troe, J., Schroeder, J., J. Chem. Phys., 97 (1992) 4827; b) Firestone, R. A., Vitale, M., J. Org. Chem., 46 (1981) 2160.

[89] The van der Waals volumes V_W can be calculated by the computer program MOLVOL: Artschwager-Perl, U., Cycloadditionen unter hohem Druck, Ph.D. Thesis, Ruhr-Universität Bochum, 1989. This program uses the cartesian coordinates of a molecular structure resulting from a force field or quantum mechanical calculation and can be obtained on request. V_W of ground states can also be calculated from tables of group contributions to the van der Waals volumes published by Bondi, A., J. Chem. Phys., 68 (1964) 441.

[90] Asano, T., le Noble, W. J., Rev. Phys. Soc. Jpn., 43 (1973) 82.

[91] Similar calculations of van der Waals volumes have been carried out: Firestone, R. A., Smith, G. M., Chem. Ber., 122 (1989) 1089.

[92] Krawczyk, B., Klärner, F.-G., unpublished results.

[93] a) Dauben, W. G., Kessel, C. R., Takemura, K. H., J. Am. Chem. Soc., 102 (1980) 6893; b) Dauben, W. G., Gerdes, J. M., Smith, D. B., J. Org. Chem., 50 (1985) 2576.

[94] Tietze, L. F., Hübsch, T., Oelze, J., Ott, C., Wörner, G., Buback, M., Chem. Ber., 125 (1992) 2249; Tietze, L. F., Hübsch, T., Voss, E., Buback, M., Tost, W., J. Am. Chem. Soc., 110 (1988) 4065.

[95] Buback, M., Tost, W., Tietze, L. F., Voss, E., Chem. Ber., 121 (1988) 781.

[96] Jurczak, J., Kozluk, T., Tkacz, M., Eugster, C. H., Helv. Chim. Acta, 66 (1983) 218.

[97] Grieco, P. A., Nunes, J. J., Gaul, M. D., J. Am. Chem. Soc., 112 (1990) 4595.

[98] Breslow, R., Maitra, U., Rideout, D., Tetrahedron Lett., 24 (1983) 1901.
[99] Grieco, P. A., Nunes, J. J., Gaul, M. D., J. Am. Chem. Soc., 112 (1990) 4595.
[100] Sauer, J., Kredel, J., Tetrahedron Lett., 7 (1966) 731, 6359.
[101] Breslow, R., Acc. Chem. Res., 24 (1991) 159.
[102] Blokzijl, W., Blandamer, M. J., Engberts, J. B. F. N., J. Am. Chem. Soc., 113 (1991) 4241.
[103] Blake, J. F., Jorgensen, W. L., J. Am. Chem. Soc., 113 (1991) 7430.
[104] Pindur, U., Lutz, G., Otto, C., Chem. Rev., 93 (1993) 741.
[105] Posner, G. H., Ishihara, Y., Tetrahedron Lett., 35 (1994) 7545.
[106] Tietze, L. F., Ott, C., Gerke, K., Buback, M., Angew. Chem., 105 (1993) 1536; Angew. Chem. Int. Ed. Engl., 32 (1993) 1485.
[107] Dauben, W. G., Kowalczyk, B. A., Nieduzak, T. R., J. Org. Chem., 55 (1990) 2391.
[108] Kozikowski, A. P., Konoike, T., Nieduzak, T. R., J. Chem. Soc., Chem. Comm., (1986) 1350.
[109] Jenner, G., Tetrahedron Lett., 35 (1994) 1189.
[110] Isaacs, N. S., Maksimovic, L., High Pressure Res., 13 (1994) 21.
[111] Fessner, W.-D., Grund, C., Prinzbach, H., Tetrahedron Lett., 30 (1989) 3133; Klärner, F.-G., Artschwager-Perl, U., Fessner, W.-D., Grund, C., Pinkos, R., Melder, J. P., Prinzbach, H., Tetrahedron Lett., 30 (1989) 3137.
[112] Dolbier, W. R., Weaver, S. L., J. Org. Chem., 55 (1990) 711.
[113] Asano, T., chapter 2 in ref. [3]; Jenner, G., chapter 6 in ref. [2].
[114] Roth, W., Doering, W. v. E., J. Am. Chem. Soc., 112 (1990) 1722.
[115] Morokuma, K., Borden, W. T., Hrovat, D. A., J. Am. Chem. Soc., 110 (1988) 4474.
[116] Walling, C., Naiman, M., J. Am. Chem. Soc., 84 (1962) 2628.
[117] Stashina, G. A., Vasylvitskaya, E. N., Gamalevich, G. D., El'yanov, B. S., Serebryakov, E. P., Zhulin, V. M., Izv. Akad. Nauk. SSSR, Ser. Khim., (1986) 329.
[118] Schulman, E. M., Merbach, A. E., Turin, M., Wedinger, R., le Noble, W. J., J. Am. Chem. Soc., 105 (1983) 3988.
[119] Zimny, B. U., Organisch-chemische Reaktionen unter hohem Druck: Der Druckeffekt auf Cycloadditionen von Cyan- und Dicyanacetylen, Cope-Umlagerungen von Phenyl-substituierten 1,5-Hexadienen, Cycloreversion und Umlagerungen von substituierten Cyclobutanen, Ph.D. Thesis, Universität Essen, 1994.
[120] Gehrke, J.-S., Der Druckeffekt auf pericyclische Reaktionen: Druckabhängigkeit der thermisch induzierten Reorganisationsprozesse von cis- und trans-1,2-Divinylcyclobutan, Diploma thesis, Universität Essen, 1993.
[121] Jansen, P., Der Druckeffekt auf die Cope-Umlagerung: Druckabhängigkeit der Umlagerung von dl-2,2-Bismethylencyclohexan zu 1,2-Di-(1-cyclohexenyl)-ethan, Diploma thesis, Universität Essen, 1994.
[122] Schulman, E. M., Merbach, A. E., Turin, M., Wedinger, R., le Noble, W. J., J. Am. Chem. Soc., 105 (1983) 3988.
[123] le Noble, W. J., Rajaona Daka, M., J. Am. Chem. Soc., 100 (1978) 5961.
[124] Schöllkopf, U., Hoppe, I., Liebigs Ann., (1972) 153.

3.9. REFERENCES

[125] Sugiyama, S., Mori, A., Takeshita, H., Chem. Lett., (1987) 1247.
[126] Sugiyama, S., Takeshita, H., Chem. Lett., (1986) 1203.
[127] Dauben, W. G., Kowalczyk, B. A., Funhoff, D. J. H., Tetrahedron Lett., 29 (1988) 3021.
[128] a) le Noble, W. J., Brower, K. R., Brower, C., Chang, S., J. Am. Chem. Soc., 104 (1982) 3150; b) le Noble, W. J., J. Chem. Educ., 44 (1967) 729.
[129] Mündnich, R., Plieninger, H., Tetrahedron, 32 (1976) 2335.
[130] Diedrich, M. K., Klärner, F.-G., unpublished results.
[131] The structural parameter necessary for the calculation of V_w was taken from an *ab initio* calculation. Baldwin, J. E., Reddy, V. P., Schaad, L. J., Hess Jr., B. A., J. Am. Chem. Soc., 110 (1988) 8554.
[132] Mündnich, R., Plieninger, H., Vogler, H., Tetrahedron, 33 (1977) 2661.
[133] Mündnich, R., Plieninger, H., Tetrahedron, 34 (1978) 887.
[134] Breitkopf, V., Hopf, H., Klärner, F.-G., Witulski, B., Zimny, B., Liebigs Ann., (1995) 613.
[135] Isaacs, N. S., van der Beeke, P. G., Tetrahedron Lett., 23 (1982) 2147.
[136] Buback, M., Gerke, K., Ott, C., Tietze, L. F., Chem. Ber., 127 (1994) 2241.
[137] Buback, M., Abeln, J., Hübsch, T., Ott, C., Tietze, L. F., Liebigs Ann., (1995) 9.
[138] Diedrich, M. K., Hochstrate, D., Klärner, F.-G., Zimny, B., Angew. Chem., 106 (1994) 1135; Angew. Chem. Int. Ed. Engl., 33 (1994) 1079.
[139] Jenner, G., Papadopoulos, M., El'yanov, B., Gonikberg, E. M., J. Org. Chem., 47 (1982) 4201.
[140] Jenner, G., ben Salem, R., J. Chem. Soc., Perkin Trans. II (1989) 1671.
[141] Stephenson, L. M., Mattern, D. L., J. Org. Chem., 41 (1976) 3614.
[142] Although the new C–C bond is only partially formed in the transition structure, the C–H–C distances are obviously both within the range of van der Waals distances, whereas in the product the non-bonding ($=H_2C$---H–CH_2–) distance is, according to the most stable geometry, larger than the van der Waals distance.
[143] Doering, W. v. E., Birladeanu, L., Sarma, K., Teles, J. H., Klärner, F.-G., Gehrke, J.-S., J. Am. Chem. Soc., 116 (1994) 4289.
[144] a) Jurczak, J., Pietraskiewicz, M., Top. Curr. Chem., 130 (1986) 185; b) Salanski, P., Ostaszewski, R., Jurczak, J., High Pressure Res., 13 (1994) 35; c) Stankiewicz, T., Tetrahedron, 49 (1993) 1471.
[145] Benkhoff, J., Boese, R., Klärner, F.-G., Wigger, A. E., Tetrahedron Lett., 35 (1994) 73.
[146] Kohnke, F. H., Slawin, A. M. Z., Stoddart, J. F., Williams, D. J., Angew. Chem., 99 (1987) 941; Angew. Chem. Int. Ed. Engl., 26 (1987) 892; Kohnke, F. H., Stoddart, J. F., Pure Appl. Chem., 61 (1989) 1581.

4

ORGANIC SYNTHESIS AT HIGH PRESSURE

Janusz Jurczak*[1,2] and Daniel T. Gryko[1]

[1] Institute of Organic Chemistry, Polish Academy of Sciences, 01-224 Warsaw, Poland
[2] Department of Chemistry, Warsaw University, 02-093 Warsaw, Poland

4.1 INTRODUCTION
4.2 FUNDAMENTAL ASPECTS OF THE HIGH PRESSURE TECHNIQUE USED IN ORGANIC SYNTHESIS
4.3 A SURVEY OF PREPARATIVELY IMPORTANT HIGH PRESSURE ORGANIC REACTIONS
4.4 CONCLUSIONS
4.5 REFERENCES

4.1 INTRODUCTION

Modern organic synthesis aims at the efficient preparation of increasingly complex compounds. This results in the dynamic development of new methods. They are mostly based on chemical modification of substrates, reagents and catalysts. Nevertheless, frequent use has been made recently of techniques involving high pressure, ultrasound, microwaves as well as photo- and electrochemical phenomena.

The effect of pressure on the rate and direction of organic reactions in solution has been known for a long time, and it has been the object of extensive physical and chemical studies. In practice, the high pressure technique has gained more attention since the 1980's, when the pioneering studies of Dauben *et al.* [1, 2] testified to the advantages of high pressure application for an effective course of the Diels-Alder reaction. During the past two decades, organic synthesis has developed greatly in this direction. Several interesting reviews [3–8] and two books [9, 10] devoted to this topic have recently been published.

The application of high pressure offers the possibility of solving some problems in organic synthesis that cannot be resolved by other more accessible methods. The rates of organic reactions characterized by a large negative activation volume substantially increase when high pressure is applied (see Chapter 3). Consequently, the reactions can be carried out under mild conditions which usually prevents adverse side effects. In many instances the reaction times are shortened and yields are enhanced. An example is shown in Scheme 4.1.

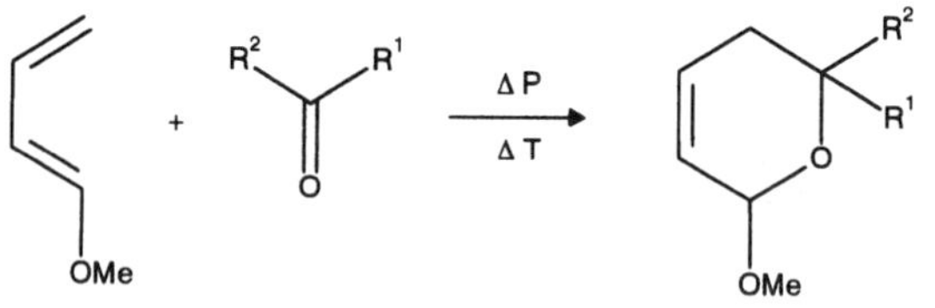

Scheme 4.1

2-Methoxy-6-substituted-5,6-dihydro-2H-pyrans (3) are versatile intermediates for the syntheses of monosaccharides and related natural compounds. A convenient synthetic route to these compounds *via* the thermal Diels-Alder reaction of 1-methoxybuta-1,3-diene (1) with carbonyl compounds (2) is limited to activated aldehydes with an electron-withdrawing substituent, *e.g.*, glyoxylic acid esters. This reaction involving non-activated carbonyl compounds fails under normal conditions, whereas upon use of high pressure it can be carried out with high yield [11]. Recently, the method was improved by the use of Eu(fod)$_3$ as a mild Lewis acid catalyst, allowing the [4+2] cycloaddition to be carried out at lower pressure [12].

Pressure as an experimental variable influences, apart from the rate constant, the regio- and stereoselectivity of organic reactions. This effect is related to the difference between activation volumes of parallel reactions leading to regio- and/or stereoisomers (Scheme 4.2).

The high pressure (196 MPa) reaction of 1-alkyl derivatives of buta-1,3-diene (4) with alkyl acrylates (5) affords a mixture of four isomers 6: regioisomers *ortho* and *meta* in two *cis* and *trans* diastereoisomer series [13]. An increase of pressure from 196 to 588 MPa promotes predominant formation of the *ortho* regioisomer in both the *cis* (86%) and *trans* (87%) diastereoisomer series. The effect of pressure on the ratio of *cis:trans* diastereoisomers in this case is much smaller, though it distinctly favors the *cis* form in both the *ortho* (53%) and *meta* (54%) series.

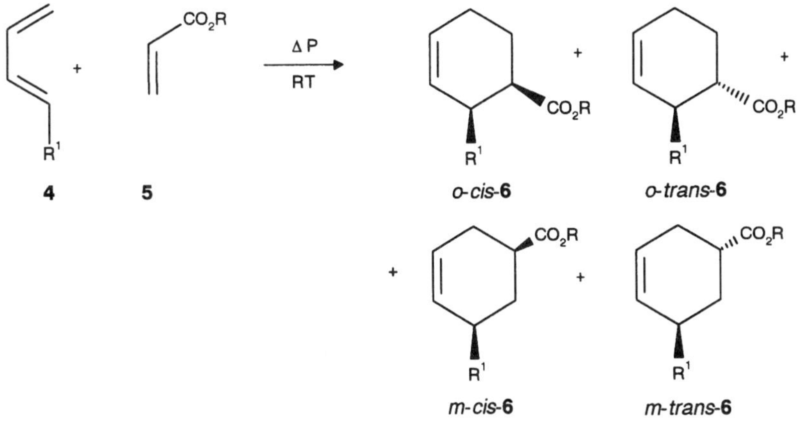

4.1. INTRODUCTION

Scheme 4.2

Furan and its derivatives form [4+2] cycloadducts with a variety of monoactivated dienophiles under high pressure, while the analogous reactions at atmospheric pressure fail to produce any cycloadducts [2, 14]. For example, furan (**7**) reacts with alkyl acrylates (**5**) under high pressure to give a 7:3 diastereoisomer mixture of *endo*-**8** and *exo*-**8**, in 62% yield (Scheme 4.3).

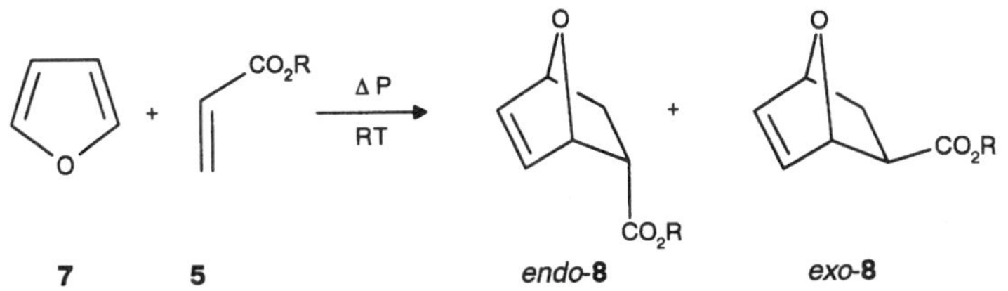

Scheme 4.3

As experimental parameters, temperature and hydrostatic pressure are generally used for a shift of the position of reaction equilibrium. It is, of course, much more common to use temperature instead of pressure for this purpose. However, application of high pressure allows us, in some cases, to obtain compounds usually thermodynamically less favored at normal pressure. For example, the high pressure [4+2] cycloaddition of furan (**7**) to *p*-benzoquinone (**9**) has been carried out successfully (Scheme 4.4) [15].

This type of Diels-Alder reaction can be expected to be accelerated by an increase in pressure, whereas the *retro* reaction would be inhibited. Consequently, it was expected that under very high pressure and at moderate temperature, the equilibrium of this cycloaddition would be shifted toward the products. Indeed, in the reaction at 2000 MPa pressure and ambient temperature, a 1:1.1 mixture of the expected *endo*-**10** and *exo*-**10** cycloadducts was obtained.

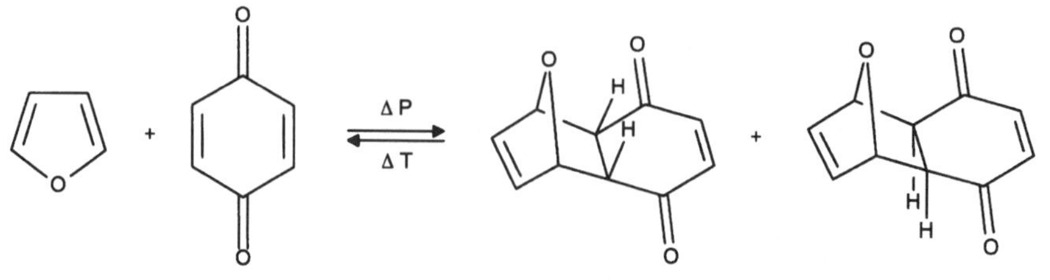

Scheme 4.4

However, after keeping the reaction mixture under normal conditions for 30 min, no cycloadducts were detected. These facts suggest that both cycloadducts are unstable at room temperature and undergo the *retro* reaction [16, 17]. These observations explain why the cycloadducts **10** could not be obtained under thermal conditions.

These are important aspects of high pressure organic synthesis and are the topic of this Chapter. A survey of high pressure reactions useful for preparative organic synthesis is presented. Moreover, some applications of high pressure methodology to syntheses of natural products are also described. First, however, some fundamental principles of the effect of pressure on organic reactions are presented.

4.2 FUNDAMENTAL ASPECTS OF THE HIGH PRESSURE TECHNIQUE USED IN ORGANIC SYNTHESIS

Pressure times volume ($p \times V$) has the dimensions of energy. Thus, the use of pressure constitutes a non-thermal means for carrying out reactions. Every reaction is characterized by a volume of activation ($\Delta V^{\neq}$) defined as the difference between the partial molar volume occupied by the transition state (V_{TS}) and that occupied by the reactant(s) (V_R) (see Chapter 3):

$$\Delta V^{\neq} = V_{TS} - V_R$$

Pressure can influence the reaction rates as shown in Figure 4.1. Pressure affects the free energy of activation ($\Delta G^{\neq}$) according to the sign and magnitude of $\Delta V^{\neq}$. If $\Delta V^{\neq} < 0$, the application of pressure lowers $\Delta G^{\neq}$ and accelerates the reaction. Conversely, if $\Delta V^{\neq} > 0$, increase in pressure retards the reaction. The volume of activation is obtained from the pressure dependence of the rate constant and may be a function of pressure:

$$\Delta V^{\neq} = -RT(\delta \ln k/\delta p)_T$$

4.2. FUNDAMENTAL ASPECTS OF THE HIGH PRESSURE TECHNIQUE

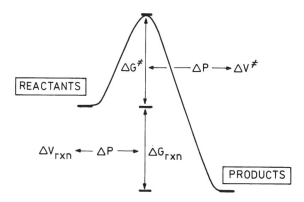

Figure 4.1 Reaction free energy profile: effect of pressure on free energies.

Pressure can also influence reaction equilibrium (Figure 4.1). For each reaction there is a volume of reaction (ΔV_{rxn}) defined as the difference in partial molar volume between products (V_P) and reactants (V_R):

$$\Delta V_{rxn} = V_P - V_R$$

Pressure affects the free energy of reaction (ΔG_{rxn}) according to the sign and magnitude of ΔV_{rxn}. The influence of pressure on the reaction equilibrium follows directly from the Le Chatelier-Braun principle. If $\Delta V_{rxn} < 0$, the application of pressure shifts the equilibrium toward the products. The volume of reaction is obtained from the pressure dependence of the equilibrium constant and may also be a function of pressure:

$$\Delta V_{rxn} = -RT(\delta \ln K/\delta p)_T$$

As mentioned above (see also Chapter 3), pressure influences, apart from the chemical equilibrium and rate constant, the regio- and stereoselectivity of organic reactions. This effect is related to the difference between activation volumes of parallel reactions leading to regio- and/or stereoisomers:

$$\Delta V^{\neq} = \Delta V_1^{\neq} - \Delta V_2^{\neq} \dots$$

Therefore, pressure proves to be an important factor for controlling asymmetric induction which is an essential process in the total synthesis of natural products.

High pressure reactions carried out in solution can be limited by solvent-related properties. This is due to the effect of pressure on the behavior of organic liquids. The property most affected is the freezing pressure (p_f), i.e., the pressure at which the liquid solidifies at a given temperature. The freezing pressures of typical organic solvents are presented in Table 4.1 [4].

To prevent the drawback involved in solvent freezing under high pressure, temperature elevation is mostly used. For the same purpose, reactions may be

4. ORGANIC SYNTHESIS AT HIGH PRESSURE

Table 4.1 Freezing pressures of common organic solvents close to ambient temperature

Solvent	p_f [MPa$(T/°C)$]
Hexane	1020 (30)
Toluene	960 (30)
Dichloromethane	1300 (25)
Chloroform	550 (25)
Methanol	3000 (25)
Acetone	800 (20)
Diethyl ether	1200 (35)
Ethyl acetate	1210 (25)

carried out in solvent mixtures, which usually exhibit a substantial increase in solidification pressure, as compared to pure components. For example, for pure benzene $p_f \approx 90$ MPa (30 °C), and for pure toluene $p_f \approx 960$ MPa (30 °C), whereas for their mixture (benzene:toluene = 2:8) $p_f \approx 1560$ MPa (30 °C).

High pressure organic reactions on a preparative scale are mostly carried out in a piston cylinder type apparatus. The main features of the typical apparatus are presented in Figure 4.2 (see also Figure 2.1).

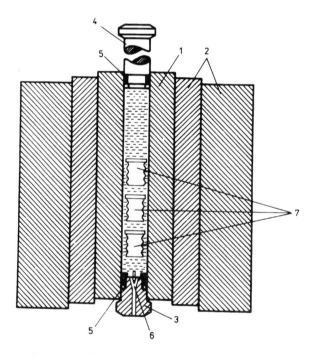

Figure 4.2 Piston-cylinder high pressure apparatus.

The high pressure apparatus comprises two (or one) external rings (2) surrounding an internal steel conical vessel (1). The cylindrical high pressure space is closed from below with a steel stopper (3). All electrical connections (manganin manometer, thermocouple, etc.) are led through a conical electrode (6) placed in the stopper. The internal vessel is closed from above by a mobile piston (4) moved by an independent hydraulic force. The piston and the stopper are sealed with resin O-rings and brass rings (5). For reactions performed at higher temperatures, an external heating jacket is used. Flexible Teflon ampoules (7) may be used as reaction containers. Special high pressure equipment for preparative organic synthesis is now commercially available; for example, the complete high pressure system shown in Figure 4.3 is available from Unipress (Warsaw, Poland).

4.3 A SURVEY OF PREPARATIVELY IMPORTANT HIGH PRESSURE ORGANIC REACTIONS

Among cycloadditions, the Diels-Alder reaction is frequently used in organic synthesis. Since there is a multitude of dienes and dienophiles known, it is possible to obtain differently functionalized adducts as starting compounds for syntheses of many important natural products. The high pressure technique is a very convenient tool for the Diels-Alder reaction, providing highly regio- and stereocontrolled syntheses of [4+2] cycloadducts, and in many instances, reaction times are shortened and yields are improved.

Naphthalene derivatives can participate in the Diels-Alder reaction as diene components. However, since naphthalene (11) itself is a rather unreactive diene, only 1% of the cycloadduct 13 is obtained from the thermal reaction with maleic anhydride (12). Application of high pressure (950 MPa, 100 °C) in this reaction affords a mixture of *endo*-13 and *exo*-13 in 78% yield, with predominant formation of the *endo* cycloadduct (Scheme 4.5).

Furan and its derivatives receive a great deal of attention as dienes and form cycloadducts with a variety of dienophiles under high pressure (*cf.* Schemes 4.3 and 4.4). Thiophene has a greater stabilization energy than furan and thus is more reluctant to form cycloaddition products. However, there is one example of a high pressure reaction (1500 MPa, 100 °C) of thiophene (14) with maleic anhydride (12) to afford a 43% yield of the [4+2] cycloadduct *exo*-15 (Scheme 4.6) [18].

It is well known that furan (7) fails to react with dimethylmaleic anhydride even at 4000 MPa [19], probably because of both the low reactivity of furan as a diene and the extremely high steric hindrance of the dienophile. The steric restrictions of the dienophile can be reduced by conversion of dimethylmaleic anhydride to its bicyclic thia-derivative 16, which reacts smoothly with furan (7) under high pressure conditions (1500 MPa, rt) to give a quantitative yield of a mixture of diastereoisomers *exo*-17 and *endo*-17 in a 85:15 ratio (Scheme 4.7) [20].

Figure 4.3 Commercially available high pressure apparatus LCP 30 (Unipress, Warsaw, Poland).

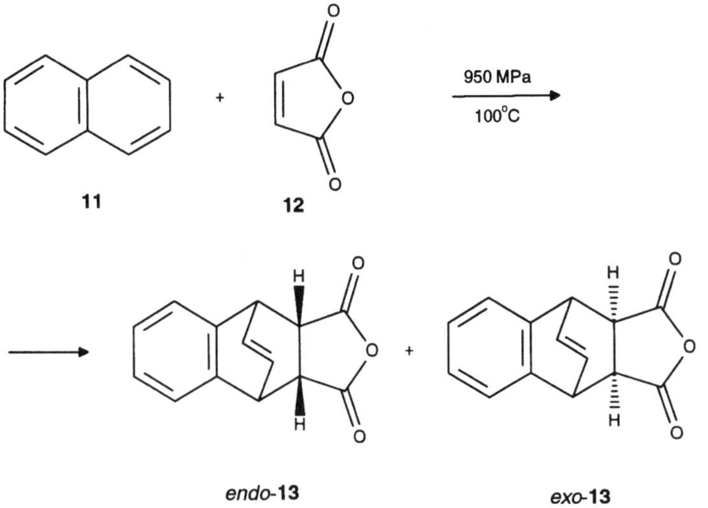

4.3. A SURVEY OF HIGH PRESSURE ORGANIC SYNTHESIS

Scheme 4.5

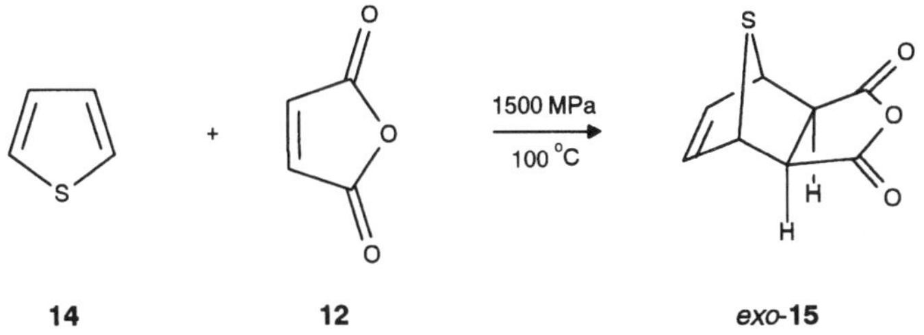

Scheme 4.6

Following chromatographic separation, pure *exo*-**17** is reduced with Raney nickel, affording the Spanish fly constituent cantharidin (**18**).

The Diels-Alder reaction of isophorone dienamine (**19**) with acrylonitrile (**20**), under thermal conditions, affords a product mixture containing, apart from the desired [4+2] cycloadduct **21** (3%), two other compounds **22** (10%) and **23** (20%) (Scheme 4.8) [1].

Under high pressure conditions, the same reaction proceeds cleanly to give, as the only product, the [4+2] cycloadduct **21**, in 90% yield. This is another feature of high pressure synthesis: pressure can change the course and product distribution of organic reactions.

Another important utilization of the high pressure Diels-Alder reaction is exemplified by the Stoddart synthesis of interesting "belt" compounds **26**, starting from diene **24** and dienophile **25** (Scheme 4.9) [21, 22].

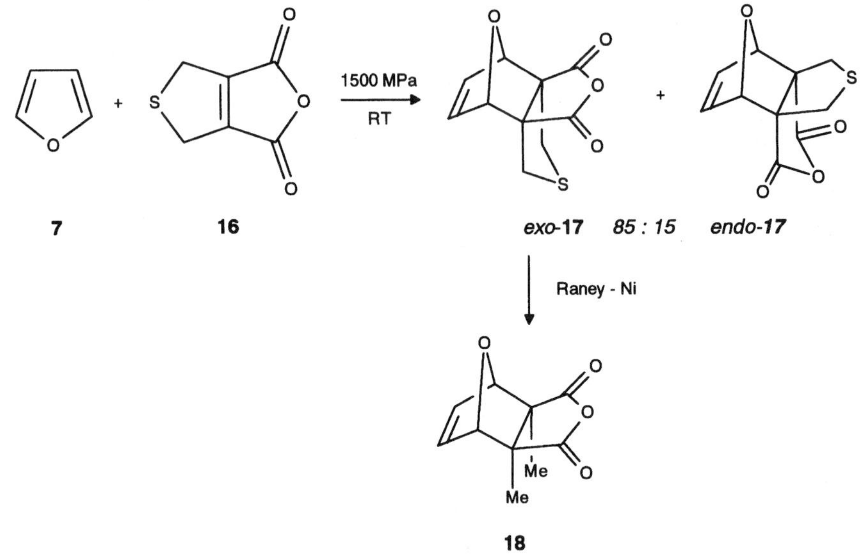

172   4. ORGANIC SYNTHESIS AT HIGH PRESSURE

Scheme 4.7

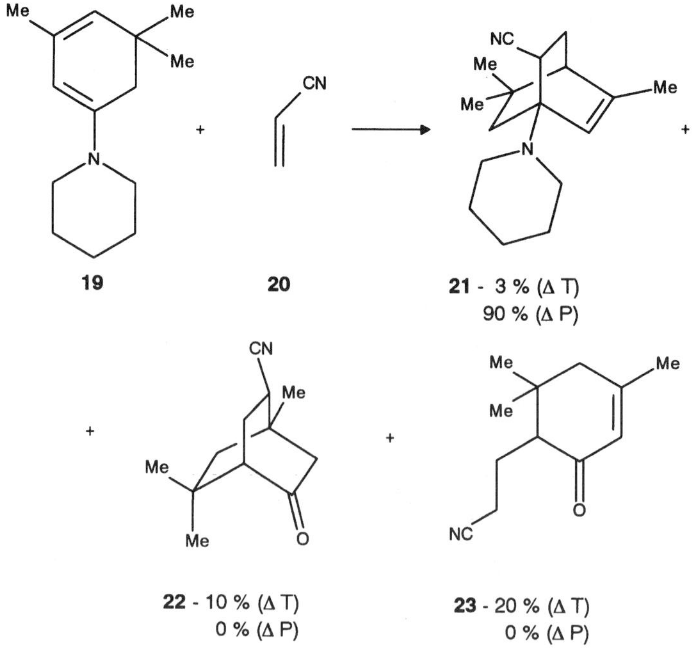

Scheme 4.8

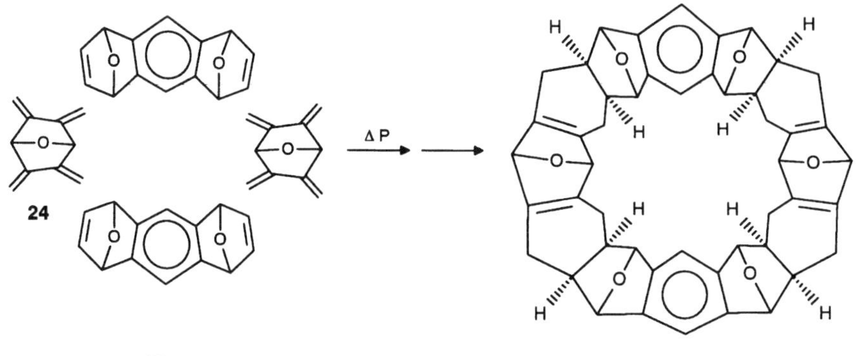

4.3. A SURVEY OF HIGH PRESSURE ORGANIC SYNTHESIS

Scheme 4.9

This synthesis required a succession of Diels-Alder reactions, followed by finally an intramolecular coupling. The 2–3% yield at atmospheric pressure was improved to 30–35% at 1000 MPa.

The high pressure hetero-Diels-Alder reaction is a convenient method for the preparation of variously substituted derivatives of 5,6-dihydro-2H-pyran [11]. Compound **28**, a precursor of the pheromone massoi lactone (**29**), can be obtained by direct high pressure [4+2] cycloaddition of 1-methoxybuta-1,3-diene (**1**) to hexanal (**27**) in 28% yield (Scheme 4.10) [23].

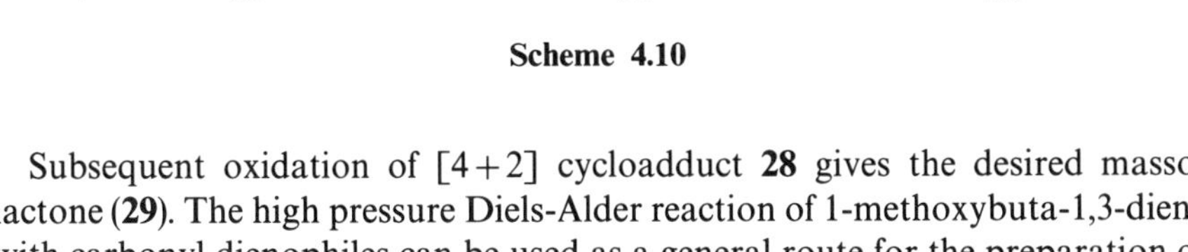

Scheme 4.10

Subsequent oxidation of [4+2] cycloadduct **28** gives the desired massoi lactone (**29**). The high pressure Diels-Alder reaction of 1-methoxybuta-1,3-diene with carbonyl dienophiles can be used as a general route for the preparation of naturally occurring δ-lactones.

[4+2] Cycloaddition of 1-methoxybuta-1,3-diene (**1**) to 2,3-O-isopropylidene-D-glyceraldehyde (**30**), bearing a chiral center located in the α-position to the formyl group, gives rise to four optically active diastereoisomeric (*cis* and *trans*) cycloadducts **31**. Acid-catalyzed isomerization at the C-2 carbon atom leads to two *trans* diastereoisomers (2S,6S)-**31** and (2R,6R)-**31** (Scheme 4.11) [24].

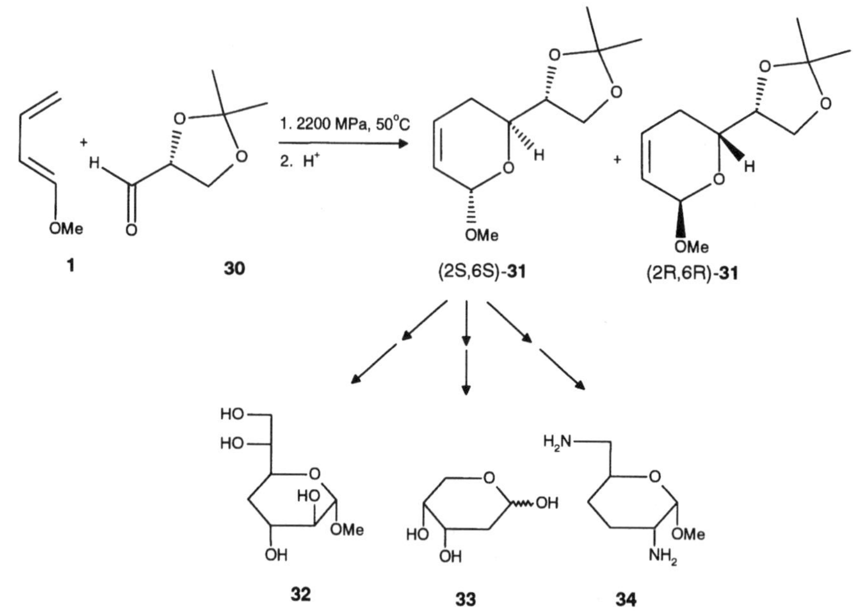

Scheme 4.11

The cycloadduct (2S,6S)-**31** can be utilized for preparation of a variety of interesting sugars. For example, oxidative functionalization of the double bond leads to methyl 4-deoxyheptoside **32** [25], whereas its ozonolysis, followed by hydrolysis, affords 2-deoxy-D-ribose **33** [26]. On the other hand, subjecting cycloadduct (2S,6S)-**31** to a sequence of simple reactions gives optically pure methyl purpurosaminide C (**34**), a component of some aminoglycosidic antibiotics [5].

A similar approach has been used for the synthesis of 6-*epi*-D-purpurosamine B (**37**), a component of the antibiotic fortimycin. The Eu(fod)$_3$-mediated high pressure [4+2] cycloaddition of 1-methoxybuta-1,3-diene (**1**) to *N*-carbo-*t*-butoxy-L-alaninal (**35**), followed by anomeric isomerization, affords a 2:1 mixture of cycloadducts (2S,6S)-**36** and (2R,6R)-**36** in 70% yield. Chromatographic separation, followed by functionalization of the optically pure compound (2S,6S)-**36**, leads to the desired product **37** (Scheme 4.12) [27].

The use of α,β-unsaturated carbonyl compounds as heterodienes provides an attractive route to 3,4-dihydro-2H-pyrans which, like the 5,6-isomers, can serve as useful starting materials for the synthesis of sugars. For example, methyl vinyl ketone (**38**) reacts with ethoxyethylene (**39**) under high pressure (1500 MPa, rt) to afford the cycloadduct **40** in 62% yield [28]. The same reaction carried out under 200 MPa pressure and at 100 °C gives only 25% yield of the product **40** and 50% yield of the dimer **41** (Scheme 4.13) [29].

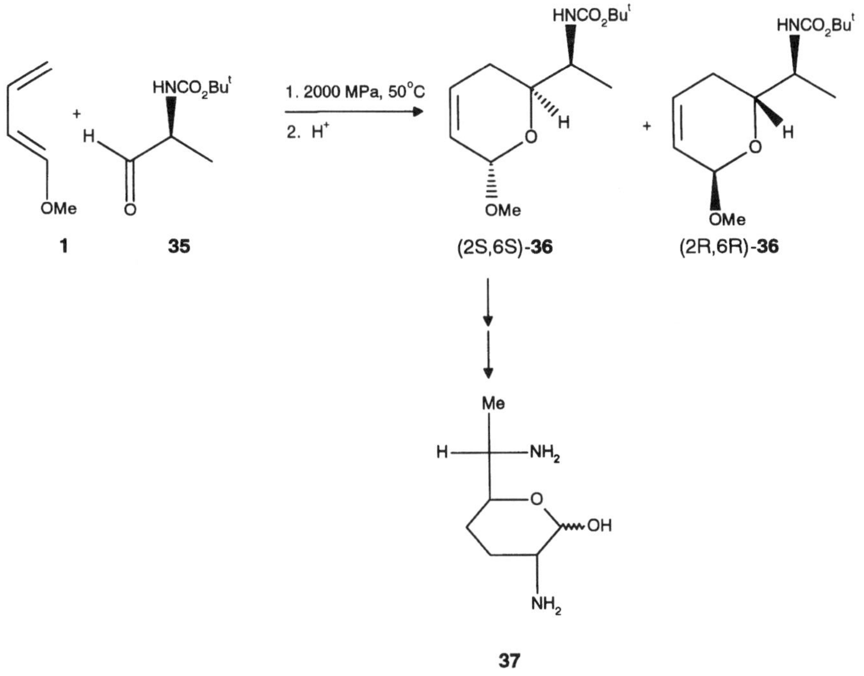

4.3. A SURVEY OF HIGH PRESSURE ORGANIC SYNTHESIS

Scheme 4.12

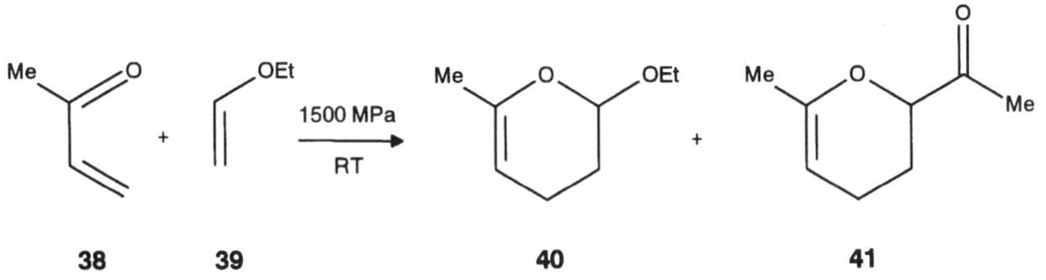

Scheme 4.13

The hetero-Diels-Alder reaction with imines as dienophiles is potentially useful for preparation of a variety of nitrogen-containing heterocycles, *e.g.*, pyridines, quinolines, and related alkaloids. The high pressure reaction (1000 MPa, 110 °C) of isoprene (**42**) with triethyl azomethinetricarboxylate (**43**) affords in 60% yield the 1,2,3,6-tetrahydropyridine derivative **44** (Scheme 4.14) [30].

[2+2] Cycloaddition is strongly promoted by pressure because it proceeds *via* a dipolar transition state [31]. Polar [2+2] cycloadditions are characterized by large negative activation volumes (see Section 3.3), which are of the same order of magnitude as those found for the Diels-Alder reactions. Application of high pressure to this cycloaddition offers a valuable route to four-membered rings, such as cyclobutanes, oxetanes, and β-lactams. For example, the cyclobutane derivative **47** can be obtained in good yield from the high pressure

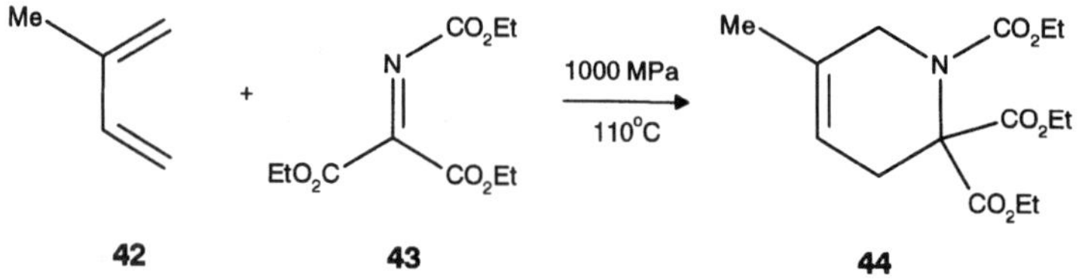

Scheme 4.14

reaction (1200 MPa, RT) of 1-ethoxycyclohexene (**45**) with dicyanostyrene (**46**), whereas no reaction is observed at atmospheric pressure and 50 °C (Scheme 4.15) [32].

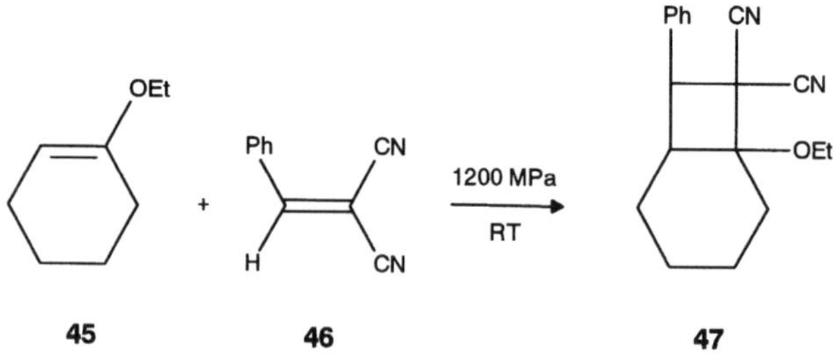

Scheme 4.15

1,1-Dimethoxy-1-propene (methylketene acetal **49**), which does not react with benzaldehyde (**48**) at room temperature in the absence of a catalyst, is completely converted into a mixture of oxetanes, *trans*-**50** and *cis*-**50**, under high pressure conditions (1200 MPa, RT); the more stable compound *trans*-**50** is the major product (Scheme 4.16) [33].

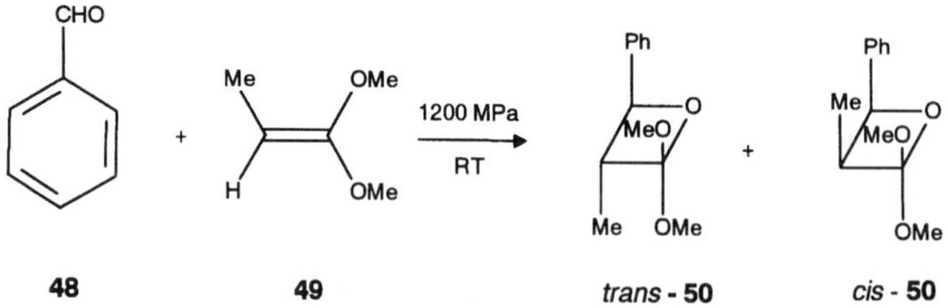

Scheme 4.16

The bicyclic [2+2] cycloadducts of isocyanates and glycals (1,2-unsaturated monosaccharides) are interesting compounds with potential biological activity. They are also valuable chiral intermediates, suitable for further transformations into known or new derivatives of β-lactams. Application of high pressure (1000 MPa, RT) promotes the [2+2] cycloaddition of tosyl isocyanate (**52**) to tri-O-acetyl-D-glucal (**51**), leading to the bicyclic adduct **53**. The reaction proceeds with high stereoselectivity to afford the four-membered β-lactam ring *anti* to the acetoxy group at the C-3 carbon atom (Scheme 4.17) [34].

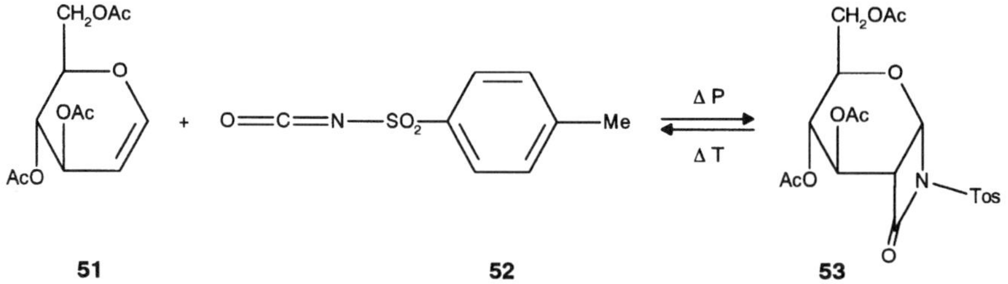

4.3. A SURVEY OF HIGH PRESSURE ORGANIC SYNTHESIS

Scheme 4.17

Upon heating or even after standing under ambient conditions, the bicyclic adduct **53** exhibits *retro*-reaction to afford starting glycal **51**. This fact explains why β-lactams of this type could not be obtained from glycals and isocyanates under thermal conditions.

Pressure also exerts a considerable effect on the rates of 1,3-dipolar cycloadditions. Application of high pressure in 1,3-dipolar cycloadditions is extremely useful when nitrone or cycloadduct products are thermally unstable. Nitrones are usually sensitive to heat, and therefore the long heating required to induce cycloaddition at atmospheric pressure often results in a poor yield of the cycloadduct. Thus the high pressure reaction (200 MPa, 50 °C) of nitrone **54** with ethyl vinyl ether (**39**) affords the cycloadduct **55** in 83% yield, whereas no cycloadduct is obtained at atmospheric pressure and 80 °C (Scheme 4.18) [35].

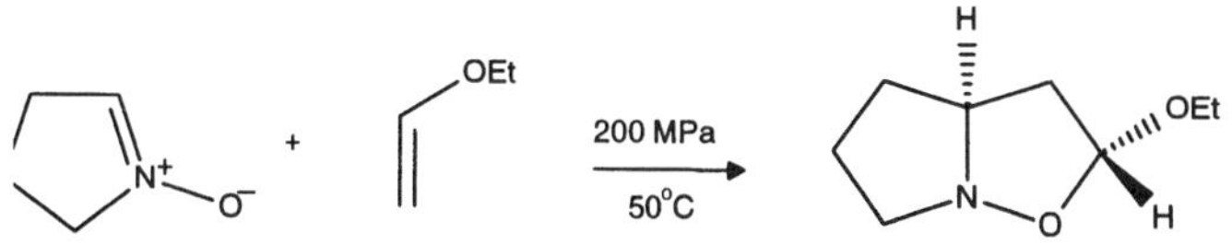

Scheme 4.18

The ene reaction, closely related to [4+2] cycloaddition, is also accelerated by pressure (see Section 3.3). The thermal uncatalyzed ene reactions are usually low yield processes requiring elevated temperatures. Application of high pressure strongly promotes ene reactions of β-pinene (**56**) with activated enophiles, which otherwise occur only at temperatures higher than 150 °C or in the presence of Lewis acid catalysts. For example, β-pinene (**56**) reacts under high pressure conditions (4000 MPa, RT) with methyl pyruvate (**57**) to give the expected product **58** in quantitative yield (Scheme 4.19) [36].

The widely used Michael reaction is an example of a pressure-accelerated polar addition. Many Michael addition reactions which were unsuccessful under thermal or Lewis acid catalyzed conditions proceed smoothly under high pressure. The high pressure approach to Michael addition reactions is especially

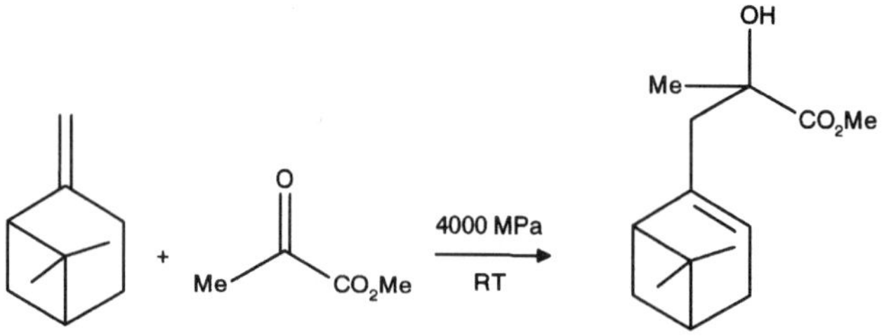

4. ORGANIC SYNTHESIS AT HIGH PRESSURE

Scheme 4.19

effective in overcoming steric hindrance. The high pressure Michael addition reaction (1500 MPa, RT) of doubly activated donor diethyl malonate (**60**) with the sterically demanding acceptor 1-acetylcyclohexene (**59**), in the presence of appropriate bases, gives the primary Michael adduct **61** in 60 % yield (Scheme 4.20) [37].

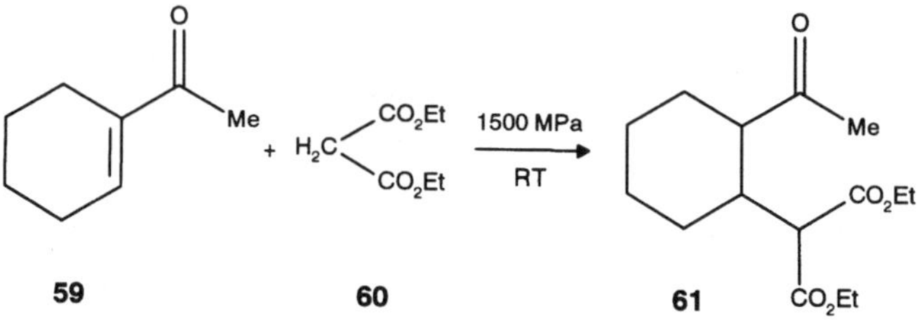

Scheme 4.20

Conjugate addition of *O*-silylated ketene acetals to α,β-unsaturated carbonyl systems is a useful and efficient method for introducing an acetate residue in the β-position with respect to the carbonyl group. These transformations are carried out thermally for relatively unhindered compounds or by Lewis acid catalysis for more sterically congested systems. It was recently demonstrated [38, 39] that the high pressure technique provides an alternative route for inducing ketene acetal additions to sensitive enones with steric constraints. The high pressure reaction (1500 MPa, RT) of the crotonic aldehyde acetal **63** with the sterically hindered cyclic enone **62**, in the absence of a catalyst, gives the Michael adduct **64** in 81% yield (Scheme 4.21) [38].

Aldol reactions usually proceed under basic or acidic conditions. At high pressure these additions can be carried out under neutral conditions. Thus it is now possible to perform aldol reactions with compounds that are acid- or base-sensitive. The trimethylsilyl enol ether of cyclohexanone (**65**) reacts with benzaldehyde (**48**), in the presence of titanium tetrachloride, to afford β-silyloxyketones (*S,S*)-**66** and (*R,S*)-**66** in a ratio of 25:75. Under high pressure

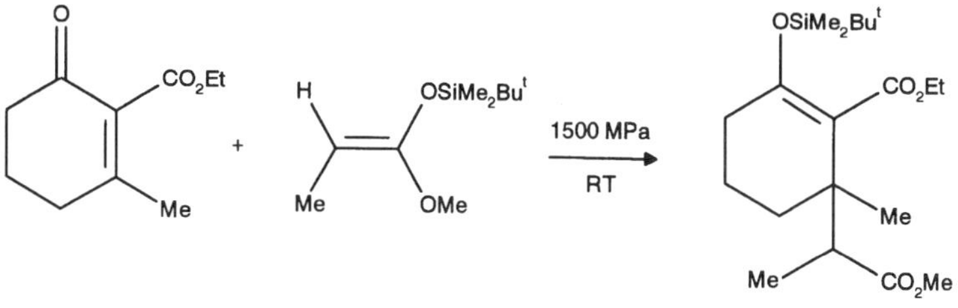

4.3. A SURVEY OF HIGH PRESSURE ORGANIC SYNTHESIS

Scheme 4.21

and neutral conditions (1000 MPa, 50–60 °C) this reaction yields the same reaction mixture but with reversed stereochemistry (Scheme 4.22) [40].

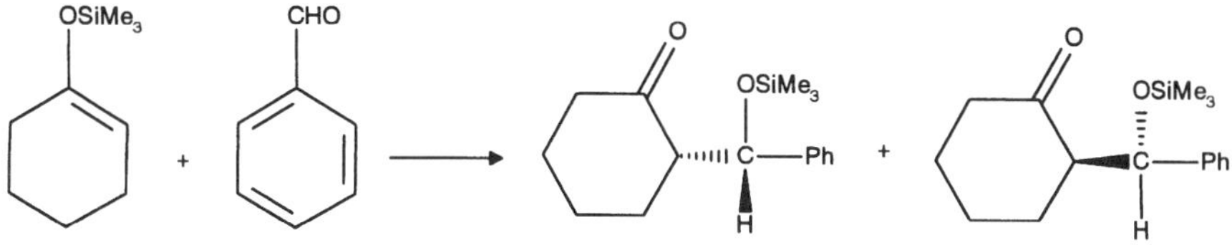

Scheme 4.22

TiCl$_4$, 0.1 MPa 25 : 75
1000 MPa 75 : 25

Base-catalyzed nitro-aldol reactions of carbonyl compounds with nitroalkanes often fail due to steric restrictions. For example, attempted additions of nitroalkanes to 2-methylcyclohexanone (**67**) under normal conditions are unsuccessful. However, nitromethane (**68**) adds to **67** under high pressure conditions (900 MPa, 30 °C), giving the nitroalcohol **69** in 50% yield (Scheme 4.23) [41].

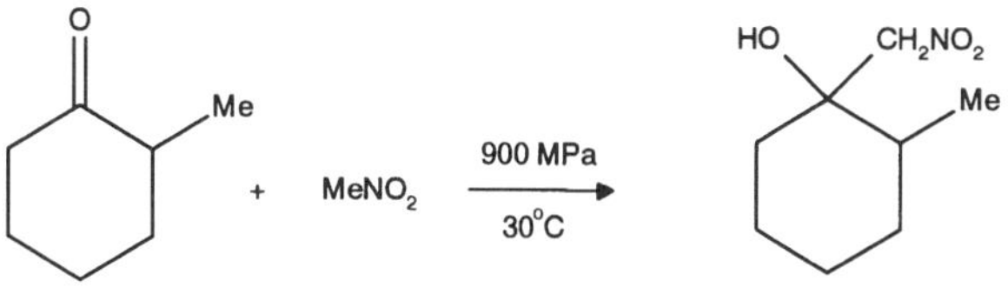

Scheme 4.23

The α-substitution of acrylonitrile (**20**) by aldehydes and ketones is one of the most pressure-sensitive reactions known. At atmospheric pressure only the most reactive aldehydes undergo addition. However, a quantitative yield of the product **71** is obtained from the reaction of acrylonitrile (**20**) with acetone (**70**) within a few minutes, under 900 MPa pressure and ambient temperature (Scheme 4.24) [42].

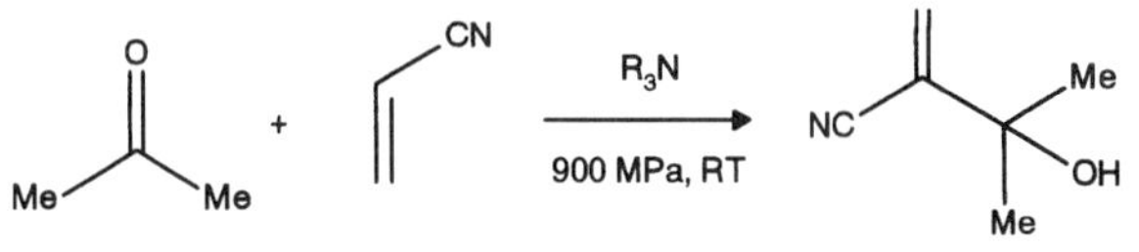

Scheme 4.24

Hydroboration is another example of an addition reaction that is highly sensitive to pressure. The atmospheric pressure hydroboration of sterically hindered olefins usually proceeds no further than the monoalkylborane stages, as in the case of dimethyl-di-*t*-butylethylene (**72**). When olefin **72** reacts with BH_3 under normal conditions, the monoalkylborane **74** is the only product formed. On the other hand, the monoalkylborane **74** reacts under high pressure (500 MPa) with two molecules of olefin **72** to give the extremely hindered trialkylborane **75** (Scheme 4.25) [43].

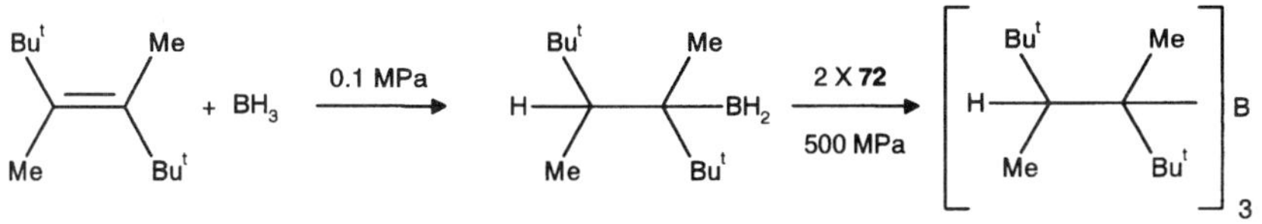

Scheme 4.25

Alcohols can be protected with acylating or silylating agents under high pressure conditions [44]. This is particularly useful when steric hindrance or lability of alcohols severely limits the scope of thermal reactions. The high pressure reaction (1500 MPa, 40 °C) of fenchone (**76**) with ethylene glycol (**77**) affords in 82% yield the acetal **78**, whereas this reaction fails to proceed on heating at atmospheric pressure (Scheme 4.26) [45].

Ionogenic reactions are characterized by a transition state that is more polar than the initial state. An increase in solvation causes a reduction of the volume, which is more significant when charge is created. Thus ionogenic reactions should be accompanied by very negative activation volumes that are strongly solvent dependent; thus these reactions should be accelerated by pressure, especially when reactions are carried out in less polar solvents. Consequently, some nucleophilic and electrophilic reactions that hardly occur at atmospheric pressure are readily facilitated under high pressure.

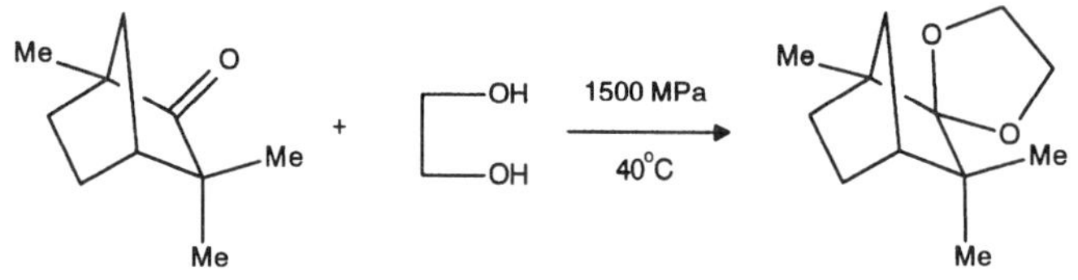

4.3. A SURVEY OF HIGH PRESSURE ORGANIC SYNTHESIS

Scheme 4.26

The formation of charged products from neutral substrates usually results in increased solvation, and consequently the volumes of activation of these reactions are often strongly negative (see Chapter 3). For example, the formation of a quaternary ammonium salt from a tertiary amine and an alkylating agent (Menshutkin reaction) has an activation volume within the range -20 to -50 cm^3 mol^{-1}. Such reactions should therefore, be dramatically accelerated by pressure. 2,6-Di-*t*-butylpyridine (**79**) reacts with methyl iodide (**80**) at high pressure (550 MPa, 60 °C) to give the methiodide **81** in 40% yield (Scheme 4.27). The reaction of **79** with methyl fluorosulfate under the same conditions affords the corresponding quaternary salt in quantitative yield [46].

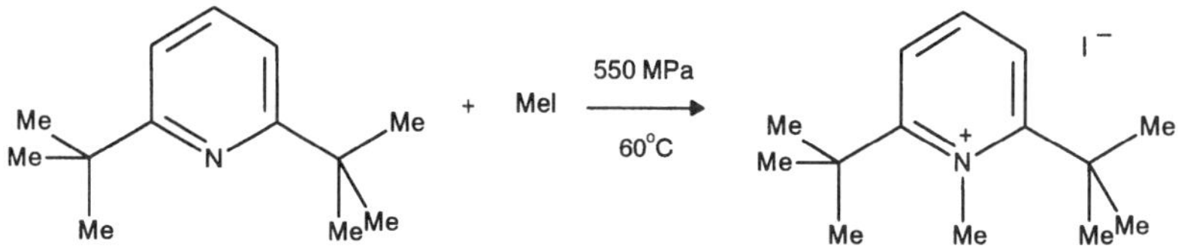

Scheme 4.27

A comparison of activation volumes of quaternization reactions of various 2,6-disubstituted pyridines with different alkyl iodides clearly indicates that the volume of activation is more negative for more sterically hindered reactions [47]. This means that the high pressure approach should be especially effective when highly hindered tertiary amines are quaternized. An analogous situation should pertain to the quaternization of tertiary amines with sterically hindered alkylating agents.

Diiodomethane (**83**) reacts under high pressure conditions (2000 MPa, 50 °C) with the alkaloid sporteine (**82**) to afford the bis-quaternary salt **84** in quantitative yield (Scheme 4.28) [48].

The latter result has inspired further investigations of the double quaternization of *N,N'*-dimethyl diazacoronands that could lead to the formation of cryptand frameworks [48]. When an equimolar mixture of *N,N'*-dimethyl diazacoronand (**86**) and bis(2-iodoethyl)ether (**85**), dissolved in acetone, is

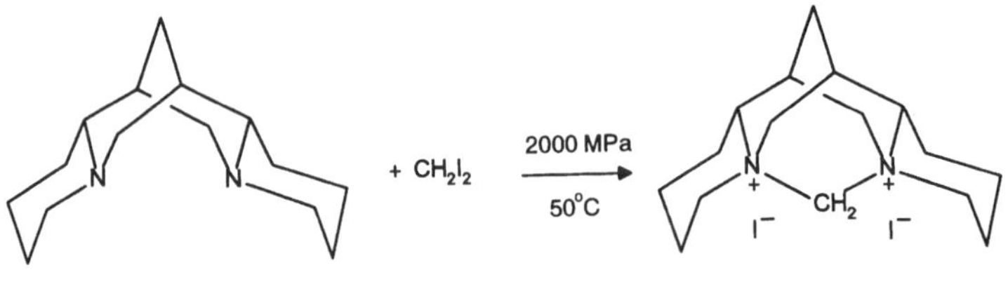

Scheme 4.28

subjected to 1000 MPa pressure at room temperature for a few hours, the bis-quaternary salt **87** precipitates quantitatively (Scheme 4.29) [49].

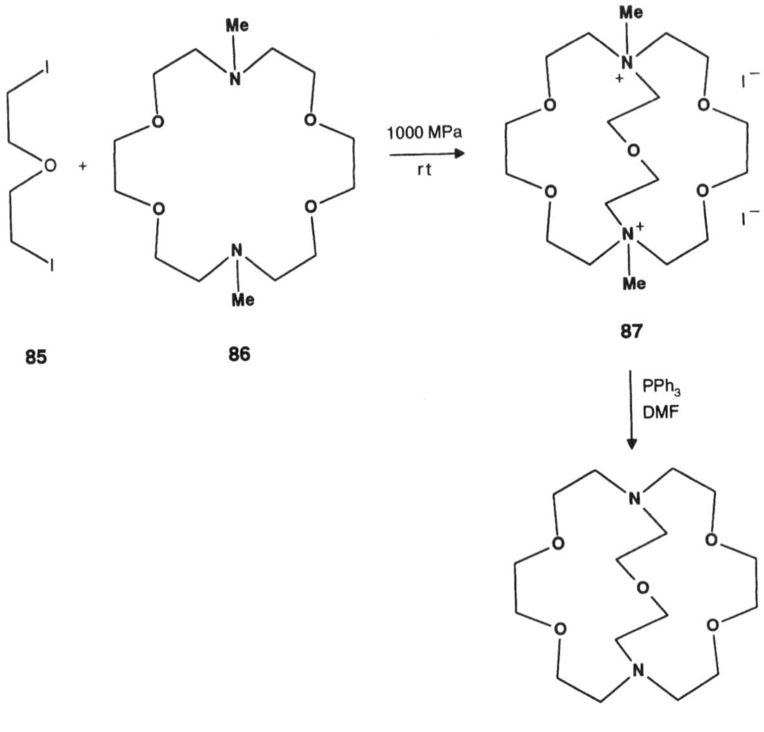

Scheme 4.29

The ultimate product, [2.2.1] cryptand (**88**), was obtained by treatment of the bis-quaternary salt **87** with triphenylphosphine in boiling dimethylformamide. In order to demonstrate the utility of the high pressure method, double quaternization reactions between other N,N'-dimethyl diazacoronands and various

4.3. A SURVEY OF HIGH PRESSURE ORGANIC SYNTHESIS

bridging components, including chiral compounds, were extensively studied [50, 51].

A convenient high pressure synthesis of a tricyclic cryptand-related compound is exemplified in Scheme 4.30. The high pressure reaction (1100 MPa, RT) between the simple N,N'-dimethyl diazacoronand (**89**) and its N,N'-diiodoamide derivative (**90**) gives in good yield the bis-quaternary salt **91**, which can readily be transformed into the desired tricyclic system [50, 52].

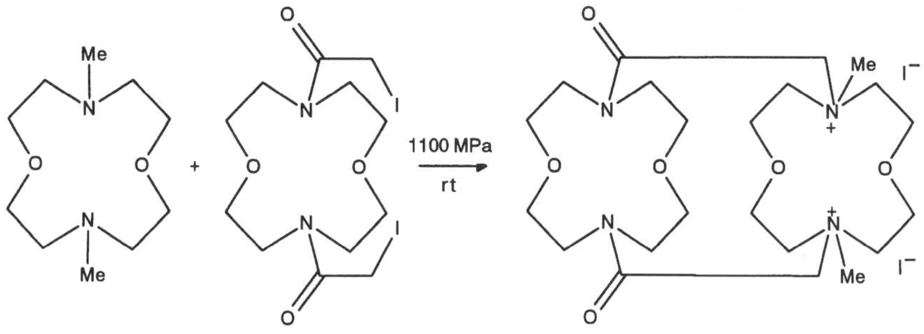

Scheme 4.30

It was expected that a tertiary open-chain α,ω-diamine would react under high pressure conditions with an appropriate α,ω-diiodo compound to form a cyclic bis-quaternary salt. Indeed, when the tertiary amine **92** reacted with the diiodo derivative **93** under 1000 MPa pressure at 30 °C, it afforded the desired bis-quaternary salt **94** in crystalline form (Scheme 4.31) [53, 54].

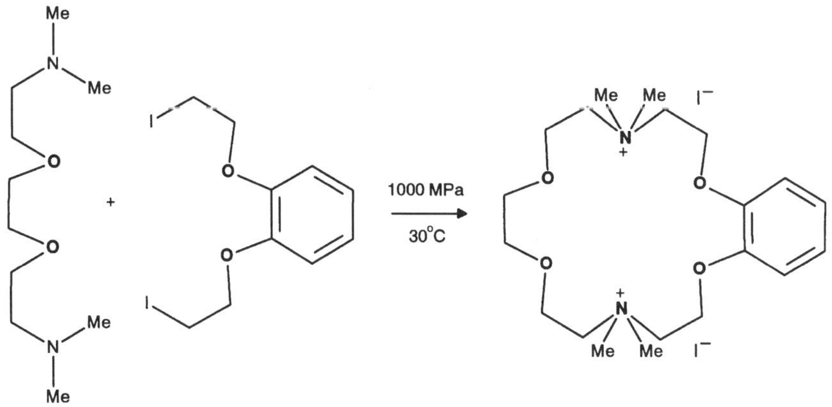

Scheme 4.31

Subsequent demethylation with L-Selectride® yielded the respective N,N'-dimethyl diazacoronand. In a similar manner several chiral diazacoronands were prepared [55]. N,N'-Dimethyl diazacoronands were also obtained by a double alkylation reaction involving secondary α,ω-diamines and typical α,ω-diiodo compounds [56].

Dichloromethane (**96**) can serve as a C_1 unit in the Mannich reaction. A number of Mannich bases were prepared under high pressure conditions from secondary amines and highly branched ketones [57]. For example, dichloromethane (**96**) reacts with piperidine (**95**) under 800 MPa pressure at 40 °C, yielding the corresponding ammonium salt, which with acetophenone (**97**) gives the desired aminomethylated product **98** in 93% yield (Scheme 4.32) [58].

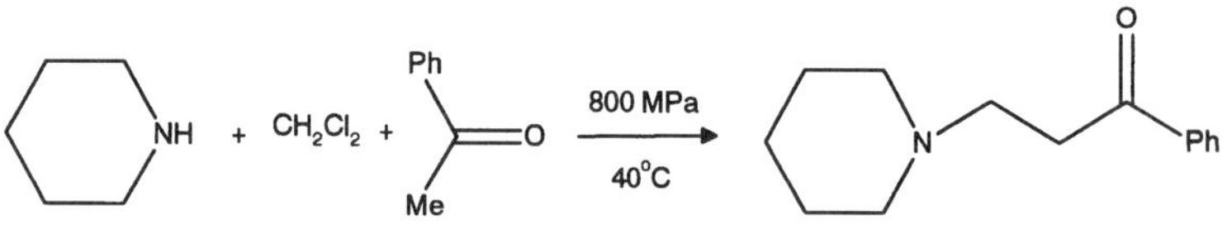

Scheme 4.32

Phosphonium salts for use in the preparation of Wittig reagents are efficiently formed under high pressure conditions [59]. The Wittig reaction itself is also accelerated by pressure. Consequently a convenient method for the synthesis of tri- and tetrasubstitued olefins is available. For example, the high pressure reaction (1000 MPa, 50 °C) of p-methoxybenzaldehyde (**100**) with triphenylphosphoranylideneacetone (**99**) affords the trisubstituted olefin **101** in 67% yield (Scheme 4.33) [60].

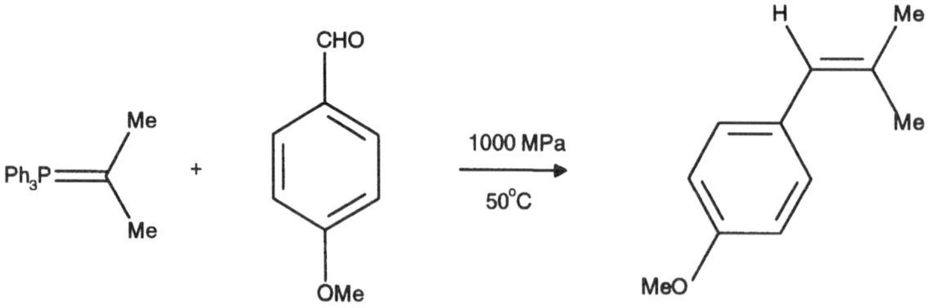

Scheme 4.33

Electrophilic aromatic substitution reactions have moderate negative activation volumes and should therefore be accelerated by pressure [3]. Unsubstituted or 2-alkylsubstituted five-membered heterocycles react under thermal and high pressure conditions with activated aldehydes and ketones, affording electrophilic aromatic substitution products [61, 62]. The high pressure reaction

(850 MPa, 70 °C) of 2-methylfuran (**102**) with methyl pyruvate (**57**) gives, in 80% yield, the expected electrophilic substitution product **103**, whereas under thermal, atmospheric pressure conditions the reaction fails (Scheme 4.34) [63].

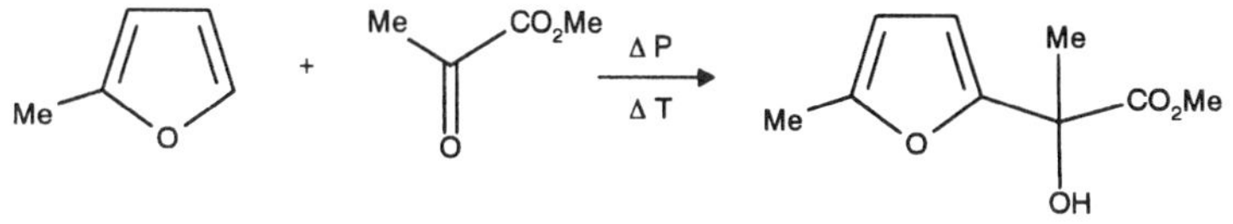

Scheme 4.34

The formation of compound **106** was interpreted earlier as following an "ene-like" reaction pathway [64]. However, recent studies [65] rule out this mechanism (Scheme 4.35). It seems that the course of this reaction is more complex, and that the first stage consists of an electrophilic substitution of a carbonyl compound at positions 2 or 3 of the furan ring. The products of this reaction constitute versatile synthons for syntheses of carbohydrates and other polyhydroxylated natural products. Thus, the asymmetric reaction of 2,5-dimethylfuran (**104**) with 2,3-O-isopropylidene-D-glyceraldehyde (**90**) was also studied successfully under high pressure [66].

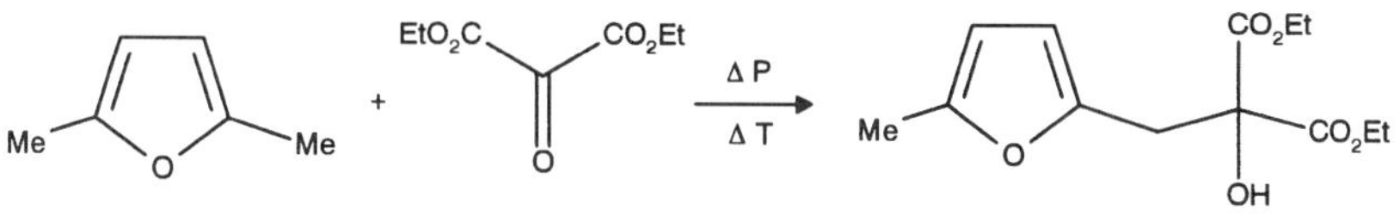

Scheme 4.35

4.4 CONCLUSIONS

The examples of application of high pressure techniques presented to illustrate the great advantages of this non-conventional method of organic synthesis. The scope of many important organic reactions can be extended by this methodology. High pressure, strongly affecting both reaction rates and equilibria, may be utilized for carrying out the syntheses of organic compounds which are difficult to obtain under either thermal or catalytic conditions. The influence of pressure on the regio- and stereoselectivity of organic reactions is extremely attractive from the point of view of total synthesis of natural products. High pressure asymmetric reactions are also of great importance, particularly because of the growing interest in optically active materials. In the light of recent results, the high pressure technique seems to be a powerful tool for the synthesis of complex organic molecules, such as cryptands, belts, collars and phthalocyanines, to mention only a few.

Further development of the potential of high pressure organic synthesis depends also on improvement of high pressure equipment. Designing larger units as well as reducing the cost of acquisition would extend the introduction of high pressure equipment for synthesis into organic laboratories. Efforts are being made to extend the applicability of the high pressure approach to organic synthesis on processes of industrial importance.

4.5 REFERENCES

[1] Dauben, W. G., Kozikowski, A. P., J. Am. Chem. Soc., 96 (1974) 3664.
[2] Dauben, W. G., Krabbenhoft, H. O., J. Am. Chem. Soc., 98 (1976) 1992.
[3] Matsumoto, K., Sera, A., Uchida, T., Synthesis, (1985) 1; Matsumoto, K., Sera, A., Synthesis, (1985) 999.
[4] Jurczak, J., Physica, 139/140B (1986) 709.
[5] Jurczak, J., (le Noble, W. J. (ed.)), Synthesis in Organic High Pressure Chemistry, p. 304, Elsevier, Amsterdam, 1988.
[6] Jurczak, J., High Press. Res., 1 (1989) 99.
[7] Isaacs, N. S., Tetrahedron, 47 (1991) 8463.
[8] Jenner, G., Polish J. Chem., 66 (1992) 1535.
[9] Jurczak, J., Baranowski, B. (eds.), High Pressure Chemical Synthesis, Elsevier, Amsterdam, 1989.
[10] Matsumoto, K., Acheson, R. M. (eds.), Organic Synthesis at High Pressures, Wiley, New York, 1991.
[11] Jurczak, J., Chmielewski, M., Filipek, S., Synthesis, (1974) 41.
[12] Jurczak, J., Golebiowski, A., Bauer, T., Synthesis, (1985) 928.
[13] El'yanov, B. S., Shakhova, S. K., Polkovnikov, B. D., Rar, L. F., J. Chem. Soc., Perkin Trans., 2 (1985) 11.
[14] Rimmelin, J., Jenner, G., Rimmelin, P., Bull. Soc. Chim. Fr., (1978) 461.
[15] Jurczak, J., Kozluk, T., Filipek, S., Eugster, C. H., Helv. Chim. Acta, 65 (1983) 222.
[16] Jurczak, J., Kozluk, T., Tkacz, M., Eugster, C. H., Helv. Chim. Acta, 66 (1983) 218.
[17] Jurczak, J., Kawczynski, A. L., J. Org. Chem., 50 (1985) 1106.
[18] Kotsuki, H., Nishizawa, H., Kitagawa, S., Ochi, M., Yamasaki, N., Matsuoka, K., Tokoroyama, T., Bull. Chem. Soc. Jpn., 52 (1979) 544.
[19] Gladysz, J. A., Chemtech., (1979) 373.
[20] Dauben, W. G., Gerdes, J. M., Smith, D. B., J. Org. Chem., 50 (1985) 2576.
[21] Ashton, P. R., Isaacs, N. S., Kohnke, F. H., Mathias, J. P., Stoddart, J. F., Angew. Chem. Int. Ed. Engl., 28 (1989) 1258.
[22] Ashton, P. R., Isaacs, N. S., Kohnke, F. H., D'Alcontres, G. S., Stoddart, J. F., Angew. Chem. Int. Ed. Engl., 28 (1989) 1261.
[23] Chmielewski, M., Jurczak, J., J. Org. Chem., 46 (1981) 2231.
[24] Jurczak, J., Bauer, T., Tetrahedron, 42 (1986) 5045.

4.5. REFERENCES

[25] Jurczak, J., Bauer, T., Kihlberg, J., J. Carbohydr. Chem., 4 (1985) 447.
[26] Chmielewski, M., Jurczak, J., J. Carbohydr. Chem., 6 (1987) 1.
[27] Golebiewski, A., Jacobsson, U., Jurczak, J., Tetrahedron, 43 (1987) 3063.
[28] Dauben, W. G., Krabbenhoft, H. O., J. Org. Chem., 42 (1977) 282.
[29] Jenner, G., Abdi-Oskoui, H., Rimmelin, J., Bull. Soc. Chim. Fr., (1979) 33.
[30] Vor der Brüch, D., Buhler, R., Plieninger, H., Tetrahedron, 28 (1972) 791.
[31] Scheeren, J. W., Recl. Trav. Chim. Pays-Bas, 105 (1986) 71.
[32] Scheeren, J. W., (Jurczak, J., Baranowski, B. (eds.)), Synthesis of Carbocyclic Systems in High Pressure Chemical Synthesis, p. 168, Elsevier, Amsterdam, 1989.
[33] Aben, R. W. M., Scheeren, J. W., Tetrahedron Lett., 24 (1983) 4613.
[34] Chmielewski, M., Kaluza, Z., Belzecki, C., Salanski, P., Jurczak, J., Adamowicz, H., Tetrahedron, 41 (1985) 2441.
[35] Dicken, C. M., DeShong, P., J. Org. Chem., 47 (1982) 2047.
[36] Gladysz, J. A., Yu, Y. S., J. Chem. Soc., Chem. Commun., (1978) 599.
[37] Dauben, W. G., Gerdes, J. M., Tetrahedron Lett., 24 (1983) 3841.
[38] Bunce, R. A., Schlecht, M. F., Dauben, W. G., Heathcock, C. H., Tetrahedron Lett., 24 (1983) 4943.
[39] Yamamoto, Y., Maruyama, K., Matsumoto, K., Tetrahedron Lett., 25 (1984) 1075.
[40] Yamamoto, Y., Maruyama, K., Matsumoto, K., J. Am. Chem. Soc., 105 (1983) 6963.
[41] Matsumoto, K., Angew. Chem. Int. Ed. Engl., 23 (1984) 619.
[42] Hill, J. S., Isaacs, N. S., Tetrahedron Lett., 27 (1986) 5007.
[43] Rice, J. E., Okamoto, Y., J. Org. Chem., 47 (1982) 4189.
[44] Dauben, W. G., Bunce, R. A., Gerdes, J. M., Henegar, K. E., Cunningham, A. C., Ottoboni, T. B., Tetrahedron Lett., 24 (1983) 5709.
[45] Dauben, W. G., Gerdes, J. M., Look, G. C., J. Org. Chem., 51 (1986) 4964.
[46] Hou, C. J., Okamoto, Y., J. Org. Chem., 47 (1982) 1977.
[47] Jenner, G., Angew. Chem. Int. Ed. Engl., 14 (1975) 137.
[48] Jurczak, J., Pietraszkiewicz, M., Top. Curr. Chem., 130 (1985) 183.
[49] Pietraszkiewicz, M., Salanski, P., Jurczak, J., Bull. Pol. Ac. Chem., 33 (1985) 433.
[50] Jurczak, J., Ostaszewski, R., J. Coord. Chem., 27 (1992) 201.
[51] Pietraszkiewicz, M., Salanski, P., Jurczak, J., Tetrahedron, 40 (1984) 2971.
[52] Jurczak, J., Ostaszewski, R., Polish J. Chem., 66 (1992) 1733.
[53] Jurczak, J., Ostaszewski, R., Salanski, P., Stankiewicz, T., Tetrahedron, 49 (1993) 1471.
[54] Stankiewicz, T., Jurczak, J., Polish J. Chem., 66 (1992) 1743.
[55] Salanski, P., Ostaszewski, R., Jurczak, J., High Press. Res., 13 (1994) 35.
[56] Jurczak, J., Ostaszewski, R., Synth. Commun., 19 (1989) 2175.
[57] Matsumoto, K., Uchida, T., Hashimoto, S., Yonezawa, Y., Iida, H., Kahedi, A., Otani, S., Heterocycles, 36 (1993) 2215.
[58] Matsumoto, K., Angew. Chem. Int. Ed. Engl., 21 (1982) 922.
[59] Dauben, W. G., Gerdes, J. M., Look, G. C., Synthesis, (1986) 532.
[60] Nonenmacher, A., Mayer, R., Plieninger, H., Ann. Chem., (1983) 2135.

[61] Jenner, G., Rimmelin, J., Bull. Soc. Chim. Fr., (1981) 65.
[62] Jurczak, J., Belniak, S., High Press. Res., 11 (1992) 119.
[63] Jenner, G., Papadopoulos, M., Jurczak, J., Kozluk, T., Tetrahedron Lett., 25 (1984) 5747.
[64] Jurczak, J., Kozluk, T., Pikul, S., Salanski, P., J. Chem. Soc., Chem. Commun., (1983) 1447.
[65] Jurczak, J., Belniak, S., Chmielewski, M., Kozluk, T., Pikul, S., Jenner, G., J. Org. Chem., 54 (1989) 4469.
[66] Jurczak, J., Pikul, S., Tetrahedron, 44 (1988) 4569.

5

INORGANIC AND RELATED CHEMICAL REACTIONS IN SUPERCRITICAL FLUIDS

Martyn Poliakoff*, Michael W. George and Steven M. Howdle
Dept. of Chemistry, University of Nottingham, Nottingham, England, NG7 2RD

5.1 INTRODUCTION
5.2 EXPERIMENTAL TECHNIQUES
5.3 "TUNABLE" DENSITY
5.4 SUPERCRITICAL FLUIDS AS SOLVENTS
    5.4.1 Solubility
    5.4.2 Rapid Precipitation
    5.4.3 Antisolvents, Micelles and Electrochemistry
    5.4.4 Unusual Solubilities
    5.4.5 High Solubilities of Gases
5.5 CHEMICAL INERTNESS AND THE ACTIVATION OF SUPERCRITICAL CO_2
5.6 SURFACE TENSION AND VISCOSITY
5.7 WATER AND "OTHER" SUPERCRITICAL FLUIDS
5.8 SUPERCRITICAL FLOW-REACTORS AND SCALE-UP
5.9 CONCLUSION
5.10 REFERENCES

Over the past decade, there have been substantial advances in the chemical applications of fluids with critical temperatures close to ambient, *i.e.*, C_2H_4, CO_2, C_2H_6, etc. Much current interest derives from the potential of supercritical (sc) CO_2 as an environmentally acceptable replacement for more conventional solvents. However, supercritical fluids have a range of unusual properties which can be exploited for reactions, which are qualitatively different from standard laboratory chemistry. This chapter aims to show how these properties have been exploited chemically and to indicate some of the avenues which are still to be explored. In particular, we give a brief introduction to these unusual fluids, outline some of the experimental problems and how they may overcome, and then describe a number of reactions in inorganic and related chemistry where the unique properties of supercritical fluids have been exploited to particular advantage.

5.1 INTRODUCTION

A supercritical fluid is any gas, the temperature and pressure of which are above their critical values, T_c and p_c respectively. In practice, the definition is restricted to fluids close to their critical points and, hence, close to their critical density ρ_c. There is a considerable number of fluids with critical temperatures close to ambient and Table 5.1 lists just a small selection, together with a few substances of chemical importance with rather higher values of T_c.

Figure 5.1 shows a generalized phase diagram for a pure substance with qualitative indications of the density of that substance in the various regions of T, p space. The most important feature of this diagram is the indication that, within the supercritical region, there is no phase boundary between the gas and liquid phases. This means that there is a continuity in physical properties of the fluid between the gas and liquid states. The consequence is that supercritical fluids have properties, which are a curious hybrid of those normally associated with liquids and gases. Thus, under most conditions, the viscosities and diffusivities are similar to those of gases while the density is closer to that of a liquid. It is this combination of properties which has continued to fascinate physical

Table 5.1 Critical data for selected substances[a]

Substance	T_c/K	p_c/MPa	ρ_c/gml
C_2H_4	282.3	5.041	0.214
Xe	289.7	5.84	1.110
C_2F_6	293.0	3.06	0.622
CHF_3	299.3	4.858	0.528
CO_2	304.1	7.375	0.468
C_2H_6	305.4	4.884	0.203
N_2O	309.6	7.255	0.452
$C_2H_4(NH_2)_2$	593	6.27	0.29
H_2O	647.1	22.06	0.322

[a] Further data can be found in [76].

5.1. INTRODUCTION

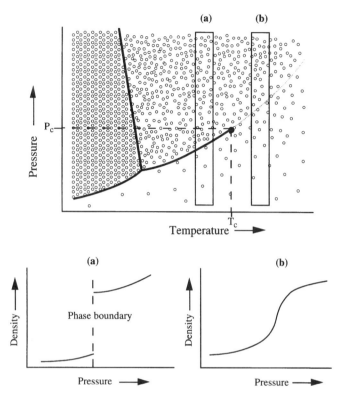

Figure 5.1 Schematic phase diagram illustrating the variable density of a supercritical fluid. The two plots show the contrasting effects of increasing pressure on the density of the fluid in the regions marked "a" and "b" on the phase diagram. In (a) increasing pressure leads to condensation of the gas as the phase boundary is crossed, with a corresponding step-increase in the fluid density. On the other hand, in the supercritical region (b), increasing the pressure increases the fluid density smoothly and continuously with the greatest rate of change in the region of the extrapolated liquid/vapor equilibrium curve. It is this behavior which allows the supercritical fluid to be exploited as a "tunable" solvent. (Adapted from [23].)

chemists for over 150 years. Any historical account of supercritical fluids is beyond the scope of this chapter and readers are strongly recommended to read the historical chapter in McHugh and Krukonis's book [1]. However, it is important to warn those new to the field that supercritical fluids were researched intensively by most of the leading physical chemists in the 19th century. Therefore, the experiment which you are planning to carry out, may have been performed already very many years ago, as in the classic work of the Scottish chemists Hannay and Hogarth [2, see below].

Supercritical fluids have a surprisingly long history as solvents for reaction chemistry, beginning with the high pressure polymerization of ethylene [1]. The success of this process contains an important lesson for contemporary supercritical chemists; the process was adopted, despite the difficulties of working at high pressures, because it promoted a reaction (the formation of polyethylene) *which at that time could not be carried out in any other way.* Thus, pioneering work [3] by Plesch in 1960 on the polymerization of isobutene in liquid CO_2 was sadly curtailed because the advantages of the process were not apparent. Today, the situation is different. Environmental concerns are creating a distinct role for supercritical solvents, particularly $scCO_2$, which are currently regarded as potentially cleaner solvents for a whole variety of chemical reactions. Nevertheless, the capital costs of supercritical equipment remain high, even on a laboratory scale, and so the need to demonstrate real chemical advantages for supercritical chemistry is as strong as ever.

Supercritical reactions have been reviewed several times over the last few years [4] and we do not intend to repeat these excellent reviews in this chapter. Instead, we will discuss some of the properties of supercritical fluids which differentiate them from more conventional solvents: we describe briefly how they have been exploited already and then indicate a few of the directions in which they might still be developed. Before discussing the chemistry, we describe some of the experimental problems and how they can be tackled.

5.2 EXPERIMENTAL TECHNIQUES

All supercritical fluid experiments involve high pressures but these are generally much lower than those associated with the high pressure experiments described in Chapters 2–4 in this book. Typically, supercritical experiments are carried out at pressures below 400 bar; nevertheless, the experimental problems can appear quite daunting to those inexperienced in the field. In reality, the difficulties are probably less than the novice would imagine but the potential dangers should never be ignored and full safety assessments should be made for all experiments. Fortunately, most of the basic items needed for supercritical experiments are available commercially and there are comprehensive texts covering the design of more specific equipment such as optical cells [5]. Here, we just give a brief indication of the equipment which is required.

Figure 5.2 shows a manual pump of a type familiar to those involved in high pressure experiments. It has been modified for use with supercritical fluids (*e.g.*, $scCO_2$, scC_2H_4, etc.) by wrapping the barrel with a copper coil through which cold N_2 gas can be passed to cool the pump sufficiently for filling with the liquefied gas. Such pumps are simple to use and ideal for a wide range of experiments.

As will become clear in the course of this chapter, we believe that spectroscopy is the key to *in situ* monitoring of chemical processes in supercritical solution so that these processes can be optimized in real time. Such optimization is considerably more important in supercritical fluids than in conventional solvents because the tunability of the fluids results in a greater number of parameters

5.2. EXPERIMENTAL TECHNIQUES

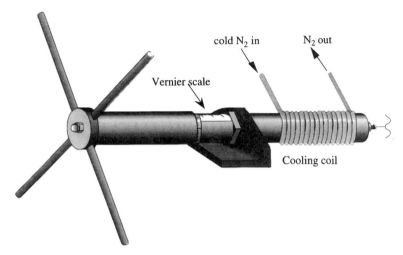

Figure 5.2 A manually operated high-pressure pump adapted for use with supercritical fluids. (Adapted from [23].)

which can affect the outcome of a reaction. Thus, the chances of hitting the optimal conditions purely by "trial and error" are much less in supercritical solution than in conventional reactions. Figure 5.3 illustrates an optical cell, developed at Nottingham [6] for use in our studies of photochemical reactions and spectroscopic investigations. The cell is specifically designed for ease of dismantling so that reaction residues can be rapidly cleaned out. The basic design can be adapted for a variety of uses which have already included

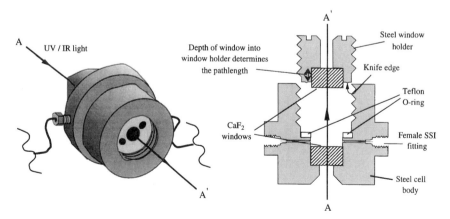

Figure 5.3 A miniature high-pressure optical cell for use with supercritical fluids. Left, overall view; right, cross section drawn on a larger scale to show the main features of the construction. (Adapted from [23].)

photochemistry, FT Raman spectroscopy, flow-reactors, time-resolved IR spectroscopy, formation of inverse micelles, H-bonding and polymerization. Further discussions of optical cells can be found in recent reviews [4b, 7].

Much of the published work on supercritical chemistry has been carried in autoclaves or similar high pressure vessels. This approach has the advantage of simplicity but spectroscopic monitoring is relatively difficult in such vessels [5b]. An interesting idea exploited by Kolis *et al.* [8] combines the advantages of the conventional test-tube with the high pressure capabilities of an autoclave. Their method involves sealing the reactants plus a solvent which is liquid at ambient temperatures, *e.g.*, ethylene diamine, into a thin-walled glass capillary and then heating this in a pressurized autoclave; the reaction mixture can then be recovered and worked up as if the whole reaction had been carried out in the thin-walled tube. A somewhat similar approach has been used to obtain very high pressures of Xe for the formation of the endohedral complex C_{60}@Xe [9]. Here, the Xe was sealed into a copper rather than glass tube, which was deliberately intended to collapse under pressure to raise the pressure of Xe into the kilobar region.

Our research group favors the use of continuous flow-reactors whenever possible because these can frequently offer significant advantages over batch reactions in sealed autoclaves. The low viscosity and good thermal and mass transport properties of supercritical fluids lend themselves particularly to the use of flow systems. In any flow-reactor, such as that shown schematically in Figure 5.4, residence times are shorter than in batch reactors [10]. Reactions can, therefore, run at higher temperatures with a reduced reaction time and a correspondingly increased throughput for a reactor of given volume. This increase in throughput is of particular advantage for supercritical chemistry, both on grounds of safety and of cost. Those who use supercritical flow-reactors on an industrial scale are understandably reticent about the precise details of their process. Pickel and coworkers at Hoffman La Roche are using flow systems for catalytic hydrogenation on a commercial scale [11]. The supercritical process has permitted a $\times 25$ reduction in the volume of the high pressure reactor compared to the previous process without any loss in throughput. The principal reaction, heterogeneous hydrogenation, benefits from the high solubility of H_2 in supercritical solvents [12], see below.

At Nottingham, we are developing miniature flow-reactors for use on a laboratory scale [13]. Our reactor is based on modular components as shown schematically in Figure 5.5; valves, sensors, optical cells, etc., are all mounted on small magnets so that they can be rapidly moved around on the steel bench top (see below and ref. [14] for a more complete description). This design gives

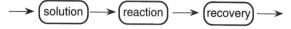

Figure 5.4 A block diagram of a flow-reactor, illustrating the spatial separation of the three processes, dissolution of the reactant, the reaction itself and, finally, the separation of product and solvent.

5.3. "TUNABLE" DENSITY

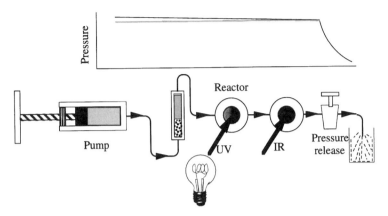

Figure 5.5 A schematic view of a supercritical flow-reactor for supercritical photochemistry. The key features are the spectroscopic monitoring for real-time optimization and the rapid precipitation of the product. The plot above the reactor emphasizes the fact that there is an extremely small pressure drop along the length of the reactor up to the point of pressure release. (Adapted from [23, 69].)

considerable flexibility in the layout of the reactor which can be rapidly modified to suit individual experiments. Safety will always be of overriding importance in supercritical experiments and the very small volume of these reactors combined with the use of commercial chromatographic fittings reduces the scale of any safety problem significantly.

Many supercritical fluids are highly flammable including scC_2H_4, scC_2H_6, $scNH_3$, etc. Thus, it is important to consider using pneumatically operated pumps in any investigation where the work extends beyond $scCO_2$. Our group at Nottingham has made extensive use of such equipment; indeed, it is a key component in the flow-reactors described above. Pumps suitable for such work are produced by NWA GmbH, Lörrach, Germany.

We now describe how the unusual properties of supercritical fluids have been exploited chemically. We begin by considering tunable density, which is held by some to be the supreme advantage of supercritical fluids. Then, we discuss solvent properties and particularly the high solubility of H_2 in supercritical solution, which already has been exploited both on laboratory and commercial scales. In subsequent sections, we describe the unusual solubility of fluorocarbons in $scCO_2$, the use of scH_2O and other high T_c fluids for synthesizing new materials and finally we discuss the scaling-up of reactions by means of continuous or semi-continuous flow-reactors.

5.3 "TUNABLE" DENSITY

A unique feature of supercritical solvents is the possibility of changing the density of the fluid by varying the applied pressure. Several other properties change in parallel with this change in density; those more relevant to reactions include viscosity, solvent power, diffusivity and dielectric constant, where the magnitude of the change varies strongly from fluid to fluid, low in $scCO_2$ but high in $scCHF_3$. The sensitivity of all of these properties to pressure is highest in the vicinity of the critical point because it is here that the fluid is most compressible [1].

The precise control of chemical reactions is one of the principal goals of chemical science. For a given set of reactants in a conventional solvent, the controllable parameters are normally restricted to concentration, temperature and pressure. Usually, the dependence on pressure is small and significant effects are only seen with very high pressures (>1 kbar). Thus, the tunable density of supercritical fluids provides an extra variable with which to "tune" chemical reactions. This possibility has inspired a sizable number of studies, most of which have centered on reactions with proven sensitivity to solvent (*e.g.*, Diels-Alder [15]), where significant changes in rate or product distribution have already been reported for different conventional solvents, as discussed in more detail in Chapter 6.

In some of these reactions, changes in density of the supercritical solvent have been accompanied by dramatic changes in rate constants. By contrast, the effect on product distribution has been surprisingly small. One of the most striking examples is the photo-chemical dimerization of isophorone in $scCO_2$ and in $scCHF_3$ (Scheme 5.1), changing the density of the fluid was found to change the relative yields of the three stereochemically distinct dimers [16]. These changes were attributed to the fact that formation of the three dimers involves different transition states which have significantly different dipole moments and, hence, interact with the solvent to different extents. Varying the solvent density varies the degree of interaction and changes the relative energies of the transition states [16].

Recent results [17] have shown that the density of $scCHF_3$ can also be used to alter the equilibrium between the *trans* and *gauche* rotamers of $(CF_3)_2CHOH$. Increasing density favors the *gauche* form, Scheme 5.2, an effect attributed to the increase in dielectric constant of CHF_3 which accompanies the increase in density. By contrast, almost no change was observed in the *trans:gauche* ratio on increasing the density of a solution of $(CF_3)_2CHOH$ in $scSF_6$, where the dielectric constant is almost unaffected by changes in supercritical density.

Ben-Amotz et al. have suggested a somewhat different effect which has potential for tuning product distribution [18]. Their suggestion involves a

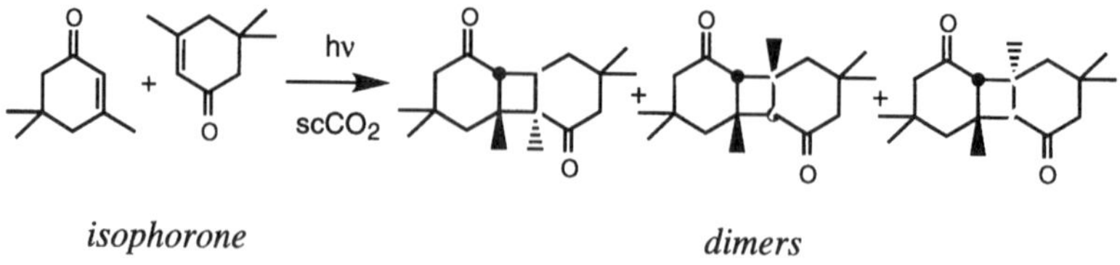

isophorone   *dimers*

Scheme 5.1

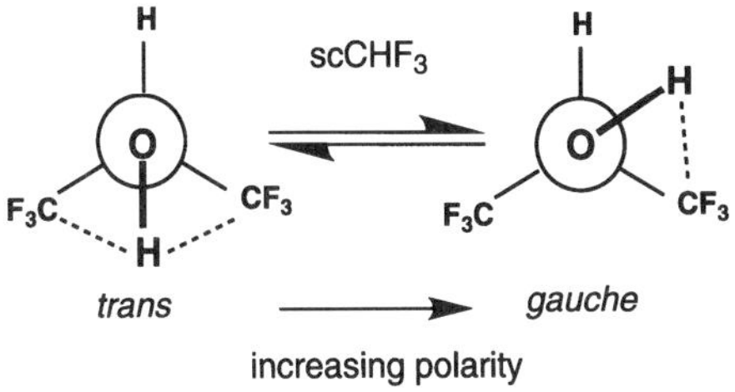

5.3. "TUNABLE" DENSITY

Scheme 5.2

reaction leading to two or more products, each of which is reached by a path involving a different vibrational mode of the same reactant molecule. They had already observed that pressure-induced shifts in the Raman bands of *near-critical* C_2H_6 were significantly different for different vibrational modes, *i.e.*, v(C–H) and v(C–C). Thus, they suggested that in such a multi-path reaction, the different paths should show a different pressure dependence, just like the modes of C_2H_6, and therefore changes in pressure should be reflected by changes in product distribution. This suggestion still remains hypothetical but clearly needs investigating.

There have been a large number of studies on the effects of solvent density on reaction kinetics but these have been discussed at length elsewhere [19]. Much of this work has relied on transient UV-visible absorption or emission spectroscopy. Such techniques have the advantages of high sensitivity but, because the bands are usually broad, it is rarely possible to monitor very many different species simultaneously in the same solution. We have recently begun to apply nanosecond time-resolved IR spectroscopy (TRIR) [20] to supercritical fluid solutions to monitor the reactions of intermediates in organometallic reactions. TRIR has a great advantage for monitoring complex reactions because IR bands, particularly in the v(C–O) region of metal carbonyls, are relatively narrow. Thus, it is possible to follow the behavior of several different chemical species without significant overlap between their IR absorption bands. Our preliminary results indicate that, in $scCO_2$, diffusion controlled reactions (*i.e.*, radical recombination) are slowed down by increasing solvent density as a consequence of the decreased diffusivity in the denser medium, see Figure 5.6. By contrast, the rates of unimolecular isomerization appear to be almost independent of solvent density under the same conditions [21].

In a related study [22] involving flash photolysis of metal carbonyl compounds the kinetics of ring closure for $M(CO)_5$L–L compounds (L–L = bidentate ligand) were measured (eq. 1).

$$M(CO)_5\text{L–L} \xrightarrow{scCO_2} M(CO)_4(\text{L–L}) + CO \qquad (1)$$

These studies in $scCO_2$ and scC_2H_6 revealed an extremely high volume of activation for the reaction, approaching $+7 \times 10^3 \, cm^3 \, mol^{-1}$ at temperatures

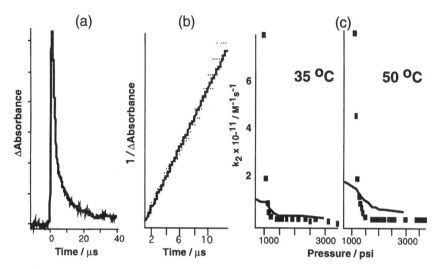

Figure 5.6 Showing (a) the Time-resolved IR decay trace of the CpMo(CO)$_3$ radical (monitored at 2011 cm^{-1}) obtained by UV photolysis (355 nm) of [CpMo(CO)$_3$]$_2$ in scCO$_2$ (37 °C, 1870 psi) and (b) the corresponding plot of the data, showing the good fit to second order kinetics; (c) pressure dependence of k_2, rate constant for radical recombination; the solid line represents the calculated diffusion controlled rate constants. Scale: 14.7 psi = 1 atm. (Adapted from [21].)

close to T_c. This high value reflects the high compressibility of these fluids close to the critical point.

Overall, the general implication of "tuning" studies is that they provide a unique method for understanding the role of the solvent in supercritical reactions. However, there are problems in making use of these properties in preparative-scale reactions. These problems arise because pressure effects are often only significant under conditions close to the critical point, where partial molar volumes have their extreme values. Any preparative scale reaction will require at least moderately concentrated solutions and the presence of dissolved reactants and products will therefore affect the critical point of the reaction mixture. As the reaction proceeds, reactants will be converted into products, the composition of the mixture will change and so will the critical parameters. Thus, without sophisticated monitoring, it will be very difficult to maintain the mixture close enough to its critical point throughout the course of the reaction to exploit the pressure effects to the full (see also Section 5.7, below).

5.4 SUPERCRITICAL FLUIDS AS SOLVENTS

5.4.1 Solubility

One of the most counter-intuitive properties of supercritical fluids is their ability to dissolve solid substances. Even those familiar with supercritical science

5.4. SUPERCRITICAL FLUIDS AS SOLVENTS

remain fascinated by the fact that solid materials can be dissolved in a gas, albeit a very dense gas. Hannay and Hogarth [2] observed the first example of supercritical solubility (cobalt chloride dissolved in supercritical ethanol) and, since then, the effect has continued to be investigated intensively [1]. As a rough guide, solubility of a given substance depends on the vapor pressure of that substance and the density of the fluid. Thus, solubility increases with temperature (at constant density) or with pressure (at constant temperature).

Figure 5.7 shows IR spectra, which demonstrate the increasing solubility of $W(CO)_6$ in C_2H_4 with increasing pressure [23]. It is clear from the plots that the solubility increases in a very non-linear manner as the pressure is increased because, like other supercritical fluids, the behavior of C_2H_4 is far from ideal. Furthermore, the changing contour of the bands of C_2H_4 in Figure 5.7 illustrates the change from a gas-like to a liquid-like fluid as the pressure is increased. Similar solubility plots are observed for a wide variety of fluids [1].

Although solubility increases with temperature *at constant density*, the same is not true for heating at constant pressure, because under those conditions the solubility can actually decrease [1]. Heating at constant pressure causes a decrease in fluid density and, at temperatures close to T_c, the effect of this drop in density more than cancels the effect of increasing temperature with a concomitant drop in solubility. This so-called *retrograde precipitation* can be exploited to deposit material from supercritical solution onto a heated surface, see Section 5.3.2. Retrograde precipitation is only observed quite close to the critical point and, above a particular temperature, the *cross-over temperature* (T_{cross}) solubility increases again with increasing temperature. For a given supercritical solvent, different solutes may well have different cross-over temperatures. Attempts have been made to use such differences in T_{cross} to separate mixtures by selective retrograde precipitation [24]. This retrograde effect should also have exploitable chemical applications (see Section 5.4.2).

Efforts have been made to quantify the solvent powers of supercritical fluids, particularly $scCO_2$. One approach has been to measure the precise value of the UV-Vis absorption maximum of a suitable probe molecule as the density of the fluid is increased [25]. Then the resulting solvatochromism is used to relate the solvent properties of $scCO_2$ at a particular temperature and pressure to that of a conventional solvent, which produces the same solvatochromic shift. However, a more practical guide for the average user of $scCO_2$ is to assume that the solvent properties are broadly similar to those of a light alkane, *e.g.*, *n*-heptane; with a few exceptions, *e.g.*, fluorocarbons (see Section 5.3.4), it is reasonable to assume that, if something will not dissolve in *n*-heptane, it is unlikely to be significantly soluble in $scCO_2$.

One method, commonly used to improve the solvent power of $scCO_2$, is to add a relatively small amount (5–10%) of an organic solvent to the $scCO_2$. Addition of these so-called modifiers can result in a dramatic enhancement of solubility. The most frequently used modifier is CH_3OH and it is possible to buy cylinders of premixed CO_2/CH_3OH [26]; addition of CH_3OH improves the solubility of polar materials in $scCO_2$, an effect which is due at least in part to

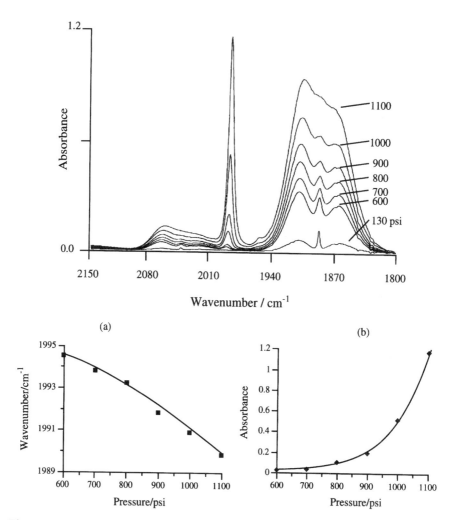

Figure 5.7 Series of IR spectra showing the increase in solubility of $W(CO)_6$ with increasing pressure of ethene. The graphs show the variation of (a) the position and (b) the intensity of the main absorption band of $W(CO)_6$ with pressure of ethene. Increasing the pressure of ethene increases the solubility of $W(CO)_6$ and also causes the band to shift to lower wavenumber. Scale: 14.7 psi = 1 atm. (Adapted from [23].)

$CH_3OH \rightarrow$ solute hydrogen-bonding [27]. Hydrogen-bonding can also be an important factor in the action of other modifiers, perhaps indeed, of most modifiers. For example, a combination of IR and UV spectroscopy has shown that hydrogen bonding is directly associated with the increased solubility of p-HOC_6H_4OH in $scCO_2$ modified with $(n$-$BuO)_3P{=}O$ [28]. In general, modifiers have not been widely used for reaction chemistry in supercritical fluids because of the possibility of the modifiers taking part in the reaction itself.

5.4. SUPERCRITICAL FLUIDS AS SOLVENTS

Furthermore, for larger-scale reactions, particularly those involving organic molecules the reactant(s) can act as their own modifiers.

5.4.2 Rapid Precipitation

Undoubtedly, the most important "tunable" property of supercritical fluids is its tunable solvent power [1]. Not only is this the basis of supercritical fluid chromatography [29] but also it features in many chemical reactions and processes, where reduction in pressure and associated expansion of the solvent is an efficient method for recovering solid products from supercritical reactions (Figure 5.8). However, this so-called "RESS" process [1, 30] (Rapid Expansion Supercritical Solution) has already proved itself to have applications much wider than just recovering reaction products, because adjustment of the expansion conditions allows the morphology of the precipitated solids to be manipulated [31].

Although there are still relatively few examples of combining RESS with chemical reactions, the work of Sievers and coworkers is particularly interesting [32]. Their experiments involved formation of thin metal-containing films on inorganic substrates such as silicon. Appropriate metallic precursors, either organometallics or coordination compounds, were deposited onto a heated substrate by RESS. Most of the precursors were too thermally labile for conventional Chemical Vapour Deposition (CVD) but could be deposited from scN_2O or scPentane *(Readers are reminded that N_2O is an extremely powerful oxidant and there have been reports* [33] *of extremely violent explosions involving scN_2O).* Pentane has a relatively high value of T_c (197 °C) and the solution was only heated above T_c for ca. 10^{-3} s, immediately prior to deposition. The solid particles, generated in this way by RESS, were then reacted by heating or by radio frequency (RF) discharge to form the film, Scheme 5.3. The film qualities achieved were very high and clearly the method has considerable potential.

Alternative approaches to thin film formation have been described by Popov and coworkers who compared both the quality of Cu films generated by RESS and other supercritical processes [34]. In particular, they examined a variant of

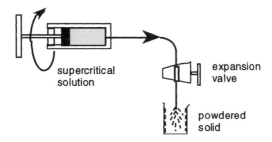

Figure 5.8 A highly schematic view of the RESS process for generating powdered material by rapid expansion.

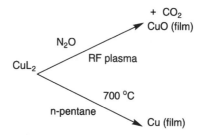

Scheme 5.3 Formation of films following RESS deposition of CuL_2 (L = 2,2,6,6-tetramethyl-3,5-heptanedionato), see [32].

retrograde precipitation (see above) whereby Cu films are deposited directly from a $scCO_2$ solution of $Cu(CF_3COCHCOCF_3)_2$ onto a heated substrate. They found that the films have better quality than those formed from the same precursor deposited by RESS but the concentrations of residual C and O in the film are higher.

5.4.3 Antisolvents, Micelles, and Electrochemistry

Although polar compounds or high molecular weight materials, such as polymers, are frequently soluble in conventional solvents, they often do not have significant solubility in supercritical fluids and are, therefore, beyond the scope of RESS. Recently, a number of groups have shown that $scCO_2$ can be used as an *antisolvent* to precipitate such materials from conventional solvents [35], Figure 5.9.

Schematically, this process looks deceptively similar to RESS but the physical principles are quite different. The role of the supercritical fluid is to dilute the conventional solvent, thus reducing its solvent-power and precipitating the solute; unlike RESS, the pressure remains constant throughout the precipitation. Indeed, the application of the so-called PCA (Precipitation superCritical Antisolvent) is similar to using hexane to induce precipitation from more polar organic solvents. The difference between PCA and hexane is that particle size

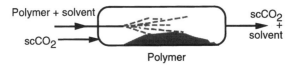

Figure 5.9 Schematic view of the PCA process, also known by the alternative acronym GAS.

5.4. SUPERCRITICAL FLUIDS AS SOLVENTS

and morphology of the precipitate can be changed in a controlled manner by varying the pressure and temperature of the CO_2 antisolvent. PCA experiments have already given results which would have been difficult, if not impossible, to obtain by conventional methods.

Supercritical fluids are potentially attractive as solvents for electrochemistry. Unfortunately, the insolubility of ionic compounds has been restrictive. Among the few successes have been an electrochemical study in scH_2O [36a], a study in $scCHClF_2$ [36b], and a number of studies in $scCO_2$ with microelectrodes, most recently involving the electrochemical activation of CO_2 itself using electrodes coated with doped polymers [37]. The scH_2O experiments [36a] are striking, firstly from the standpoint of technical virtuosity (making any measurements under such conditions would be a triumph) but also because it is one of the few studies to probe the effects of pressure on a redox reaction. The reaction in this case was the I_2/I^- couple and it was found that increasing the pressure by a modest amount, ×1.5, was sufficient to *reverse the sign* of the couple. As with studies described above, this result was attributed to the high compressibility of the fluid which results in high volumes of activation and leads to a substantial change in dielectric constant with pressure

A very recent paper [38] describes the use of microelectrodes in $scCO_2$ as a highly selective detector for supercritical fluid chromatography. This work was an imaginative development of that in ref. [37] and allows substances such as ferrocene to be detected with very high sensitivity even in fluids as non-polar as $scCO_2$. These electrodes rely on partition of the analyte between $scCO_2$ and the polymer coating of the microelectrode. Thus, there is the interesting consequence that, for a given analyte concentration, the electrochemical response becomes *less* as the pressure of $scCO_2$ is raised and hence its solvent power is increased.

Before leaving the subject of electrochemistry, it is tempting to suggest that interesting effects might be observed in more exotic fluids, *e.g.*, $scCF_3CN$ ($T_c = 38\,°C$) which may have greater solvent power than $scCO_2$ for polar compounds.

A huge number of inorganic and organometallic reactions involve ionic species. Thus, there is a continuing research effort aimed at dissolving these ionic compounds in supercritical fluids. It is possible to generate ionic species transiently in supercritical solution, for example, by pulse radiolysis [39] or by protonation [40] (see eq. 2). However, such solutions appear to be unstable and precipitation occurs rapidly.

$$[(C_5Me_5)Ir(CO)_2] + HCl \xrightarrow{scXe} [(C_5Me_5)Ir(CO)_2]^+ + Cl^- \rightarrow \text{precipitate} \quad (2)$$

Very recently, Tumas and coworkers have shown [41] that organometallic compounds with bulky *fluorinated* anions, *e.g.*, $[B(3,5-C_6H_4(CF_3)_2)_4]^-$ or $[CF_3SO_3]^-$ can be dissolved in unmodified $scCO_2$ to an extent sufficient for their use as homogeneous hydrogenation catalysts. This seemingly simple discovery promises to open up a very wide range of new reactions in $scCO_2$, for example, catalytic asymmetric hydrogenation [41].

An alternative method of dissolving ionic species involves aqueous inverse micelles. Much work has been concerned with the search for suitable surfactants for stabilizing reverse micelles in supercritical solution. Considerable success has been achieved employing scXe, scC_2H_6, and scC_3H_8, particularly by Smith, Fulton and coworkers [42]. By contrast, it has proved much harder to generate reverse micelles in $scCO_2$, possibly because CO_2 generates an acidic solution in H_2O. Recently, there has been some success with fluorinated surfactants and Howdle and coworkers have succeeded in generating micelles which can be shown spectroscopically to contain water rather than merely being aggregates of surfactant molecules [43a]. They have used these micelles to solubilize *real inorganic* compounds, *e.g.*, $KMnO_4$, in $scCO_2$ thereby creating "purple CO_2" [43b]. So far, there has been little chemical opportunity to exploit these reverse micelles in supercritical solution but clearly there are extensive possibilities.

5.4.4 Unusual Solubilities

Sadly, supercritical scientists often manage to create the impression that supercritical fluids are "wonder solvents" with almost magical powers to dissolve even the most recalcitrant solids. In reality, supercritical fluids usually have poorer solvent properties than equivalent conventional solvents. In a few cases, however, supercritical fluids are *better* solvents; most striking is the ability of $scCO_2$ to dissolve a wide range of fluorocarbons which are notorious for their low solubility in conventional solvents other than CFCs. The precise reason for this enhanced solubility of fluorinated compounds in $scCO_2$ is still the subject of debate but the effect has considerable importance [1]. For example, DeSimone and coworkers have exploited this solubility for polymerization processes [44], Scheme 5.4.

A somewhat different application involves the use of fluorinated β-diketonates which have been used both as precursors for film deposition (see Section 5.4.2), and by Wai [45] and others [46] as complexing agents for SF extraction of divalent metal ions not only on an analytical scale but also potentially for large scale remediation of environmental pollution. Equally, fluorination of the alkyl groups on conventional surfactants has been one of the more promising

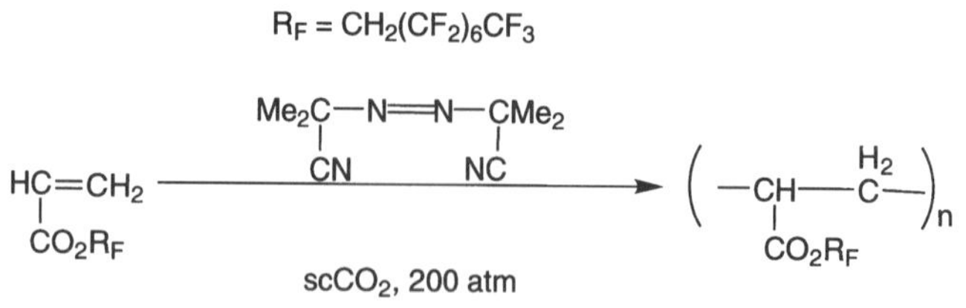

$R_F = CH_2(CF_2)_6CF_3$

$$HC{=}CH_2 \atop |\atop CO_2R_F \quad \xrightarrow{\substack{Me_2C-N{=}\!\!=\!\!N-CMe_2 \\ | \qquad\quad | \\ CN \qquad\quad NC \\ \\ scCO_2,\ 200\ atm}} \quad \left(-CH{-}\overset{H_2}{C}- \atop |\atop CO_2R_F \right)_n$$

Scheme 5.4

5.4. SUPERCRITICAL FLUIDS AS SOLVENTS

attempts at tailoring surfactants for generating reverse micelles in $scCO_2$ but, as explained above, the ideal surfactant has yet to be found.

A rather different reaction involving unusual solubility in a supercritical fluid is already being exploited industrially on a significant scale (ca. 40 kton p.a.) by the Idemitsu Petroleum Company [47]. This reaction involves the acid catalysed hydration of butene to generate butan-2-ol, Scheme 5.5.

$$EtCH=CH_2 + H_2O \rightleftharpoons EtCH(OH)-CH_3$$

n-butene: $T_c = 146\,°C$; $P_c = 39.7$ atm.

Scheme 5.5

Normally, the equilibrium between butene and butan-2-ol lies largely in favor of butene. However, the solubility of butan-2-ol is higher in scbutene at 200 °C than in the aqueous reaction mixture. Extraction of butan-2-ol into scbutene can therefore be used to shift the equilibrium in favor of butan-2-ol, which can then be recovered from the scbutene by a reduction in pressure. Unreacted butene is then recycled. The use of a supercritical solvent to shift an equilibrium in this way will, we believe, have much wider applications in the future. Already a two-phase system has been exploited for the formation of dimethyl formamide, DMF, see below.

5.4.5 High Solubility of Gases

A property of particular chemical significance is the almost complete miscibility of "permanent" gases, *e.g.*, H_2, N_2, etc., with supercritical fluids. H_2 has a relatively low solubility in common organic solvents and so this miscibility allows higher concentrations of dissolved gas to be achieved in supercritical solution than in conventional solvents (Figure 5.10). The effect is large; for a given

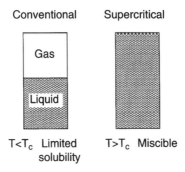

Figure 5.10 Schematic representation of the enhanced solubility of gases in supercritical solution. This effect only occurs at medium pressures (<1 kbar). At higher pressures (>1 kbar) there is the possibility in some systems of gas/gas phase separations, *e.g.*, in He/Xe [77].

pressure of gas, the concentration of H_2 can be over an order of magnitude higher in supercritical solution [48]. This property has been exploited in several areas of organometallic chemistry, both for the synthesis of new compounds and also for enhancing existing reactions (*e.g.*, hydrogenation and hydroformylation).

This effect was a key factor in the synthesis of $CpRe(N_2)_3$ ($Cp = \eta^5\text{-}C_5H_5$) in scXe, Scheme 5.6, where the high concentration of dissolved N_2 permitted the substitution of all three CO groups in the starting material $CpRe(CO)_3$ [12a]. At first sight, the reaction is so simple that it appears improbable to synthetic chemists more familiar with working in conventional solvents. The products, $CpRe(CO)(N_2)_2$ and $CpRe(N_2)_3$, had only been observed previously at cryogenic temperatures and had been presumed to be so reactive that their synthesis had not been attempted before. However, the supercritical synthesis revealed them to be relatively stable even at room temperature. The products could all be identified spectroscopically, because scXe has no absorption in a region stretching from far UV to far IR.

Although the reaction has only been carried out on a very small scale, the combination of spectroscopic techniques and supercritical fluids allows even tiny amounts of material to be handled with sufficient precision to permit determination of some properties of the products. There is sufficient chemical interest in the products for scale-up to be desirable (see Section 5.7). The multiple substitution of CO by N_2 in supercritical solution is not unique to $CpRe(CO)_3$ and formation of a number of $(\eta^n\text{-}C_nR_n)M(CO)(N_2)_2$ species has been observed for other metals (*e.g.*, Mn, Fe, Cr, etc.) [12b, 49]. However, Re remains the only metal for which significant amounts of a tris-dinitrogen compound can be formed.

The high solubility of H_2 in sc fluids has also played a key role in a number of organometallic reactions, analogous to those in Scheme 5.6. This has led to the identification of several previously unknown compounds containing the $\eta^2\text{-}H_2$ ligand, discussed in more detail below [12b, c]. Rathke and Klingler have also exploited this effect in a series of elegant high pressure NMR experiments

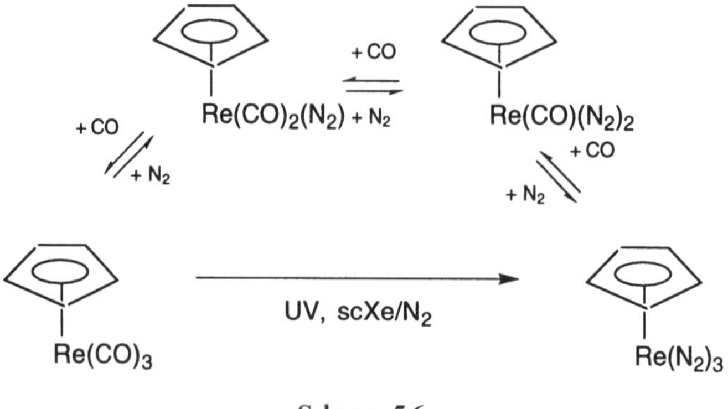

Scheme 5.6

5.4. SUPERCRITICAL FLUIDS AS SOLVENTS

involving $Co_2(CO)_8$ or $Mn_2(CO)_{10}$ as catalysts in *thermal* hydroformylation reactions [50], eq. (3).

$$RCH=CH_2 \xrightarrow[Co_2(CO)_8]{H_2/CO/scCO_2} RCH_2CH_2CH=O + RCH(CHO)CH_3 \qquad (3)$$

In this case, it was the distribution of the products in the supercritical reaction rather than the products themselves, which were different from the conventional reaction. A US Patent has been granted on the reaction with propene, $CH_3CH=CH_2$ [50]. The NMR monitoring of these reactions was aided considerably by the very low viscosity of $scCO_2$ because nuclei with spin ≥ 1 have shorter correlation times in supercritical solution and correspondingly sharper resonances. The challenge is to devise an apparatus which is small enough to fit inside an NMR spectrometer yet robust enough to avoid catastrophic failure under pressure. Klingler and Rathke used a high pressure autoclave with a RF coil located *inside* the autoclave, thus combining NMR detection with relatively conventional high pressure techniques (see also Chapter 2). Recent publications have described two radically different approaches to supercritical NMR spectroscopy. The first method involves a specially constructed flow-cell/probe which has been successfully interfaced to supercritical chromatography [51a] and extraction [51b]. On the other hand, Yonker and coworkers have used a very narrow bore capillary tube folded in a concertina fashion so as to fit inside a standard 5 mm NMR tube, which was filled with a conventional deuterated solvent (D_2O) to act as the NMR lock, Figure 5.11. Such a capillary arrangement can be used up to pressures as high as 4 kbar [52].

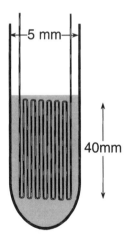

Figure 5.11 Illustration of the method of Yonker and coworkers for recording NMR spectra of supercritical fluids in a narrow bore capillary tube folded within a conventional NMR tube. (Redrawn from [52].)

Very recently, Tumas and coworkers have exploited the same effect, namely the high concentration of dissolved H_2, to carry out asymmetric hydrogenation in $scCO_2$ using a homogeneous Rh(I) catalyst [41]. The enantiomeric excess was somewhat higher than when carrying out the same reaction in conventional solvents, presumably as the result of the higher H_2 concentration.

5.5 CHEMICAL INERTNESS AND THE ACTIVATION OF $scCO_2$

The synthesis of $CpRe(N_2)_3$ in Scheme 5.6 was originally carried out in scXe because of its spectroscopic transparency [12a]. However, the chemical inertness of scXe was also of advantage since the possibility of the occurrence of side reactions was avoided. Frequently, chemists are unaware that $scCO_2$ is also an extremely inert solvent. Indeed, recent time-resolved IR experiments indicate that CO_2 interacts with unsaturated metal centers less strongly than even Xe [21, 53]. This inertness of $scCO_2$ can be turned to advantage, particularly because there are surprisingly few solvents available, which can be considered totally inert. $scCO_2$ has been exploited in this way by Jobling et al. [48] for the photochemical activation of C–H bonds of alkanes, e.g., C_2H_6, by $Cp^*Ir(CO)_2$ dissolved in $scCO_2$, eq. (4).

$$Cp^*Ir(CO)_2 + C_2H_6 \xrightarrow{UV, scCO_2} Cp^*Ir(CO)(Et)H + CO \qquad (4)$$

Clearly, such reactions could not be carried out satisfactorily in hydrocarbon solvents because $Cp^*Ir(CO)_2$ would react with the C–H bonds of the solvent in preference to those of the alkane substrate. These reactions [48] confirmed the inertness of $scCO_2$ but eventually it transpired that the reactions occurred more efficiently in pure scC_2H_6 in the absence of CO_2! A particularly interesting feature of the reaction was that C–H activation of C_2H_6 was very much more effective if H_2, at high pressure, was applied to the solution; under this condition, $Cp^*Ir(CO)_2$ was initially converted to $Cp^*Ir(CO)(H)_2$ which then reacted photochemically with C_2H_6 [48].

By contrast, relatively few catalysts manage to induce CO_2 to react with other substrates. Jessop et al. have recently reported [54a] the catalytic hydrogenation of $scCO_2$ itself, eq. (5).

$$CO_2 + H_2 (85\ atm.) \xrightarrow[scCO_2,\ p_{total}\ 205\ atm.,\ 50\ ^\circ C]{Ru(PMe_3)_4H_2 + NEt_3} HCOOH \qquad (5)$$

This experiment aroused widespread interest, not because the product, formic acid, was in any way different from that generated by the same catalyst in THF solution but because the reaction is over an order of magnitude faster in $scCO_2$. This acceleration has been attributed to the greater concentration of H_2 in the

5.5. CHEMICAL INERTNESS AND THE ACTIVATION OF scCO$_2$

supercritical reaction, see Section 5.4.5. The work has now been extended to the synthesis of dimethylformamide [54b] and, most recently, methyl formate [54c] by combining the initial hydrogenation of CO$_2$ with subsequent reaction with Me$_2$NH or MeOH, respectively. These reactions operate with a two phase system. The hydrogenation occurs in the scCO$_2$ phase, and the second step takes place in an organic phase rich in Me$_2$NH or MeOH. The turnover numbers for these reactions are much higher than those previously observed in subcritical systems, possibly because of the high concentration of H$_2$ (see above) but perhaps also because of high rates of diffusion in scCO$_2$ or lack of significant solvation of the catalytic intermediates by scCO$_2$ [54d].

Another recent example, from Reetz and coworkers [55], involves the catalyzed reaction of CO$_2$ with hex-3-yne, Scheme 5.7.

As explained in Chapter 6, there is now some doubt whether this reaction occurred in the supercritical phase because the catalyst appears to be virtually insoluble in scCO$_2$ [56]. Nevertheless, the reaction makes an important point, namely that processes where CO$_2$ itself acts as a reagent are obviously suited to scCO$_2$. Clearly, efforts should be made to discover further reactions of this type. The action of the catalyst presumably involves coordination of CO$_2$ to the Ni center. However, it is not easy to isolate compounds where CO$_2$ is coordinated to a metal in a stable manner. Some years ago, Ibers and coworkers reported an attempt to generate such CO$_2$ complexes of transition metals by reactions in *liquid* CO$_2$ solution [57]. Although the desired reactions did not occur, a number of unwanted products were formed *via* reaction with H$_2$O adventitiously present in the CO$_2$.

It is sometimes forgotten that H$_2$O is relatively more soluble in scCO$_2$ than in hydrocarbon solvents and thus H$_2$O may play a significant (and possibly unanticipated) role in the chemistry involved. For example, we have recently found [14] that photolysis of CpMn(CO)$_3$ in wet scCO$_2$ leads to formation of a white solid, most probably manganese(II) carbonate, MnCO$_3$. The mechanism of formation of carbonates in the presence of low pressure CO$_2$ has been discussed in some detail elsewhere [58] and the mechanism in wet scCO$_2$ is probably similar.

The presence of water is a crucial factor in the supercritical decaffeination process [1] and also may be important in enzymatic reactions in scCO$_2$ [59]. The advantages of scCO$_2$ for such reactions are similar to those for

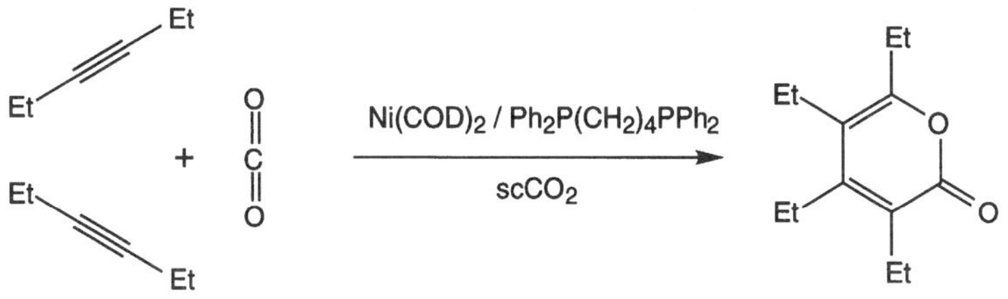

Scheme 5.7

non-enzymatic reactions; good mass transport, high diffusivity and low viscosity, and ease of recovering the product. In addition, there is the possibility of optimizing the enantiomeric purity of the products *via* solvent density.

5.6 SURFACE TENSION AND VISCOSITY

The absence of surface tension in supercritical fluids has been widely exploited in the removal of solvents, particularly from aerogels. Recently, Lynch and coworkers extended this technique to the drying of thin films of porous silicon [60]. Such films can easily be prepared by electrochemical etching in aqueous solvent, but attempts to remove the solvent by conventional means invariably led to disintegration of the film. By contrast, use of scCO$_2$ allowed the solvent to be removed without any damage; the final film had properties far superior to those prepared by other routes.

The low viscosity of supercritical fluids is a key factor in their use for extraction. The work of Wai and Beckman, see Section 5.4.4 above, indicates how a physical and a chemical role of a supercritical fluid can be combined into a single process; in their case, the process was reactive extraction, in which a supercritical fluid is used both to derivatize and to extract an otherwise insoluble analyte. Low viscosity is also important for impregnation of substances into porous materials. As with extraction, this impregnation is achieved without the solvent residues which usually remain after impregnation with conventional solvents [28, 61]. Howdle *et al.* describe the impregnation of organometallic compounds into polyethylene and their subsequent reaction with the polymer [62]. This technique has recently been extended to zeolites and to porous glasses (*e.g.*, Vycor) [63] and, clearly, the approach has potential for new materials.

5.7 WATER AND "OTHER" SUPERCRITICAL FLUIDS

The critical parameters of H$_2$O ($T_c = 374\,°C$; $p_c = 218$ atm.) are very much higher than those of CO$_2$ with the result that experiments in scH$_2$O place much more stringent requirements on apparatus [64]. Much of the current research effort has been aimed at oxidation of toxic wastes and chemical munitions (see Chapter 7). Indeed, a whole conference was recently devoted to these topics [65]. It is therefore important to remind readers that scH$_2$O also offers opportunities as a medium for chemical reactions and that the corrosion problems associated with small-scale reaction chemistry under these conditions may well be less serious. Scheme 5.8 shows a simple reaction, recently reported by Myrick *et al.* [66], who used Raman spectroscopy to follow the kinetics. Organic chemistry in scH$_2$O appears ripe for exploration but the field is somewhat outside the scope of this chapter.

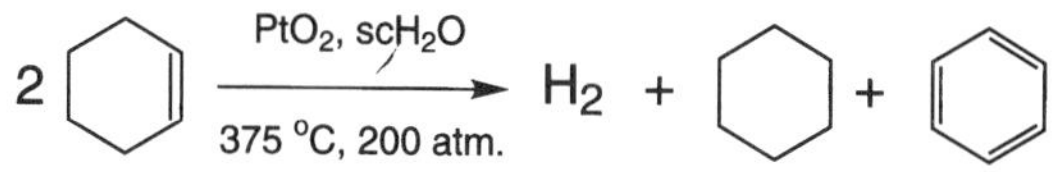

5.7. WATER AND "OTHER" SUPERCRITICAL FLUIDS

$$2\,\text{C}_6\text{H}_{10} \xrightarrow[375\,°C,\ 200\,atm.]{PtO_2,\ scH_2O} H_2 + \text{C}_6\text{H}_{12} + \text{C}_6\text{H}_6$$

Scheme 5.8

The critical temperature of D_2O (371.5 °C) is 2.6 °C lower than T_c of H_2O and it is tempting to speculate that reactions carried out in this temperature region may well show quite significant solvent isotopic differences between D_2O and H_2O.

An interesting inorganic application of scH_2O was reported by Arai and coworkers [67], who devised a flow-reactor for preparing powdered metal oxides in scH_2O. In their method, aqueous solutions of metal salts, *e.g.*, Fe^{3+}, Al^{3+}, etc., were pumped rapidly through a reactor held above the critical temperature and pressure of H_2O. Under these conditions, hydrolysis of the metal ions appeared to be very rapid and metal oxide particles were formed, presumably *via* precipitation of $M(OH)_3$, which was then dehydrated very rapidly. In the case of Fe^{3+}, Fe_2O_3 particles were formed even when the residence time in the reactor was as short as 1 s, but the particle size, 20 μm, was smaller than produced using longer residence times. It is clear that much of the impact of supercritical fluids over the next decade will be in the area of materials and these experiments underline the possibilities of using sc fluids to control the physical form of a material as well as its chemical composition.

There have been comparatively few reports of reactions in supercritical fluids with critical temperatures between those of say NH_3 ($T_c = 132\,°C$) and H_2O. Of course, Hannay and Hogarth conducted their classic experiments on solubility in scC_2H_5OH ($T_c = 243.5\,°C$). They also tried an extremely interesting experiment in $scNH_3$ [2], which potentially explores the highly topical electrical conductor/insulator transition; they attempted to heat a solution of Na metal in NH_3, which contains a high concentration of ammoniated electrons, through its critical point. Sadly, the solution always decomposed, presumably as in eq. (6).

$$2e^-_{NH_3} + 2NH_3 \rightarrow 2[NH_2]^- + H_2 \qquad (6)$$

Very recently, Kolis, Wood and coworkers have started using supercritical amines, particularly $scC_2H_4(NH_2)_2$, for the synthesis of new solid state compounds [8]. As explained in Section 5.2, their experiments have a simple elegance; suitable precursors are sealed into a quartz tube containing a measured quantity of amine, a number of tubes are loaded into a conventional autoclave which is pressurized with Ar gas and then heated to the required temperature. At the end of the reaction, the bomb is cooled and the products are recovered from their individual tubes. Scheme 5.9 shows one of these reactions, the generation of a complex anion from simpler precursors [8a]. The reaction has some interesting features: the anion contains a tetrahedral core of four W atoms with bridging S atoms between them, while the precursor molecule $W(CO)_6$ is only mononuclear; each W atom loses all six of its original CO

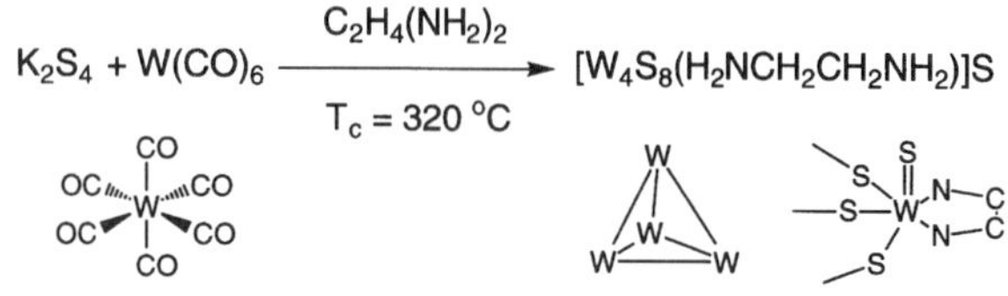

$$K_2S_4 + W(CO)_6 \xrightarrow[T_c = 320\,°C]{C_2H_4(NH_2)_2} [W_4S_8(H_2NCH_2CH_2NH_2)]S$$

Scheme 5.9

ligands in the course of the reaction and, at the end, has one diamine and four S ligands; the product is a completely new compound.

The experimental conditions of these reactions are reminiscent of those used for many reactions with conventional solvents in high pressure autoclaves. Indeed, it is quite possible that a significant number of such reactions are being carried out under supercritical conditions, albeit unwittingly. If such cryptocritical reactions could be identified, there might well be opportunities to modify the chemistry by altering the reaction conditions in a way more appropriate to supercritical fluids (*i.e.*, varying the pressure of added reactant gas would alter T_c of the reaction mixture, etc.). A study of one such reaction, the radical chain addition of an alkane to an olefin, was recently reported [68].

5.8 SUPERCRITICAL FLOW-REACTORS AND SCALE-UP

We briefly outlined in Section 5.2 some of the advantages of flow-reactors in scaling up supercritical reactions even on a very small scale. Figure 5.12 illustrates schematically two reactors, one for reactions in scC_2H_4 and the other for reactions in $H_2/scCO_2$ mixtures. Both reactors have been used successfully to isolate organometallic compounds which were previously considered to be too labile to be isolated [14, 69]. The isolation of $CpMn(CO)_2(\eta^2\text{-}H_2)$ (Scheme 5.10) is interesting because attempts to separate it from conventional solvents also removed the labile $\eta^2\text{-}H_2$ ligand [70]. By contrast, RESS precipitation of the product is extremely rapid and takes place under a high pressure of H_2. Separation of solid product does, however, necessitate the use of a sophisticated back-pressure regulator to avoid clogging [14]. This problem does not arise in Roche reactions which are exothermic and where the products are liquids [11]. A similar route can also be used for a straightforward synthesis of the known complex $CpMn(CO)_2(N_2)$, Scheme 5.10.

Flow-reactors have been used quite widely for other supercritical reactions. Apart from Arais work in scH_2O [67] described above, Gourinchas *et al.* [71] recently reported a flow-reactor for the pilot-scale synthesis of powdered TiO_2 in $scCH_3CH(OH)CH_3$. However, the supercritical fluid has the same advantage as in the organometallic reactions; the product is delivered as a dry and finely divided solid. In the field of organic chemistry, Buback and coworkers have developed flow-reactors for C_2H_4 polymerisation [72] and for producing estrone (a hormone of pharmaceutical importance) in supercritical tetralin [73].

5.9. CONCLUSIONS

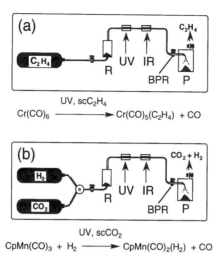

Figure 5.12 Schematic diagrams of flow-reactors for supercritical organometallic photochemistry. (a) for preparation of $Cr(CO)_5(C_2H_4)$ [14, 69] and (b) for synthesis of $CpMn(CO)_2(H_2)$ [14]. The parts are labelled as follows: R, reactant; P, product; BPR, back pressure regulator; UV, photolysis cell; IR, spectroscopic monitoring. (Adapted from [14].)

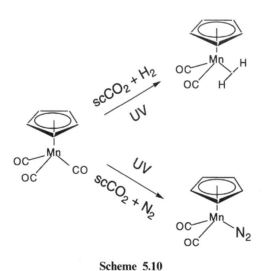

Scheme 5.10

5.9 CONCLUSIONS

This chapter serves in part as a review and has highlighted only a small proportion of the supercritical chemistry which has already been published.

Exciting new work is in progress, including the use of phase transfer catalysis shortly to be published by Eckert's group [74] and major developments in polymer chemistry and processing. Nevertheless, there are still huge areas of chemistry ripe for exploration. The techniques of supercritical chemistry need to become more widespread and to leave the specialist laboratory.

The program of the Third International Symposium on Supercritical Fluids held in Strasbourg in 1994 was very broad and served as testimony to the interdisciplinary nature of the subject. Unfortunately, there are already signs of compartmentalization into analytical, process, spectroscopy, etc. It seems not only desirable but essential that this fragmentation should be resisted. Supercritical science must remain "single phase". The full development of reaction chemistry in supercritical fluids requires not only the full range of existing supercritical expertise but also a much greater participation from synthetic chemists, who can highlight those chemical problems best tackled by application of supercritical fluids. One should always bear in mind one of the key ideas to emerge from the First International Symposium in Nice (1988), encapsulated by Krukonis [75], who said:

"There is no point in doing something in a supercritical fluid just because its neat. Using the fluids must have some real advantage."

ACKNOWLEDGEMENTS

We thank our coworkers, colleagues and technical staff at Nottingham and our UK and overseas collaborators for their contribution to our work described within this chapter. We gratefully acknowledge financial support for this work from EPSRC Chemistry and Clean Technology (Grant nos. GR/H95464 & GR/J95065), from EC Human Capital and Mobility and COST programs, Perkin-Elmer Ltd, BP International, BP Chemicals, Zeneca plc, NATO Laboratory Linkage Grant 920570, and particularly to the EPSRC Clean Technology Unit for funding a fellowship for MP and to the Royal Society for granting a University Research Fellowship to SMH. This chapter is based in part on the Review Paper delivered by MP at the 3rd International Symposium on Supercritical Fluids, Strasbourg [75]; we thank Professor M. Perrut for encouraging us to publish this material for a wider audience.

5.10 REFERENCES

[1] For an excellent introduction see, McHugh, M. A., Krukonis, V. J., Supercritical Fluid Extraction: Principles and Practice, Butterworth-Heinemann, Boston, 1994.
[2] Hannay, J. B., Hogarth, J., Proc. Roy. Soc., 30 (1880) 178.
[3] Biddulph, R. H., Plesch, P. H., J. Chem. Soc., (1960) 3913.
[4] a) Subramaniam, B., McHugh, M. A., Ind. Eng. Chem. Process Des. Dev., 25 (1986) 1; b) Buback, M., Angew. Chem. Int. Ed. Engl., 30 (1991) 641; c) Caralp, M. H. M., Clifford, A. A., Colby, S. E., (King, M. B., Bott, T. R. (eds.)), Extraction of Natural

Products using Near-critical Solvents, Chapman and Hall, Glasgow, 1993, Chapter 3; d) Boock, L., Wu, B., Lamarca, C., Klein, M., Paspek, S., CHEMTECH, (1992) 718; e) Clifford, A. A., Coleby S. E., (King, M. B., Bott, T. R., (eds.)), Extraction of Natural Products using Near-critical Solvents, Chapman and Hall, Glasgow, 1993.

[5] See for example, a) Sherman, W. F., Stadtmuller, A. A., Experimental Techniques in High-Pressure Research, John Wiley and Sons Ltd., London, 1987; b) Whyman, R., (Willis, H. L. (ed.)), Laboratory Methods in Vibrational Spectroscopy, John Wiley and Sons Ltd., London, 1987, pp 281–307.

[6] Howdle, S. M., Poliakoff, M., (Kiran, E., Sengers, J. M. H. (eds.)), Supercritical Fluids: Fundamentals and Applications, NATO ASI Series, Kluwer Academic Publications, Dordrecht, 273 (1994) 527.

[7] Poliakoff, M., Howdle, S. M., Kazarian, S. G, Angew. Chem., Intl Ed, English, 34 (1995) 1275.

[8] a) Wood, P. T., Pennington, W. T., Kolis, J. W., Wu, B., O'Connor, C. J., Inorg. Chem., 32 (1993) 129; b) Wood, P. T., Pennington, W. T., Kolis, J. M., J. Chem. Soc., Chem. Commun., (1993) 235; c) Wood, P. T., Pennington, W. T., Kolis, J. W., Inorg. Chem., 33 (1994) 1556.

[9] Jiménez-Vázquez, H. A., Cross, R. J., Mroczkowski, S., Gross, M. L., Giblin, D. E., Poreda, R. J., J. Am. Chem. Soc., 116 (1994) 2193.

[10] Tundo, P., Continuous Flow Methods in Organic Synthesis, Ellis Horwood, Chichester, 1991.

[11] Roche Magazin, 41 (1992) 2.

[12] a) Howdle, S. M., Grebenik, P., Perutz, R. N., Poliakoff, M., J. Chem. Soc., Chem. Commun., (1989) 1517; b) Howdle, S. M., Healy, M. A., Poliakoff, M., J. Am. Chem. Soc., 112 (1990) 4804; c) Howdle, S. M., Poliakoff, M., J. Chem. Soc., Chem. Commun., (1989) 1099.

[13] Banister, J. A., Howdle, S. M., Poliakoff, M., J. Chem. Soc., Chem. Commun., (1993) 1814.

[14] Banister, J. A., Lee, P. D., Poliakoff, M., Organometallics, 14 (1995) 3876.

[15] Paulaitis, M. E., Alexander, G. C., Pure and Appl. Chem., 59 (1987) 61; Isaacs, N. S., Keating, N., J. Chem. Soc., Chem. Commun., (1992) 876; Ikushima, Y., Saito, N., Arai, M., Bull. Chem. Soc. Japan., 64 (1991) 282; Ikushima, Y., Saito, N., Arai, M., J. Phys. Chem., 96 (1992) 2293.

[16] Hrnjez, B. J., Mehta, A. J., Fox, M. A., Johnston, K. P., J. Am. Chem. Soc., 111 (1989) 2662.

[17] Kazarian, S. G., Poliakoff, M., J. Phys. Chem., 99 (1995) 8624.

[18] Ben-Amotz, D., LaPlant, F., Shea, D., Gardecki, J., List, D., (Brennecke, J. F., Kiran, E. (eds.)), ACS Symposium Series, 488 (1992) 18.

[19] See *e.g.*, Betts, T. A., Bright, F. V., Applied Spectroscopy, 44 (1990) 1203; Sun, Y.-P., Fox, M. A., J. Am. Chem. Soc., 115 (1993) 747.

[20] For a description of TRIR techniques, see George, M. W., Poliakoff, M., Turner, J. J., Analyst, 119 (1994) 551.

[21] George, M. W., Sun, X.-Z., Poliakoff, M., Proc. 7th Conference, Time-resolved Vibration Spectroscopy, Santa Fe, New Mexico, in press.

[22] Ji, Q., Eyring, E. M., van Eldik, R., Johnston, K. P., Goates, S. R., Lee, M. L., J. Phys. Chem., 99 (1995) 13461.

[23] Banister, J. A., Ph.D. Thesis, University of Nottingham, Nottingham, UK, 1994.

[24] Chimowitz, E., Pennisi, K. P., AIChE J., 32 (1986) 1665; Chimowitz, E., (Perrut, M. (ed.)), 1st International Symposium on Supercritical Fluids, Soc. Chim. France, 1988, p 845.

[25] See for example, Kim, S., Johnston, K. P., Ind. Eng. Chem. Res., 26 (1987) 1206.

[26] Via, J., Taylor, L. T., Schweighardt, F. K., Anal. Chem., 66 (1994) 1459.

[27] See for example, Gupta, R. B., Combes, J. R., Johnston, K. P., J. Phys. Chem., 97 (1993) 70.

[28] Cooper, A. I., Howdle, S. M., Hughes, C., Jobling, M., Kazarian, S. G., Poliakoff, M., Shepherd, L. A., Johnston, K. P., Analyst, 118 (1993) 1111.

[29] Lee, M. L., Markides, K. E. (eds.), Analytical Supercritical Fluid Chromatography and Extraction, Chromatography Conferences, Provo, Utah, 1990.

[30] For a recent review on RESS, see Tom, J. W., Debenedetti, P. G., J. Aerosol. Science, 22 (1991) 555.

[31] Tom, J. W., Debenedetti, P. G., Biotechnol. Prog., 7 (1991) 403.

[32] Hansen, B. N., Hybertson, B. M., Barkley, R. M., Sievers, R. E., Chem. Mater., 4 (1992) 749.

[33] Raynie, D. E., Anal. Chem., 65 (1993) 3127.

[34] Howdle, S. M., Antonov, E. N., Bagratashvili, V. N., et al., Proceedings 3rd International Symposium on Supercritical Fluids, Strasbourg, 3 (1994) 369.

[35] Yeo, S.-D., Lim, G.-B., Debenedetti, P. G., Bernstein, H., Biotech & Bioeng., 41 (1993) 341; Dixon, D. J., Johnston, K. P., J. Applied Polymer Science, 50 (1993) 1929; Randolph, T. W., Randolph, A. D., Mebes, M., Yeung, S., Biotechnol. Prog., 9 (1993) 429.

[36] a) Flarsheim, W. M., Bard, A. J., Johnston, K. P., J. Phys. Chem., 93 (1989) 4234; b) Olsen, S. A., Tallman, D. E., Anal. Chem., 66 (1994) 503.

[37] Sullenberger, E. V., Dressman, S. F., Michael, A. C., J. Phys. Chem., 98 (1994) 5347.

[38] Dressman, S. F., Michael, A. C., Anal. Chem., 67 (1995) 1339.

[39] Zhang, J., Connery, K., Chateauneuf, J. E., Brennecke, J. F., Proceedings 3rd International Symposium on Supercritical Fluids, Strasbourg, 3 (1994) 369.

[40] Kazarian, S. G., Jobling, M., Poliakoff, M., Mendeleev Commun., (1993) 148.

[41] Feng, S., Gross, M. F., Burk, M. J., Tumas, W., ACS Div. Inorganic Chem. Abs. No. INOR 572, Anaheim, 1995; Burk, M. J., Feng, S., Gross, M. F., Tumas, W., J. Am. Chem. Soc., 117 (1995) 8277.

[42] Kaler, E. W., Billman, J. P., Fulton, J. L., Smith, R. D., J. Phys. Chem., 95 (1991) 458; Fulton, J. L., Blitz, J. P., Tingey, J. M., Smith, R. D., J. Phys. Chem., 93 (1989) 4198; Smith, R. D., Fulton, J. L., Blitz, J. P., Tingey, J. M., J. Phys. Chem., 94 (1990) 781.

[43] a) McFann, G. J., Johnston, K. P., Howdle, S. M., AIChE J., 40 (1994) 543; b) Clarke, M. J., Howdle, S. M., Johnston, K. P., to be published; c) Johnston, K. P., Harrison, K. L., Clarke, M. J., Howdle, S. M., Heitz, M. P., Bright, F. V., Carlier, C., Randolph, T. M., Science 271 (1996) 624.

[44] a) DeSimone, J. M., Guan, Z., Elsbernd, C. S., Science, 257 (1992) 945; b) Combes, J. R., Guan, Z., DeSimone, J. M., Macromolecules, 27 (1994) 865.

[45] Laintz, K. E., Wai, C. M., Yonker, C. R., Smith, R. D., J. Supercrit. Fluids, 4 (1991) 194; Lin, Y., Brauer, R. D., Laintz, K. E., Wai, C. M., Anal. Chem., 65 (1993) 2549; Lin, Y., Wai, C. M., Anal. Chem., 66 (1994) 1971; for a broad review of Wai's work, see Lin, Y., Smart, N. G., Wai, C. M., Trends in Analytical Chem., 14 (1995) 123.

[46] Yadzi, A., Beckman, E. J., Proceedings 3rd International Symposium on Supercritical Fluids, Strasbourg, 2 (1994) 283.

[47] Fukuzato, R., (McHugh, M. A. (ed.)), Proceedings 2nd International Symposium on Supercritical Fluids, Johns Hopkins University, Baltimore, p. 196, 1991.

[48] Jobling, M., Howdle, S. M., Healy, M. A., Poliakoff, M., J. Chem. Soc., Chem. Commun., (1990) 1287; Banister, J. A., Cooper, A. I., Howdle, S. M., Jobling, M., Poliakoff, M., Organometallics, 15 (1996) 1804.

[49] Banister, J. A., George, M. W., Grubert, S., Howdle, S. M., Jobling, M., Johnston, F. P. A., Morrison, S. L., Poliakoff, M., Schubert, U., Westwell, J. R., J. Organomet. Chem., 484 (1994) 129.

[50] Rathke, J. W., Klingler, R. J., Krause, T. R., Organometallics, 10 (1991) 1350; Klingler, R. J., Rathke, J. W., Inorg. Chem., 31 (1992) 804; Rathke, J. W., Klingler, R. J., U.S. Patent 5198589, 1994.

[51] a) Albert, K., Braumann, U., Tseng, L.-H., Nicholson, G., Bayer, E., Spraul, M., Hofmann, M., Dowle, C. J., Chippendale, M., Anal. Chem., 66 (1994) 3042; b) Braumann, U., Händel, H., Albert, K., Anal. Chem., 67 (1995) 930.

[52] Pfund, D. M., Zemanian, T. S., Linehan, J. C., Fulton, J. L., Yonker, C. R., J. Phys. Chem., 98 (1994) 11846.

[53] Kazarian, S. G., Gupta, R. B., Johnston, K. P., Clarke, M. J., George, M. W., Poliakoff, M., 3rd International Symposium on Supercritical Fluids, Strasbourg, France, 1994, p 343.

[54] a) Jessop, P. G., Ikariya, T., Noyori, R. Nature, 368 (1994) 231; b) Jessop, P. G., Hsaio, Y., Ikariya, T., Noyori, R., J. Am. Chem. Soc., 116 (1994) 8851; c) Jessop, P. G., Hsaio, Y., Ikariya, T., Noyori, R., J. Chem. Soc., Chem. Commun., (1995) 707; d) Jessop, P. G., Ikariya, T., Noyori, R., Chem. Rev., 95 (1995) 273; e) Jessop, P. G., Hsaio, Y., Ikariya, T., Noyori, R., J. Am. Chem. Soc., 118 (1996) 344.

[55] Reetz, M. T., Konen, W., Strack, T., Chimia, 47 (1993) 493.

[56] Dinjus, E., COST/Dechema Workshop, Lahnstein, Germany, April 1995.

[57] Mason, M. G., Ibers, J. A., J. Am. Chem. Soc., 104 (1982) 5153.

[58] Alvarez, R., Carmona, E., Galindo, A., Gutierrez, E., Monge, A., Poveda, M. L., Ruiz, C., Savariault, J. M., Organometallics, 8 (1989) 2430.

[59] See for example, Russell, A. J., Beckman, E. J., Chaudhary, A. K., Chemtech, (1994) 33; Kamat, S. V., Iwaskewycz, B., Beckman, E. J., Russell, A. J., Proc. Natl. Acad. Sci. U.S.A., 90 (1993) 2940; Kamat, S. V., Beckman, E. J., Russell, A. J., J. Am. Chem. Soc., 115 (1993) 8845; Randolph, T. W., Clark, D. S., Blanch, H. W., Prausnitz, J. M., Science, 34 (1988) 1354; Castillo, E., Marty, A., Combes, D., Condoret, J. S., Biotechnology Letters, 16 (1994) 169.

[60] Canham, L. T., Cullis, A. G., Pickering, C., Dosser, O. D., Cox, T. I., Lynch, T. P., Nature, 368 (1994) 133.

[61] Jobling, M., Howdle, S. M., Poliakoff, M., J. Chem. Soc., Chem. Commun., (1990) 1762.

[62] Clarke, M. J., Howdle, S. M., Jobling, M., Poliakoff, M., J. Am. Chem. Soc., 116 (1994) 8621.

[63] Howdle, S. M., Popov, V. K., unpublished results.
[64] For a good introduction to scH_2O, see Shaw, R. W., Brill, T. B., Clifford, A. A., Eckert, C. A., Franck, E. U., Chem. Eng. News, 69 (1991) 26. Savage, P. E., Gopalan, S., Mizan, T. I., Martino, C. J., Brock, E. E., AIChE J. 41 (1995) 1725.
[65] First International Conference on Supercritical Water Oxidation, Jacksonville, Florida, U.S.A., Feb. 1995.
[66] Myrick, M. L., Kolis, J., Parsons, E., Chilke, K., Lovelace, M., Scrivens, W., J. Raman Spectrosc., 25 (1994) 59.
[67] Adschiri, T., Kanazawa, K., Arai, K., J. Am. Ceram. Soc., 75 (1992) 1019.
[68] Metzger, J. O., Proceedings 3rd International Symposium on Supercritical Fluids, Strasbourg, 3 (1994) 99.
[69] Banister, J. A., Howdle, S. M., Poliakoff, M., J. Chem. Soc., Chem. Commun., (1993) 1814.
[70] Leong, V. S., Cooper, N. J., Organometallics, 7 (1988) 2080.
[71] Gourinchas, V., Bocquet, J. F., Chhor, K., Tufeu, R., Pommier, C., Proceedings 3rd International Symposium on Supercritical Fluids, Strasbourg, 3 (1994) 315.
[72] Brackermann, H., Buback, M., Makromol. Chem., Rapid Commun., 10 (1989) 283; Buback, M., Busch, M., Lovis, K., Mühling, F.-O., Chem.-Ing.-Techn., 66 (1994) 510.
[73] Buback, M., Proceedings 3rd International Symposium on Supercritical Fluids, Strasbourg, 3 (1994) 93; Hanke, A., Dechema Workshop, Applications of Supercritical Fluids, Frankfurt, 1995.
[74] Eckert, C. A., Suleiman, D., Liotta, C. L., Boatwright, D. L, 1995, in press.
[75] Poliakoff, M., Howdle, S. M., Proceedings 3rd International Symposium on Supercritical Fluids, Strasbourg, 3 (1994) 81.
[76] Reid, R. C., Prausnitz, J. M., Poling, B. E., The Properties of Gases and Liquids, 4th edition, McGraw-Hill Inc., New York, 1986.
[77] De Swaan Arons, J., Diepen, G. A. M., J. Chem. Phys., 44 (1965) 2322.

6

ORGANIC CHEMISTRY IN SUPERCRITICAL FLUIDS

E. Dinjus*, R. Fornika and M. Scholz
*Max-Planck-Gesellschaft zur Förderung der Wissenschaften e.V.,
Arbeitsgruppe 'CO$_2$-Chemie' an der Friedrich-Schiller-Universität Jena, Lessingstraße 12,
D-07743 Jena, Germany*

6.1 INTRODUCTION
6.2 FUNDAMENTALS
6.3 ORGANIC REACTIONS IN SUPERCRITICAL FLUIDS
    6.3.1 Diels-Alder Reactions
    6.3.2 Heterogeneous Catalysis and Fischer-Tropsch Synthesis
    6.3.3 Polymerization, Depolymerization and Dimerization Reactions
    6.3.4 Esterification and Carboxy-Inversion Reactions
    6.3.5 Electrochemical Reactions
    6.3.6 Biochemical and Enzymatic Reactions
    6.3.7 Photochemical Investigations of Organic Reactions in Supercritical Fluids
    6.3.8 Supercritical Fluids as Solvents and Substrates in Homogeneous Transition Metal Catalyzed Reactions
6.4 FURTHER INVESTIGATIONS
6.5 REFERENCES

6.1 INTRODUCTION

Supercritical fluids (SCF) are current research topics both in the field of fundamental research and in the application in reactions and material separation (extraction processes) [1–12]. Apart from carbon dioxide, a multitude of other compounds which are gaseous (ethene, perchlorinated and fluorinated hydrocarbons) or liquid (acetone, hexane, pentane) under normal conditions are used. Thus, supercritical and hypercritical fluids have been successfully used in the large-scale extraction of ingredients in food chemistry and pharmaceutical industry, the extraction of coffeine from coffee or tea, the extraction of nicotine from tobacco as well as obtaining the extract from hops being well known examples. The extraction of highly polluting organic contaminants from waste waters is also being investigated [3–5]. SCF could be used in the textile industry. When synthetic fibers are dyed with aqueous based coloring substances (see Chapter 7), the application of supercritical carbon dioxide (sc CO_2) as a solvent could simplify several processes or make them unnecessary and thus reduce the pollution of waste waters [13]. Furthermore, SCF are used to solubilize and purify polymers [14], and the synthesis and fractionation of polymers are also investigated in sc solvents [15]. For analytical purposes, the application of SCF in the form of Supercritical Fluid Chromatography (SFC) has proved efficient in the extraction and separation of non-volatile and temperature-sensitive substances [16, 17].

The advantageous physicochemical properties of SCF have led to their increased use in other fields of chemistry during the last few years (see Chapters 5 and 7). This is demonstrated by the growth in the number of conferences on the subject of SCF and the intensified research activities in universities, institutes and in industry. Increasing global environmental exposure to atmospheric pollutants requires the reduced utilization of solvents and alternative reaction media. Comprehensive efforts are being made to avoid organic solvents in chemical synthesis or to carry out organic syntheses in SCF, *e.g.*, in sc CO_2. This reaction medium is toxicologically harmless, incombustible and inexpensive. The advantage of ecological harmlessness, however, is offset by the disadvantage of high costs in the development of high pressure apparatus.

In general, SCF have many advantages as solvents. They are often extremely good solvents and their solvent power can be adjusted by the variation of pressure and temperature. Moreover, they can differ in their effect on solutes; for example, CO_2 exercises a selectivity of almost 100 in relation to cholesterol 1 and ergosterol 2 (cf. Figure 6.1) though they have similar structures. Both substances have very low vapor pressures (order of 10^{-10} bar) but the careful

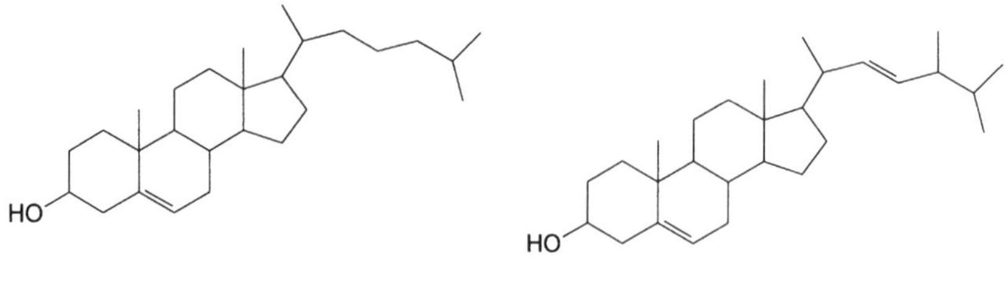

Figure 6.1 Molecular structures of cholesterol 1 and ergosterol 2.

measurement of vapor pressures reveals a parallel behavior of their solubility in sc CO_2. The examination of such systems in basic physical chemistry is necessary for understanding the nature of solubility in SCF.

6.2 FUNDAMENTALS

Almost any chemical substance can be transferred into the three conditions of matter (gaseous, liquid, solid) by the variation of pressure (p) and temperature (T). The p-T diagram of a compound (phase diagram of CO_2 in Figure 6.2) can be divided into three phases in which the substance in question exists as gaseous, liquid or solid matter. At the phase transitions, the phases are in equilibrium. Therefore, at the triple point, TP (crossing point of all 3 lines) all three phases are in equilibrium. When a substance is heated, the critical point (CP) is reached on the boiling line (phase transition liquid/gaseous) at which intermolecular forces

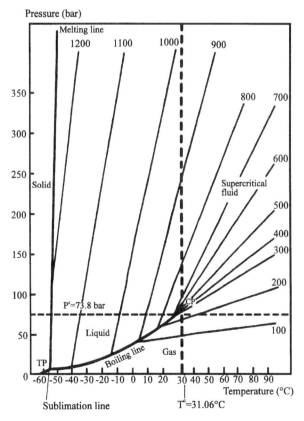

Figure 6.2 Phase diagram of carbon dioxide with density (in kg m^{-3}) as the third dimension.

and thermal energy become equal. SCF are compressed and heated substances beyond the CP which is characteristic of each substance (see Table 6.1) (for instance at the reduced temperature $T_r = T/T_c = 1.01$ and the reduced pressure $p_r = p/p_c = 1.01$). The CP for carbon dioxide is at $p_c = 73.83$ bar and at $T_c = 31.06\,°C$. The critical density ρ_c is $0.467\,\text{g mL}^{-1}$.

Supercritical fluids have lower viscosities than conventional solvents. The solubilities of the substances to be extracted can be widely varied by the selection of the parameters pressure and temperature in liquid-like densities. The possibility of the complete separation from the extract is a considerable advantage compared to conventional solvents. A comparison of the orders of magnitude of density ρ, the binary diffusion coefficient D_{12} and the dynamic viscosity for the different ranges is indicated in Table 6.2.

A unique property of SCF is their pressure dependent density. If the temperature is constant, density can be adjusted from that of a vapor to that of a liquid without any discontinuity. It is common to focus on the region where the reduced temperature ($T_r = T/T_c$) and the reduced pressure ($p_r = p/p_c$) are of the order of unity. In this region, considerable changes in fluid density and related properties, such as the solubility of the material, are observed with small changes in pressure. These characteristics make SCF very attractive as tunable process solvents or reaction media. The pressure dependence of some important

Table 6.1 T_c, p_c and μ values of some selected solvents [18].

Solvent	T_c [°C]	p_c [atm]	μ [debye]
Acetone	236	47.0	2.9
Acetonitrile	275	48.3	3.5
n-butane	152	37.5	0.0
Carbon dioxide	31	72.9	0.0
Carbon tetrafluoride	45	37.4	0.0
Chlorodifluoromethane	96	49.7	1.4
Chlorotrifluoromethane	29	38.7	0.5
Dichlorodifluoromethane	112	41.4	0.5
Diethylamine	223	36.6	1.1
Diethyl ether	193	35.6	1.3
Ethanol	241	61.4	1.7
Hexane	234	29.9	0.0
Methanol	240	80.9	1.7
Methylene chloride	237	63.0	1.8
Nitrous oxide	37	72.4	0.2
Pentane	197	33.7	0.0
Pyridine	347	56.3	2.3
Sulfur hexafluoride	45.5	37.6	0.0
Triethylamine	259	30	0.9
Water	374	221	1.8

6.2. FUNDAMENTALS

Table 6.2 Orders of magnitude of density ρ, diffusion coefficient D_{12} and viscosity η in gases, liquids, and supercritical fluids.

Property	Symbol	Unit	Gas	Liquid	SCF
Density	ρ	[g cm^{-3}]	10^{-3}	1	0.2–0.8
Diffusion coefficient	D_{12}	[cm^2 s^{-1}]	10^{-1}	10^{-6}	10^{-3}
Dynamic viscosity	η	[g cm^{-1} s^{-1}]	10^{-4}	10^{-2}	10^{-4}

solvent properties of supercritical carbon dioxide at 40 °C is presented in Figure 6.3.

There are several ways of comparing the properties of SCF with those of conventional solvents. If the temperature is constant, the solubility is a complicated function with respect to external pressure. Proceeding from the solvent as a lightly compressed gas (in which the solubility of any substance is very low), the solubility will increase to a maximum with increasing pressure and then decrease again. In contrast to that, the solubility (x) increases at constant pressure and increasing temperature according to the ideal law (ln $x \propto T^{-1}$) if the pressure is continuously adjusted in order to keep the solvent at constant density. Figure 6.4 shows the solubility of fluoranthrene in sc CO_2 at different densities as a function of the reciprocal absolute temperature T. The change of

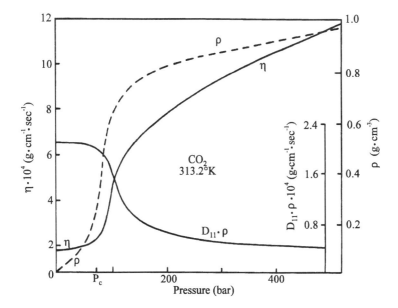

Figure 6.3 Density ρ, viscosity η and $D_{11} \times \rho$ for pure CO_2 as a function of pressure at 40 °C (D_{11} = self-diffusion coefficient) [19].

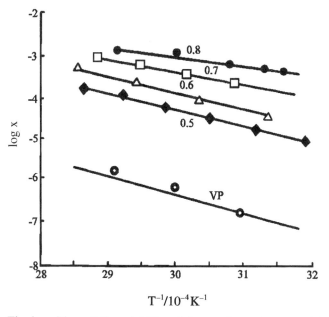

Figure 6.4 The logarithm of the solubility of fluoranthrene in supercritical CO_2 as a function of T^{-1} at various CO_2 densities (g L^{-1}) [20].

the vapor pressure of fluoranthrene depending on T^{-1}, which follows the empirical law (eq. 1), can also be seen in this figure.

$$\ln x = \frac{\Delta H_{melt}}{R}\left[\frac{1}{T_m} - \frac{1}{T}\right] \quad (1)$$

ΔH_{melt} is the melting enthalpy and T_m is the melting point.

Careful examination of the available data shows that the solubility of a substance in a supercritical fluid is an extremely sensitive function of density. $\partial p/\partial V_m = 0$ (V_m = partial molar volume) at the CP. Since the density ρ is proportional to V_m^{-1}, the term $\partial p/\partial \rho$ is also zero under the same conditions or $\partial \rho/\partial p = \infty$. Thus, the density is extremely sensitive to pressure at and above the critical point and hence it follows that the solubility of a substance in a SCF is very sensitive to pressure in this region. If, for instance, pressure is increased to 400 bar, the solubility of squalene in CO_2 increases by 10 orders of magnitude from 10^{-11} to 10^{-1} g cm^{-3}. This property of markedly temperature and pressure dependent density allows a considerable control of the capacity of a solvent to dissolve a particular substance or to distinguish between different solutes.

An empirical description proceeding from thermodynamic principles was proposed by Chrastil [21] and is presented in eq. (2), in which the density and

6.2. FUNDAMENTALS

the temperature-dependent solubility in SCF are related to each other:

$$c = \rho^k \exp\left(\frac{a}{T} + b\right) \quad (2)$$

where c is the solubility in g L^{-1}, ρ is the density in g L^{-1} and k, a and b are characteristic constants for the solvent and the solute. k is related to the association number and corresponds to the average number of solvent molecules which are associated with a molecule of a dissolved substance. For sc CO_2, k has a considerable range from 1.58 for H_2O to 12.1 for cholesterol. Generally k increases with an increase in the size of the molecules of the solute, a fact associated with the clustering of solvent molecules around the dissolved molecule; this assumption is supported by spectroscopic measurements. Dissolved molecules can induce a very big increase in the density of the solvent in the very easily compressible region near the CP (if the pressure is constant). This effect is formally related to a large, negative partial molar volume of the solute.

One way of relating solvent properties is as follows: an aromatic substance with a chromophoric group (*e.g.*, p-nitroanisole) is added to the solvent and afterwards the shift of the absorption band of the compound is determined for the assessment of the solvent properties. By means of the $p \to \pi^*$ and $\pi \to \pi^*$ electronic transitions in the UV-Vis spectra, Kamlet et al. [22] were able to relate the effects of the polarities and polarizabilities of conventional solvents. When several parameters are neglected, the π^* value for any solvent can be determined by the simplified eq. (3).

$$v_{max} = v_0 + s\pi^* \quad (3)$$

SCF can be compared with conventional solvents *via* the shift of absorption maxima (see Table 6.3) [23].

The E_T values introduced by Dimroth and Reichardt [24] can be used as a further spectroscopic method for the comparison of solvent properties. The

Table 6.3 π^*-values for solvents and supercritical fluids.

Substance	T	Pressure p [bar]	π^*-value	Ref.
sc CClF$_3$	303	91.2	-0.29 ± 0.03	[23]
liq. CClF$_3$	296	273.5	-0.21 ± 0.04	[23]
sc N$_2$O	323	182.3	-0.12 ± 0.03	[23]
liq. N$_2$O	296	273.5	-0.03 ± 0.02	[23]
sc CO$_2$	323	182.3	-0.07 ± 0.04	[23]
liq. CO$_2$	296	273.5	0.04	[22]
sc NH$_3$	413	172.2	0.34 ± 0.07	[23]
liq. NH$_3$	296	273.5	0.80 ± 0.02	[23]

molar excitation energy of the charge-transfer band of a highly solvatochromic dye (*e.g.*, N-phenylpyridine-betaine has the formula number 30 in the literature), the E_T values, are determined in the solvent in question (see Table 6.4). The $E_T(30)$ value is a measure of the strength of the interactions between solvent and solute.

With respect to polarity, sc CO_2 approximately corresponds to carbon tetrachloride and *n*-hexane. The main difference between solvent strengths of sc CO_2 and liquid hexane resides in the Lewis-acid/Lewis-base properties. Furthermore, sc CO_2 can form hydrogen bonds under suitable conditions. Both effects may be responsible for the extraordinary solvent properties of sc CO_2.

6.3 ORGANIC REACTIONS IN SCF

6.3.1 Diels-Alder Reactions

The Diels-Alder reaction ($[4+2]$ cycloaddition) is one of the most important and most interesting reactions in Organic Chemistry (see Chapter 3). As schematically shown in Figure 6.5, an electron-rich conjugated diene **3** with an electron-poor dienophile **4** (molecule with a double or a triple bond) is usually transformed to an unsaturated six-membered ring **5**. The positions 1 and 4 in the diene are the centers of the reaction. The usefulness of the Diels-Alder reaction is based on the high regio- and stereoselectivity. Diels-Alder reactions have been extensively utilized for the preparation of insecticides [27], fragrances [28, 29], plasticizers [30] and dyes [31].

The influence of solvents is generally small in Diels-Alder reactions. Apart from other hints, this fact further suggests a concerted mechanism. However, the

Table 6.4 $E_T(30)$ values for different solvents.

Solvent	$E_T(30)$ value/kcal mol^{-1}	Ref.
Water	63.1	[24]
Methanol	55.5	[24]
Acetonitrile	46.0	[24]
Acetone	42.2	[24]
Dichloromethane	41.1	[24]
Pyridine	40.2	[24]
Chloroform	39.1	[24]
THF	37.4	[24]
Benzene	34.5	[24]
liq./sc carbon dioxide	33.8	[25]
Carbon tetrachloride	32.5	[24]
sc carbon dioxide ($p = p_c$)	31.5	[26]
n-hexane	30.9	[24]

6.3. ORGANIC REACTIONS IN SCF

Figure 6.5 Reaction scheme of the Diels-Alder reaction ([4 + 2] cycloaddition).

dependence on pressure, which is used for the increase of stereoselectivity, can be considerable (see Section 3.3).

Comprehensive studies of the [4+2] cycloaddition were carried out in the liquid phase. Thus, the Diels-Alder reaction is a useful probe for studying the effects of SCF on reaction kinetics. Paulaitis and Alexander [32], Kim and Johnston [33] and Ikushima et al. [34] investigated the cycloaddition of isoprene and maleic anhydride in sc CO_2. Paulaitis and Alexander [32] studied the influence of pressure on the reaction rate in supercritical reaction mixtures. The authors reported an increase in the second order rate constant with increasing pressure. This effect is most distinct near the CP where the activation volume also reaches high negative values. The rate constants in sc CO_2 at very high reduced pressures resemble those values obtained in three different conventional solvents.

The results are in accordance with the predictions of Kim and Johnston [33] with regard to the effect of pressure on the rate constant. Combining the data for solvatochromic shifts of phenol blue (E_T values) and the kinetic data for this reaction in other solvents, they developed a kinetic correlation according to eq. (4):

$$\ln k_x = -a E_T + b \quad (4)$$

The reaction rate constant k_x is a linear function with respect to the transition energies E_T, a and b are parameters independent of pressure. This correlation was used for the prediction of the rate constant in sc CO_2. The authors calculated the activation volume $\Delta V^{\neq}$ of this reaction with the help of eq. (5):

$$\Delta V^{\neq} = aRT \left(\frac{\partial E_T}{\partial \rho} \right) T \rho \kappa_T \quad (5)$$

Thus, the activation volume amounted to $-4000 \text{ cm}^3 \text{ mol}^{-1}$ at 35 °C and 75 bar. The activation volume is only $-55 \text{ cm}^3 \text{ mol}^{-1}$ at 300 bar and the same reaction temperature. The value of the activation volume for this reaction in liquid ethyl acetate at 35 °C was indicated to be $-37.4 \text{ cm}^3 \text{ mol}^{-1}$, by way of comparison. The large negative activation volumes near the critical pressure are attributed to the partial molar volumes which reach highly negative values near the critical point. Eckert *et al.* [35] measured large negative partial molar volumes near the CP for the systems naphthalene/sc CO_2, naphthalene/sc C_2H_4, CBr_4/sc C_2H_4 and camphor/sc C_2H_4.

Paulaitis and Alexander [32] also developed phase diagrams for the systems CO_2/maleic anhydride and a CO_2 Diels-Alder adduct. They presented a general analysis of the reaction kinetics under near-critical conditions by relating geometric properties of species near the critical point with transition state theory.

Ikushima *et al.* [34] used high pressure FT-IR-spectroscopy for the *in situ* study of the $AlCl_3$-catalyzed Diels-Alder reaction of isoprene and maleic anhydride. The results obtained led the authors to propose a two-step mechanism under these conditions (see Figure 6.6).

Kim and Johnston investigated the cycloaddition of methyl acrylate and cyclopentadiene in sc CO_2 (see Figure 6.7) [36]. They could demonstrate experimentally that the selectivity of a parallel reaction network can be controlled by pressure. The authors reported on the effect of pressure with respect to the selectivity of the *endo* and *exo*-addition compound for reactions in the

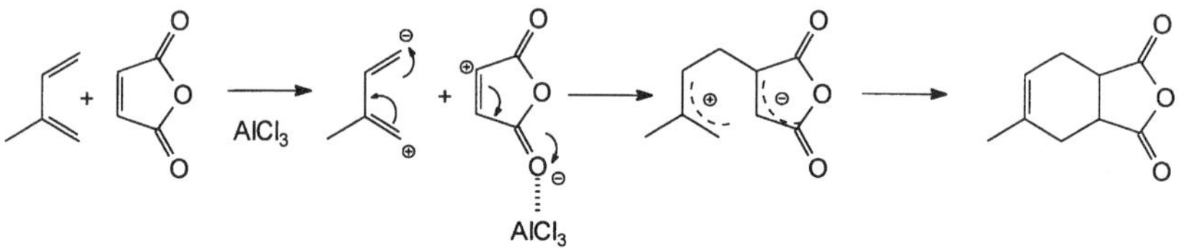

Figure 6.6 $AlCl_3$-catalyzed Diels-Alder reaction of isoprene and maleic anhydride.

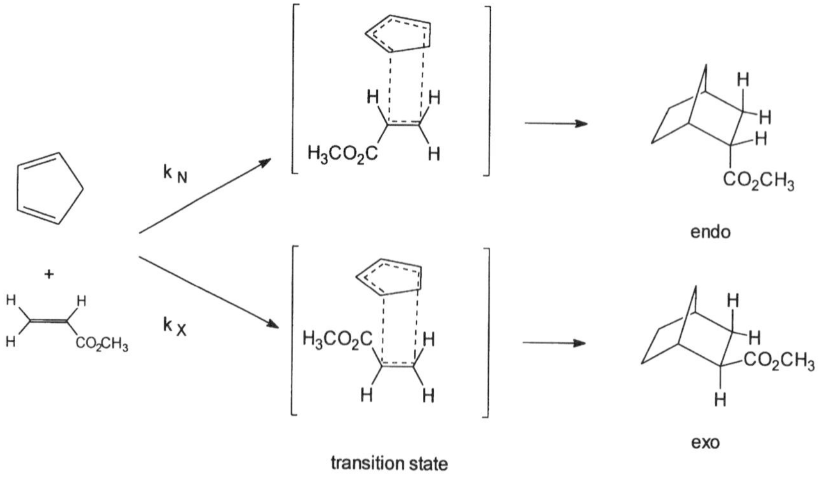

Figure 6.7 Cycloaddition of methylacrylate and cyclopentadiene in supercritical CO_2.

6.3. ORGANIC REACTIONS IN SCF

temperature range of 35 to 45 °C and a pressure range of 80 to 300 bar. The *endo/exo* selectivity slightly increased with increasing pressure. The difference between the highest and lowest selectivities observed amounted to about 2.5% at a given temperature, with an uncertainty in the determination of the selectivity of about 1%. This slight increase was explained by the effect of pressure, with the comment that the *endo* transition state has a larger dipole moment than the *exo* transition state. Hence it follows that the *endo* transition state will have greater affinity for the solvent and it will be favored by increasing density and polarity of the solvent. This pressure effect on selectivity is described by eq. (6). The increase in selectivity results from

$$\frac{RT \partial \ln\left(\frac{k_N}{k_X}\right)}{\partial p} = -(\overline{V_N^t} - \overline{V_X^t}) \tag{6}$$

where k_N is the reaction rate constant for the *endo*-product and k_X is the rate constant for the *exo*-product. The change in selectivity corresponds to the difference between the volumes for the *endo* $\overline{V_N^t}$ and *exo* transition states $\overline{V_X^t}$. The *endo* and *exo* selectivities were also correlated with the transition energies of phenol blue (E_T values). This correlation was used to predict selectivity near the critical region.

Ikushima et al. [25, 37] published kinetic data for the Diels-Alder addition of methyl acrylate to isoprene in sc CO_2 at 323 K and pressures of 4.9 to 20.6 MPa (Figure 6.8). The rate constant increased with increase in pressure. The selectivity in the formation of the two isomers **6** and **7** also depended on pressure. The occurrence of the second product shows a maximum near the critical pressure of CO_2. This observation was attributed to the local solvent density around the dissolved molecule which exceeds the bulk density in this case. This kind of clustering of solvent molecules around the transition state complex in the critical region is held responsible for the generation of steric coercion so that one isomer is mainly formed.

The authors found a correlation between the transition energies of pyridine-N-phenoxide ($E_T(30)$ values) and the rate constants at 323 K, according to eq. (4). The activation volume was calculated at 323 K, near the critical pressure, and amounted to $-700 \text{ cm}^3 \text{ mol}^{-1}$. The absolute value of the activation

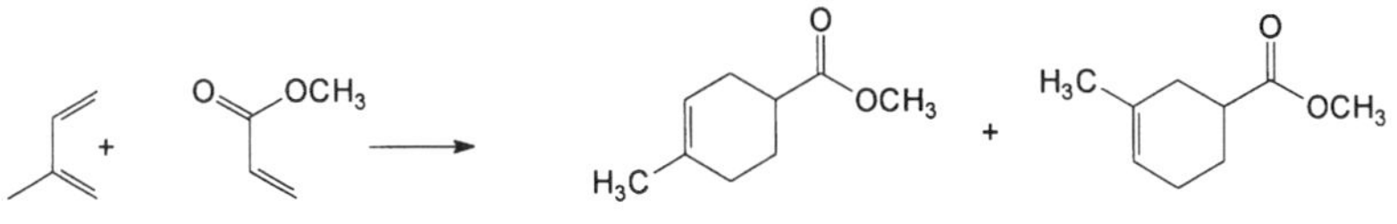

Figure 6.8 Diels-Alder addition of methylacrylate to isoprene in sc CO_2.

volume is about one order of magnitude smaller at pressures on either side of the critical point.

The authors used the method developed by Debenedetti [38] to determine the number of solvent molecules associated with the dissolved molecule. The cluster size near the CP is about 10, and that for pressures on either side of p_c is lower than 2. This confirms previous investigations which showed the potential use of pressure as a suitable parameter for the manipulation of reaction kinetics and the selectivity near the CP [38]. These effects of pressure are most distinct near the critical region.

Isaacs and Keating [39a] reported kinetic data for the cycloaddition of p-benzoquinone and cyclopentadiene in CO_2 at temperatures between 25 and 40 °C and at pressures between 60 and 240 bar. In this case, the rate constant also increases with increasing pressure and is about 20% higher than in diethyl ether. The plot of the rate constant as a function of temperature at constant density does not show any discontinuity at the critical temperature, *i.e.*, no anomalies occur near the CP. (For another recent kinetic study see [39b].)

To sum up, Diels-Alder reactions do not show any discontinuity of the reaction rate in the transition from fluid to supercritical CO_2 at constant density. The rates are approximately the same as in solvents under normal pressure [40].

6.3.2 Heterogeneous catalysis and Fischer-Tropsch synthesis

The potential usefulness of SCF as media for heterogeneous catalytic reactions was first detected by comparing the results obtained for identical reactions in the gaseous, liquid and SCF phase and dates back to the work of Tiltscher *et al.* [41, 42] in 1981 and 1984. They were able to show that catalyst deactivation occurred during isomerization of hexene in the gaseous phase because non-volatile hexene oligomers deposited on the surface of the catalyst. If, however, the reaction is carried out by maintaining a constant reaction temperature, but above the critical pressure, the sedimentation of oligomers is avoided and catalyst deactivation prevented. An essential advantage for the performance of heterogeneous reactions under sc conditions can be derived therefrom, *i.e.*, the possibility of an *in situ* extraction of coke precursors.

On the other hand, Tiltscher *et al.* [43] found in investigations of 1-hexene isomerization in a low activity macroporous Al_2O_3 that the initial ratio of cis/trans-2-hexene formed is not affected by the variation of temperature in the gaseous phase. In the liquid phase, only a slight effect on this ratio is observed due to the variation of pressure and temperature; this effect is much more distinct in the sc phase. The investigation of the 1-hexene isomerization was extended to include high activity microporous industrial Pt/Al_2O_3 catalysts, and use of CO_2 as a diluent [44–50].

Saim and Subramanian [44–47] analyzed the phase and reaction equilibria to establish the thermodynamic constraints in this system, and they also determined the equilibrium conversions and the monophase range. Continuous runs

6.3. ORGANIC REACTIONS IN SCF

were used in order to assess the catalyst deactivation in hexene/CO_2 mixtures. In accordance with Tiltscher's investigations of 1-hexene, catalyst deactivation (probably caused by carbonization) was observed at subcritical pressure whereas, at the same temperature, but supercritical pressure, the catalyst activity did not decline. That precipitation of coke precursors occurs was concluded from the brown coloring of the reaction solution in the supercritical range, whereas the solution remained clear at subcritical conditions. Advantages of the SCF technology compared to conventional processes may be derived therefrom for other heterogeneously catalyzed, technologically relevant processes and applications.

The following example illustrates the complications and complexities of such applications. The isomerization rates towards the end of the run decline in the case of an isothermal increase of pressure in the subcritical range, but they increase with increasing pressure in the supercritical range. In contrast to the almost constant activity found when macroporous catalysts were used [41], Saim and Subramanian [44–47] observed a deactivation of the applied microporous Pt/Al_2O_3 even under supercritical conditions.

The loss of activity is apparently due to the formation of inextractable coke in the pores of the catalyst under the subcritical conditions during the initial phase. Due to the coke desorption rates and the solubility of model coke compounds in SCF, Manos and Hofmann [48] concluded that a complete *in situ* reactivation of a microporous zeolite catalyst by SCF is not possible. Only a reduced deactivation rate under sc conditions can be expected because the newly formed coke precursors can be dissolved by the SCF reaction medium.

Model observations on a single-pore model (coke formation and *in situ* coke extraction) showed that when pressure is increased along a near-critical isotherm, a pressure and density optimum exists at which the catalyst activity reaches a maximum [49]. At low densities, a catalyst deactivation due to insufficient coke extraction is assumed to occur. The catalyst activity is reduced by pore-diffusion limitations in the liquid-like reaction mixture at a density higher than the optimal (see Figure 6.9). Ginosar and Subramanian [50] were able to confirm these qualitative findings for the isomerization of 1-hexene in real systems.

These investigations of heterogeneous isomerization catalysts under supercritical conditions allow the following conclusions. The blocking of catalyst pores, which leads to catalyst deactivation at subcritical conditions, was avoided by the *in situ* extraction of the coke compounds by near-critical and supercritical mixtures. In spite of the reduced coke deposits under supercritical conditions, the isomerization rate was lower and the deactivation rate was higher due to the limited diffusion through the pores in the liquid-like reaction mixture.

It is concluded that near-critical reaction mixtures produce an optimum combination of solvent and transport properties, which, in comparison with subcritical (gas-like) or dense supercritical (liquid-like) mixtures, is best suited for a maximum isomerization rate and a minimum catalyst deactivation rate. Applications of sc conditions for technologically relevant chemical processes can be expected in the near future.

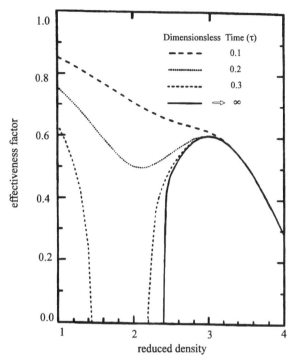

Figure 6.9 Effectiveness factor map for 1-hexene isomerization at supercritical conditions [49].

Fischer-Tropsch Synthesis One of the technologically relevant synthetic processes, in which the effects of supercritical phases have been intensively investigated, is the Fischer-Tropsch synthesis. The Fischer-Tropsch synthesis (FTS), in which liquid hydrocarbons are produced from synthesis gas (CO and H_2), is a heterogeneously catalyzed gas-phase reaction. The catalysts used are typically based on iron or cobalt in the presence of additional promotors such as lanthanum and ruthenium. Since heat dissipation is often difficult in exothermal reactions, local superheating on the catalyst surface easily occurs which can lead to the deactivation of the catalyst. Furthermore, high molecular weight waxes which do not vaporize under reaction conditions and block the micropores of the catalyst are formed at the same time.

In order to improve the characteristics of the heat transport and to extract the wax (as an unwanted by-product from the catalyst surface), a slurry phase FTS process was developed in which a slurry of the catalyst and mineral oil were used as a reaction medium [51–53]. It was shown that the diffusion of synthesis gas into the micropores of the catalyst is so slow in the liquid phase that the overall reaction rate is slower than in the gaseous phase [54, 55] These factors led to investigations in SCF phases [56]. Fujimato *et al.* [57, 58a, b] studied the Fischer-Tropsch synthesis in a supercritical *n*-hexane phase. They compared the

6.3. ORGANIC REACTIONS IN SCF

performance of silica-supported cobalt/lanthanum and/or alumina supported ruthenium catalysts in the gaseous phase, the liquid phase (trickle bed operation with n-hexadecane and nitrogen) and supercritical phase FTS in a fixed-bed reactor. The following changes occurred in the supercritical phase compared to the other phases:

- the heat transport was faster than in the gaseous phase, but slower than in the liquid phase;
- the mass transfer was more efficient than in the liquid phase, but less efficient than in the gaseous phase;
- *in situ* extraction of high molecular weight products from the pores of the catalyst;
- a higher olefin selectivity than in the two other phases.

Furthermore, less methane was formed, probably due to the better heat distribution in the reactor. The more distinct formation of long-chain olefins is apparently due to the improved solubility of these higher hydrocarbons in SCF. During the gas phase synthesis, the olefins remain in the pores of the catalyst and can be hydrogenated there. Interesting effects of the size of catalyst pores were found with regard to reaction rate and product distribution. For a given total pore volume, as the pore diameter decreases, the surface area increases, but the diffusivity within the pores decreases. These two competing effects lead to an optimum of the pore size in different catalytic reactions. The optimum pore diameter for CO transformation was found to be about 10 nm for the alumina supported ruthenium catalyst.

Yokota and Fujimoto [56, 57] found that smaller catalyst pores led to higher rates, but noted another distribution of products. Catalysts with smaller pores produced higher proportions of short olefins instead of the more desired longer chain olefins. (See also [58c, d] for effects on product substitution.)

Burkur *et al.* [59] studied the steady state FTS in supercritical propane and were able to confirm that olefin hydrogenation and isomerization reactions are diffusion controlled. The observed increase of olefin selectivity during the supercritical FTS is in accordance with the results obtained by Fujimoto *et al.* [56, 58] in tests of short duration with a cobalt- and ruthenium-based Fischer-Tropsch catalyst in supercritical n-hexane.

Compared to both the traditional gas phase reaction and the experiments in the liquid or slurry phase, the Fischer-Tropsch synthesis in a SCF has advantages which are due to diffusion, heat transfer and solubility effects. The higher selectivity and the reduced formation of unwanted by-products are other advantages of FTS in the supercritical phase.

6.3.3 Polymerization, Depolymerization and Dimerization Reactions

Polymerization Reactions in Inert Supercritical Fluids The attractive characteristics of performing polymerization reactions under supercritical conditions are the same as for other chemical reaction systems under supercritical conditions. High pressures favor addition polymerizations significantly since they typically have negative intrinsic activation volumes.

Furthermore, the properties of SCF such as solvent parameters are well defined functions of density, temperature, and pressure near the critical region. This possibility of the "fine tuning" of sc fluids allows, on the one hand, the control of the solubility of polymers with a given molecular weight in the reaction mixture and, on the other, control of the molecular weight distribution when there are only slight changes in the reaction conditions.

In addition, the fact that diffusion coefficients in SCF are higher allows higher net reaction rates than in conventional solvents. This is particularly true when diffusion restrictions are of importance. Finally, the lower viscosity in sc fluids can reduce solvent cage effects which impede the efficiency of a free radical initiated reaction. The solvent can also affect the transition state to such an extent that a change in the kinetics of the polymerization reaction occurs.

Almost all papers dealing with polymerization in SCF can be divided into two groups. In one case, the reaction conditions exceed the CP of the monomer (or comonomer), or in the other case, the reaction conditions exceed the CP of an inert solvent. The latter method has lately been developed much further. The idea of using inert SCF for polymerization reactions has only been tested during the last ten years. Therefore, the synthesis of polymers in an inert sc medium has not yet been commercialized.

Styrene was polymerized in highly volatile hydrocarbons such as sc ethane, sc propane or sc n-butane [60–63]. Kumar et al. [60, 61] described precipitation polymerization reactions in sc fluids. They added the radical starter AIBN (azodiisobutyronitrile) to styrene in an optical high-pressure cell which was put under pressure in the presence of ethane and this led to a single, homogeneous phase. Afterwards, the reaction medium was mixed using a circulating pump or a shaking bath and the samples were analyzed by means of size-exclusion chromatography (SEC). The precipitation polymerization occurred within 30 minutes when the monomer concentration in the reactor was kept below 20 mass%. A higher monomer concentration probably increases the solubility parameters of the solvent and this prevents the occurrence of precipitation. After another 30 minutes, the reaction medium is clear and the precipitated polymer has settled on the wall of the reactor. If the reaction is carried out in a pressure range of 120 to 250 atm and a temperature range of 60 to 70 °C, oligomers of styrene with a molecular weight of between 1000 and 6000 are formed. The authors assume that the equilibrium solubility of a polymer chain in an SCF is independent of the composition of the original polymer and that it only depends on the molecular mass of the chain.

Kiran and Saraf [62, 63] also investigated styrene polymerization in sc ethane, sc propane and sc n-butane. The reactions, carried out in a cell in which the volume could be varied and the contents could be viewed, were initiated by means of the radical starters AIBN or t-butyl peroxide and butyl peroxybenzoate. Temperature as well as pressure have decisive effects on the molecular

6.3. ORGANIC REACTIONS IN SCF 235

weight and size distribution of the products. In ethane, reduction in pressure from 272 atm to 136 atm causes an increase in the polydispersity from 6.29 to 75.9, with a number average molecular weight M_n of about 1870. If the pressure is maintained at 272 atm, the polydispersity increases from 1.66 to 6.29 in the temperature range from 70 to 100 °C and the number average molecular weight decreases from 5446 to 1877.

Polymerization of Ethylene The use of CO_2 as an inert reaction medium in the polymerization of ethylene was described by Hagiwara *et al.* [64–66]. A specially designed autoclave with a quartz-glass inspection window permitting the visual observation of the polymerization reaction was used. The ethylene/ CO_2 mixtures exist as a single, homogeneous phase and the polymer precipitates from the solution as slightly puffy particles, which deposit on the bottom of the reactor. This result can be compared with the block polymerization of ethylene. The IR analysis also indicated that the polymer formed in sc CO_2 and the block polymer are comparable materials. The polymerization rate of ethylene as well as the molecular weight and the polymer yield were lower in sc CO_2 than in the block-polymerized product. This observation is explained by a reduced propagation rate due to the lower fugacities of ethylene in the reaction mixture.

The reduction of the molecular weight was explained by the increased chain termination at higher CO_2 concentrations. The main advantage of performing the ethylene polymerization in sc CO_2 is that sc CO_2 can serve as a relatively inert solvent which does not tend to decompose in the radiation field. Furthermore, CO_2 can be separated easily from the product and prevents the adhesion of the polymer to the reactor walls, an important prerequisite for a successful continuous polymerization process. It is, however, not yet clear if CO_2 functions as an inert or reactive solvent, since the copolymerization of ethylene and CO_2 is also known to occur.

Terry *et al.* [67] polymerized light olefins in sc CO_2 and investigated the feasibility of conducting polymerization reactions in CO_2 *in situ* since CO_2 is increasingly being used in the recycling of oil. Increased viscosity in CO_2 could lead to a better sedimentation of the oil in the fluid. These authors also reported the polymerization of ethylene and octene in sc CO_2, but they did not observe the expected increase in viscosity. The same idea of transporting a monomer by means of SCF into a porous material and then conducting the polymerization at supercritical conditions was described by Sunol [68] in a patent application. The patent describes the impregnation of wood and other porous materials with polymers, which leads to a modification of the mechanical properties of the material.

Polymerization of Acryl Monomers Hartmann and Denzinger [69] devised a method of synthesizing cross-linked acryl-polymers in sc CO_2. The reactions were carried out with a semi-batch procedure in an agitated autoclave at temperatures between 31 and 150 °C and pressures between 80 and 300 bar. The reaction mixtures contained several unsaturated olefinic monomers, some were bifunctional and could act as a suitable initiator for free radicals. The weight ratio monomer:CO_2 varied between 1:1 and 1:15. The primary advantage of this process, in contrast to the reaction in most other solvents, is the easy separability of CO_2 from the polymer powder produced. The particle diameter was found to be between 0.5 and 5 μm. The fact that cross-linked polymers occur indicates that the polymerization within the precipitated polymer particles takes place more quickly than in a homogeneous solution.

DeSimone et al. [70] examined the homogeneous free radical polymerization of fluorocarbon polymers with a higher molecular weight in sc CO_2. Homopolymers of 1,1-dihydroperfluorooctylacrylate (FOA) were polymerized in a window cell at 59.4 °C and a pressure of 207 atm, and AIBN was used as radical starter (see Figure 6.10). This allowed the visual verification of homogeneous polymerization. The polymer remaining in the cell was removed with Freon-113. The product was a viscous, transparent liquid and the yield amounted to 65%. The average molecular weight of 270,000 was determined by means of gel permeation chromatography. Mixtures of FOA and other monomers, such as methylmethacrylate, butylacrylate, styrene and ethylene, also copolymerize homogeneously in sc CO_2 under these conditions. The primary aim was to replace chlorofluorocarbons (CFC), which are usually used for the production of materials of this type; this could bring about important ecological advantages since CFC are considered to be responsible for the depletion of ozone in the earth's upper atmosphere.

Guan et al. [71] studied the kinetics of the initial step, the decomposition of 2,2'-azodiisobutyronitrile in sc CO_2. The activation energy for the apparent first order rate constant was found to be 135 kJ mol^{-1} which is within the range quoted for liquid solvents (120 to 140 kJ mol^{-1}). A lower reaction rate than in benzene was found in sc CO_2, an observation which is attributed to the lower dielectric constant of CO_2. The kinetic data found in CO_2 and those published for liquid solvents could be compared with the Kirkwood correlation and supported this interpretation. Experiments with benzene led to the determination of the intrinsic activation volume (solvation effects excluded) of 23.6 cm^3 mol^{-1}. The authors also developed a complete correlation of pressure effects with the observed rate constant for AIBN decomposition. They combined the correlation from the Kirkwood plot with the pressure effect on the rate constant resulting from the intrinsic activation volume.

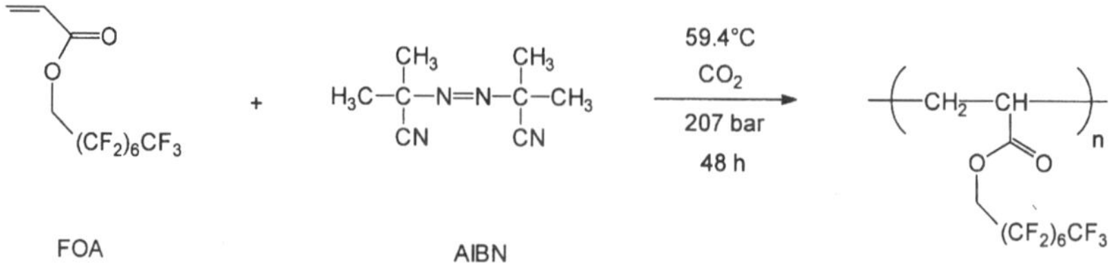

Figure 6.10 Polymerization of 1,1-dihydroperfluorooctylacrylate.

6.3. ORGANIC REACTIONS IN SCF

Inverse Emulsion Polymerization Beckmann *et al.* [72–74] described the inverse emulsion polymerization of acrylamide in supercritical mixtures of ethane and propane. This reaction can be regarded as heterogeneous, in which an aqueous phase can be dispersed in a continuous, supercritical hydrocarbon phase.

A non-ionic surfactant, acrylamide and water were put in a window cell and mixed with a magnetic stirrer. The propane/ethane mixture was added with a syringe pump. After the equilibration was reached at the required temperature (up to 65 °C) and pressure (up to 440 atm), it was possible to produce a homogeneous dispersion of the aqueous phase in the continuous, supercritical phase with the aid of a circulating pump. The circulating pump was switched on for 30-min intervals during the 6 hour reaction in order to maintain the reaction mixture in a disperse state. The window cell was depressurized at the end of the reaction and water, surfactant and the unreacted material were flushed out of the reaction container with a chloroform/water mixture. The polymer retained in the cell was dissolved by a mixture of chloroform, water and acetone.

The molecular weight of the polymeric product was between 483,000 and 864,000. The conversion rates (5–30%) are low at reaction temperatures between 35 and 50 °C, but increase to more than 95% at 57.5 °C.

Elliott and Cheung [75] and Srinivasan and Elliott [76] investigated the possibility of producing a polymeric, microcellular foam (organic aerogel) *via* polymerization in SCF. They polymerized methylmethacrylate with ethylene glycol dimethacrylate in propane and Freon-22, and succeeded in combining the polymerization step and the subsequent drying step to a single integrated process. The foams obtained, however, had undesired high bulk densities.

A further use of SCF as reaction media for polymerization is their application as a form of a reaction suppression system. Lee and Hoy describe a method for suppressing the reactivity by adding SCF (mainly CO_2) to a reaction mixture [77]. This effect might be due to the SCF-induced dilution of the reaction mixture and the associated decrease in the reaction rate it causes.

Depolymerization Dhawan *et al.* [15, 78a] described a further application of SCF, namely depolymerization reactions. Depolymerization of synthetic wastes is a possible method of treating these materials in an environmentally safe way and of recycling monomers or other useful products. Pyrolytic depolymerization has some of the same disadvantages as the pyrolytic decomposition of coal or biomass:

- low yield of the wanted products;
- excess formation of soot and gas;
- insufficient control of the reaction.

Dhawan *et al.* [15, 78a] investigated the thermolysis of cinnamic alcohol-butadiene copolymers in sc toluene and sc tetralin and the thermolysis of cis-polyisoprene and scrap rubber in sc toluene. Their results showed that polymers in SCF were reduced to lower molecular weight products and demonstrated that secondary reactions to form polycondensates were suppressed in contrast to pyrolytic depolymerization. It was possible to identify more than 100 different products and many of them were produced from reactions of the polymer with SCF. Chen *et al.* [78b] depolymerized post-consumer fire and natural rubber using sc H_2O and sc CO_2.

Dimerization Saito *et al.* [79] reported the dimerization of benzoic acid *via* hydrogen bonding in sc CO_2 studied by FT-IR spectroscopy. The absorption ratio between dimer and monomer is highly affected by the solvent density and reaches a maximum in the density region of the medium where the isothermal compressibility also has a maximum. This indicates that solute-solute interactions should be taken into consideration even in very highly diluted solutions for the development of a predictable model to describe the thermodynamic properties of SCF. Further studies are necessary for the clarification of the dimerization mechanism of benzoic acid.

6.3.4 Esterification and Carboxy-Inversion Reactions

Ellington and Brennecke [80, 81] studied the non-catalytic esterification of phthalic anhydride with methanol in sc CO_2 (see Figure 6.11) in order to test if an increase in the local concentration in SCF increases the reaction rate [82]. The values of the solvatochromic shifts of cosolvents in sc CO_2 indicate that the local composition of the cosolvent around the dissolved sample is a strict function of the pressure in the compressible region and that it can be up to seven times higher than the bulk concentration.

As shown schematically in Figure 6.11, the reaction consists of the addition of a methanol molecule per molecule of phthalic anhydride; the addition of a second methanol molecule would require a catalyst under the prevailing reaction conditions. The experiments both in conventional solvents and in SCF confirm a first-order reaction in phthalic anhydride and in methanol concentration. The reactants were mixed in a high pressure vessel (750 ml) and then transferred to an optical high pressure cell [83].The rate constants were determined by UV-Vis spectroscopy at 50 °C and at pressures of 166.5 to 97.5 bar. The concentrations of reactants were 3×10^{-4} mol dm^{-3} (phthalic anhydride) and 3.6×10^{-2} to 0.36 mol dm^{-3} (methanol). Based on the bulk concentrations, the second order rate constant reached values of between 1.42×10^{-3} dm^3 mol^{-1} min^{-1} at 166.5 bar and 4.27×10^{-2} dm^3 mol^{-1} min^{-1} at 97.5 bar. This 30 fold increase in the rate constants is a noteworthy and

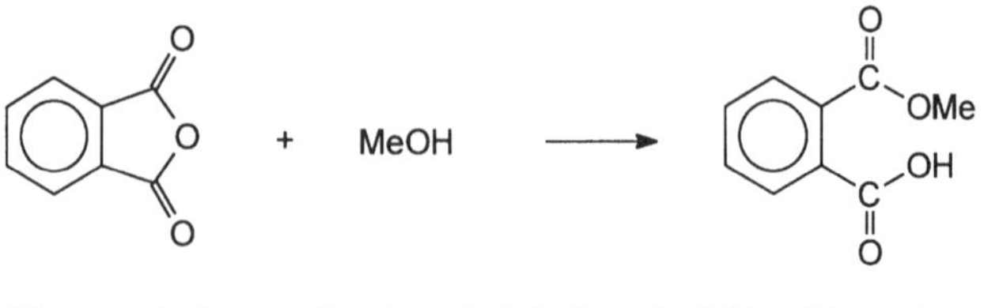

Figure 6.11 Non-catalytic esterification of phthalic anhydride with methanol in sc CO_2.

6.3. ORGANIC REACTIONS IN SCF

important change for a reaction in SCF. The results demonstrate that a chemical reaction can be accelerated when carried out in an sc fluid without limitation to a small range of conditions above the CP.

The thermodynamic pressure effect on the rate constant as it is determined by means of the transition state theory can be calculated with eq. (7) [84].

$$\frac{\partial \ln k}{\partial p} = -\frac{\Delta V^{\neq}}{RT} - k_T \qquad (7)$$

$\Delta V^{\neq}$ designates the activation volume of the reaction (difference of the partial molar volumes of transition state and reactants) and k_T designates the isothermal compressibility of the solvent. The assessment of $\Delta V^{\neq}$ and k_T with the Peng–Robinson equation of state leads to negative values for the term $\partial \ln k/\partial p$, i.e., the rate constant should, as it is observed, increase at lower pressures [85]. Therefore, the rate constant might only increase by 1.2 fold over the pressure range from 166.5 to 97.5 bar.

The results obtained by the authors cannot be completely explained by the effect of pressure on the rate constant. They assume that an increase of the local concentration of methanol by the dissolved phthalic anhydride has an effect on the reaction. The measured rate constants reflect the increase in the methanol concentration, which is more distinct at lower pressures near the CP.

Moulougi et al. investigated the esterification of oleic acid with methanol [86]. This reaction is catalyzed by p-toluenesulfonic acid (p-TSA) or by the sulfonated cation-exchange resins (styrene polymers with a cross-linking level of 8% divinyl benzene) K 2411 and K 1481. Following pretreatment with 2 N hydrochloric acid (transformation into the H^+ form) and the subsequent neutralization with water, the exchange resins were dried at 313 K and put into a sapphire reactor together with oleic acid and methanol. Previous results showed that a condensation of oleic acid and methanol in sc CO_2 at 313 K and 135 bar did not occur without the presence of a catalyst.

When the catalyst p-TSA was used, very good yields of methyl oleate in a homogeneous reaction mixture were obtained at relatively short reaction times (160 min). Thus, p-TSA is an excellent catalyst for this esterification in sc CO_2 and the yield obtained was higher than in the alcoholic medium [87, 88]. The kinetics of the esterification are best described by a first-order model with a rate constant of 8.64×10^{-3} min^{-1} at a pressure of 165 bar and in which a molar ratio p-TSA to oleic acid of $r = 1/15$ is reached. The rate constant was found to be 1.43×10^{-2} min^{-1} at 195 bar and $r = 1/5$.

In the case of the sulfonated cation-exchange resins K 2411 and K 1481, heterogeneous mixtures were obtained in sc CO_2 and the reaction rate is controlled by the rates of adsorption of oleic acid and of desorption of methyl oleate. The authors found a significant influence of r (ratio of the moles of the groups of sulfonic acids in the resin to the moles of oleic acid) on the reaction in the heterogeneous phase. High values for r reduced both adsorption and desorption. The pellet size of the resin, however, was not affected. They concluded that the reaction could be limited by external diffusion [89]. The esterification reaction, however, only takes place on the surface *via* the adsorption of oleic acid, the reaction itself and the succeeding desorption of methyl oleate. The external groups of sulfonic acid are poisoned by chemisorption since they are occupied by bulky molecules [90]. The initial catalytic activity can be restored by simply flushing out the resin with methanol or ether.

The esterification of oleic acid by methanol in sc CO_2 proceeds similarly to the corresponding reaction in conventional solvents with similar hydrophobic properties. Further investigations could aim at development of polymer resins with specific functional groups that can be used as heterogeneous catalysts in SCF [91, 92].

Leffler *et al.* [93] investigated the decomposition of diacyl peroxide in sc CO_2, sc CCl_4 and sc $CHCl_3$ and argued in favor of an ionic reaction. For bisisobutyryl peroxide (see Figure 6.12), the rate constant k is 3.6×10^{-5} s^{-1} in sc CO_2 (40 °C, density: 0.93 g mL^{-1}) and thus, it is lower than in CCl_4 (7.72×10^{-5} s^{-1}) and in $CHCl_3$ (42.5×10^{-5} s^{-1}). Apart from the carboxy-inversion product isopropyl-isobutyryl carbonate (yield 17%), isopropyl isobutanoate (5%) and isobutyric acid (17%) were formed.

There is no exchange between the inverted CO_2 groups in isopropyl isobutyrylcarbonate and the CO_2 from the reaction medium. Cyclobutane carbonyl-*m*-chlorobenzoyl peroxide also shows a slower decomposition rate in sc CO_2 (55 °C, density: 0.81 g mL^{-1}, $k = 2.2 \times 10^{-5}$ s^{-1}) than in sc CCl_4 ($k = 2.93 \times 10^{-5}$ s^{-1}) and sc $CHCl_3$ ($k = 27.2 \times 10^{-5}$ s^{-1}).

A partial rearrangement of alkyl groups in cyclopropylmethyl and 3-butenyl groups takes place in alkyl-*m*-chlorobenzoic acid esters and in the carboxy-inversion product alkyl-*m*-chlorobenzoic carbonate. Experiments using ^{13}C-labelled CO_2 resulted in a 12% exchange of the carbonyl-carbonate carbon in cyclopropylmethyl-*m*-chlorobenzoic carbonate. A rearrangement does not take place in the carboxy-inversion product with unshifted alkyl groups.

The authors concluded that the unshifted carboxy-inversion product is formed in a separate and isolated way, which does not go *via* the ion-pair intermediate step R$^+$CO$_2$$^-$OOCR′, in which R$^+$ (R = cyclopropylmethyl) is not associated with CO_2.

6.3.5 Electrochemical reactions

The first investigations of the use of the non-aqueous supercritical systems sc CO_2, sc CF_3Br, sc HCl and sc NH_3 in electrochemical reactions were performed

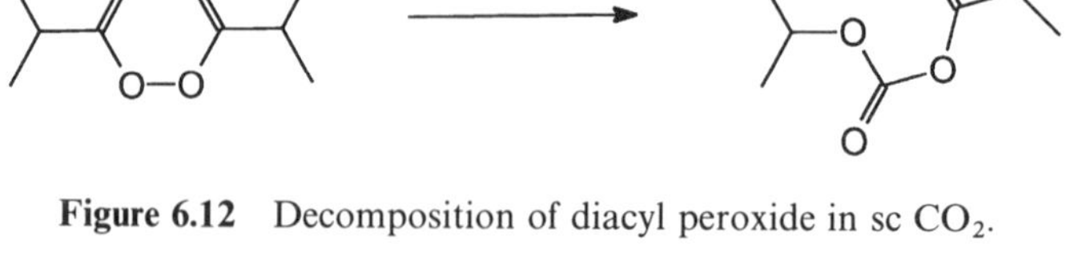

Figure 6.12 Decomposition of diacyl peroxide in sc CO_2.

6.3. ORGANIC REACTIONS IN SCF

by Silvestri et al. [94]. sc CO_2 and sc CF_3Br with $(n\text{-Bu})_4NI$ as supporting electrolyte are not very good conductors and therefore, they are not suited for electrochemical syntheses. In the sc HCl/KI system, iodide was oxidized to elementary iodine at the anode. The corrosiveness of HCl inhibited the reactions. In the sc NH_3/NaCl system, silver was dissolved at the silver anode and precipitated as powder at the cathode. The anode current efficiencies were between 82% and 95%. The current efficiency of iron anodes was 45%, which is lower due to the formation of Fe(III) and the oxidation of ammonia. The electrical resistance of the supercritical medium was one to two orders of magnitude higher than in the liquid medium. Thus, only very narrow cells with a very small distance between the electrodes are suitable for electrochemical syntheses.

Bard et al. [95–103] investigated electrochemical reactions of different molecules in sc NH_3, sc H_2O, sc SO_2 and sc acetonitrile by means of cyclic voltammetry, chronocoulometry and chronoamperometry. The measuring methods are described thoroughly in the literature [104, 105]. A constant potential, which is sufficiently high to initiate reaction (8), is applied to the working electrode in chronocoulometry and chronoamperometry.

$$\text{Ox} + n e^- \rightarrow \text{Red} \tag{8}$$

After the reaction has begun, the charge (chronocoulometry) or current (chronoamperometry) is measured as a function of time. The dependence of current on time, a criterion for the reaction progress, can be described by the Cottrell equation (eq. 9) for linear diffusion of the reactants:

$$i = \frac{nFAD_{Ox}^{0.5}c_{Ox}}{\pi^{0.5}t^{0.5}} \tag{9}$$

where n is the number of electrons taking part in the reaction, F the Faraday constant, c_{Ox} the starting concentration of the reactant and t the time. The diffusion coefficient of the reactant, $D_{Ox}^{0.5}$, can now be obtained from a plot of i versus $t^{0.5}$ according to eq. (9). Bard et al. [95–103] obtained the diffusion coefficients with this method. Furthermore, they tested whether the molecules investigated follow the Stokes-Einstein relation. This can be represented in effect by a model of a macroscopic sphere which moves in an incompressible continuous fluid. The following relationship (eq. 10) exists between the diffusion coefficient D of the reactants and the viscosity η of the fluid:

$$D = \frac{kT}{A\pi\eta r} \quad \text{with} \quad A = \frac{6\left(1 + \frac{2\eta}{\beta r}\right)}{\left(1 + \frac{2\eta}{\beta r}\right)} \tag{10}$$

where k is the Boltzmann constant, r the radius of the diffusing particle (molecule or solvated ion). A lies between 4, free-slip condition ($\beta = 0$), and 6, non-slip particle-solvent boundary condition ($\beta = \infty$) [102]. The diffusion coefficients measured according to the procedures described above at different temperatures, were correlated to the known viscosities according to eq. (10) for almost all molecules investigated. In reference [100], the viscosity of acetonitrile was determined according to eq. (10). It can be summarized that the Stokes-Einstein relation can also be used in SCF and that the basic structure of the fluid and the solvent sphere of the molecules are not essentially changed during the transition from the liquid to the supercritical medium.

Only Cu(II) in sc H_2O [96], hydroquinone in sc H_2O [97], pyrazine in sc NH_3 [98] and ferrocene in sc acetonitrile [102] showed higher diffusion coefficients than expected according to eq. (10). The authors assume that the radius of the diffusing species (r in eq. (10)) decreases with increasing temperature. In the first case the effective solvated ion radius of Cu(II) changes [96] and in the second case a breakdown of hydrogen bonds between water and hydroquinone is possible [97] and decreases the effective radius.

Wightman et al. [106–110] performed electrochemical reactions in sc CO_2 by adding water, tetracyclohexylammonium hexafluorophosphate or -nitrate as electrolytes and by applying microelectrodes with diameters between 10 and 250 μm. These reactions were investigated voltammetrically. When large amounts of the electrolyte were added, the electrolyte formed a second phase since it did not dissolve in sc CO_2. Furthermore, the electrolyte coated the electrode. Microelectrodes coated with ion-conducting Nafion film were also tested [108, 109].

Others have tested electrodes coated with conducting polymers [111–114] and enzymes (horseradish peroxidase) [115] in sc CO_2 and sc $HCClF_2$ [116] with the aim of obtaining electrochemical detectors (ECD) for SCF.

Dombro et al. [117] have produced dimethyl carbonate electrochemically from CO and methanol at reaction temperatures of 90 °C. CO_2 was used as a cosolvent with methanol in order to carry out the reactions near the CP at lower temperatures than in aqueous electrolytes. Tetrabutylammonium bromide (TBAB) (0–5%) served as electrolyte and bromide source (see Figure 6.13).

Table 6.5 provides a summary of the voltammetric studies described.

Better current efficiencies from 40 to 110% (current efficiencies exceeding 100% can be attributed to errors in sample recovery and analysis [117]) were obtained by Crisp et al. [118] who suggested the reaction mechanism presented in Figure 6.13.

Current efficiencies increase with the CO and TBAB content. Only dimethoxymethane was formed at a TBAB content of 1%. This is the only preparative electro-organic synthesis in an SCF.

6.3.6 Biochemical and Enzymatic Reactions

SCF as solvents in enzyme-catalyzed reactions have some advantages compared to other solvents. Diffusion-controlled reactions can be accelerated since the

6.3. ORGANIC REACTIONS IN SCF

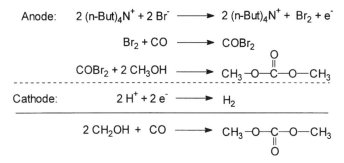

Figure 6.13 Electrochemical production of dimethylcarbonate.

transfer coefficients of the dissolved substances are increased by the lower viscosity, the lower surface tension and the high diffusion capacity. In 1991 Russel and Beckmann [119] postulated that the higher diffusivities in SCF give rise to an increase in the reaction rate of biocatalytic processes.

Further advantages are desirable separation properties and the insolubility of enzymes in SCF, which makes the reprocessing of the reaction products following the reaction and the recovery of enzymes much easier. The high solubility of gases like oxygen and hydrogen in SCF also facilitates oxidation and hydrogenation reactions. Furthermore, the non-toxicity and incombustibility of most SCF and the possibility of simply removing the solvent without residue after the reaction by decompression, are remarkable properties of SCF's.

Only SCF with critical temperatures of 10 to 80 °C are suitable for enzyme-catalyzed reactions since the enzymes denature at higher temperatures. Furthermore, the enzymes must be stable under reaction conditions. Aaltonen and Rantakylä [120] listed all enzymes which were investigated until 1990 and which are stable in sc CO_2, in a review article.

sc CO_2 was used as a solvent in almost all cases since it has advantages which are comprehensively described [120]. Furthermore, Hammand et al. [121], Kamat et al. [122, 123], and Russell et al. [124, 125] also employed fluoroform. Kamat et al. [122, 123] tested SF_6, propane, ethane, ethylene, fluoroform and CO_2 as SCF in the transesterification of methyl methacrylate with different enzymes as catalysts. Interestingly, the inorganic and hydrophobic SF_6 was found to be an excellent solvent for this reaction. The hydrophobic solvents were superior to the sc CO_2 under all conditions. This was explained by invoking unfavorable interactions of the CO_2 molecules with the free amino groups of the enzymes.

The first enzymatic catalyzed reactions in SCF were the hydrolysis of di-sodium *p*-nitrophenylphosphate to *p*-nitrophenol [126] catalyzed by the enzyme alkaline phosphatase, the oxidation of *p*-cresol to methyl-substituted catechol and *o*-quinone with the enzyme polyphenol oxidase [121], and the

Table 6.5 Voltammetric investigations.

SCF	Investigated molecules	Electrodes (diameter)	Electrolyte	Remarks	Ref.
NH_3	m-chloronitrobenzene	W (1 mm)	KI	Radical-anion is stable on the voltammetric time scale	[95]
H_2O	Cu, Cu(I), Cu(II)	Pt (1 mm)	Na_2SO_4, KCl	D of Cu(II) larger than predicted by Stokes-Einstein (S-E)	[96]
H_2O	H_2O, I^-, Br^- hydroquinone	Pt (0.5-1.3 mm)	$NaHSO_4$	D of hydroquinone larger than predicted by S-E	[97]
H_2O	I^-	Pt (25 μm)	$NaHSO_4$	Pressure effects on the redox potential, clustering of SCF, solute-solvent interactions	[103]
NH_3	Pyrazine, quinoxaline, phenazine, solvated e^-, nitrobenzene	W (1 mm)	CF_3SO_3K	Radical anions and solvated electrons stable on the voltammetric time scale	[98]
NH_3	Dimerization of quinoline and acridine radical	W or Pt (25-100 μm)	$N(n-Bu)_4$ CF_3SO_3	Dianions not stable on the voltammetric time scale, kinetic study	[99]
MeCN	Ferrocene, phenazine	W (1 mm)	CF_3SO_3Na	Viscosity of MeCN is estimated using measured D and the S-E equation	[100]
SO_2	$Fe(bpy)_3^{2+}$	Pt (10-25 μm)	Et_4NAsF_6	Electrochemical studies in sc SO_2 are possible	[101]
MeCN	Methyl viologen $Fe(Cp)_2$, $Fe(Cp^*)_2$, $Os(bpy)_3^{2+}$	W (1 mm)	CF_3SO_3Na	Only ferrocene and decamethylferrocene have a well-defined behavior under supercritical conditions, D values larger than computed with the S-E relationship	

CO_2	Ferrocene	Pt (10–50 µm)	$(C_6H_{12})_4NPF_6$	Water must be added, electrolyte forms a second liquid phase and coats the electrode	[106], [107]
CO_2 N_2O	Ferrocene, anthracene, 9,10-diphenyl-anthracene	Pt (10–250 µm)	$(C_6H_{12})_4NNO_3$	Added water forms a film on the surface of the electrodes, $[(C_6H_{12})_4N]NO_3$ film electrode as detectors for SCF limit 0.1 ng ferrocene	[110]
CO_2	Ferrocene, 4-methyl-catechol, 3,4-dihydroxy-benzyl amine	Pt (10 µm) Nafion coated	H_2O	Electrochemical detection in sc CO_2 for supercritical fluid chromatography (SCF)	[108], [109]
CO_2	Ferrocene, anthracene, 9,10-diphenyl-anthracene	Pt (10µm) coated with PEO	none	Electrochemical detection in sc CO_2 for SCF compared with a FID PEO = poly(ethylene oxide)	[111]
CO_2	p-benzoquinone, anthracene,	see above	none	Electrochemical detection in sc CO_2 for SCF, electrocarboxylation	[114]
$HCClF_2$	Ferrocene	Pt (10–25 µm)	$(n\text{-Bu})_4NBF_4$	Electrochemical detection in sc CO_2 for SCF	[116]

interesterification of triolein with stearic acid and a *Rhizopus delemar* enzyme [127]. According to Krukonis and Hammond [128], the catalyzed hydrolysis of p-nitrophenylphosphate [126] takes place in a moisture film between sc CO_2 and the enzyme, since p-nitrophenol is insoluble in sc CO_2. This moisture film is formed by the adsorption of the water dissolved in sc CO_2 onto the enzyme.

The enzyme catalyzed reactions which were frequently investigated are transesterifications as described in Figure 6.14. Esterifications were also investigated ($R_1 = H$ in Figure 6.14). In an interesterification or acidolysis, the groups of aliphatic acids are exchanged, *e.g.*, in the reaction of a triglyceride with an aliphatic acid as shown in Figure 6.15.

Table 6.6 summarizes the enzyme catalyzed reactions; SCF are presented in the first column, the enzyme in the second, the reaction in the third, remarks on the reaction in the fourth and references in the fifth.

Figure 6.16 shows the transesterification of N-acetyl-L-phenylalanine chloroethyl ester and ethanol with the enzyme *Subtilisin Carlsberg* as catalyst in sc CO_2 investigated by Pasta et al. [129]. They found a considerably higher reaction rate in sc CO_2 than in the hitherto best organic solvent, tetra-amyl alcohol (92% yield after 30 min. compared to 89% after 2 hours). This effect was attributed to the higher mass transfer and the higher diffusion rate in sc CO_2 (see also [119]).

In other reports, SCF only showed their advantages in comparison with other solvents, mostly n-hexane, when the reaction solutions were separated following the preparative step(s). It was found that it was essential for water to be present in the reaction solution for the enzyme to have activity. In order to reach its optimum activity and conformation the enzyme must be properly hydrated and, therefore, it needs a certain minimum water content [130]. If the water content is higher, a hydration sphere is formed around the enzyme, which obstructs the diffusion of the hydrophobic substrate to the active site of the enzyme. If there is

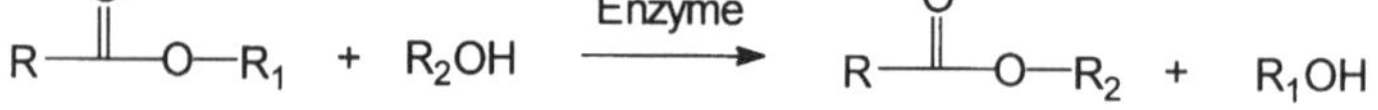

Figure 6.14 Transesterification by enzymes.

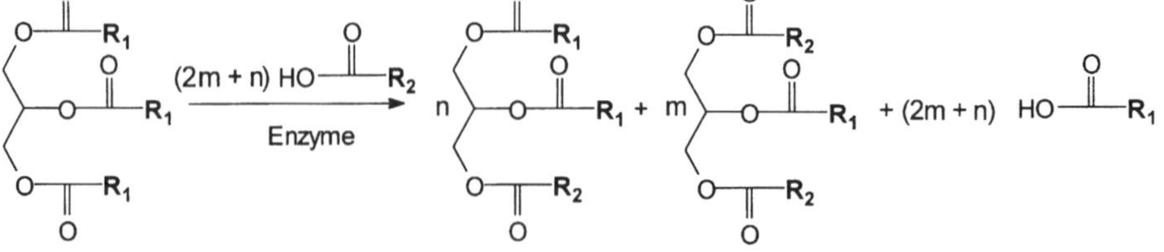

Figure 6.15 Transesterification of triglycerides with fatty acids catalyzed by enzymes.

6.3. ORGANIC REACTIONS IN SCF

a high excess of water, the enzyme can also denature [131]. Marty et al. [132, 133] investigated in detail the effect of the water content on the reaction rate. The esterification of oleic acid with ethanol served as a model reaction. An immobilized lipase of *Mucor miehei* (lypozyme) served as catalyst. The adsorption isotherms for water were measured and the dependence of the water content of the solid enzymatic phase on the water content in sc CO_2 was determined. Increasing temperature (33–50 °C) and increasing pressure (11–17 MPa), as well as an increase in ethanol content (0–450 mM), cause a negative effect on the water adsorption by the solid phase.

The maximum enzyme activity occurs at a water content of the solid phase of about 8–10% of the dry weight, independent of the reaction conditions and the solvent. Since more water dissolves in sc CO_2 than in *n*-hexane, more water must be added to sc CO_2 than to *n*-hexane to maintain the water content of the enzyme.

The first complete reaction kinetics study of an enzymatic reaction was carried out in sc CO_2. A ping-pong bi-bi mechanism with inhibition by ethanol substrate was derived and tested for the above reaction in *n*-hexane by Chulalaksananukul et al. [134]. Figure 6.17 shows this mechanism schematically. The acid is bound by the enzyme. The complex formed loses water after a rearrangement. Then the alcohol is bound and, following a further rearrangement, the ester is separated, with the enzyme being regenerated. In a side reaction the enzyme can become inactive by binding an alcohol molecule. The experimental values are in good agreement with the calculated ones. If the optimum water content of the solid phase is adjusted in both solvents, the kinetics of both phases can be compared. The values of the kinetic constants are within the same range. If only the reaction is considered, there is no clear advantage of sc CO_2 over *n*-hexane. With respect to the stability and the kinetics of the enzyme, sc CO_2 is very similar to an organic solvent if the water content of the enzyme is optimal. The advantage of sc CO_2 resides in the post-reaction separation process. The investigations by Marty et al. [135] were extended to a continuous reaction-separation process in which the reaction solution was processed after the reactor by means of four separation vessels. CO_2, emerging from the last separation vessel and after being manipulated to obtain appropriate reaction conditions, could be led back to the reactor. Water must always be added to sc CO_2 during this continuous process in order to avoid the drying out of the enzyme. The water content of the enzyme was adjusted to a optimum value of 10% by means of the adsorption isotherms [132].

In the case of sc CO_2, theoretical calculations were in good agreement with the experimentally measured values whereas, in the case of *n*-hexane, the calculated yields were higher than the experimental values. In the experiments with sc CO_2, the purity and concentration of the product were twice as high as in the experiments with *n*-hexane. This result confirms that sc CO_2 can only be used as a solvent in the reaction described above when the reaction and subsequent processing are combined. Only in this way can the real advantage of SCF, the superior separation performance, be used optimally.

Table 6.6 Some enzyme catalyzed reactions

Fluid	Enzyme	Reaction	Remarks	Ref.
CO_2	Subtilisin Carlsberg	Transesterification of N-acetyl-L-phenylalanine chloroethyl ester with ethanol (see Figure 6.16)	The reaction is faster in sc CO_2 than in organic solvents	[129]
CO_2	Lipase (Mucor miehei)	(isoamyl alcohol) –OH + (acetate ester) → (isoamyl acetate) + EtOH	Kinetic, lower yields than calculated for n-hexane	[143a]
CO_2	Lipase (Mucor miehei)	$C_9H_{19}OH$ + (acetate ester) → $C_9H_{19}O$–(acetate) + EtOH	Continuous process with integrated product recovery	[143b]
Several	Lipase (Candida cylidracea)	(methacrylate ester) + HO–CH$_2$–CH(C$_2$H$_5$)–C$_4$H$_9$ $\xrightarrow{\text{Lipase}}$ (2-ethylhexyl methacrylate) + MeOH	Test of several SCF: CO_2, SF_6, C_3H_8, C_2H_6, C_2H_4, HCF_3	[123]
HCF_3	See above	See reaction above	"Solvent engineering" (see text)	[122], [125]
HCF_3	See above	Polytransesterification of bis(2,2,2-trichloroethyl) adipate and 1,4-butanediol (see Fig. 6.18) eq. (4)	Control of polymer dispersity and molecular weight of polyester	[124], [125]
CO_2	Lipase (Mucor miehei)	Esterification of myristic acid (tetradecane acid) with ethanol	Continuous process	[131]
CO_2	Lipase (Mucor miehei)	C_8H_{17}–CH=CH–C_7H_{15}–COOH + EtOH $\xrightleftharpoons[\text{Enzyme}]{}$ C_8H_{17}–CH=CH–C_7H_{15}–COO–Et + H_2O	Effect of water, continuous process (see text)	[132], [133], [135]

Solvent	Enzyme	Reaction	Notes	Ref.
CO_2	Lipase (*Candida Cylindracea*)	Stereospecific esterification of oleic acid with ($\pm$) citronellol (see Figure 6.22)	CO_2-aggregation (see text)	[140], [141]
CO_2	Lipase (*Mucor miehei*)	Stereospecific esterification of ibuprofene with propanol (see Figure 6.20)	70% S(+) form	[138]
CO_2	Lipase (*porcine pancreatic*)	Stereospecific esterification of glycidol with butyric acid (see Figure 6.21)	Separation after reaction	[139], [143c]
CO_2	Lipase (*Rhizopus delemar*)	Figure 6.15 with $R_1 = C_7H_{14}-CH=CH-C_8H_{17}$, $R_2 = C_{13}H_{27}$	One of the first enzyme catalyzed reactions	[127]
CO_2	Lipase (*Rhizopus arrhizius*)	Figure 6.15 with $R_1 = C_{11}H_{23}$, $R_2 = C_{15}H_{31}$	Pressure effect	[137]
CO_2, C_2H_6	Lipase (*Rhizopus arrhizus*)	Figure 6.15 with $R_1 = C_{11}H_{23}$, $R_2 = C_{13}H_{27}$	Kinetic	[143d]
CO_2	Cholesterol oxidase	Oxidation of Cholesterol (see Figure 6.23)	Cholesterol aggregation	[142]
CO_2, HCF_3	Polyphenol oxidase	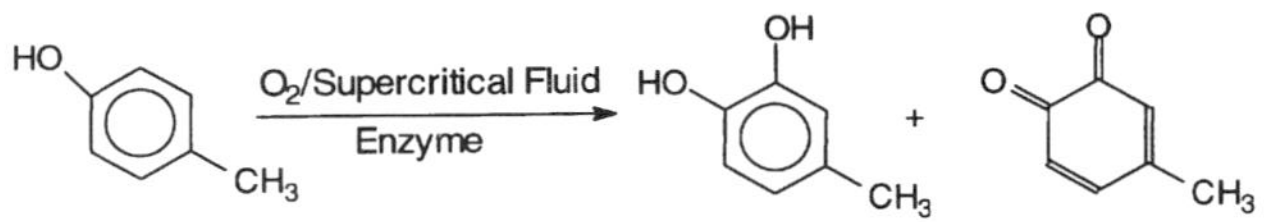 ![reaction with O$_2$/Supercritical Fluid over Enzyme converting p-cresol to catechol product + quinone]	One of the first enzyme-catalyzed reactions	[121]
CO_2	Alkaline phosphatase	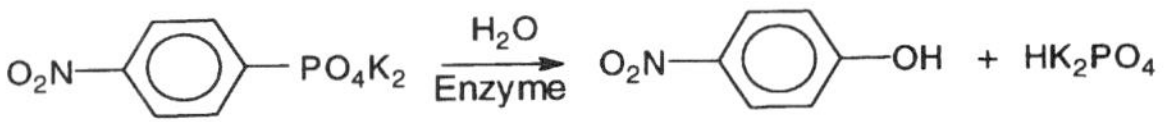 $O_2N-C_6H_4-O-PO_4K_2 \xrightarrow[\text{Enzyme}]{H_2O} O_2N-C_6H_4-OH + HK_2PO_4$	One of the first enzyme-catalyzed reactions	[128]

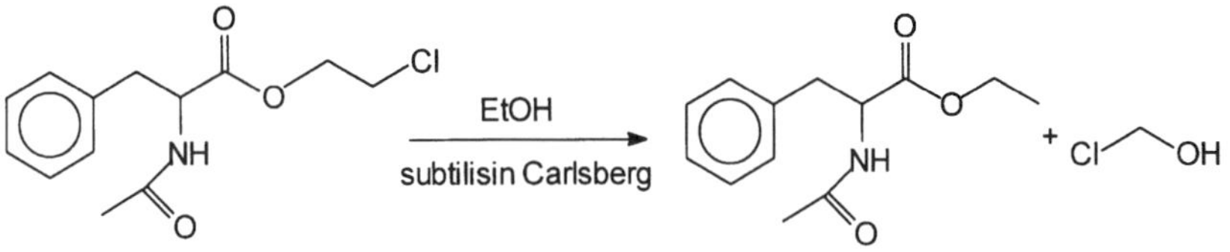

Figure 6.16 Transesterification of N-acetyl-L-phenylalanine chloroethyl ester and ethanol with the enzyme *Subtilisin Carlsberg* as catalyst.

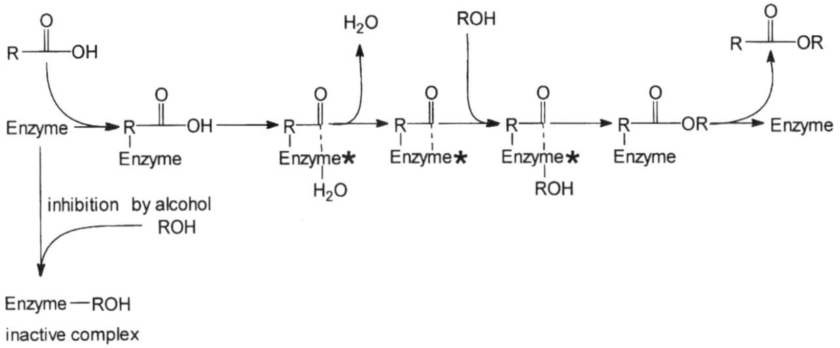

Figure 6.17 The ping-pong bi-bi mechanism with inhibition by alcohol.

Dummont et al. [136a] and Bernard et al. [136b] studied the kinetics on the esterification of myristic acid (tetradecane acid) with ethanol and the immobilized lipase of *Mucor miehei* as catalyst. The reaction mechanism in sc CO_2 as well as in n-hexane corresponds to the ping-pong bi-bi mechanism with inhibition by alcohol as explained in Figure 6.17. The maximum reaction rate was 1.5 fold higher in sc CO_2 than in n-hexane. The influence of the water content of the enzyme on its activity, as described earlier, was found here, too. For the same reaction, Dumont et al. [131] optimized a continuously running reactor. The optimum water content in sc CO_2 was about 0.25%, which means that sc CO_2 is saturated with water. A higher water content led to the denaturation of the enzyme. Under optimum conditions, it was possible to produce 24 g ethyl myristate within 6 hours, which corresponds to a yield of 89%.

Erickson et al. [137] determined the influence of pressure on reaction rates in SCF. They found that the rate of interesterification of trilaurin and palmitic acid in sc CO_2 and in sc ethane catalyzed by the lipase of *Rhizopus arrhizius* was inversely proportional to pressure. This was confirmed by other authors [122, 125]. Since this behavior was found both in sc CO_2, which is slightly acidic and can change the pH value at active sites of the enzyme, and in the neutral sc ethane, a pH effect could be excluded. The authors proposed the influence of pressure on the distribution of the reactants between the supercritical phase and the enzyme as a possible explanation of this pressure effect.

6.3. ORGANIC REACTIONS IN SCF

Kamat et al. [122] attributed the change in the reaction rate to the change in the dielectric constant of SCF with pressure. Thus, the initial rate of the lipase catalyzed transesterification of methylmethacrylate with 2-ethylhexanol decreased with increasing dielectric constant. sc C_3H_8, sc C_2H_6, sc HCF_3 and sc SF_6 were used as solvents. Since the dielectric constant of sc HCF_3 depends markedly on pressure (1 to 8 at 5.9 to 28 MPa), it is an ideal solvent with which to control the enzymatic reaction *via* the variation of pressure. Such directed control *via* a physical property of the solvent is called "solvent engineering".

Russell et al. described a polyesterification between bis (2,2,2-trichloroethyl) adipate [124, 125a] or divinyl adipate [125b] and 1,4-butanediol with porcine pancreatic lipase as catalyst in supercritical fluoroform as an example of "solvent engineering" (see Figure 6.18). The molecular weight and the distribution of the polyester formed were controlled by pressure during the synthesis. Furthermore, polymers of lower molecular weight distribution than obtained in a conventional step condensation method were produced.

The enantioselectivity of the transesterification of N-acetyl-(L or D)-phenylalanine ethylester and methanol catalyzed by *Subtilisin Carlsberg* or *Aspergillus* protease was directed by variation of pressure [125]. The reaction scheme is shown in Figure 6.19.

The enantioselective esterification reaction of ibuprofen catalyzed by the immobilized lipase of *Mucor miehei* investigated by Rantakylä and Aaltonen [138] is shown in Figure 6.20. The reaction rate in sc CO_2 was comparable

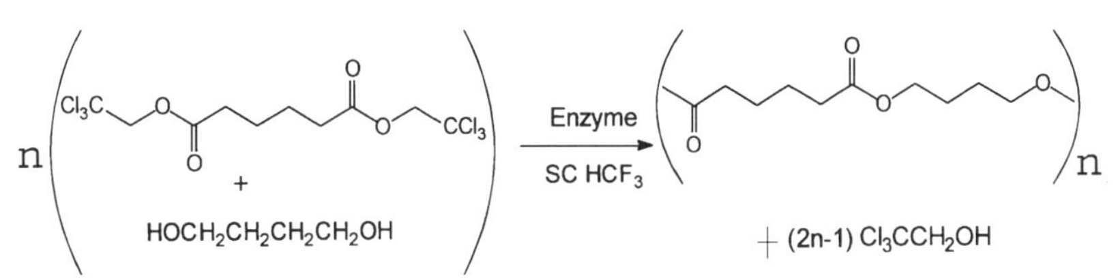

Figure 6.18 Polytransesterification of bis(2,2,2-trichloroethyl)adipate and 1,4-butanediol with porcine pancreatic lipase as catalyst

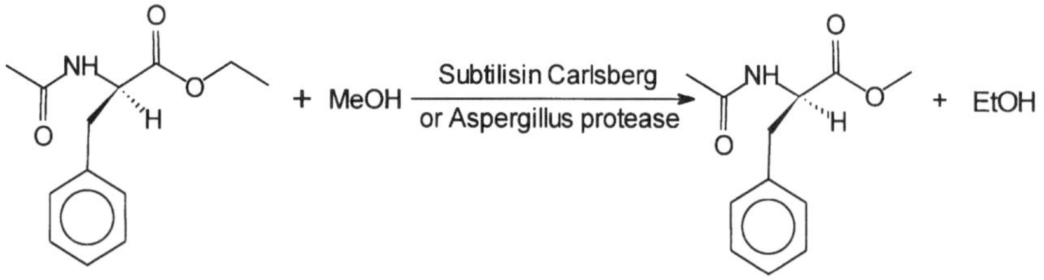

Figure 6.19 The enantioselectivity of the transesterification of N-acetyl-(L or D)-phenylalanine ethylester and methanol catalyzed by *Subtilisin Carlsberg* or *Aspergillus* protease.

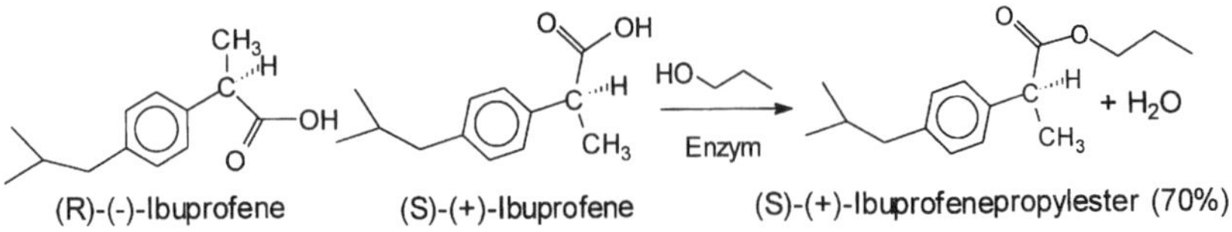

Figure 6.20 The enantioselective esterification of ibuprofen catalyzed by the immobilized lipase of *Mucor miehei*.

to that achieved in *n*-hexane. Pressure affects the reaction rate, but not the enantioselectivity.

Martins [139] investigated the esterification of glycidol with butyric acid and porcine pancreatic lipase as catalyst (Figure 6.21). The catalytic performance of the immobilized enzyme with 20–25% hydration was three times higher than that of the optimized free enzyme containing 10% water. Yields and the selectivity of the reaction in sc CO_2 were comparable with the most favorable results obtained in organic solvents. The optimum compromise between conversion and selectivity was at 25% yield and at 85% enantiomeric purity of (*S*)-glycidyl butyrate.

Ikushima et al. [140] investigated the esterification of oleic acid with ($\pm$)-citronellol catalyzed by *Candida cylindracea* enzyme (CCL) as shown in Figure 6.22. At a pressure of 8.41 MPa and a temperature of 304.1 K, 98.9% of the (*S*)-($-$)-ester was formed in a yield of 3.6%. The reaction rate was proportional to the $E_T(30)$-values which can be varied by the reaction conditions. This could be explained by an activated complex which is more polar than the reactants and stabilized by the increase of polarity taking part in the reaction. The stereoselectivity of the reaction did not show any correlation with the $E_T(30)$-values, but was very sensitive to pressure and temperature and even decreased with slight changes of these parameters. The authors explained this effect as a result of the aggregation of CO_2 molecules around the activated complex. From an aggregation of more than twenty CO_2 molecules, the activated complex can deform the enzyme into a steric organization which allows

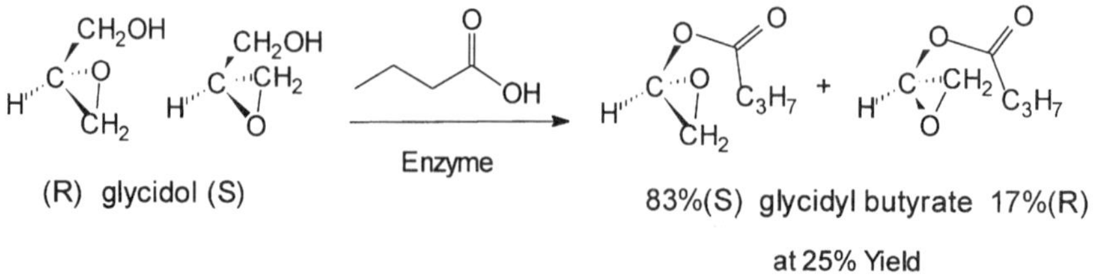

Figure 6.21 The esterification of glycidol with butyric acid and porcine pancreatic lipase as catalyst.

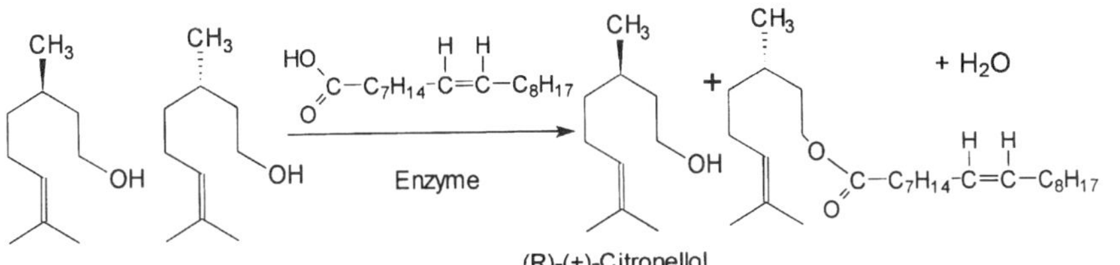

6.3. ORGANIC REACTIONS IN SCF

Figure 6.22 The esterification of oleic acid by (±)-citronellol catalyzed by an enzyme of *Candida cylindracea*.

only a reaction between (S)-(−)-citronellol and the activated complex. The extent of the aggregation, which is also called cluster size, is calculated, as in reference [25], with the fluctuation theory by Debenedetti [38].

Ikushima et al.[141] investigated the interactions between sc CO_2 and CCL in the near-critical region. They found by means of microgravimetric measurements that the mass of the enzyme increased considerably in the pressure range of 7.7 to 8.7 MPa, to an extent which suggested an aggregation of up to 3000 CO_2 molecules per CCL molecule. Within this pressure range, infrared spectroscopy showed a reversible formation of diester groups (1722, 1760, 2963, and 2860 cm^{-1}) between the alkyl residues of different chains of the amino acids of CCL. This leads to a structural change in CCL (explained in detail in the paper) which is stereospecific for the catalysis to (S)-(−) esters.

Randolph et al. [142] investigated the oxidation of cholesterol to cholest-4-ene-3-one catalyzed by cholesterol oxidase of *Gloeocysticum chrysocreas* (Figure 6.23). In order to increase the solubility of the cholesterol in sc CO_2, 2% of a cosolvent was added to sc CO_2. The reaction rate increased fourfold when 2% of *t*-butanol were added, whereas the rate decreased slightly when methanol was added, although methanol increases the solubility of cholesterol in sc CO_2 more so than does *t*-butanol. Since no considerable changes in the structure of the enzyme were found from electron paramagnetic resonance (EPR) spectroscopy after changes in pressure, as well as after the addition of different

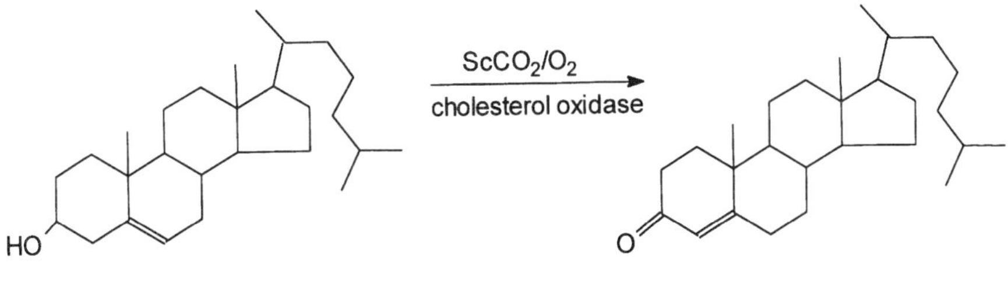

Figure 6.23 The oxidation of cholesterol to cholest-4-ene-3-one catalyzed by cholesterol oxidase.

cosolvents, the authors explained these results by proposing that there is an aggregation of cholesterol molecules. Thus larger hydrophobic surfaces are formed, and a stronger binding of the aggregates to the hydrophobic enzyme results. By means of EPR spectroscopy of spin-labelled 3-doxyl-5-alpha cholestane, the formation of aggregates due to the addition of cosolvents was detected. The reaction rate was proportional to the formation of aggregates.

Since the increase of pressure promotes the formation of aggregates and since the solubility of the cholesterol in sc CO_2 increases with pressure, the reaction rate was proportional to pressure in these experiments; in contrast to the findings of others. Further enzymatic reactions carried out in sc CO_2 are described in [144–146].

6.3.7 Photochemical Investigations of Organic Reactions in Supercritical Fluids

Fox et al. [149] investigated the [2+2] photodimerization of isophorone in sc CO_2 and sc HCF_3 (Figure 6.24, with R = CH_3). The product ratio of **9** to the sum (**9+10**) (see Figure 6.24) is higher in the more polar sc HCF_3 (0.74 at 57.6 bar to 1.00 at 482.8 bar) at all pressures than in the non-polar sc CO_2 (0.09 at 89 bar to 0.1 at 281.4 bar). Chapman et al. [150] who performed this reaction in different solvents, explained this result by considering the higher polarity of the head to head product **8** (dipole moment = 5.08D) compared to the head to tail isomers **9** and **10** (1.03 and 1.09 D, respectively). It was possible to vary the ratio from 4:1 in polar to 1:4 in non-polar solvents.

Since the dielectric constant in sc HCF_3 increases markedly with pressure [122], a more drastic increase in this ratio with pressure increase would be

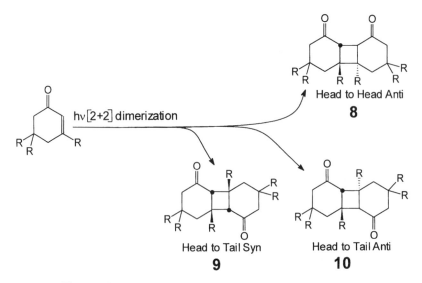

Figure 6.24 The [2+2] photodimerization of isophorone.

6.3. ORGANIC REACTIONS IN SCF

expected than was observed. This can be explained as due to a density difference between the bulk and the higher local densities adjacent to the solute, which is highest at the CP and decreases with increasing pressure. The local dielectric constant is already very high at the CP and increases less markedly than the bulk dielectric constant.

The verification of the existence of local densities is seen in the decrease in the partial molar volume of a substance near the CP, as was measured by Eckert *et al.* [35]. The systems naphthalene/sc CO_2, naphthalene/sc C_2H_4, CBr_4/sc C_2H_4 and camphor/sc C_2H_4 were investigated. The substantial increase in the local solvent density around solute molecules due to intermolecular forces, which is also known as solvent clustering, is confirmed by several authors [103, 151–154].

In contrast to the reactions in liquid solvents, in which the same amounts of the head to tail products **9** and **10** were formed [150], two to four times more **10** than **9** was formed both in sc HCF_3 and in sc CO_2. The formation of the *anti* configuration **10** requires less desolvation from the reactant dissolved in SCF than does the formation of the *syn* configuration **9**. This reorganization effect of the solvent becomes more significant with increasing pressure and increasing solvation and also explains the observed increase of the **10:9** ratio with pressure.

In the [2+2] photodimerization of 2-cyclohexene-1-one in sc ethane (Figure 6.24, with R = H), Combes *et al.* [151] found that the ratio of **10:9** increases with pressure. Between 3 and 5 times more **9** and **10** than **8** were formed at high pressures (65 to 250 bar). At lower pressures near the CP, the regioselectivity of **9** and **10** suddenly decreased to about 1.3. The authors attribute this to the dramatic increase in the local solute concentration in the solute-solute clusters formed.

An exact understanding of these solvent reorganization effects (solute-solvent repulsive interactions) and the solvent polarity effects (solute-cosolvent attractive interactions) in reactions in SCF would certainly be valuable for designing and performing regio- and stereoselective organic syntheses.

O'Shea *et al.* [155] investigated solvent cage effects by irradiating 1-phenyl-3-(4-methyl-phenyl)-acetone with UV-light ($v = 313$ nm) (Figure 6.25, R = CH_3). If the radical recombination takes place in a cage after the decarbonylation, 100% of the cage product **CG** must be produced. If the recombination takes place outside of the cage, a statistical distribution of **CG** and **ESC** (cage escape products) as shown in Figure 6.25 (far right column) is the result of the reaction. Since the **CG** yield was between 49.3 and 50.7%, it was possible to exclude the cage effect in this reaction.

Chateauneuf *et al.* [156, 157] were also not able to demonstrate a cage effect when they irradiated 1,3-diphenylacetone in sc CO_2, sc HCF_3 and sc ethane with light. The reason for the absence of a cage effect is the short lifetime (a few picoseconds [157]) of the solvent cage compared to the decarbonylation, which needs about 100,000 picoseconds [155].

Weedon *et al.* [158] investigated the Photo-Fries rearrangement of naphthyl acetate in sc CO_2 (Figure 6.26). The singlet radical pair produced has a lifetime

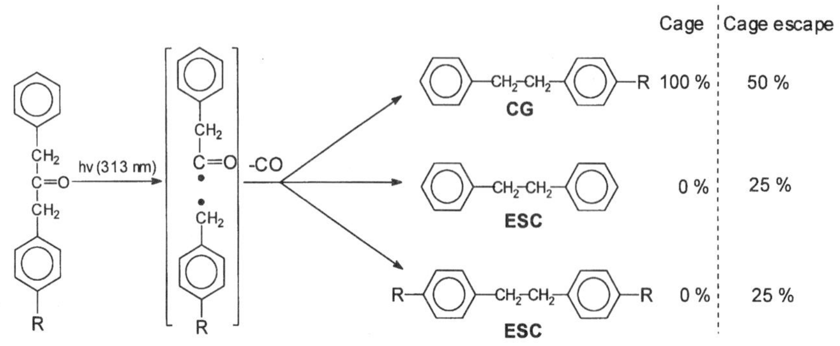

Figure 6.25 Investigation of the cage effect (irradiation of 1-phenyl-3-(4-methyl-phenyl)-acetone).

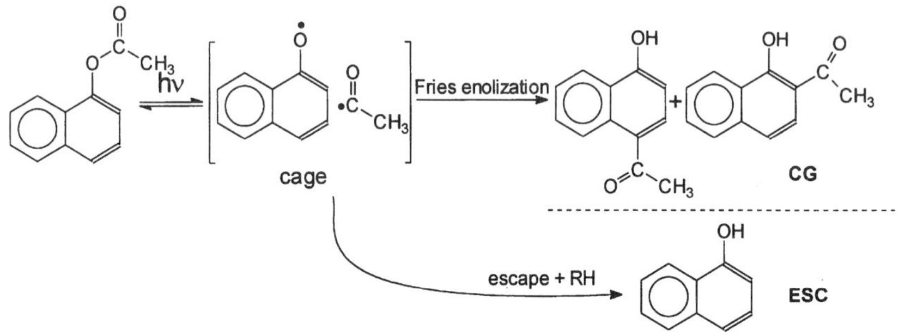

Figure 6.26 Photo-Fries rearrangement of naphthyl acetate in sc CO_2.

of only 25 picoseconds [159], *i.e.*, a lifetime comparable to that of the cage. When the reaction takes place in the cage, 4-acetylnaphthol (**CG**) is formed. Outside the cage, the naphthyloxy radical abstracts a hydrogen atom from a suitable donor (the cosolvent **RH**) and naphthol (**ESC**) is produced. The ratio of **CG** to **ESC** produced was constant at high pressures. Independent of the addition and the nature of the cosolvent, the ratio increased dramatically near the CP. Since it was possible to correlate well the change of the product distribution with pressure with the change in the partial molar volume with pressure for the system naphthalene/sc CO_2 measured by Eckert *et al.* [35], solvent-solute clusters probably exist with a lifetime which is comparable or higher than that of the singlet radical pair. This also explains why the cage products **CG** do not increase with increasing pressure (and decreasing clustering), but at reduced pressure near the CP.

Polar solvents such as methanol and isopropanol increase both the **CG/ESC** ratio and the *ortho:para* ratio of **CG**. In this case, the reaction could be influenced by solute-solute interactions involving hydrogen bonding within

6.3. ORGANIC REACTIONS IN SCF

solute-solvent clusters. In the case of slightly polar cosolvents, solute-solvent interactions have more effect on the reaction than solute-solute interactions.

6.3.8 Supercritical Fluids as Solvents and Substrates in Homogeneous Transition Metal Catalyzed Reactions

The SCF reactions presented in this chapter show that, apart from the well-known and currently used applications of extraction [160] and chromatography [161], there are other possibilities of achieving advantages in chemical reactions with the application of SCF; the special properties of SCF such as variable density between gas and liquid, high fluidity, miscibility with other gases [26] contribute to this potential. Research interest in SCF was not only inspired by the advantages of improved selectivity, the desired product distribution and facile preparation, but also by the valuable progress from an ecological point of view, for example by the substitution of organic solvents such as halogenated hydrocarbons by sc CO_2 (see Figure 6.27).

Furthermore, some of the first examples in which SCF did not only function as solvent but also as reaction partner, *i.e.*, substrate, in homogeneously catalyzed reactions have been described [162].

In 1993, Reetz *et al.* [162] reported a 2-pyron synthesis in sc CO_2 catalyzed by $(Ni(COD)_2/Ph_2P(CH)_2)_4PPh_2)$ a combination introduced by Inoue [163, 164] in 1977. In sc CO_2 the turnover number (TON) was virtually the same as reported by Inoue.

A synthesis which showed improvement in selectivity and TON was described in 1987 [165]. Monodentate phosphanes of high basic strength such as PMe_3 and $PPhMe_2$ were used as control ligands. In optimization experiments designed to increase the TON and selectivity of this $[2+2+2]$ cycloaddition of two moles of hex-3-yne with one mol of CO_2 to tetraethyl pyrone, a pronounced dependence on CO_2 pressure was found using $PPhMe_2/Ni(COD)_2$ in acetonitrile [166].

Just as in the heterogeneously catalyzed hexene isomerization in SCF in which the heating phase can be a decisive factor regarding catalyst activation, this latter feature must be taken into consideration in the 2-pyrone synthesis. Several multiphase systems exist one after the other, so it is difficult to determine which phase has the most favorable preconditions for the course of reaction.

When the catalyst has a $PMe_3/Ni(COD)_2 = 2:1$ ratio, 2-pyrone formation takes place more quickly than with dppb in sc CO_2 (the other conditions being

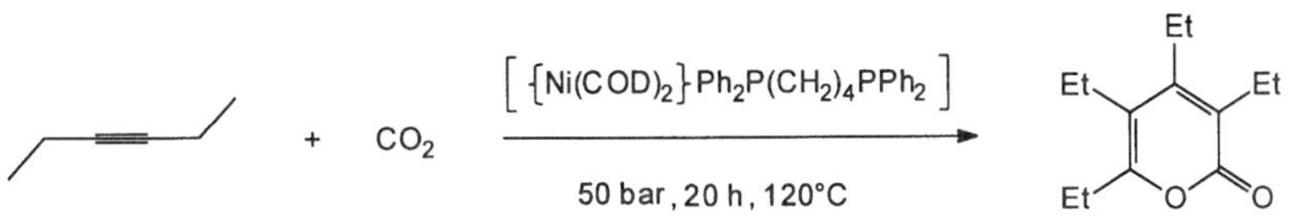

Figure 6.27 Synthesis of 2-pyrone from alkynes and CO_2.

the same). In contrast to the reaction control by dppb, an induction period is not observed with PMe_3, *i.e.*, the formation of the catalytically active species takes place much more quickly, apparently due to solubility reasons. Catalyst formation as well as catalyst solubility are obviously determined by the currently existing phase [167].

Figure 6.28 shows that in the application of PMe_3 instead of dppb the higher reaction rate is due to the omission of the induction period, *i.e.*, the formation of the active species as the first step. In the case of dppb, induction takes place so slowly that the maximum rate is only reached after 7 hours. In this time period, the reaction in the presence of PMe_3 has virtually led to complete conversion.

This can be explained by the fact that PMe_3, in contrast to dppb, forms a homogeneous solution with sc CO_2 and therefore, reacts more rapidly with $Ni(COD)_2$ to produce the catalytically active species. The selectivity of 2-pyrone formation in sc CO_2 with the catalyst PMe_3 reaches values far beyond 90% and thus, the sc medium has a clear advantage for the reaction compared with conventional solvents such as THF/acetonitrile. The reaction rate observed in THF/acetonitrile can also be achieved in sc CO_2 although much more drastic reaction conditions are required.

This has illustrated that, when SCF (here sc CO_2) are used as solvents and reaction partners in catalytic reactions, there are more complicated phase transitions and phase equilibria than in conventional solvents, and the modification of the catalyst system must be taken into consideration. In principle, it can be expected that further applications of sc CO_2 as an environmentally safe solvent and as a C_1-building block will be found when suitable catalysts are available.

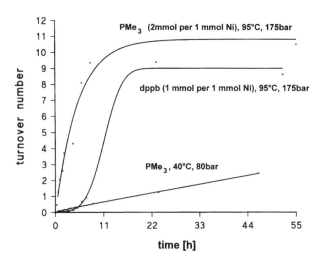

Figure 6.28 Kinetic data for the formation of 2-pyrone from hex-3-yne and CO_2 with $Ni(COD)_2$/phosphine under various conditions [167].

6.3. ORGANIC REACTIONS IN SCF

Hydrogenation of CO_2 to Formic Acid At present, formic acid is only synthesized *via* the reaction of NaOH with CO under pressure at 210 °C. The first synthesis of formic acid or the formate anion by the hydrogenation of CO_2 was discovered by Farlow and Adkins by the use of Raney nickel as catalyst in 1935 [168].

The first example of a homogeneously catalyzed reaction was reported by Inoue *et al.* in 1976 [169]. Rhodium(I) phosphane complexes such as Wilkinson's catalyst [Rh(PPh$_3$)$_3$Cl] were employed for the catalytic hydrogenation of CO_2 in benzene solution in the presence of tertiary amines.

The addition of H_2 to CO_2 is an endergonic but exothermic process under standard conditions [170] (see Figure 6.29). The equilibrium can be shifted to the right by suitable choice of the reaction conditions (elevated pressure, base addition, etc.) [171–173] or by trapping formic acid as formate esters or formamides [174–177]. In addition, the high kinetic barrier can be overcome by suitable catalysts [178] (see Figure 6.29).

Inoue's catalyst showed a better performance when small amounts of water were added, but the TON did not reach more than 150 even employing drastic reaction conditions. Other investigations showed that it was possible to obtain higher yields when an isopropanol/amine mixture which contained up to 20% of water was used [179]. Gaßner and Leitner [180] reported TONs up to 3440 catalytic cycles in 12 hours under very mild conditions with a water-soluble rhodium catalyst in water/amine solutions.

For certain reactions aqueous media often facilitate reaction rates and produce higher yields than do organic solvents. The accelerating effect of small amounts of water in organic solvents [169, 172] has several mechanistic explanations. It is possible that a donor interaction from water to the carbon atom of CO_2 increases the nucleophilicity of the oxygen atoms of CO_2 and increases its capacity to bind to a metal center. Calculations by *ab initio SCF methods* confirm that a CO_2-water interaction of the character described is more stable than the two species alone [181].

Up to 2200 moles of HCO_2H per mole of rhodium with turnover frequencies as high as 374 h^{-1} can be achieved with the *in situ* catalyst [Rh(COD)H]$_4$/dppb [182]. Isolated complexes of the type [(P$_2$)Rh(hfacac)] as catalyst precursors used by Fornika *et al.* [183] have shown a higher catalytic activity than the *in situ* catalysts. Higher TONs than 3000 with turnover frequencies up to 1335 per hour were achieved under the following reaction conditions: 2.5×10^{-3} mmol of (Cy$_2$P(CH$_2$)$_4$PCy$_2$)Rh(hfacac) as catalyst, DMSO/NEt$_3$ (5:1) as solvent, 40 atm

$$CO_2 (g) + H_2 (g) \underset{}{\overset{\text{catalyst}}{\rightleftharpoons}} HCO_2H (l)$$

$\Delta H° = -31.6$ kJ/mol
$\Delta G° = +32.9$ kJ/mol

Figure 6.29 Hydrogenation of CO_2 to formic acid.

($H_2/CO_2 = 1:1$) and 25 °C. This is the highest reported catalytic activity in water free organic solvents. Noyori et al. [184] reported that ruthenium(II)-phosphine complex catalysts are highly active for the hydrogenation of CO_2 to formic acid in sc CO_2. Figure 6.30 shows the reaction conditions and TON.

The catalysts were selected due to their favorable solubility. The solubilities are similar to those in hexane. The reaction conditions (pressure and temperature) are more drastic than those described earlier. It is also interesting that traces of water are indicated to be necessary for reaching a high reaction rate. Typically, reaction conditions were a Ru-complex/triethylamine/water ratio of 1:3100:40, dissolved in a supercritical mixture of H_2 (85 atm) and CO_2 at a total pressure of 200 to 220 atm at 50 °C. In order to avoid the formation of liquid triethylamine, the amount of amine was kept at $\leq 1/3$ of the quantity of amine which could be completely dissolved in supercritical CO_2 (see [185]). The absence of a liquid phase was controlled by a reactor equipped with sapphire windows.

The high performance of the formic acid production in the case of ruthenium complex catalysis is a consequence of the presence of the supercritical CO_2 phase. If the reaction in liquid CO_2 is carried out with comparable CO_2 concentrations at a temperature of 15 °C instead of 50 °C, but otherwise the same reaction conditions, the TON decreases from 7200 to 20 and TOF from 1400 to 1.3 per hour. This means that the characteristic properties of sc CO_2, such as the extremely high miscibility with hydrogen [185–187] and a good mass-transfer capability, are responsible for the remarkable catalytic activity and efficiency.

Under these conditions, sc CO_2 becomes an excellent medium for its own hydrogenation (see Figure 6.31). If it is possible to develop a continuous-flow

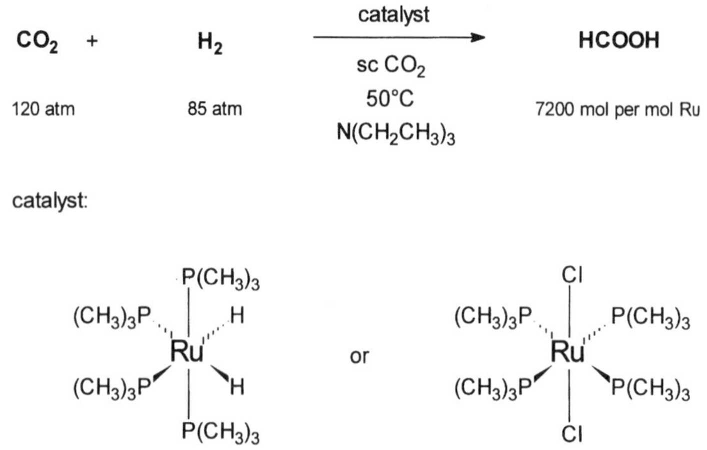

Figure 6.30 Ruthenium catalyzed hydrogenation of CO_2 in the supercritical phase (sc CO_2).

6.3. ORGANIC REACTIONS IN SCF

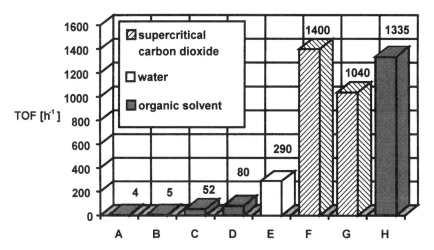

Figure 6.31 Rates of formic acid production by hydrogenation of CO_2 under various conditions and in different solvents. For reaction conditions, see (A) [164b], (B) [172], (C) [184], (D) [185], (E) [186], (F, G) [185], (H) [183].

system, to solve the problems of extraction and of recycling the amine and the catalyst following the release of formic acid, the industrial realization of this high pressure process can be expected.

Synthesis of Methyl Formate The synthesis of methyl formate in sc CO_2 is similarly successful in the presence of methanol and ruthenium(II) catalyst systems (Figure 6.32) [188]. Typical reaction conditions are: 80 °C, 5 mmol NEt_3, 13 mmol MeOH, 3 mmol catalyst dissolved in a supercritical mixture of H_2 (80 atm at reaction temperature) and CO_2 (total pressure 200–210 atm). The reaction mixture forms a single supercritical phase. The time dependence of the product formation shows that formic acid is formed first followed by

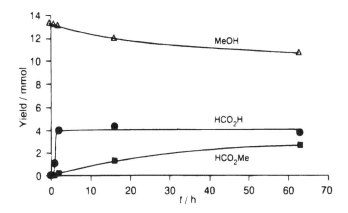

Figure 6.32 Time dependence of the product yields [188].

esterification which must take place thermally. The amine is an inhibitor of esterification, but its presence is required to reach reasonable yields in the hydrogenation step.

The use of sc conditions for the homogeneous hydrogenation of CO_2 with subsequent thermal esterification leads to high yields of methyl formate under mild conditions. Therefore, it might be possible to replace the carbonization of methanol presently used, with an industrial process developed for the synthesis of methyl formate from CO_2 as a C_1 building block once catalysts and reaction conditions are optimized.

Catalytic Production of Dimethylformamide (DMF) from sc CO_2   DMF is a useful polar solvent and is produced industrially (250,000 tons/year) by carbonylation of dimethylamine in the presence of methanol [189]. Using Raney nickel as catalyst the synthesis of DMF from dimethylamine, CO_2 and hydrogen was first reported by Adkins and Farlow [168]. The formation of DMF from dimethylamine, H_2 and CO_2 is only barely thermodynamically favorable at standard conditions, eq. (11). Thermodynamic data are available for aqueous reactants and liquid products [177].

$$CO_2 + NH(CH_3)_2 + H_2 \rightarrow HCON(CH_3)_2 + H_2O \qquad (11)$$

$$\Delta G^0 = -0.75 \text{ kJ/mol}$$

$$\Delta H^0 = -56.5 \text{ kJ/mol}$$

$$\Delta S^0 = -119 \text{ J/mol K}^{-1}$$

The enthalpy change accompanying DMF production (-56.5 kJ/mol) is more favorable than that for methyl formate formation (-15.3 kJ/mol).

Homogeneous catalysis of the former reaction was first reported by Haynes *et al.* in 1970 [190], using a palladium catalyst with benzene as a solvent and a pressure of 28/28 atm (H_2/CO_2) and 100 °C. A TON of 120 was realized. A ruthenium complex with dppe as a chelating phosphine ligand as described by Kiso and Saeki [191] was more efficient with a TON of 3400 in hexane under similar conditions (pressure 29/29 atm (H_2/CO_2), 130 °C). Jessop *et al.* [192] found that in the presence of a catalytic amount of $RuCl_2(PMe_3)_4$ as a catalyst precursor, sc CO_2 reacts with H_2 and dimethylamine to give DMF with TONs of up to 370,000. As a source of dimethylamine they used liquid dimethylammonium dimethylcarbonate, although dimethylamine gave the same results. The reaction conditions were: 100 °C, 80 atm H_2 and with CO_2 a total pressure of 210 atm. The conversion of dimethylamine was 94%, and the selectivity for DMF was 99%. The TON of 370,000 is superior to the largest TON of 3400 for DMF formation from CO_2 in a conventional liquid solvent. The authors discuss the production of DMF from sc CO_2 as proceeding in two steps on the basis of the composition of the product as a function of reaction time (see Figure 6.33).

6.3. ORGANIC REACTIONS IN SCF

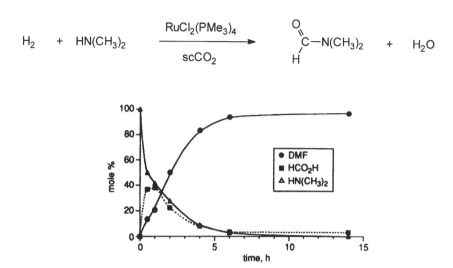

Figure 6.33 The composition of the product as a function of reaction time for the reaction of dimethylammonium dimethylcarbamate (5.0 mmol), H_2, and sc CO_2 (130 atm) at 100 °C catalyzed by $RuCl_2(PMe_3)_4$ (2.5 µmol) [192].

The fast Ru-catalyzed hydrogenation of CO_2 to formic acid is followed by the slower thermal condensation of formic acid and dimethylamine. Dimethylamine acts as a base to stabilize the formic acid in the first step and serves as a reactant in the second step.

The driving force for this process is embodied in the existence of a two phase system, a supercritical phase and a liquid phase. The combination of the steps in a one pot procedure is also responsible for the high rate of DMF production (see Figure 6.34). With this improved catalytic efficiency and the lower toxicity of CO_2 compared to CO, the reaction of CO_2 with hydrogen and dimethylamine could become competitive with the carbonylation of dimethylamine as an industrial method for the production of DMF.

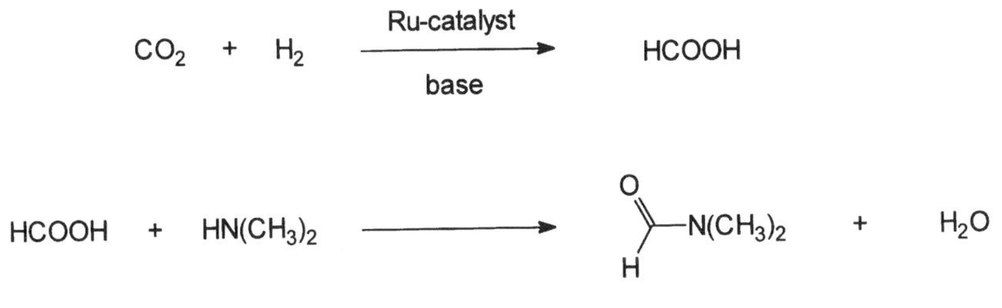

Figure 6.34 Two step mechanism for DMF production.

6.4 FURTHER INVESTIGATIONS

In this chapter we have discussed various types of reactions which have been performed under supercritical conditions and the advantages offered by SCF as reaction media. Relatively few experiments have been performed to date attempting to utilize the unique properties exhibited by substrates and solvents in the supercritical-fluid region. A more detailed investigation of the fundamental behavior of reactivity in the SCF region appears warranted, especially where small changes in physical properties of the fluid produce the most pronounced reactivity changes.

The potential use of sc CO_2 in combination as solvent and substrate is a particularly intriguing prospect. Increased governmental and environmental restrictions on solvent emissions make SCF more attractive as reaction media. Often they can be easily separated from the product and recycled more efficiently than conventional liquid solvents. Knowledge developed so far has illustrated the multi-purpose use of sc CO_2 as a reaction medium and suggests strongly that it has great promise as an ecologically harmless reaction medium.

The examples provided in Sections 6.3.1 to 6.3.8 show very clearly the principal advantages (miscibility, transport properties, etc.) of using SCF as solvents and substrates. The utilization of SCF as a C_1-building block, especially in the case of CO_2, will be achieved satisfactorily when suitable catalysts are developed. A catalyst which is efficient in conventional solvents will not necessarily function well in SCF.

Recent results (126, 147, 148] demonstrate the potential of supercritical media for homogenous catalytic hydrogenation and, in general, for asymmetric catalysis.

The special properties of SCF require the development of modified catalysts. Catalysts on supports may play a particular role in this field. Product extraction by means of depressurizing the supercritical phase and subsequent compression of the CO_2 (solvent/substrate) should permit the development of a profitable continuous process, employing a regenerated catalyst/support.

6.5 REFERENCES

[1] Paulaitis, M. E., Krukonis, V., Kurnik, R. T., Reid, R. C., Rev. Chem. Eng., 1 (1983) 179.

[2] Schneider, G. M., Angew. Chem., 90 (1978) 762; Angew. Chem. Intl. Ed. Engl. 17 (1978) 716.

[3] Gupta, S., Ghonasgi, D., Dooley, K. M., Knopf, F. C., J. Supercrit. Fluids, 4 (1991) 181.

[4] Hess, R. K., Erkey, C., Akgerman, A., Dexter, B. J., Irvin, T. R., J. Supercrit. Fluids, 2 (1989) 47.

[5] Roop, R. K., Akgerman, A., Dexter, B. J., Irvin, T. R., J. Supercrit. Fluids, 2 (1989) 51.

6.5. REFERENCES

[6] Kiran, E., Brennecke, J. F., Supercritical Fluid Engeneering Science, ACS Symposium Series, 514 (1993) 1 and literature cited therein.

[7] McNally, M. E. P., Bright, F. V., (Bright, F. V., McNally, M. E. (eds.)), Supercritical Fluid Technology, ACS Symposium Series, 488 (1992) 1 and literature cited therein.

[8] Subramaniam, B., McHugh, M. A., Ind. Eng. Chem. Process Des. Dev., 25 (1986) 1.

[9] Squires, T. G., Venier, C. G., Aida, T., Fluid Phase Equilibria, 10 (1983) 261.

[10] Boock, L., Wu, B., LaMarca, C., Klein, M., Paspek, S., CHEMTECH, 12 (1992) 719.

[11] Scholsky, K. M., J. Supercrit. Fluids, 6 (1993) 103 and references cited therein.

[12] a) Savage, P. E., Gopalan, S., Mizan, T. I., Martino, C. J., Brock, E. E., AIChE J., 41 (1995) 1723 and references cited therein; b) Morgenstern, D. A., LeLacheur, R. M., Morita, D. K.; Borkowsky, S. L.; Feng, S., Brown, G. H.; Luan, L., Gross, M. F., Burk, M. J., Tumas, W., ACS Symp. Ser., 626 (1996) 626, 132.

[13] Knittel, D., Schollmeyer, E., GIT Fachz. Lab., 10 (1992) 993 and references cited therein.

[14] Saraf, V. P., Kiran, E., J. Supercrit. Fluids, 1 (1988) 37 and references cited therein.

[15] Dhawan, J. C., Legendre, R. C., Bancsath, A. F., Davis, R. M., J. Supercrit. Fluids, 4 (1991) 160 and references cited therein.

[16] Wright, B. W., Smith, R. D., Chromatographia, 18 (1984) 542.

[17] Jonker, C. R., Wright, B. W., Udseth, H. R., Smith, R. D., Ber. Bunsenges. Phys. Chem., 88 (1984) 908.

[18] Reid, R., Poling, B., The Properties of Gases and Liquids, McGraw Hill, New York, 1987, 656.

[19] Bartmann, D., Schneider, G. M., J. Chromatogr., 83 (1973) 135.

[20] Burk, R., Kruus, P., Can. J. Chem. Eng., 70 (1992) 403.

[21] Chrastil, J., J. Phys. Chem., 86 (1982) 3016.

[22] Kamlet, M. J., Abboud, J. L., Abraham, M. H., Taft, R. W., J. Org. Chem., 48 (1983) 2877.

[23] Yonker, C. R., Frye, S. L., Kalkwarf, D. R., Smith, R. D., J. Phys. Chem., 90 (1986) 3022.

[24] Reichardt, C., Solvents and Solvent Effects in Organic Chemistry, VCH, Weinheim, Germany, 1990.

[25] Ikushima, Y., Saito, N., Arai, M., J. Phys. Chem., 96 (1988) 2293.

[26] Hyatt, J. A., J. Org. Chem., 49 (1984) 5097 and references cited therein.

[27] Dawson, G. W., Griffiths, D. C., Pickett, J. A., PCT int. Appl., WO 82/04 249 (Dec. 9, 1982).

[28] Takasago Perfumery Company, Ltd., Jpn. Kokai Tokyo Koho JP 57183728 A2 (1982).

[29] Upadek, H., Bruns, K., Eur. Pat. Appl. EP 53717A1 (1982).

[30] Matsushita Electrical Industrial Company, Ltd., Jpn. Kokai Tokyo Koho JP 57109830A2 (1982).

[31] Greenhalgh, C. W., Birch, A. M., Mercer, A. J. H., British Patent GB 1601945 (1981).

[32] Paulaitis, M. E., Alexander, G. C., Pure Appl. Chem., 59 (1987) 61.
[33] Kim, S., Johnston, K. P., (Squires, T. G., Paulaitis, M. E. (eds.)), Effects of Supercritical Solvents on the Rates of Homogeneous Chemical Reactions, ACS Symposium Series, 329 (1987) 42.
[34] Ikushima, Y., Saito, N., Arai, M., Bull. Chem. Soc. Japan, 64 (1991) 282.
[35] Eckert, C. A., Ziegler, D. H., Johnston, K. P., Kim, S., J. Phys. Chem., 90 (1986) 2738.
[36] Kim, S., Johnston, K. P., Chem. Eng. Commun., 63 (1988) 49.
[37] Ikushima, Y., Ito, S., Asano, T., Yokoyama, T., Saito, N., Hatakeda, K., Goto, T., J. Chem. Eng. Japan, 23 (1990) 96.
[38] Debenedetti, P. G., Chem. Eng. Sci., 42 (1987) 2203.
[39] a) Issacs, N., Keating, N., J. Chem. Soc. Chem. Commun., (1992) 876; b) Knutson, B. L., Dillow, A. K., Liotta, C. L., Eckert, C. A., ACS Symp. Ser., 608 (1995) 166.
[40] Kaupp, G., Angew. Chem., 106 (1994) 1519. Angew. Chem. Intl. Ed. Engl. 33 (1994) 1452.
[41] Tiltscher, H., Wolf, H., Schelchshorn, J., Angew. Chem., Int. Ed. Engl., 20 (1981) 892.
[42] Tiltscher, H., Hofmann H., Chem. Eng. Sci., 42 (1987) 959.
[43] Tiltscher, H., Wolf, H., Schelchshorn, J., Ber. Bunsenges. Phys. Chem., 88 (1984) 897.
[44] Saim, S., Subramaniam, B., Chem. Eng. Sci., 43 (1988) 1837.
[45] Saim, S., Ginosar, D. M., Subramaniam, B., (Johnston, K. P. (ed.)), ACS Symposium Series, 406 (1989) 301.
[46] Saim, S., Subramaniam, B., J. Supercrit. Fluids, 3 (1990) 214.
[47] Saim, S., Subramaniam, B., J. Catal., 131 (1991) 445.
[48] Manos, G., Hofmann, H., Chem. Eng. Technol., 14 (1991) 73.
[49] Baptist-Ngyen, S., Subramaniam, B., AIChE J., 38 (1992) 1027.
[50] Ginosar, D. M., Subramaniam, S., (Delmon, B., Froment, G. F. (eds.)), Catalyst Deactivation, Proc. Int. Symp. on Catalyst Deactivation, Elsevier, Amsterdam, 1994.
[51] Kölbel, H., Ralek, M., Catal. Rev.-Sci. Eng., 21 (1980) 225.
[52] Satterfield, C. N., Stenger, Jr. H. G., Ind. Eng. Chem. Process Des. Dev., 24 (1985) 407.
[53] Deckwer, W. N., Kokoun, R., Sander, E., Ledakowicz, S., Ind. Eng. Chem. Process. Res. Rev., 25 (1986) 643.
[54] Stern, D., Bell, A. T., Heinemann, H., Chem. Eng. Sci., 38 (1983) 597.
[55] Huang, S. H., Lin, H. M., Tsai, F. N., Chao, K. L., Ind. Eng. Chem. Res., 27 (1985) 102.
[56] Yokota, K., Fujimoto, K., Fuel, 68 (1989) 255.
[57] Yokota, K., Fujimoto, K., Ind. Eng. Chem. Res., 30 (1991) 95.
[58] a) Fan, L., Yokota, K., Fujimoto, K., AIChE J., 38 (1992) 1639; b) Fan, L.; Yokota, K.; Fujimoto, K., Top. Catal., 2 (1995) 267–83; c) Fujimoto, K.; Fan, L.; Yoshii, K., Top. Catal. 2 (1995) 259; d) Fan, L.; Yokota, K.; Fujimoto, K.; Sekiyu Gakkaishi 38 (1995) 71.

6.5. REFERENCES

[59] Lang, X., Akgerman, A., Bukur, D. B., Ind. Eng. Chem. Res., 34 (1995) 72.
[60] Kumar, S., Reid, R., Suter, U., Polym. Prepr., 27 (1986) 224.
[61] Kumar, S., Suter, U., Polym. Prepr., 28 (1987) 286.
[62] Saraf, V., Kiran, E., Polym. Prepr., 31 (1990) 687.
[63] Kiran, E., Saraf, V., J. Supercrit. Fluids, 3 (1990) 198.
[64] Hagiwara, M., Mitsui, H., Machi, S., Kagiya, T., J. Polym. Sci, A-1, 6 (1968) 603.
[65] Hagiwara, M., Mitsui, H., Machi, S., Kagiya, T., J. Polym. Sci, A-1, 6 (1968) 609.
[66] Kagiya, T., Machi, S., Hagiwara, M., Hosaki, Y., US Patent 3 516 912 (1970).
[67] Terry, R. E., Ziad, A., Angelos, C., Whitman, D. L., Energy Prog., 8 (1988) 48.
[68] Sunol, A. K., Int. Patent Appl. WO 90/02612 (1992).
[69] Hartmann, H., Denzinger, W., German Patent 3609826A (1987).
[70] DeSimone, J., Guan, Z., Elsbernd, C., Science, 257 (1992) 945.
[71] Guan, Z., Combes, J. R., Menceloglu, Y. Z., DeSimone, J. M., Macromolecules, 26 (1993) 2663.
[72] Beckmann, E., Fulton, J., Matson, D., Smith, R., (Johnston, K., Penninger, J. (eds.)), Supercritical Fluid Science and Technology, ACS Symposium Series, 406 (1989) 184.
[73] Beckmann, E. J., Smith, R. D., J. Supercrit. Fluids, 3 (1990) 205.
[74] Beckmann, E. J., Smith, R. D., J. Phys. Chem., 94 (1990) 345.
[75] Elliott, J. R. Jr., Cheung, H. M., Supercritical Fluid Engeneering Science, ACS Symposium Series, 514 (1993) 271.
[76] Srinivasan, G., Elliot, J. R. Jr., Ind. Eng. Chem. Res., 31 (1992) 1414.
[77] Lee, C., Hoy, K. L., Eur. Patent Appl. EP506041A2 (1992).
[78] a) Dhawan, J. C.; Bencsath, A. F.; Legendre, R. C., Supercritical Fluid Engineering Science. A, ACS Symposium Series, 514 (1993) 380; b) Chen, D. T.; Perman, C. A.; Riechert, M. E.; Hoven, J., J. Hazard. Mater. 44 (1995) 53.
[79] Tsugane, H., Yagi, Y., Inomata, H., Saito, S., J. Chem. Eng. Jpn., 25 (1992) 351.
[80] Kim, S., Johnston, K. P., AIChE J., 33 (1987) 1603.
[81] Kim, S., Johnston, K. P., Ind. Eng. Chem. Res., 26 (1987) 1206.
[82] Ellington, J. B., Brennecke, J. F., J. Chem. Soc., Chem. Commun., (1993) 1094.
[83] Roberts, C. B., Chateauneuf, J. E., Brennecke, J. F., J. Am. Chem. Soc., 114 (1992) 8455.
[84] Evans, M. G., Polanyi, M., Trans. Faraday Soc., 311 (1935) 875.
[85] Peng, D.-Y., Robinson, D. B., Ind. Eng. Chem. Fundam., 15 (1976) 59.
[86] Vieville, C., Mouloungui, Z., Gaset, A., Ind. Eng. Chem. Res., 32 (1993) 2065.
[87] Kita, H.,Tanaka, K., Okamoto, K. I., Yamamoto, M., Chem. Lett., (1987) 2053.
[88] Kita, H., Sasaki, S., Tanaka, K., Okamoto, K. I., Yamamoto, M., Chem. Lett., (1988) 2025.
[89] Boyer, J. L., Gilot, B., Guiraud, R., Bull. Soc. Chim. Fr., (1989) 260.
[90] Chou, T. C., Yeh, H. J., Ind. Eng. Chem. Res., 31 (1992) 130.
[91] Riad, A., Mouloungui, Z., Delmas, M., Gaset, A., Synth. Commun., 19 (1989) 3169.
[92] Asdih, M., Marchal, P., Mouloungui, Z., Gharbi, R. L., Delmas, M., Gaset, A., Rev. Fr. Corps Gras, 37 (1990) 240.

[93] Sigman, M. E., Barbas, J. T., Leffler, J. E., J. Org. Chem., 52 (1987) 1754.
[94] Silvestri, G., Gambino, S., Filardo, G., Cuccia, C., Guarino, E., Angew. Chem., 93 (1981) 131; Angew. Chem. Int. Ed. Engl., 20 (1981) 101.
[95] Crooks, R. M., Fan, F.-R. F., Bard, A. J., J. Am. Chem. Soc., 106 (1984) 6851.
[96] McDonald, A. C., Fan, F.-R. F., Bard, A. J., J. Phys. Chem., 90 (1986) 196.
[97] Flarsheim, W. M., Tsou, Y.-M., Trachtenberg, I., Johnston, K. P., Fan, F.-R. F., Bard, A. J., J. Phys. Chem., 90 (1986) 3857.
[98] Crooks, R. M., Bard, A. J., J. Phys. Chem., 91 (1987) 1274.
[99] Crooks, R. M., Bard, A. J., J. Electroanal., 240 (1988) 253.
[100] Crooks, R. M., Bard, A. J., J. Electroanal., 243 (1988) 117.
[101] Cabrera, C. R., Garcia, E., Bard, A. J., J. Electroanal., 260 (1989) 457.
[102] Cabrera, C. R., Garcia, E., Bard, A. J., J. Electroanal., 273 (1989) 147.
[103] Flarsheim, W. M., Bard, A. J., Johnston, K. P., Fan, F.-R. F., J. Phys. Chem., 93 (1989) 4234.
[104] Galus, Z., (Chalmers, R. A., Bryce, W. A. J. (eds.)), Fundamentals of Electrochemical Analysis, Coedition between Ellis Horwood Limited, Chichester, England and Polish Scientific Publishers PWN, Warsaw, Poland, 1994.
[105] Bard, A. J., Faulker, L. R., Electrochemical Methods, Wiley, New York, 1980.
[106] Philips, M. E., Deakiln, M. R., Novotny, M. V., Wightman, R. M., J. Phys. Chem., 91 (1987) 3934.
[107] Niehaus, D., Philips, M. E., Adrean, M., Wightman, R. M., J. Phys. Chem., 93 (1989) 6232.
[108] Adrean, M., Wightman, R. M., Anal. Chem., 61 (1989) 270.
[109] Adrean, M., Wightman, R. M., Anal. Chem., 61 (1989) 2193.
[110] Niehaus, D., Wightman, R. M., Flowers, P. A., Anal. Chem., 63 (1991) 1728.
[111] Dressman, S. F., Michael, A. C., Anal. Chem., 67 (1995) 1339.
[112] Sullenberger, E. F., Michael, A. C., Anal. Chem., 65 (1993) 2304.
[113] Sullenberger, E. F., Michael, A. C., Anal. Chem., 65 (1993) 3417.
[114] Sullenberger, E. F., Dressman, S. F., Michael, A. C., J. Phys. Chem., 98 (1994) 5347.
[115] Dressman, S. F., Garguilo, M. G., Sullenberger, E. F., Michael, A. C., J. Am. Chem. Soc., 115 (1993) 7541.
[116] Olsen, S. A., Tallman, D. E., Anal. Chem., 66 (1994) 503.
[117] Dombro, R. A., Prentice, G. A., McHugh, M. A., J. Electrochem. Soc., 135 (1988) 2219.
[118] Cipris, D., Mador, I. L., J. Electrochem. Soc., 125 (1978) 1954.
[119] Russell, A. J., Beckman, E., Enzyme Microb. Technol., 13 (1991) 1007.
[120] Aaltonen, O., Rantakylä, M., CHEMTECH, (1991) 240.
[121] Hammond, D. A., Karel, M., Klibanov, A. M., Krukonis, V. J., Appl. Biochem. Biotech., 11 (1985) 393.
[122] Kamat, S. V., Iwaskewycz, B., Beckman, E. J., Russell, A. K., Proc. Natl. Acad. Sci. U.S.A., 90 (1993) 2940.
[123] Kamat, S. V., Barrera, J., Beckman, E. J., Russel, A. K., Biotechnol. Bioeng., 40 (1992) 158.

6.5. REFERENCES

[124] Chaudhary, A. K., Beckman, E. J., Russel, A. K., J. Am. Chem. Soc., 117 (1995) 3728.

[125] a) Russell, A. J., Beckman, E., Chaudhary, A. K., CHEMTECH, (1994) 33; b) Russell, A. J.; Beckman, E. J., Abderrahmare, D., Chaudhary, A. K., US 5478910 A (1995).

[126] Randolph, T. W., Blanch, H. W., Prausnitz, J. M., Wilke, C. R., Biotechnol. Lett., 7 (1985) 325.

[127] Nakamura, K., Chi, Y. M., Yamada, Y., Yano, T., Chem. Eng. Commun., 45 (1986) 207.

[128] Krukonid, V. J., Hammond, D. A., Biotechnol. Lett., 10 (1988) 837.

[129] Pasta, P., Mazzola, G., Carrea, G., Riva, S., Biotechnol. Lett., 11 (1989) 643.

[130] Dordick, J. S., Enzyme Microb. Technol., 11 (1989) 194.

[131] Dumont, T., Barth, D., Perrut, M., J. Supercrit. Fluids, 6 (1993) 85.

[132] Marty, A., Chulalaksananukul, W., Willemot, R. M., Condoret, J. S., Biotechnol. Bioeng., 39 (1992) 273.

[133] Marty, A., Chulalaksananukul, W., Condoret, J. S., Willemot, R. M., Durand, G., Biotechnol. Lett., 12 (1990) 11.

[134] Chulalaksananukul, W., Deloreme, P., Willemot, R. M., FEBS Lett., 276 (1990) 181.

[135] a) Marty, A., Combes, D., Condoret, J. S., Biotechnol. Bioeng., 43 (1994) 497; b) Marty, A.; Manon, S.; Ju, D. P.; Combes, D.; Condoret, J.-S., Ann. N. Y. Acad. Sci., 750 (1995) 408.

[136] a) Dummont, T., Barth, D., Coebier, C., Branlant, G., Perrut, M., Biotechnol. Bioeng., 39 (1992) 329; b) Bernard, P., Barth, D., Biocatal. Biotransform. 12 (1995) 299.

[137] Erickson, J. C., Schyns, P., Cooney, C. L., AIChE J., 16 (1990) 299.

[138] Rantakylä, M., Aaltonen, O., Biotechnol. Lett., 16 (1994) 824.

[139] Martins, J. F., Borges de Carvalho, I., Correa de Sampaio, T., Barreiros S., Enzyme Microb. Technol., 16 (1994) 785.

[140] a) Ikushima, Y., Saito, N., Yokoyama, T., Hatakeda, K.; Ito, S.; Arai, M.; Blanch, H. W., Chem. Lett., (1993) 109; b) Ikushima, Y., Saito, N., Hatakeda, K., Ito, S., US 5403739 A (1995).

[141] Ikushima, Y., Saito, N., Arai, M., Blanch, H. W., J. Phys. Chem., 99 (1995) 8941.

[142] a) Randolph, T. W.; Blanch, H. W., Prausnitz, J. M., AIChE J., 34(8) (1988) 1354; b) Randolph, T. W.; Clark, D. S., Blanch, H. W.; Prausnitz, J. M., Science, 238 (1988) 387; c) Randolph, T. W.; Clark, D. S.; Blanch, H. W.; Prausnitz, J. M., Proc. Natl. Acad. Sci. USA, 85 (1988) 2979.

[143] a) Bolz, U., Stephan, K., Stylos, P., Riek, A., Rizzi, M., Reuss, M., (Reuss, M., (ed.)), Proc. 2nd Int. Symp. Biochem. Eng., Fischer, Stuttgart (1991) 82; b) Jong, J. P. J., Doddema, H. J., Janssens, R. J. J., van der Lugt, J. P., Oostrom, H. H., (Christiansenc, C., Munck, L. Villadsen, J. (Eds.)), Proc. 5th Eur. Congr. Biotechnol. 239, Munksgrad Copenhagen, 1990; c) Shen, X. M., de Loos, T. W., de Swaan Arons, S. J., (Tramper J. et al. (eds)), Biocatalysis in Non-Conventional Media, 417–423, Elsevier, Amsterdam, 1992; d) Miller, D. A.; Blanch, H. W.; Prausnitz, J. M., Ind. Eng. Chem. Res., 30 (1991) 939.

[144] Noritomi, H., Miyata, M.; Kato, S.; Nagahama, K., Biotechnol. Lett. 17 (1995) 1323.
[145] Gunnlaughsdottir, H.; Sivik, B., J. Am. Oil Chem. Soc., 72 (1995) 399.
[146] Knez, Z.; Rizner, V., Habulin, M.; Bauman, D., J. Am. Oil Chem. Soc., 72 (1995) 1345.
[147] Burk, M. J.; Feng, S.; Gross, F. M., Tumas, W., J. Am. Chem. Soc. 117 (1995) 8277.
[148] Xiao, J., Nefkens, S. C. A., Jessop, P. G., Ikariya, T., Noyori, R., Tetrahedron Letters 37 (1996) 2813.
[149] Hrnjez, B. J., Metha, A. J., Fox, M. A., Johnston, P. K., J. Am. Chem. Soc., 111 (1989) 2662.
[150] Chapman, O. L., Nelson, P. J., King, R. W., Trecker, D. J., Griswold, A., Rec. Chem. Prog., 28 (1967) 167.
[151] Combes, J. R., Johnston, K. P., O'Shea, K. E., Fox, M. A., (Bright, F. V., McNally, M. E. (eds.)) Supercritical Fluid Technology, ACS Symposium Series, 488 (1992) 31.
[152] Knutson, B. L., Tomasko, D. L., Eckert, C. A., Debenedetti, P. G., Chialvo, A. A., (Bright, F. V., McNally, M. E (eds.)), Supercritical Fluid Technology, ACS Symposium Series, 488 (1992) 60 and references cited therein.
[153] Debenedetti, P. G., Mohamed, R. S., J. Chem. Phys., 90 (1989) 4528.
[154] Petsche, I. B., Debenedetti, P. G., J. Phys. Chem., 95 (1991) 386.
[155] O'Shea, K. E., Combes, J. R., Fox, M. A., Johnston, K. P., Photochem. Photobiol., 54 (1991) 571.
[156] Roberts, C. B., Zhang, J., Chateauneuf, J. E., Brenneke, J. F., J. Phys. Chem., 97 (1993) 5618.
[157] Roberts, C. B., Zhang, J., Chateauneuf, J. E., Brenneke, J. F., J. Am. Chem. Soc., 115 (1993) 9576.
[158] Andrew, D., Des Islet, B. T., Margaritis, A., Weedon, A. C., J. Am. Chem. Soc., 117 (1995) 6132.
[159] Nakasgaki, R., Hiramatsu, M., Watanabe, T., Tanimoto, S., J. Phys. Chem., 89 (1985) 3222.
[160] Zosel, K., Angew. Chem., 90 (1978) 748. Angew. Chem. Intl. Ed. Engl., 17 (1978) 702.
[161] Klesper, E., Angew. Chem., 90 (1978) 785.
[162] Reetz, M. T., Könen, W., Strack, T., Chimia, 47 (1993) 493.
[163] Inoue, Y., Itoh, Y., Hashimoto, H., Chem. Lett., (1978) 633.
[164] a) Inoue, Y., Itoh, Y.; Kazama H.; Hashimoto, H., Bull. Chem. Soc. Japan, 53 (1980) 3329; b) Sasahi, Y., Inoue, H., Hashimoto, H., J. Chem. Soc., Chem. Commun. (1976) 605.
[165] Walter, D., Dinjus, E., Schönberg, H., Sieler, J., J. Organomet. Chem., 334 (1987) 377.
[166] Plenz, F., Diplomarbeit, Jena (1992).
[167] Dinjus, E., Geyer, C., Plenz, F., unpublished results.
[168] Farlow, M. W., Adkins, H., J. Am. Chem. Soc., 57 (1935) 2272.
[169] Inoue, Y., Izumida, H., Sasaki, Y., Hashimoto, H., Chem. Lett., (1976) 863.

6.5. REFERENCES

[170] Weast, R. C. (ed.), Handbook of Chemistry and Physics, GSTh ed., CRC Press, Boca Raton, 1984.
[171] Khan, M. M., Halligudi, S. B., Shuky, S., J. Mol. Catal., 51 (1989) 407.
[172] Tsai, J.-C., Nicholas, K. M., J. Am. Chem. Soc., 114 (1992) 5117.
[173] Graf, E., Leitner, W., J. Chem. Soc., Chem. Commun., (1992) 623.
[174] Inoue, Y., Sasaki, Y., Hashimoto, H., J. Chem. Soc., Chem Commun., (1975) 718.
[175] Darensbourg, D., Ovalles, C., Pala, M., J. Am. Chem. Soc., 105 (1983) 5937.
[176] Darensbourg, D., Ovalles, C., J. Am. Chem. Soc., 106 (1984) 3750.
[177] Schreiner, S., Yu, J. Y., Vaska, L., J. Chem. Soc., Chem Commun., (1988) 602.
[178] Burgemeiser, T., Kastner, F., Leitner, W., Angew. Chem., 105 (1993) 781.
[179] Drury, D. J., Hamlin, J. E., Eur. Pat. Appl. EB 95, 321 (1983).
[180] Gaßner, F., Leitner, W., J. Chem. Soc., Chem. Commun., (1993) 1465.
[181] Ngyen, M. T., Ha, T.-K., J. Am. Chem. Soc., 106 (1984) 599.
[182] Leitner, W., Dinjus, E., Gaßner, F., J. Organomet. Chem., 475 (1994) 257.
[183] Fornika, R., Görls, H., Seemann, R., Leitner, W., J. Chem. Soc., Chem. Commun., (1995) 1479.
[184] a) Jessop, P. G., Ikariya, T., Noyori, R., Nature, 368 (1994) 231; b) Jessop, P. G., Hsiao, Y., Ikariya, T., Noyori, R., J. Am. Chem. Soc. 118 (1996) 344; c) Jessop, P. G., Ikariya, T., Noyori, R., Science, 269 (1995) 1065; d) Ikariya, T.; Hsiao, Y.; Jessop, P. G.; Noyori, R., Eur. Pat. Appl., EP 652202 A1 (1995).
[185] Tsang, C. Y., Streett, N. B., Chem. Eng. Scie., 36 (1989) 993.
[186] Howdle, S. M., Poliakoff, M., J. Chem. Soc., Chem. Commun., (1989) 1099.
[187] Howdle, S. M., Healy, M. A., Poliakoff, M., J. Am. Chem. Soc., 112 (1990) 4804.
[188] Jessop, P. G., Hsiao, Y., Ikariya, T., Noyori, R., J. Chem. Soc., Chem. Commun., (1995) 707.
[189] Bipp, H., Kiczcka, U. K., Ullmann's Encyclopedia of Industrial Chemistry, VCH Weinheim, Vol. A 12, 1–12, 1989.
[190] Haynes, P., Slaugh, H., Kohnle, J. F., Tetrahedron Lett., (1970) 365.
[191] Kiso, Y., Saeki, K., Jpn. Kokai Tokyo Koho, JP 52036617 (1977).
[192] Jessop, P. G., Hsiao, Y., Ikariya, T., Noyori, R., J. Am. Chem. Soc., 116 (1994) 8851.

7

INDUSTRIAL AND ENVIRONMENTAL APPLICATIONS OF SUPERCRITICAL FLUIDS

H. Schmieder*, N. Dahmen, J. Schön and G. Wiegand
Forschungszentrum Karlsruhe, Institut für Technische Chemie CPV, D-76021 Karlsruhe, Germany

7.1 SURVEY
7.2 CURRENT PROCESS DEVELOPMENTS
    7.2.1 Food and Pharmaceutical Industries
    7.2.2 Textile Industry
    7.2.3 Nanopowder Production
    7.2.4 Polymers
7.3 ENVIRONMENTAL TECHNOLOGY
    7.3.1 Extractive Treatment of Solids
    7.3.2 Separation of Aqueous Residues
    7.3.3 Supercritical Water: Oxidation and Conversion Processes
7.4 REFERENCES

This chapter contains a brief survey of the state of the art and an outline of major ongoing process developments in which supercritical fluids are used. Under the heading of development work, special mention is made of those processes which are concerned with the disposal of waste and the recycling of residues. In these cases, too, supercritical carbon dioxide and water are the dominating media under study, because they are non-toxic, non-flammable, and environmentally compatible.

7.1 SURVEY

It took approximately a hundred years from the discovery of the unusual solvent power of supercritical fluids, especially for substances of low volatility, to their industrial use as extraction agents. In 1822, Cagniard de la Tourin [1] described his observation of the liquid phase disappearing when various liquids were heated in closed vessels. Upon cooling, the liquids were restored. That study is considered to be the first publication describing the phenomenon of the supercritical state.

The definition of the critical point was introduced by Andrews in 1869 [2]. Even at that time it was possible to determine very precisely the critical temperature of 30.9 °C and the critical pressure of 73.0 atm (7.30 MPa) for carbon dioxide. In the following decades, numerous studies were published, especially about the solubility of inorganic and organic substances in condensed and supercritical gases [3–5].

The first industrial use of compressed gases as solvents for separation is considered to be the deasphalting of heavy mineral oil fractions by means of dense propane in the petrochemical industry in the late thirties [6, 7]. Since the fifties, studies and development efforts have been focused on new ways of separating substances by making use of the extraordinary properties of supercritical fluids. In this context, mention must be made of the excellent solvent power for a wide variety of organic substances resulting from the fluid-like densities of supercritical fluids and their favorable transport properties (gas-like viscosity and much higher diffusion coefficients than in liquids); the activities of Zosel at the Max Planck Institute for Coal Research [8, 9] are noteworthy with respect to successful applications in extraction processes.

An extractive separation process using compressed gases (see, *e.g.*, monographs by Brunner [10] and by McHugh and Krukonis [11]) differs from the usual solid–liquid and liquid–liquid extractions mainly by the need to design the process for high pressures, which also results in capital costs being higher than in the traditional extraction methods.

For the purpose of designing and planning the process, the frequently complex phase equilibria are needed as a function of pressure and temperature. Such phase equilibria have been measured in large numbers for binary systems and, to a lesser extent, also for ternary systems and other multicomponent mixtures [12]. Equilibrium studies are considerably more expensive than those for conventional extraction under normal conditions. This is one of the reasons why mathematical modeling studies are of great significance in process development. The thermodynamic conditions underlying phase equilibria have been known since the last century, and can be calculated by means of so-called equations of state for any given fluid mixed system. Mixing rules can be employed to describe the phase equilibria with satisfactory accuracy for multicomponent systems, too [13]. As far as carbon dioxide and water are concerned, reference is made here to only some of the extensive studies by Schneider and Franck [14–17]. The

7.1. A SURVEY

solubility of the component to be extracted determines the necessary amount of fluid (mass flow of the extraction solvent). Adding a cosolvent, also called entrainer or modifier, may increase solubility, sometimes drastically so, thereby reducing the required extracting agent stream and, consequently, also the energy requirement of the process.

The kinetic data regarding mass transfer, which are needed in addition to the equilibrium data, have so far been measured and studied less extensively. In this respect, and in basic fluid dynamics for extractor design (extractor type, packages, internals, hold-up, flooding limits, coalescent, etc.), considerable gaps in our knowledge exist about the design and optimization of liquid-fluid countercurrent flow processes. In solid-fluid extractions, which have found the broadest industrial application so far, continuous process control should be sought with a sufficiently high mass transfer rate. The state of the art is represented by a quasi-continuous feed and discharge of solids through controlled pressure locks and by a cascade of several batch extractors [11, 18].

The motivation for developing extractive separation techniques with supercritical fluids can be summarized as follows:

- Low temperature, mild conditions.
- Residue-free extracts due to simple and complete separation of the solvent.
- Substitution of problematic traditional solvents (halogenated hydrocarbons, aromatic compounds) as a result of stronger restrictions regarding solvent residues in food, and more restrictive environmental standards.

Moreover, some authors expect power consumption [11] and even the total cost to be lower in liquid-fluid extraction [19, 20] than in distillation.

Supercritical carbon dioxide is an extracting agent used in many industrial processes, in a large scale primarily those for decaffeinating coffee and tea, and for producing hops extract. In 1978, the first decaffeination plant was commissioned by Hag AG in Bremen with a capacity of 10,000 t/a, which has been increased considerably in the meantime. Following this, in 1982, again in Germany, a plant for hops extraction was constructed. In 1985 and 1988, facilities for hops (Pfizer) and coffee (General Foods) were commissioned in the U.S.. The latter facility is said to have an annual throughput of 25,000 t and uses extractors more than 20 m high and fed in a quasi-continuous mode (pressure lock) [11, 18]. In addition, many industrial applications on a smaller scale have been reported [11, 21–23]. These include:

- Spice extraction (pepper, red pepper),
- Aromatic substances (aniseed, citrus fruit),
- Fragrances (perfumes),
- Pharmaceuticals (natural substances),

- Seed oils (rapeseed, soybeans, olives, etc.),
- Nicotine removal.

Especially in the U.S., processes for producing low-fat and low-cholesterol food (from high-cholesterol eggs, butter, etc.) are pursued on a large scale.

Only very few industrial applications in waste treatment are known so far. For liquid effluent treatment, for instance, a countercurrent flow extraction plant with a column of 0.5 m diameter and 10 m height is operated by Clean Harbors [18].

In addition to extraction processes, there are a number of non-extractive industrial applications, including the fractionation of polymers, coating processes, crystallization of pharmaceutical products and explosives, and spray paint. In this latter process, which is commercialized by Union Carbide, approximately half of the conventional solvent is said to be replaced by carbon dioxide [18].

Another successful application of supercritical fluids is in extraction and chromatography for analytical purposes; thermally unstable and less volatile substances can be separated by this method [24]. In addition, numerous development activities have been reported for preparative chromatographic techniques in the pharmaceutical sector [25].

The industrial use of highly compressed gases in the supercritical state for synthetic purposes began early this century in the high-pressure process of ammonia production, and was continued in the Fischer-Tropsch synthesis, in polymer production and other processes (see Chapter 6).

In the literature, especially in the past decade, many more far reaching proposals and studies can be found of reactions with supercritical fluids, ranging from hydrogenation or the formation of organometallic compounds to enzymatic reactions. In a review, Clifford [26] discusses the basic physical chemistry and the benefits to be expected from such reactions in which the fluid may be the reaction medium and/or the reactant.

A promising process development, whose technical application is being studied in the United States and in Europe, is the low-temperature incineration of waste and pollutants in supercritical water, to which more detailed reference will be made below. The basic principles of the process were already elaborated by Modell at MIT [27], in the late seventies. While organic substances are completely oxidized to carbon dioxide by the addition of oxygen or other oxidants in this process, a different line of process development makes use of near-critical water to convert organic matter in the presence of catalysts [28].

7.2 CURRENT PROCESS DEVELOPMENTS

Supercritical Fluid Extraction (SFE), also referred to as Gas Extraction, was conducted mainly with light hydrocarbons as fluids up until the early seventies [6, 7, 29]. The most important processes currently under development, in which supercritical carbon dioxide is used as a fluid, will be presented below.

7.2. CURRENT PROCESS DEVELOPMENTS

Compared to other fluids, carbon dioxide is characterized by advantageous critical data ($p_c = 7.3$ MPa and $T_c = 31.2\,°C$), which result in relatively simple technical handling of the fluid, and by it being non-flammable and not corrosive. As a natural constituent of air (0.04 vol.%) CO_2 is not toxic in low concentrations and is environmentally compatible; its maximum work place concentration (MAK) level can be 5000 ppm [30] (9.2 g CO_2/m^3 = 0.5 vol.%). Moreover, CO_2 is available at a low price in technical quantities of the necessary quality. Further important information on CO_2 can be found in the literature [10, 11, 21, 31].

7.2.1 Food and Pharmaceutical Industries

In the food and luxury food industries, but also in the production of flavors, fragrances, spices, and essential oils, use of the extracting agents employed up to that time became more restricted from the seventies onward due to more stringent legislation regarding food products [32, 33]. In the mid-eighties, the European Community and the U.S. Food and Drug Administration redefined the permitted extracting agent residue level in foodstuffs [32], with the result that many conventional solvents were barred from further use. This triggered the technical application of supercritical carbon dioxide as an alternative extracting agent.

One of the first summaries of technical applications of supercritical CO_2 in the field of natural substances was published by Stahl [34]. This publication also contains a discussion of the solubilities in dense CO_2 of 49 organic compounds in the substance categories of polyaromatic compounds, phenols, carboxylic acids, anthraquinones, pyrones, hydrocarbons and lipids, as a function of the polarity and structure of substances. In addition, the solubilities of various alkaloids in compressed CO_2 were given [35].

A first overview of the state of the art of extraction and chromatography with supercritical fluids was presented in 1978 at a symposium organized at the Haus der Technik in Essen, Germany [36, 37]. This was followed by a meeting of the SCI Food Group in London in 1982 [38], and the subject was also reviewed by Williams [29], Paulitis et al. [39], and by Randall et al. [40], whose article also contained a list of extraction processes arranged according to the supercritical fluids employed.

As mentioned earlier, this technology has found its broadest application in the food industry [31, 41]. In more recent literature, non-extraction applications are also listed; supercritical CO_2 is proposed as a medium to increase the yield of enzymatic reactions, such as dephosphonation, esterification or enzymatic oxidation of cholesterol to cholesterol oxidases [42] (see Chapter 6). The RESS process (Rapid Expansion from Supercritical Solutions) [22, 42–44], the GAS process (Gas Antisolvent Crystallization) [22, 42, 43, 45], or the PGSS method (Particles from Gas Saturated Solutions) [46] allow microcrystalline powders to be produced (see also Section 7.2.3) which are also of interest to the food and pharmaceutical industries (see Chapter 5). Highly soluble food additives (polylactic acid) [42], and pharmaceutical powders, such as phenacetin [42], steroids [47], and coprecipitates for direct encapsulation of drugs [22, 48, 49] were prepared by RESS from supercritical CO_2 solutions. In the GAS process, CO_2 acts as an anti-solvent in the production of protein powder (insulin, catalase) [43, 49], and the PGSS method was used to produce glyceride and nifedipin powder [46].

A comprehensive recent description of the technical application of supercritical fluids in the food flavoring and pharmaceutical industries, as well as information about the technical equipment and the process cost, is given by King and Bott [21]. The report by Palmer [23] contains a list of patented, commercially used processes using CO_2 under supercritical conditions.

7.2.2 Textile Industry

Increasingly more restrictive waste water legislation in Germany in the late eighties resulted in the development, at the German Textile Research Center (DTNW) in Krefeld, of a dyeing process free from water [50], which was demonstrated in a pilot plant for the first time in 1991 [51]. In this alternative dyeing process, the dyes are dissolved in supercritical CO_2 and applied to the swelled textile fiber from the fluid phase, as in conventional impregnation (see Chapter 5). In comparison with the conventional dyeing process, there is no waste water, and the energy requirement is lower because no drying steps are included; hazardous air emissions are avoided because CO_2 is recycled, and surplus dyestuff can be recovered.

Studies of the solubility of dyestuffs in dense CO_2 were first conducted at the University of Bochum [52, 53]. In dyeing polyester fibers [51], the dyestuffs in supercritical CO_2 act on the fiber for ten minutes at 25 MPa and 130 °C; subsequently, the compressed CO_2 is expanded to atmospheric pressure by way of a separator to allow the excess dyestuff to be recovered. Wool and cellulose fibers, such as cotton or viscose, do not swell in supercritical CO_2. Before dyeing, these fibers therefore must undergo a preliminary treatment step to improve their affinity to the dyestuff and its penetration into the fiber. These fibers are then treated with the dyestuff for 15 minutes at 30 MPa and 160 °C, followed by an expansion of the fluid to ambient pressure at 1 MPa/min [54].

In addition to dyeing, cleaning of textiles by supercritical CO_2 was studied as an alternative to the use of perchloroethylene as a solvent [55]. A good washing effect was achieved for oil soiled textiles, which were treated at pressures of 20–25 MPa and temperatures of 50–70 °C. Standardized soiled fabric with pigment-like impurities has not been washed satisfactorily before, probably because of the absence of mechanical movement of the textiles in those experiments. It is hoped that the washing effect can be improved by addition of entrainers to the supercritical CO_2 which may then also be applicable for the washing of hydrophilic soiled fabric [55].

7.2.3 Production of Nanopowders

The term "Supercritical Fluid Nucleation" [22] stands for a special application of supercritical fluids in making very fine particles, so-called "nanopowders". These powders are characterized by their highly uniform grain size. Depending on the process conditions and the substances involved, grains are produced in which the diameter ranges between a few and several tens of micrometers, characterized by large surfaces. Microparticles are becoming increasingly important in the pharmaceutical industry for controlled drug delivery applications [42, 49], and in the production of ceramics and plastics [22, 44, 56]. Three methods of making microparticles are currently available [22, 43–46].

The RESS process is used for substances which are soluble in supercritical fluids [57]. Precipitation as submicroparticles occurs during very fast expansion (in microseconds) of the supercritical solutions through a nozzle, which abruptly reduces the dissolving power of supercritical fluids and leaves no time for the particles to grow larger. During expansion, the pressure can be reduced partly or completely to atmospheric level, and the mechanism of particle formation is based on free jet expansion. The schematic diagram of the equipment, and particle formation as a function of pressure, temperature, and expansion rate, were presented [57] for the SiO_2 and GeO_2 systems dissolved in supercritical water, and for polystyrene dissolved in supercritical pentane. If CO_2 is used as a fluid, and if the system is expanded completely, the residual precipitate is a solvent-free powder as described for the phenanthrene/supercritical CO_2 system [44] and the salicylic acid/supercritical CO_2 system [58]. Other applications of the RESS process with supercritical CO_2 can be found [42, 43, 47–49].

Gas Antisolvent Crystallization (GAS) [45], also referred to as SAS by some authors [43, 44], is used on substances which are insoluble in supercritical fluids. The substance to be crystallized is dissolved in a suitable solvent, and then a cosolvent is added to reduce the solvent power of the solvent to such a level that the dissolved substance is rapidly precipitated. If a highly soluble gas or fluid is used as a cosolvent, injection (within a few seconds to a few minutes) will cause pronounced expansion of the initial solution. This rapid increase in volume is associated with a reduction in the solvent power of the solvent. As a consequence, the dissolved substance is precipitated as very small particles from the supersaturated solution. The particle size obtained is influenced by the concentration of the initial solution, the injection as a function of time, the expansion of the initial solution, and the temperature. This method, with CO_2 as the anti-solvent, was used for the first time in 1988 to prepare very fine explosive powders [45, 56, 59]. In that case, nitroguanidine was precipitated from a dimethylformamide solution at 5.5 MPa [45], and cyclotrimethylene trinitroamine was recrystallized from cyclohexanone at CO_2 pressures of 7–16 MPa [49]. Citric acid was precipitated as microparticles from acetone-water solutions (0–30% H_2O) with CO_2 at 1.5–4.5 MPa by the GAS process [60]. Polystyrene powder was produced by precipitation of the polymer from a toluene solution at 6–22 MPa [61]. Further information about polymer powders can be found [22, 43, 56, 62], while the preparation of microparticles for various drugs is reported [43, 49].

Another process for generating microparticles, the PGSS process, is based on a melt saturated with CO_2, which is expanded through a nozzle. The cooling effect during expansion causes the melt to solidify in the form of very small particles which are several tens of micrometers in diameter. The process, the equipment, and the mechanism of particle formation are reported [46]. In order to produce glyceride powders, the melt was saturated with CO_2 in the range of 8–20 MPa and 65–100 °C, while pressures of 10–20 MPa and temperatures of 170–200 °C were applied in the case of nifedine [46].

7.2.4 Polymers

The increasingly more stringent requirements to be met by plastics in terms of purity and material properties, especially in medicine, in food and packaging applications, and in microelectronics, in many cases necessitated modifications of the traditional manufacturing processes. As a consequence, supercritical fluids are being used as alternative solvents in polymer production [56, 62–66]. Early studies by Zosel [8] dealt with improvements in polymer quality.

In a large number of high pressure polymerization processes, most of which occur as free-radical polymerizations, supercritical light hydrocarbons are used as solubility promoters [62–65]. As a consequence of the selective solvent power of the fluid, the monomer and the initiator dissolve in the supercritical fluid, whereas the polymer is no longer soluble beyond a specific molecular mass, and precipitates. This selective solvent power of supercritical light hydrocarbons and a number of chlorofluorohydrocarbons has been applied to the fractionation and purification of polymers from the feed product monomers and oligomers by molar mass separation [62–64]. This application of supercritical fluids has been covered extensively [64].

The mechanical, chemical, thermal and other physico-chemical properties of polymers are determined mainly by their structure and molecular weight distribution. The use of fractionated polymers with a narrow molecular weight distribution has greatly improved the quality of polyethylene and ethylene-based copolymers [62–67]. Because of its lower solvent power, supercritical CO_2 has found only limited application in the fractionation and purification of polyethyleneglycols, polydimethylsiloxane, and perfluoroalkylpolyethers [64]. Studies of the solubility of polymers in supercritical CO_2 [64, 68, 69] have shown that the exchange of hydrogen atoms for fluorine atoms in the hydrocarbon chain of the polymer increases considerably the solubility of the polymer in dense CO_2. This effect is used in homogeneous free radical polymerization of 1,1-dihydroperfluorooctyl acrylate, using azobisisobutyronitrile (AIBN) as the initiator. The monomer, as a 25% w/v solution in supercritical CO_2, was polymerized at 60 °C and 20.7 MPa [62, 70, 71]. Over a polymerization period of 48 h, fluoropolymers with molar masses of up to 27,000 were obtained. Moreover, in supercritical CO_2 this fluorine-bearing monomer enabled the production of copolymers with methyl methacrylate, styrene, and ethylene [62].

Ethylene as a supercritical fluid acted both as a solvent and a reactant in the production of polyethylene [62–65]. Several polymerization reactions are described [63], in which CO_2 in the supercritical state is also a reactant at the same time: polyethercarbonates were produced from ethylene oxide and CO_2 at 150–240 °C and 13.5 MPa. Diamines react with CO_2 at 200–250 °C and 50–200 MPa to produce polyurethanes. The copolymerization of ethylene and CO_2 proceeds in the presence of a catalyst at 30–100 MPa and 140 °C. The reaction of alkene oxide with CO_2 yields polycarbonates, methylvinylether and CO_2 react to produce polyvinyl polymers, and acrylonitriles and CO_2 form polyacrylonitrile at temperatures of 120–160 °C. In the polymerization of ethylene, induced by UV radiation, at pressures above 38.5 MPa, CO_2 was employed as solvent.

Polyacrylic compounds are soluble in compressed CO_2 at temperatures of 31–150 °C and pressures of 8–30 MPa, and thus allow polymer powders with diameters of 0.5–5 µm to be produced by the RESS process [44, 62, 63]. Recently, the solubility of polysilanes and novolak resins in supercritical CO_2 has been exploited to make microelectronic components. In this photolithographic procedure the unexposed resin remains as monomer and is removed by supercritical CO_2 (80 °C, 41.5 MPa) from exposed photoresist resins [56].

When supercritical CO_2 and other fluids are dissolved in plastics, there may be changes in the morphological properties of the polymers (plasticization); in this way, the dissolved fluids act like plasticizers [62], aiding the processing of plastics. Moreover, efforts are made to improve the solubility of polymers in dense CO_2 by surface modification and by the attachment of "CO_2-philic groups" (fluorocarbon or polydimethylsiloxane segments) [62–68, 72].

7.3 ENVIRONMENTAL TECHNOLOGY

7.3.1 Extractive Treatment of Solids

The beneficial properties of supercritical fluids open up new possibilities of treating solid, slurry and liquid residues in environmental technology [73, 74]. One of the first applications of supercritical carbon dioxide in this area can be considered to be the regeneration of activated carbon used to clean polluted liquid or gaseous effluent streams [75]. Regeneration by fluid CO_2 allowed the adsorber material to be recycled many times without suffering any major decline in the adsorbing capacity. The regeneration behavior is determined not only by the solubility in the fluid but, especially, by the desorption characteristics of the organic substances.

The sorption behavior of pollutants toward the matrix is important especially in the treatment of organic-polluted soils by supercritical fluids. Consequently, the more recent literature lists a number of studies, *e.g.*, on the adsorption of polychlorinated aromatic hydrocarbons (PAH) [76, 78], of chlorobenzene and chlorophenol [78], and of phenol [76, 79, 80] on soils, in the presence of supercritical carbon dioxide. The use of supercritical fluids to separate pollutants from real soil samples for analytical purposes is described in a number of reports [41, 81–86], although there is little literature on the use of supercritical fluids for the clean up of soil on a technical scale.

On an extended laboratory scale, contaminated soil samples were treated with supercritical CO_2 in 0.2 dm^3 autoclaves to remove PAH (polycyclic aromatic hydrocarbons) [87] and PCB (polychlorinated biphenyls) [80, 88, 89]. At only 14 MPa and 40 °C, most of the PCB were removed, and construction of a mobile facility with 1 m^3 extractors has been proposed [88]. In order to isolate natural long-chained aliphatic compounds from soils, soil samples of up to 120 g were used and extracted for up to 4 h at 37.5 MPa and 80 °C [90], whereas samples of 100–1000 g were used to clean up oil-bearing soils with fluid CO_2 [91]. Schulz and co-workers [92–95] used an extractor volume of 4 dm^3 for SFE of soils contaminated with PAH; within that study, the extraction was optimized [92, 93], and basic design data were determined for a larger plant and for the assessment of soil clean up costs [94, 95]. In Australia, a mobile facility for extraction with condensed CO_2 was developed for cleaning soils contaminated with petroleum [96]. The facility is 12 m long and has been designed for a throughput of 2 t of soil/h. It comprises four extractors, each of 500 kg capacity [97]. No report is available regarding operating experience with this plant.

Now that the SFE technique has achieved the status of a standard analytical procedure for assays for hydrocarbons and contaminants in soils in the U.S. with the introduction of EPA method 3560, it may be assumed there will be increased use of SFE in environmental technology. Furthermore costs of DM 200–500/t ($140–350/t) are expected to arise from this clean up technique, depending on plant size and operation mode [94, 95]. In addition to carbon dioxide, supercritical and near-critical water ($T_c = 374$ °C; $p_c = 22.1$ MPa) is used for pollutant removal. In this case, the water can be used as an extracting agent, and also as a reactive solvent for further treatment of its pollutant contents (see Section 7.2.1) [98, 99].

The technique of supercritical extraction is not restricted to the treatment of solid residues in the form of soils, but can also be applied to other kinds of waste, such as industrial waste. Once pollutants have been removed by SFE, a concentrate is obtained which can be treated for the decomposition of the pollutants. The treatment of residues resulting in reusable fractions after separation is a process of particular economic interest. This has been demonstrated with industrial waste in the form of oil-contaminated residues from glass and metal processing on the basis of laboratory analyses and semi-technical scale extraction in a bench plant [100] up to experiments in a plant of industrial size. This kind of material arises in metal work industries in various forms, ranging from mechanical compacts of chips, grits and grinds to loose beds of sludge with up to 50 wt. % of oil.

Before the supercritical extraction of grinding residues can be implemented on a technical scale, a number of laboratory studies must be conducted. An example involving glass and metal working residues (grits and grinds) that are contaminated with oil will be described below. The dissolution behavior of the components to be extracted is of interest. Often they consist of complex mixtures mainly of aliphatic hydrocarbons with, typically, a few percent by weight of different chemical additives with different solubilities in supercritical CO_2. The oils used in metal processing may be straight petroleum based oils, but synthetic and native oils are also used. Complete separation and subsequent quantitative determination of each individual component is not necessary, since the separation effect can be characterized sufficiently well by integral parameters. The organic-bonded carbon (Total Carbon, TC) was determined in the solid residue. Some of the samples contained considerable amounts of organics in addition to the oil, *i.e.*, filtering aids. For these samples another method of analysis was applied to prove that the carbon content remaining in a sample after extraction was not oil. Here the EPA method no. 3560 was used. This method describes the determination of total recoverable petroleum hydrocarbons by SFE. The extract was collected in trichlorotrifluoroethane ($C_2Cl_3F_3$), the concentration of oil was then determined by IR-spectroscopy.

Figure 7.1 shows the solubility behavior in supercritical CO_2, as determined in a phase equilibrium apparatus, of a typical petroleum grade which is a mixture of mainly aliphatic hydrocarbons with up to 32 carbon atoms. The assay was conducted by a synthetic method in which the petroleum is dissolved homogeneously in fluid CO_2, and subsequently the pressure is reduced gradually down to the level of demixing, which can be observed optically through sapphire windows. In contrast to the fast and easily recognized phase separation in pure substances, gradual demixing in this case occurs as a result of the oil components precipitating one by one, beginning with the sparingly soluble fractions. To characterize demixing, an optical system with a photocell was

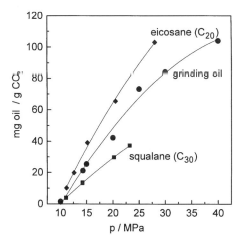

Figure 7.1 Solubility of a grinding oil in supercritical carbon dioxide as a function of pressure at 50 °C.

designed which traces clouding in the demixing process. The figure shows that the petroleum studied has a solubility of only a few mg/g of CO_2 at a pressure of approx. 10 MPa. The solubility can be increased approximately by a factor of 100 by raising the pressure to 40 MPa. Measurements were conducted at 50 °C. For comparison purposes, the diagram also shows the solubilities of two pure aliphatic hydrocarbons, namely squalane ($C_{30}H_{62}$) and eicosane ($C_{20}H_{42}$), in supercritical carbon dioxide [101].

To determine optimum values for pressure, temperature and the CO_2 flow and throughput as process parameters, experiments were run in an analytical SFE apparatus with sample amounts of a few grams. The carbon dioxide is fed to the top of the extraction cell (1 ml) by means of a piston pump. Extraction may be a dynamic or a steady-state process brought about by stopping the flow, or it can be achieved by a combination of both methods. In the last step, the loaded CO_2 stream is expanded through a restrictor into a solvent trap thus dissolving the precipitated organic material. As a result of the pressure decrease at the restrictor, the components may precipitate at this end and can either change or completely disrupt the CO_2 flow. In a number of facilities tested, the best solution was found to be a variable restrictor whose cross section can be adapted to the requirements in each case. An in-line monitor was used for tracing the course of extraction; downstream of the extraction cell, a small part of the CO_2 stream was diverted and passed over an FID (flame ionization detector). This allowed the hydrocarbons in the fluid phase to be monitored up to complete extraction. The residual carbon content was determined by TC measurements on the solid residue. Figure 7.2 shows the residual petroleum content plotted against the CO_2 throughput expressed in g of CO_2/g of material for the extraction tests of glass grinding sludge. At 10 MPa, the residual content approaches the set level of 1% only gradually,

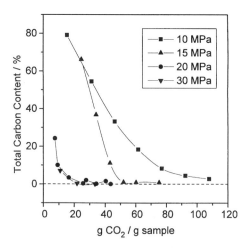

Figure 7.2 Residual carbon content as a function of CO_2 throughput at pressures between 10 and 30 MPa, and at 50 °C.

7.3. ENVIRONMENTAL TECHNOLOGY

while at 20 MPa, complete extraction is achieved already at 30 g of CO_2/g of material. The degree of extraction is a function not only of external parameters, but also of the oil used.

Figure 7.3 is a schematic diagram of the bench-scale test facility used for supercritical extraction of larger quantities of grinding sludge. The fluid reservoir holds liquefied carbon dioxide at approx. 0 °C, which can be raised to the desired operating pressure (50 MPa max.) by a membrane pump with a maximum capacity of 30 kg/h. The condensed CO_2 passes through the solid-fluid extractor (4 dm^3 volume), which contains the basket loaded with the extraction material. The basket is closed at the top and the bottom with sintered metal plates to allow the bed to be permeated by the extracting agent without removing the solid from the container. Temperatures of up to 100 °C can be set by means of a heat exchanger. The carbon dioxide loaded with the extracted components is passed through a relief valve into the phase separator in which a gaseous and a liquid phase are coexisting at a pressure of approx. 6 MPa, with the pollutants becoming enriched in the liquid phase. Given a sufficient degree of purity, the gas can be returned to the reactor *via* the fluid tank. In the studies described here, which were conducted with petroleum samples, the residual content of oil in the gas phase was always below 1 ppm. At higher concentrations, these components can be collected in an activated carbon adsorption column upstream from the condensation unit of the system.

Unlike the laboratory experiments, the bench-scale plant used a steady-state-dynamic extraction mode, *i.e.*, the extractor filled with the grit and grind residue was first exposed to CO_2, then left standing for a certain period of time to equilibrate, and then the oil enriched carbon dioxide was flushed out. In this way, the grit and grind are purified in batches with a minimum of extracting agent. Experiments were run with a CO_2 flow, in one case, from bottom to top (test No. 1, Table 7.1) and, in the other cases, from top to bottom.

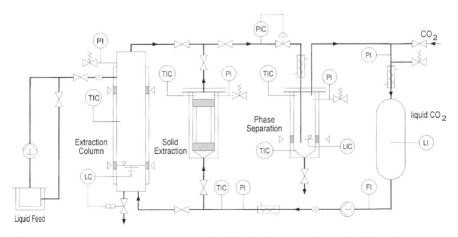

Figure 7.3 Schematic diagram of a test facility for the extraction of solids and aqueous effluents.

Table 7.1 Test results obtained in the bench-scale plant.

No.	Material	Oil content	CO_2/Material	Residual TC/%
1	Glass Sludge 1	28–32	18.4	0.53–0.57
2	Glass Sludge 2	28–32	14.8	0.51–0.61
3	Metal-Compact 1	6–7	21.1	0.43
4	Metal-Compact 2	—	13.1	≈ 4
5	Metal-Compact 3	≈ 3	11.4	0.15
6	Metal Sludge 1	26	15.0	< 1.0
7	Metal Sludge 2	48–55	26	< 1.0
8	Brass Compact	2–3	10.0	0.4
9	Al-Compact	15–16	26.8	0.1
10	Cu rolling scale	3–4	11.7	< 0.2

The successful treatment of metal work residues can be seen from Table 7.1. The table is a summary of the residual carbon contents (TC) of all the samples, together with the initial oil content of untreated materials and the specific CO_2 throughput required for clean up. The compacts listed in the table in some cases consist only of compacted metal chips, while others contain metal chips and grit and grind, with the grit and grind reaching up to 40 wt. %. In principle, it was possible to attain residual oil contents of less than 1% in all tests. In one case (test No. 4) a constant carbon fraction remained and could not be further reduced, even after repeated treatment. This residual TC was not removed even by washing of a sample of the compact in hexane, acetone or methanol. Consequently, it cannot be petroleum. What kind of organical bonded carbon is involved has not yet been determined. In test No. 7, a non homogeneous mixture of materials existed after extraction, and this could be separated by simple magnetic separation into a metal fraction with a TC of less than 1%, and a fine grained non-metallic fraction with a residual carbon content of 4–5%. The latter was caused by organic filter contaminants in the residue.

Materials of a variety of compositions have the common property that their solid constituents, ranging from coarse metal chips to very small abraded fines, are essentially inert in the extraction process. The composition, at least of the metallic components, exhibited no influence on the extraction parameters. Both the oil extract obtained in separation and the solid residue can be treated further by conventional means. Depending on purity, the oils can be either recycled directly into the grinding process or subjected to used oil processing. The metal components can be passed on to foundries or steel mills in the form of oil-free compacts. Residues from metal working, of which some 150,000 t per year [102] arise in Germany at the present time, are mostly disposed of as hazardous waste in thermal processes, or are used in the cement industry, for instance, as iron additives. It is seen that detailed knowledge of the composition of any waste to be treated is desirable, both to optimize oil removal by supercritical extraction and also for further recycling purposes. Moreover, reliable data on the

7.3. ENVIRONMENTAL TECHNOLOGY

operating, capital, and manpower costs of the process are required, because the attractiveness of the process will depend decisively on its economics compared to those of competing processes.

Moreover, not only oil removal from residues left over in various processing steps is a possibility, but also degreasing of the work pieces could be carried out as far as necessary for further processing. In this case, supercritical CO_2 could be used as a substitute for the organic solvents or aqueous emulsions employed so far [103].

Although none of the processes mentioned in this chapter have been employed on a technical scale, the extent of activity in this field shows that these processes are thought to hold a considerable potential for the solution of current and future problems.

7.3.2 Separation of Aqueous Residues

Especially in the presence of less volatile organic substances in low concentrations in an aqueous solution, liquid-fluid extraction may be a meaningful alternative to other separating processes, such as distillation or membrane separation, for concentrating the organic components. In a countercurrent flow process, in which the aqueous phase is fed to the top of a column, the organic substances are continuously dissolved in supercritical CO_2 and discharged (see Figure 7.3). Expansion of this extract in a separator can produce the pollutant concentrate, while the water, depleted in organic substances, is removed at the bottom of the column. Increasing attention has been devoted to the separation of organic substances from aqueous solutions (which often constitute complex mixtures of chemically different substances) for environmental purposes. This requires knowledge not only of the hydrodynamic characteristics in the water/CO_2 system, but also of the mass transfer properties and phase equilibria, as characterized by the partition coefficient, K_x, of the organic substances involved. This is a necessary prerequisite for the technical design of liquid-fluid countercurrent flow extraction. The partition coefficient of an organic substance, which is defined in this context as the ratio of its mole fractions in the aqueous and in the fluid CO_2 phases, is a function of pressure and temperature. Partition coefficients for the binary water/CO_2 system are shown in Table 7.2 for several organic substances. Figure 7.4 shows the dependence of the K_x-values of phenol on pressure for various temperatures. In modeling, the dependence of the partition coefficients on concentration plays a major role. In the ideal case, there is a linear relationship between the concentrations in the two coexisting phases. This is true in the case of phenol (Figure 7.5).

The application of supercritical gases for separating substances in liquid products has already attained technical maturity in the food and pharmaceutical industries (see Section 7.1.1); as yet, only a few applications exist in environmental technology [18]. Since Elgin and Weinstock drew attention to a possible application of compressed gases to remove organic substances from aqueous solutions in 1959 [104], a number of studies have been conducted, especially in

Table 7.2 Partition coefficients K_x of selected organic substances at 10 and 30 MPa, and 50 °C.

Compound	Chem. Formula	10 MPa	30 MPa
Phenol	C_6H_5OH	0.43	1.31
Benzaldehyde	C_6H_5CHO	12.7	46.4
Benzylalcohol	$C_6H_5CH_2OH$	0.48	2.34
Benzoic acid	$C_6H_5CO_2H$	0.55	1.70
Cyclohexanone	$C_6H_{10}O$	6.1	20.1
Coffeine	$C_8H_{10}N_4O_2$	0.11	0.23

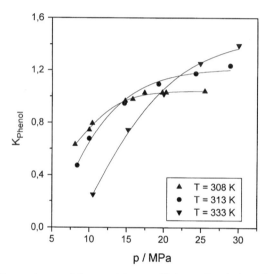

Figure 7.4 Dependence of the partition coefficients K_x of phenol on pressure.

recent years, on the supercritical extraction of organic substances from aqueous solutions. The removal of ethanol and other simple alcohols from aqueous solutions has been a subject of particular interest. This was motivated by the production of alcohols as fuels or chemical feedstocks from biomass fermentation [105–107]. Many basic studies were conducted on the phase equilibria of these systems [108]. Separation experiments were also run in countercurrent columns in laboratory scale and pilot plants [109] to examine the suitability of the process as an energy-saving alternative to distillation. Using an aqueous ethanol solution of up to 20%, separation was conducted in a perforated tray column (inner diameter: 10 cm; length: 20 m); subsequently, the CO_2 solvent was removed from the extract by distillation. In connection with the possibility of purifying drinking water by means of compressed gases, Ehntholt studied the extraction capability of a number of substances typically occurring in low concentrations in drinking water [110, 111].

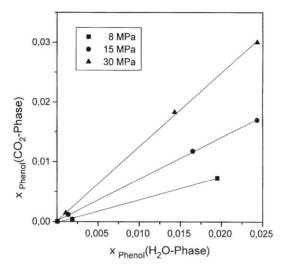

Figure 7.5 Dependence of the K_x-value of phenol on concentration.

Tests, mainly conducted as laboratory scale experiments, have been performed to examine the suitability of supercritical extraction for removal of pollutants from liquid effluents polluted with organic substances. The extraction of phenol was studied in great detail [112–114], especially with respect to partition characteristics. The dependence on pressure and temperature of the partition coefficients of other substances likely to occur as pollutants, such as aromatic compounds, pentachlorophenol [115, 116], cresol, benzene [117], toluene and naphthalene [118], or halogenated hydrocarbons [119], were also determined by measurements on individual substances and mixtures.

Mass transfer parameters for ethanol and benzyl alcohol were studied by Knez in a batch apparatus in which the aqueous solution of the two substances was present in the steady-state [120]. Experiments on a larger scale were performed by McGovern and Moses in a countercurrent flow column with a throughput of 473 L/min. The aqueous solution contained not only acetonitrile and acrylonitrile, but also 2 wt. % ammonium sulfate in order to simulate a typical liquid effluent of an acrylonitrile production plant. On the basis of the experimental data, the separation performance of a 25-tray column for the extraction of ten organic substances was calculated [121].

Although effectiveness is improved by the use of entrainers, mostly acetone, methanol or ethanol, of which up to 20 wt. % are added to the carbon dioxide extracting agent [122, 123]. This adds to the expense of technical implementation because of the later need to remove the entrainer.

7.3.3 Supercritical Water: Oxidation and Conversion Processes

Water is an ecologically safe substance widespread throughout nature. At standard conditions it shows many anomalies when compared with other solvents. In the subcritical region, the vapor pressure curve, which ends at the critical point, separates liquid and gaseous phases and a continuous density variation is not possible. Upon approaching and passing the critical point ($T_c = 647$ K, $p_c = 22.1$ MPa, $\rho_c = 320$ kg/m^3), water exhibits unusual and unexpected behavior. In the region of the supercritical fluid, the density can be varied without any liquid-like to gas-like phase transition over a wide range of conditions (see Chapter 6).

Liquid water at standard conditions ($T = 298$ K, $p = 0.1$ MPa) is poorly miscible with hydrocarbons and gases. In contrast, it is a good solvent for salts because of its high density (997 kg/m^3) and high dielectric constant (78.5). The density and dielectric constant decrease simultaneously with increasing temperature above 377 K. At near-supercritical temperature and pressure the dielectric constant is in the range of 10 [127, 128]. Near the critical point, the dissociation constant of supercritical water for complete ionization is comparable with the values at standard conditions (10^{-14}), but it is remarkably higher at 1273 K (10^{-6}), where the density increases to 1000 kg/m^3 [129]. Therefore, supercritical water at low densities, is a poor solvent for ionic species such as inorganic salts. Calculations of the solubility of NaCl and KCl in supercritical water have been carried out by Pitzer and co-workers [130–132]. Supercritical water, however, is completely miscible with many organic compounds and gases [133–136]. Diffusion rates are high and viscosity is low in a supercritical aqueous mixture. A summary of the density, dielectric constant, and viscosity of water as a function of temperature at constant pressure of 30 MPa is given in Figure 7.6.

Transport properties and miscibility are important parameters which determine the rate and homogeneity of chemical reactions. High diffusion rates and low viscosity, together with the near complete miscibility with many substances, make supercritical water an excellent medium for homogeneous, fast, and efficient reactions. An example of such a reaction is oxidation with or without catalyst.

Supercritical water can also be used as an extraction medium because of its extraordinary properties. Because it is a good solvent for hydrocarbons, it is used in the purification of soil contaminated with hydrocarbons (Section 7.3.2). Initially, the hydrocarbons are removed from the soil by dissolution into supercritical water and are then oxidized [25].

Supercritical Water Oxidation Supercritical Water Oxidation (SCWO) technology may be a suitable alternative method for destroying hazardous waste. Hazardous waste destruction processes are subject to very strict emission regulations. SCWO results in harmless natural products such as carbon dioxide and water and therefore, there is a great impetus to develop methods like SCWO which can meet the strict environmental laws. Hazardous wastes include:

- Process effluents from the chemical, pharmaceutical, and food industry
- Warfare agents

7.3. ENVIRONMENTAL TECHNOLOGY

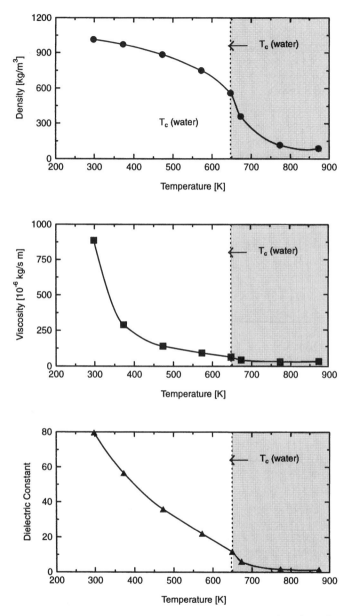

Figure 7.6 Water density, viscosity, and dielectric constant as a function of temperature at 30 MPa.

- Propellants
- Ammunitions
- Nuclear waste
- Wastes generated from pulp and paper industries
- Municipal sewage.

The first research on SCWO was reported in the early 1980s [137, 138]. Since then several groups have initiated research projects which have varying emphases. There are obvious problems to be solved and issues to be addressed:

- A lack of data on thermodynamic phase behavior for supercritical fluids, including those containing salts
- Unknown chemical reaction kinetics and mechanisms
- Materials research related to corrosion problems
- Engineering tasks (*e.g.*, reactor design, pressurization, mass and heat transfer, separation techniques, handling of solids) and process control at high pressure and temperature
- Safety considerations for safe operation.

Many research projects have been carried out successfully. The first commercial scale use of the SCWO process has been reported recently [139].

The principle of the oxidation reaction is simple (Figure 7.7). Pressurized water, oxygen (from air or from H_2O_2) and organic waste are mixed and heated to the supercritical state, with the temperature ranging between 700 K and 900 K, and pressure ranging between 24 MPa and 30 MPa. Because the chemical reaction is homogeneous, fast, and efficient, the required residence time is short, generally less than one minute, and often less than 30 seconds. Heteroatoms such as Cl and S which are present are converted to mineral acids or inorganic salts which precipitate at low concentrations. After the single phase mixture has left the reactor, temperature and pressure are reduced to standard conditions and the supercritical mixture separates into a liquid phase and a gas phase.

The hydrothermal oxidation under SCWO conditions involves free radical chain mechanisms. The products are carbon dioxide and water in the ideal case. The schematic reaction equation for the complete oxidation of hydrocarbons is given by:

$$C_xH_y + (x + y/4)O_2 \rightarrow x\,CO_2 + y/2\,H_2O.$$

In reality, more byproducts can be formed, such as carbon monoxide in the gas phase and, for example, aldehydes, ketones, and organic acids in the liquid phase. Nitrogen is usually found as nitrogen gas or dinitrogen oxide [140–142].

Kinetic Studies For a practical SCWO process, it is very important to achieve high destruction efficiencies; this requires an excess of oxygen. Kinetic and

7.3. ENVIRONMENTAL TECHNOLOGY

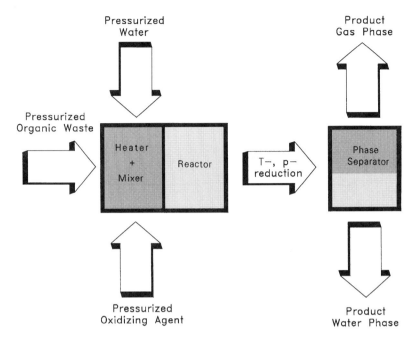

Figure 7.7 Principle for Supercritical Water Oxidation.

mechanistic research on the SCWO process has been carried out for model substances like carbon monoxide, hydrogen, ammonia, methane, methanol, ethanol, acetic acid, and phenol. Comprehensive reviews are available [140, 143, 143a]. Research groups at the University of Illinois/Georgia [136, 144], MIT [145–151], the University of Michigan [152, 153], and the University of Texas [140, 154–156] have made efforts to collect and evaluate experimental kinetic data. Los Alamos National Laboratory has contributed data [157, 158]. At the University of Karlsruhe, kinetic investigations have been performed [159, 160].

Reaction rates are usually treated as global rate expressions. Pseudo-first-order kinetics with respect to the organic substrate (or waste) are applied for the oxidation in supercritical water. The activation energies, calculated from an Arrhenius plot, are in the range of 30 to 480 kJ/mol for different model substances (Table 7.3). Elementary reaction models have been applied to the different experimental results to determine the appropriate reaction mechansims. Some models cannot reproduce the experimental data, while some models fit very well. It is evident that more effort into modelling of these reactions is necessary.

Salt Formation Another topic requiring fundamental research is the formation of salt in SCWO processes. In the supercritical region, at low water densities, salt solubility is very low. Organic substances, containing heteroatoms such as Cl and S,

Table 7.3 Examples of kinetic data for SCWO processes [140]. E_a is the activation energy, A is the Arrhenius pre-exponential factor.

Compound	E_a/kJ mol^{-1}	A	T in K	p/MPa	Refs.
Methane	150^a	$1.5\ 10^7$	833–973	24.5	[158, 195]
Methanol	395	$2.5\ 10^{24}$	723–823	24.5	[158]
Ethanol	340	$6.5\ 10^{21}$	755–814	24.1	[145]
Formic acid	≈ 96	–	683–691	40.8–43.2	[196]
Acetic acid	167.1	$2.63\ 10^{10}$	673–803	24–35	[155]
Phenol	63.8	$2.61\ 10^5$	557–702	24.3	[196]
2,4-Dichlorophenol	71.9	$1.92\ 10^4$	683–788	27.6	[147]
Acetamide	80	10^5	673–803	24–35	[155]
Ammonia	157	$3.16\ 10^6$	913–973	24	[149]
Carbon monoxide	134	$3.16\ 10^8$	693–844	24	[197]

^{a} Average value.

form mineral acids or salts during the oxidation. These cannot be completely dissolved in supercritical water. This may cause problems during the overall process due to the possibility of pipe blockage. Investigations of phase equilibria of salts in supercritical water at low densities are thus very important and needed, since only few data are known [132, 161, 162]. At MIT and NIST [163] a research program for measuring such solubility data, up to 873 K and 40 MPa, was initiated in 1988. These studies have been carried out in small flow reactors in order to determine the particle size and the precipitation behavior of sodium chloride and sodium sulfate in supercritical water [164].

Corrosion The problem of corrosion is linked to the formation of salts and mineral acids. Because supercritical fluid permits at least a partial ionization of electrolytes, the presence of oxygen and hydrochloric acid can cause electrochemical corrosion in an aggressive manner [165]. Consequently, it is necessary to find materials for SCWO reactors which can withstand corrosion and still provide sufficient strength at elevated temperature and pressure. High performance steels and nickel base alloys are sufficiently strong for high pressure and high temperature conditions and are corrosion resistant in pure supercritical water, but not in the presence of inorganic corroding ions. Nickel alloys like Inconel 625, Nimonic 90, and Hastalloy are rapidly corroded when exposed to hydrofluoric acid, hydrochloric acid, sulfuric acid, phosphoric acid, and mixtures thereof [166, 167]. Chlorinated cutting oils, which were oxidized in supercritical water, also caused corrosion [168]. Platinum and titanium alloys promise a good corrosion resistance at SCWO conditions [167, 169], but their cost is a limiting factor. Corrosion protective ceramic layers (*e.g.*, aluminium oxide, zirconium oxide) have been investigated, as well as coatings or liners from glass, Teflon (PTFE), titanium, hafnium, and zirconium [165]. Titania/titanium coating/metal base systems are resistant to chloride exposure [168]. Several research

7.3. ENVIRONMENTAL TECHNOLOGY

groups propose extended effort to develop improved corrosion resistant materials on the basis of systematic corrosion tests.

Deep-well Technology   One idea suggested for the process design of SCWO is to use a deep-well system, referred to as the "Oxydyne Process". Drill holes in the earth could be used at depths of 1500 m to 3000 m. Deep-shaft wet-air oxidation is proposed for subcritical or supercritical applications. The reactor consists of two concentric vertical tubes (Figure 7.8). Waste water is pumped down the outer tube, and the effluent returns in the annular space. Oxygen is added via the center tube to the reaction zone at the deep end of the drill hole. The tube diameter and length are designed to enable the reactor to attain sufficient reaction time and pressure. The reactor can be constructed from conventional oilfield equipment. Excess heat from exothermal oxidation of waste can be recovered at the surface. More details on the deep-well wet-air oxidation are given in the literature [171].

Reactors for SCWO Oxidation   In the early 1980s MODAR developed a process, which is known as the patented "MODAR Process" [172]. The reactor is a high pressure vessel (Figure 7.9) consisting of two zones. The high temperature zone, in the upper part of the vessel, is where the supercritical mixture is injected downward through a nozzle. In this supercritical region oxidation takes place. The gaseous effluents undergo a reversal in flow direction and leave the reaction vessel at the top. Insoluble material, *e.g.*, inorganic salts, precipitates in the low temperature zone at the bottom of the high pressure vessel. Cold water is injected to dissolve soluble salts and form a slurry of the insoluble materials. The brine solution or slurry is removed through a discharge nozzle. A "cold" falling

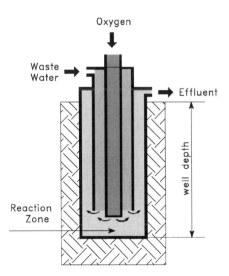

Figure 7.8   Scheme of deep-shaft wet-air oxidation reactor.

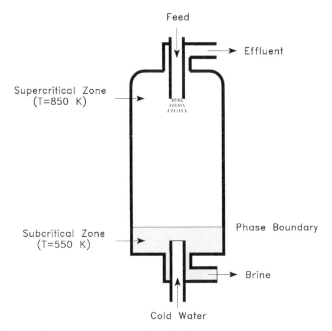

Figure 7.9 Scheme of the MODAR high pressure reactor vessel.

water film (approximately 590 K) is used to prevent salts from precipitating on the walls of the reaction vessel and to reduce corrosion. The applied temperatures are up to a maximum of 893 K, and the pressure is about 23.5 MPa. Reactors have been built on the laboratory scale (internal diameter 71 mm) and on the pilot plant scale (internal diameter 254 mm). These have been applied to a wide range of model substances and wastes, especially those forming salts and solid particles.

The simplest method for SCWO is to use a tubular (pipe) reactor with good flow properties. The length and the inner diameter of the tube are variable. Pipe reactors are suitable for all wastes which do not form solid accumulations on the walls of the tube like sticky salts, which may plug the reactor. If necessary it is very easy to replace a tube and because of the easy handling, tubular setups are widespread.

A new development is a transpiring wall reactor (or platelet reactor) which can overcome corrosion and heat transfer problems (Figure 7.10). Salt is prevented from adhering to the reactor wall and supercritical water provides a protective boundary layer against corrosion through a permeable reactor wall [173, 174]. Transpiring walls have been used for many years in gas turbine engines and rocket engines to obtain high combustion temperatures of about 3500 K. For SCWO application, the waste/water/oxygen mixture is injected at the top of the reactor inside the transpiring wall. Water enters the annular space between the reactor wall and the transpiring wall, and then passes into the inner reaction chamber through the pores of the transpiring wall. It provides cooling

7.3. ENVIRONMENTAL TECHNOLOGY

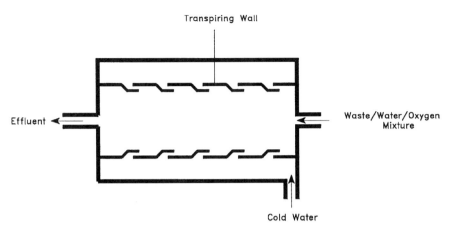

Figure 7.10 Scheme of the transpiring wall reactor.

of the transpiring wall and flushes away solid particles preventing them from sticking to the wall. Because of the lower wall temperatures, corrosive products are diluted and therefore corrosion is reduced. The hot reaction products leave the transpiring wall reactor and are quenched in a transpiring wall cooler. The brines are diluted and the quenched stream is separated into two phases. The advantage is that the reactor walls are protected from corroding substances and also from high temperatures. Therefore, cheaper materials, for example stainless steel, can be used for construction purposes. The transpiring pipe may be made of porous ceramic material or of platelet devices (*e.g.*, ceramics, stainless steel, inconel) by diffusion bonding of a stack of thin plates. A new technical concept is now being developed at the Forschungszentrum Karlsruhe für Technik und Umwelt, Institut für Technische Chemie [175].

Another approach to solve corrosion problems has been reported by Batelle, Pacific Northwest Laboratories [176]. A dual shell pressure balanced vessel is used with a high pressure wall outside and a second thin shell inside for a pressure barrier made from stainless steel. Sensors in the pressure balancing fluid between wall and shell detect by conductivity measurements when the shell is breached. A quick replacement is possible.

Burners for Flame Forming Hydrothermal Oxidation Another approach to SCWO is flame forming hydrothermal combustion of organic compounds in supercritical water. The first attempt to make a hydrothermal burner was made by Franck and Schilling at the University of Karlsruhe. Oxygen-methane flames in a supercritical water environment were formed batchwise in a burner consisting of a slit nozzle of 0.1 mm to 0.5 mm width. Oxygen is injected into a methane-water mixture at high pressures and temperatures forming a hydrothermal flame. Another possibility is to use a concentric slit nozzle where methane is injected through an inner capillary, and oxygen is injected through an outer tube

[135, 177, 177a]. In these experiments, the highest applied pressure was 200 MPa, the preheating temperature of the mixture was 770 K, and the flame temperatures were between 3000 and 4500 K, determined theoretically and experimentally [177a, 178, 179].

The same principle has been employed at Sandia National Laboratories, Livermore, California, to investigate methanol and methane flames formed by injecting oxygen into the preheated hydrocarbon-water mixture using Raman Spectroscopy. The temperatures ranged between 650 and 780 K and the pressure was kept constant at 27.5 MPa [180].

A new film-cooled coaxial hydrothermal burner for flame forming oxidation reactions is being used at the Institute for Process Engineering and Cryogenics, Swiss Federal Institute of Technology, for continuous flow operation over several hours [181]. Two coaxial tubes form the burner, with waste water entering through the center tube, and the oxidizing agent entering through the outer tube. Initiated by self-ignition, a hydrothermal flame forms, oxidizing the organic contents of the waste water. Similar to the operation of rocket engines, the walls of the combustion reactor are cooled by a thin water film. A circulation zone stabilizes the reaction by back-mixing a part of the hot reaction products with the cold reactants (Figure 7.11). The burner is installed in a high pressure optical cell with an inner diameter of 20 mm and sapphire windows with a diameter of 18 mm. The autoclave is designed to operate at 42 MPa and 873 K.

Process Design and Development In order to study a practical SCWO process, it is necessary to build a facility which includes all units for its operation, *i.e.*,

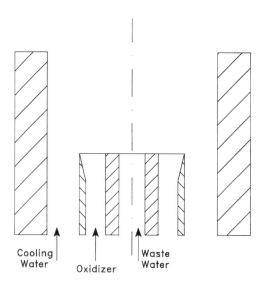

Figure 7.11 Scheme of film-cooled coaxial hydrothermal burner.

7.3. ENVIRONMENTAL TECHNOLOGY

feeding, pressurization, heat transfer, phase separation, process control and monitoring, and safety controls. A schematic presentation of the SCWO plants operated at the Forschungszentrum Karlsruhe für Technik und Umwelt [182] is shown in Figure 7.12 as an example of a complete arrangement for SCWO. In this example, SCWO occurs in a tube reactor heated by a fluidized sand bath. Similar plants are operated at Los Alamos National Laboratories [157, 158] and MIT [151]. The setup can be taken as a general example, where the central reactor unit is exchangeable. For a flexible operation, three feed devices are installed. The gas is supplied by a high pressure compressor fed with atmospheric air. Water and "fuel" (organic feed) are fed into the plant by high pressure pumps. The fluids can be preheated and are mixed in a first step, before the oxidizing agent (oxygen from air) is added in a second mixer at the entrance to the tube reactor. The mixer steps may be exchanged, *e.g.*, first mixing of gas and water, then mixing with waste. Exothermal oxidation reactions of organic substances commence immediately, as indicated by a rise in temperature along the tubular reactor together with formation of carbon monoxide and carbon dioxide in the product gas phase. The product phase is cooled after leaving the tube reactor and the pressure is reduced to standard conditions by passage through a back pressure regulator. The composition of the gas phase (oxygen, carbon dioxide, carbon monoxide) can be analyzed by an on-line spectrometer. The liquid (water) phase is sampled to determine residual total organic carbon (TOC). Special devices must be added for salt separation or neutralization.

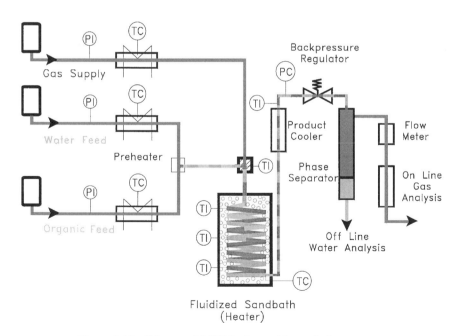

Figure 7.12 Scheme of SCWO plant with a tubular reactor.

A computer system monitors and controls the process and stores all necessary data.

Current SCWO Activities Several plants of different sizes have been designed, constructed, developed, and partly operated. The 1st International Workshop on Supercritical Water Oxidation was held at Amelia Island, Florida, in February 1995 in order to give an overview of research activities concerning SCWO. The present state of knowledge on SCWO may be summarized as follows:

In Europe several projects concerning the destruction of hazardous waste by SCWO processes are underway. In Germany, at the Institut für Technische Chemie, Forschungszentrum Karlsruhe für Technik und Umwelt, two plants are being operated for supercritical water oxidation experiments [183]. They are based on the same principles (Figure 7.12). One is a bench-scale while the other is a lab-scale setup. The bench-scale plant is scaled up by a factor of 10 over the lab-scale plant with a capacity of 1.5 L/h, several model substances such as ethanol, n-hexane, cyclohexane, benzene, toluene, nitrobenzene and xylene have been oxidized successfully with a destruction efficiency of 99.5% or higher. In order to find the optimum operation conditions, temperature, flow rate, and oxygen content were varied. The oxidizing agent is air from the atmosphere. The efficiency of the oxidation was determined by the analysis of TOC in the product water phase, and by an on-line analysis of carbon dioxide and carbon monoxide in the product gas phase [182].

Experiments with model substances are very important as a first step to achieve the efficient destruction of actual organic wastes. These usually consist of many components of different types, and they can behave differently during the course of oxidation. Two examples, ethanol and n-hexane, are discussed below in order to demonstrate differences in their processing.

During the ethanol oxidation experiments, the following parameters were varied: temperature (623–823 K); organic feed (30–50 g/h at constant air feed of 430 L/h, implying a stoichiometry of between 1.2 and 2.0); and flow rate of the water (300–1750 g/h). The pressure was held constant at 24 MPa. Figure 7.13 shows the carbon monoxide and carbon dioxide production and Figure 7.14 the TOC values as a function of bath temperature for an ethanol feed of 30 g/h. Under these conditions the formation of carbon monoxide could be completely suppressed. The lowest TOC-values of 0.02% in the water phase were found at 673 K and low water flows of 300–450 g/h at a pressure of 24 MPa. Thus, the percentage of ethanol oxidation is at least 99.98%. For 40 g/h and 50 g/h ethanol, higher carbon monoxide values are found and the residual TOC is higher. The qualitative GC-MS analysis of TOC in the water phase indicated the presence of acetone, acetic acid and propionic acid and the absence of ethanol.

n-hexane exhibits a completely different behavior from ethanol. Figures 7.15 and 7.16 show the results for carbon monoxide and carbon dioxide formation and TOC for an air supply of 430 L/h and an organic feed of 30 g/h. The curves show reverse slopes in contrast to those for ethanol. Much higher temperatures (more than 773 K) are required to obtain a percentage of

7.3. ENVIRONMENTAL TECHNOLOGY

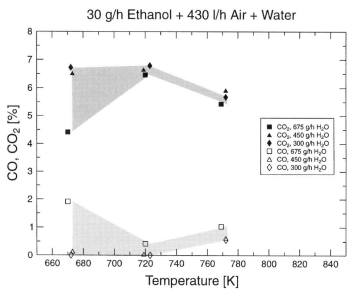

Figure 7.13 Temperature dependence of carbon dioxide and carbon monoxide formation for ethanol oxidation; trends are shaded.

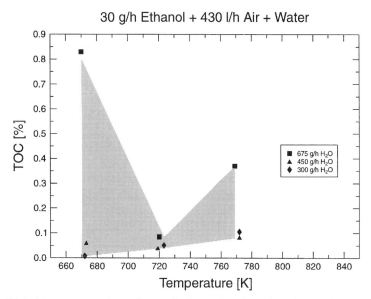

Figure 7.14 Temperature dependence of TOC values for ethanol oxidation; trends are shaded.

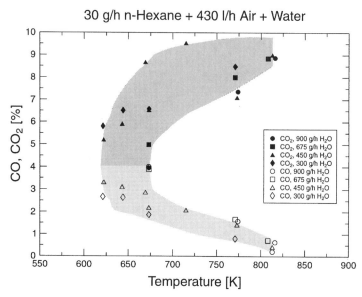

Figure 7.15 Temperature dependence of carbon dioxide and carbon monoxide formation for *n*-hexane oxidation; trends are shaded.

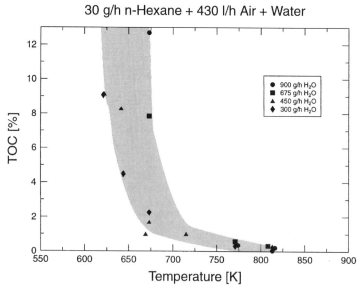

Figure 7.16 Temperature dependence of TOC values for *n*-hexane oxidation; trends are shaded.

7.3. ENVIRONMENTAL TECHNOLOGY

oxidation better than 99.5%. The lowest TOC (best conversion) was found at a bath temperature of 813 K, and carbon monoxide formation was never completely suppressed. The lowest carbon monoxide values together with lowest TOC residues were found for a water feed of 900 g/h. High TOC values in the liquid phase were the result of the formation of soot, phenols and polyaromatic compounds.

The difference in behavior of these two examples demonstrates that it is crucial to know how individual substances react during SCWO in order to apply the optimal conditions for the oxidation of real wastes.

The tube reactor used in these experiments has a length of about 6 meters (inner diameter 2.1 mm). The fluidized sandbath provides a good high temperature thermostat and is controlled by a thermocouple at the end of the tube reactor. Temperature measurement and control is made by thermocouples along the tube. It is possible to observe the progress of the exothermal oxidation reaction by a temperature rise along the tube. Two typical axial temperature profiles are shown in Figure 7.17 for n-hexane and in Figure 7.18 for toluene. The highest temperature peaks correspond to low TOC values and hence to a high percentage of conversion. For n-hexane oxidation (Figure 7.17), the reaction at a bath temperature of 813 K starts immediately after mixing of all three components. The temperature peaks are located 5 cm behind the mixer. The lowest water feed rates generate the highest reaction temperature and the lowest residual TOC value. For higher water flow rates, the peak of the temperature profile decreases and TOC increases. The same pattern is found for

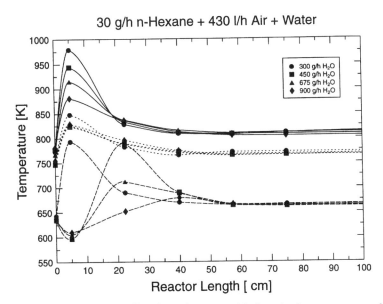

Figure 7.17 Temperature profiles for n-hexane oxidation, bath temperatures 813 K, 773 K, and 673 K.

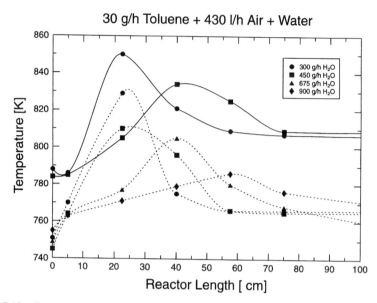

Figure 7.18 Temperature profiles for toluene oxidation, bath temperatures 813 K, 773 K, and 673 K.

bath temperatures of 733 and 673 K, at a water feed rate of 300 g/h. With increasing water flow rate, the maximum in temperature shifts further downstream along the tubular reactor [182].

In the case of toluene (Figure 7.18), with an air supply of 430 L/h, all temperature peaks are shifted downstream when compared with the observations for n-hexane. For higher water feed rates, the temperature peaks are flattened and also shifted further downstream.

The bench-scale plant (15 L/h) commenced operation in early 1995. First experiments with model substances are now completed and confirm the results obtained from the lab-scale plant. Real wastes from pharmaceutical and other industries are now being treated. The bench-scale pipe reactor with a length of 15 m exhibits an adequate oxidation efficiency, but modifications of the reactor design are necessary to overcome problems related to salt precipitation [175].

Another approach being investigated experimentally at the Research Center Karlsruhe is to combine the Supercritical Fluid Extraction (SFE) of wastes (Section 7.3.2) with the SCWO. In the case of pollutants, which are not suitable for re-use, it could be advantageous to destroy them by an integrated SFE/oxidation process. Initial experiments have been carried out using carbon dioxide instead of water as the extraction and reaction medium. The oxidation reaction proceeds readily and the destruction efficiencies are comparable with those in supercritical water [100, 184].

At the Fraunhofer Institut für Chemische Technologie (Pfinztal, Germany), experiments using batch reactors with a capacity of 250 ml are being undertaken

7.3. ENVIRONMENTAL TECHNOLOGY

to destroy explosives such as TNT, DNT, and NDT-3. Destruction efficiencies of > 97% are achieved at residence times of 30 min in the reactor [183, 185, 186]. The MODEC Company has constructed a SCWO facility consisting of six single modules at the Fraunhofer Institute site. Operation was begun in 1994. The straight tube reactor, with a length of 200 meters, is operated adiabatically in a temperature range of 823–873 K at a pressure of 25 MPa. No additional energy input is necessary to maintain the reaction once it has commenced. The oxidizing agent is pure oxygen. The reaction energy is recovered to compensate for heat loss along the reactor tube. High flow rates keep particles moving and prevent their settling on the walls. In the colder parts of the reactor, salts dissolve again as the temperature decreases. A consortium of pharmaceutical companies provides organic process wastes containing carbon, hydrogen, and sulfur. Products from the oxidation are water, carbon dioxide, completely oxidized metals, and inorganic salts. The carbon dioxide produced by the oxidation reaction is separated in a last step. The capacity is 90 L/h, corresponding to 10 kg of dry waste. Pulp and paper wastes, pharmaceutical effluents, and municipal wastes can be treated in the MODEC plant [139, 187].

At the Institut für Technische Chemie, University of München, Tiltscher and coworkers have designed a continuously-operated bench-scale SCWO plant to treat waste that cannot be destroyed by biological systems. The plant capacity is 50 L/h. Oxidation studies have been performed for pyridine, phenol, triethylamine, diethylsulfide, nitrobenzene, 4-chlorophenol, and thiophene-2-carbonic acid with hydrogen peroxide as the oxidizing agent. Destruction efficiencies are between 80% and almost 100% [188, 189].

At the Institute for Process Engineering and Cryogenics, Swiss Federal Institute of Technology, the film-cooled coaxial hydrothermal burner setup (which has been described above) is operated in a high pressure plant. Oxygen as oxidizing agent and water for cooling and a supercritical medium are used. Methane and methanol are oxidized as organic model substances by flame formation. The feed concentrations of methane or methanol are a maximum of 25%. The mixtures are heated until self-ignition occurs. The flames can be observed through windows and are recorded on video. The maximum reaction temperatures are above 1300 K. In the case of methane oxidation, the flames die when the temperature drops below the ternary phase boundary of the water/methane/oxygen system at about 370 K. With methanol injection (25%), ignition temperatures below 370 K are possible. Oxidizing efficencies between 80 and 99% have been achieved [181].

SCWO activities in the United States are widespread. Department of Energy (DOE) Institutions generate mixed nuclear and chemical waste containing organic substances at 30 sites. Oxidation in supercritical water provides a possible way to destroy organic materials in the mixed waste. At the Idaho National Engineering Laboratory, for example, a pilot plant test bed has been set up to test different reactor types for a throughput of about 200 L/h with a maximum temperature of 900 K and a pressure of 34.4 MPa. Simulated wastes have been oxidized successfully [190]. Further research and development activities are planned in order to optimize the SCWO pilot plant at Idaho Falls [183, 191]. The large scale operation of the plant for treatment of organic wastes is scheduled for mid 1996, and an operation for the treatment of a mixed waste plant for 1999. In the interim, more basic research on corrosion, electrochemical sensor development, *in situ* neutralization and system modeling are planned, in collaboration with several companies and research institutes [191].

The DOE (Department of Energy) and the DOD (Department of Defence) have a joint project to treat the U.S. Navy hazardous waste such as paint, solvent degreasing waste, adhesives, oil, halogenated organic solvents, and PCB at the U.S. Navy Facilities Engineering Service Center, Port Hueneme, California. This facility has a capacity of about 10 L/h with an operating temperature of 923 K and a pressure of 23 MPa. The residence time for the reaction is about 10 seconds [140, 183]. Another collaboration between the DOE, the DOD, the GNI-Group Inc., MODAR, and Stone & Webster aims to destroy propellants and munitions by SCWO. At Deer Park, Texas, a pilot plant which has a capacity of about 800 L/h is under construction [139, 183]. The DOD has unneeded chemical warfare agents and solid propellants which can also be oxidized by the SCWO process. Two pilot plants have been constructed, with General Atomics as partner, with throughputs of 200 L/h and 100 L/h, respectively, and a temperature of 923 K and pressure of 30 MPa. The water feed contains 5 weight percent of organic materials and 12 weight percent of inert solids. The tube reactor is lined with titanium to avoid corrosion [167, 192]. In another joint project, Sandia National Laboratory, Livermore, California, and Foster Wheeler Development Corporation, Livingston, New Jersey, have planned a new plant with a feed capacity of about 3500 L/h at Picatinny Arsenal, New Jersey. Experiments, which are necessary for the development of the new plant, will be carried out in a pilot plant with a throughput of about 350 L/h (scale-down factor of 10) at Pine Bluff Arsenal, Arkansas. The objective is the destruction of pyrotechnic colored smoke and dyes. A platelet reactor is used with a reaction temperature of 823 K, a pressure of 27 MPa, and a residence time of about 15 s in the transpiring wall tube. Operation will start in late 1996, when all necessary preparatory experiments have been completed. The DOE jointly funds the development of the platelet reactor [173, 183]. The NIST is providing support in the form of thermophysical data, optical measurements for chemical reactions, process control and monitoring, tests on multiple reactor configurations, and destruction efficiency maps for customers such as the DOD, the DOE, the MIT and Huntsman Chemical [163].

The first commercial use of SCWO technology has recently commenced, following many years of research activities at the University of Texas, in the group of E. F. Gloyna, in Austin, Texas. The partners are ECO Waste Technologies, Huntsman Chemical, and the University of Texas [139, 183, 193]. The throughput capacity is about 1150 L/h. Organic wastes of long-chain alcohols, glycols, and amines at a concentration of 10–15 wt. % in the water feed are oxidized. The product phase contains water, oxygen, carbon monoxide, and carbon dioxide. The destruction efficiencies are > 99.5% and the liquid effluent can be released to the municipal sewer. The technology is not suited to waste water containing inorganic components [139, 183].

In Canada, NORAM Engineering and Constructors in Vancouver produce pulp and paper waste which they plan to destroy using a SCWO plant with a capacity of about 200 L/h and a feed containing 10% solids, in collaboration with MODEC. The costs are estimated to be 10% lower than for usual disposal [183]. The destruction efficiencies by SCWO are about 99.999% instead of 95% by subcritical wet oxidation [139].

In Japan, the first SCWO pilot plant has been constructed by Organo Corporation for the purpose of oxidizing halogenated organic substances, detergents, developing solutions, and wastes from chemical processes. A MODAR vessel concept is used, and the start of operation was expected for mid 1995 [194].

Combined Extraction and Oxidation in Supercritical Water Supercritical water can also be utilized for the decontamination of soil. Organic pollutants (*e.g.*, PAH, oil) can be washed out of the soil by supercritical water as solvent. In a second step the extracted organic substances can be destroyed by a SCWO process. At the University of Hamburg-Harburg [25] and at the Department of Chemical Engineering of the University of Akron, Ohio [183], exploratory experiments have begun.

In conclusion, when considering the present state of SCWO technology development, two major problems have still to be solved in order to realize a reliable SCWO process. It is necessary to find corrosion resistant materials for the treatment of feeds containing corrosive compounds and plugging of the high pressure process equipment by solid precipitation must be prevented. There are many positive approaches to solve these problems. The advantages of SCWO processes are:

- Complete oxidation with a high space-time-yield.
- No NO_x formation at the SCWO reaction temperatures.
- No expensive off-gas treatment, since heteroatom compounds are precipitated as salts.
- Efficient heat recovery.
- Easy separation of carbon dioxide.

Considering these advantages, it is clear that continued research and development into SCWO technology as an alternative method for destroying hazardous wastes is desirable and necessary. Numerous other attractive projects have been proposed for environmentally compatible processes using near critical or supercritical water as reaction medium. Catalytic gasification is one of these processes described in the following section.

Conversion of Biomass and Waste Successful processes have been described in the literature in which subcritical and supercritical water are used as reaction media and as a reactant for the conversion (liquefaction and gasification, *i.e.*, for partial oxidation) of biomass and waste. This procedure differs from the classical (coal) gasification process mainly in the following characteristics:

- Higher pressure, above 20 MPa instead of 8 MPa or less.
- Lower temperature, 350–600 °C instead of 1000 °C and above.
- No addition of oxygen.
- Use of catalysts.

Surprisingly high reactivities of organic material, especially carbohydrates, are observed in high-pressure water at temperatures above 250 °C. Syskin and Katritzky [124], in a review in 1991, summarized the findings in a number of papers on this subject as several conclusions:

- Water can act as a highly efficient acid or base catalyst. The reactions considered may be accelerated further by additions of acidic or basic materials, such as clay or carbonates.
- Ion chemistry seems to dominate, opening up reaction pathways which differ from those in thermal radical reactions.
- The reactions may be autocatalyzed by dissolved reaction products.

The early experiments include studies by Modell [125], in which glucose and wood flour were thermolyzed in supercritical water. High carbon conversions, but only low gas yields (methane etc.), were found under these conditions.

Since the late seventies, many pertinent experiments have been conducted at the Pacific Northwest Laboratory; the results were published in a review [28]. A large number of model substances (ethanol, aromatic and polyaromatic compounds, chlorinated hydrocarbons, and inorganic compounds) as well as real wastes from the chemical and food industries, and wood were used as reactants. The experiments were conducted in the batch mode, but also in the continuous flow mode, mostly at subcritical temperatures and at pressures mainly between 17 and 23 MPa. In the case of cellulose and similar materials, mostly liquid reaction products arise in the low temperature range (between 250 and 300 °C) in the absence of a catalyst and at moderate carbon conversion, while, at higher temperatures and especially with the assistance of catalysts (alkali carbonates), more than 99% conversion at high gas yields (methane, carbon dioxide, and small quantities of hydrogen) can be achieved.

During pyrolysis, alkali metals seem to display catalytic activities in the liquid phase. This was also found for the steam reforming reaction ($C + H_2O \rightarrow CO + H_2$) and for the water-gas shift reaction ($CO + H_2O \rightarrow CO_2 + H_2$). On the other hand, in steam reforming and in methane production ($CO + 3H_2 \rightarrow CH_4 + H_2O$), nickel and ruthenium catalysts display some activity in hot high-pressure water, while platinum, palladium, and copper cannot be used under these conditions. Aluminium oxide and zirconium oxide may be used as catalyst supports. For *p*-cresol and methylisobutylketone as model substances conversion rates in excess of 99% were found in a continuous flow experiment at a methane concentration of roughly 60% in the product gas, and a residence time of 10 minutes. The methane yield was higher at 350 than at 400 °C. Carbon and tar formation was not observed in most experiments, or observed only to a small extent in some experiments. The least reactive material was found to be lignin and lignin-containing material (wood).

Two problems for which no satisfactory solutions seem to have been found yet are the lifetime of the catalysts studied (including commercial catalysts), and the relatively low space-time yield. In most experiments, the residence times were between ten minutes and approximately two hours. However, even without the application of catalysts, high conversions and gas yields (hydrogen, carbon dioxide, carbon monoxide, and methane) can be achieved for carbohydrates in supercritical water at temperatures of around 600 °C and a pressure of 34.5 MPa at a residence time of only 30 s [126]. In the experiments conducted with miniature reactors, however, the reactor wall seems to play the role of a catalyst.

Note added in proof:

In order to demonstrate the scale-up of laboratory and bench scale parameters to larger dimensions, experiments in an industrial plant of Messer–Griesheim in Krefeld, Germany, were performed. The plant consists of two 200 l extraction vessels continuously streamed through by carbon dioxide. In two separation vessels of 20 and 50 dm^3 volume the loaded CO_2 flow is depressurized totally to gas and the oil is precipitated. Since the flow rate in the laboratory scale experiments was found to have no effect on the extraction efficiency, the runs were performed with high flow rates of about 800 to 1200 kg/h at pressures up to 40 MPa. Because of the large amount of oil, the extraction had to be interrupted from time to time to discharge the oil from the vessels. Several batches of different materials were tested at pressures between 30 and 40 MPa. In all cases the extraction parameters required for complete de-oiling, obtained from the laboratory experiments, were found also to be valid for a plant of industrial size.

7.4 REFERENCES

[1] Cagniard de la Tour, Annales de Chimie et de Physique, 21 (1822) 127.
[2] Andrews, T., Phil. Trans. Royal Soc., 159 (1869) 575.
[3] Hannay, J. B., Proc. Royal Soc., 29 (1879) 324.
[4] Villard, P., J. Chem. Soc., 68, part II (1895) 255.
[5] Büchner, E. H., Z. f. Phys. Chemie, 54 (1906) 665.
[6] Paul, P. F. M., Wise, W. S., The Principles of Gas Extraction, Mills and Boon, London, 1971.

[7] Irani, C. A., Funk, E. W., Separations using Supercritical Gases, in Recent Developments in Separation Science, CRC-Press, Palm Beach, Vol. 3, Part A (1977) 171–193.
[8] Zosel, K., Angew. Chem., 90 (1978) 748; Angew. Chem. Intl. Ed., 17 (1978) 702.
[9] Zosel, K., German Patent, 200 52 93 (1970).
[10] Brunner, G., Gas Extraction, Steinkopf Darmstadt - Springer New York, 1994.
[11] McHugh, M. A., Krukonis, V. J., Supercritical Fluid Extraction, Butterworths, Boston, 1986.
[12] Dohrn, R. G., Fluid Phase Equilibria, 106 (1995) 213.
[13] Sadus, R. J., High Pressure Phase Behavior of Multicomponent Fluid Mixtures, Elsevier, Amsterdam, 1992.
[14] Lentz, H., Franck, E. U., Angew. Chem. Int. Ed. Engl., 17 (1978) 728.
[15] Schneider, G. M., Angew. Chem. Int. Ed. Engl., 17 (1978) 716.
[16] Franck, E. U., Fluid Phase Equilibria, 10 (1983) 211.
[17] Schneider, G. M., Fluid Phase Equilibria, 10 (1983) 141.
[18] Krukonis, V., Brunner, G., Perrut, M., Proceedings 3rd International Symposium on Supercritical Fluids, Vol. 1, 1–22, Strasbourg, Oct. 17–19, 1994.
[19] Moses, J. M., de Filippi, R. P., Proceedings International Solvent Extraction Conf., Denver, Col., 1983.
[20] Coenen, H., Rinza, P., Techn. Mitteil. Krupp, Werksber. 39 (1981).
[21] King, M. B., Bott, T. R. (eds.), Extraction of Natural Products using Near-Critical Solvents, Blackie Academic and Professional, Glasgow, 1993.
[22] McHugh, M., in [11], pp 333–365.
[23] Palmer, M. V., Ting, S .S. T., Food Chemistry, 52 (1995) 345.
[24] Schneider, G. M., (Kiran, E., Sengers, J. M. H. (eds.)), Supercritical Fluids, page 739, Kluwer Academic Publishers, Dordrecht, 1994.
[25] Brunner, G. H., (Kiran, E., Sengers, J. M. H. (eds.)), Supercritical Fluids, page 653, Kluwer Academic Publishers, 1994.
[26] Clifford, A. A., (Kiran, E., Sengers, J. M. H. (eds.)), Supercritical Fluids, page 449, Kluwer Academic Publishers, 1994.
[27] Modell, M., US-Patent, 4,338,199 (1982).
[28] Sealock, L. J. et al., Ind. Eng. Chem. Res., 32 (1993) 1535.
[29] Williams, D. F., Chem. Eng. Sci., 36 (1981) 1769.
[30] Kühn, R., Birett, K., Merkblätter gefährlicher Arbeitsstoffe, Blatt K-31, Kohlendioxid, Ecomed Verlag, München, 1994.
[31] Stahl, E., Quirin, K. W., Gerard, D. (eds.), Verdichtete Gase zur Extraktion und Raffination, Springer-Verlag, Heidelberg, 1987.
[32] Sanders, N., in [21], page 34.
[33] Kiran, E., Sengers, J. M. H. (eds.), Supercritical Fluids, Kluwer Academic Publishers, Dordrecht, 1994.
[34] Stahl, E., Schilz, W., Chem.-Ing.-Tech., 48 (1976) 773.
[35] Stahl, E., Schilz, W., Schütz, E., Willing, E., in [36], page 93.
[36] Schneider, G. M., Stahl, W., Wilke, G. (eds.), Extraction with Supercritical Gases, Verlag Chemie, Weinheim, 1980.

7.4. REFERENCES

[37] Wilke, G., Angew. Chem., 90 (1978) 747; Angew. Chem. Intl. Ed., 17 (1978) 701.
[38] Meeting of the SCI Food Group Food Engineering Panel, London, Feb. 4., 1982, Chemistry and Industry, London, June 19., 1982, 385.
[39] Paulitis, M. E., Krukonis, V. J., Kurnik, R. T., Reid, R. C., Review in Chemical Engineering, 1 (1982) 179.
[40] Randall, L. G., Bowman, L. M., Supercritical Gases in Extraction and Chromatography, Sep. Sci. Tech., 17 (1982) 1.
[41] Proceedings 3rd International Symposium on Supercritical Fluids, Oct. 17–19, 1994, Strasbourg, France.
[42] Caralp, M. H. M., Clifford, A. A., Coleby, S. E., in [21], page 50.
[43] Debenedetti, P. G., in [41], Vol. 3, page 213.
[44] Debenedetti, P., in [33], page 719.
[45] Gallagher, P. M. et al., (Johnston, K. P., Penninger, J. M. L. (eds.)), Supercritical Fluid Science and Technology, ACS Symposium Series, 406 (1989) 334.
[46] Weidner, E. et al., in [41], Vol. 3, page 229.
[47] Larson, K. A., King, M. L., Biotechnology Progress, 2 (1986) 73.
[48] Frederiksen, L. et al., in [41], Vol. 3, page 235.
[49] Tom, J. W., Lim, G.-B., Debenedetti, P. G., Prud'homme, R. K., Supercritical Fluid Engineering Science, ACS Symposium Series, 514 (1993) 238.
[50] Schollmeyer, E. et al., patent DP 3906724 A1 (1990).
[51] Saus, W., Knittel, D., Schollmeyer, E., Textil Praxis International, 47 (1992) 1052.
[52] Poulakis, K., Spee, M., Schneider, G. M., Knittel, G., Buschmann, H. J., Schollmeyer, E., Chemiefasern/Textilindustrie, 41./93. Jahrgang (1991) 142.
[53] Swidersky, P., Haarhaus, U., Tuma, D., Schneider, G. M., in [41], Vol. 1, 191.
[54] Saus, W., Hoger, S., Knittel, D., Schollmeyer, E., Textilveredelung, 28 (1993) 38.
[55] Gebert, B., Knittel, D., Schollmeyer, E., Melliand Textilberichte, 2 (1993) 151.
[56] Gallagher-Wetmore, P., in [41], Vol. 3, 253.
[57] Matson, D. W., Fulton, J. L., Petersen, R. C., Smith, R. D., Ind. Eng. Chem. Res., 26 (1987) 2298.
[58] Reverchon, E., Dons, G., Gorgoglione, D., J. Supercritical Fluids, 6 (1993) 241.
[59] Gallagher, P. M., Coffey, M. P., Krukonis, V. J., Hillstrom, W. W., J. Supercritical Fluids, 5 (1992) 130.
[60] Shishikura, A.,Takahashi, H., J. Supercritical Fluids, 5 (1992) 303.
[61] Dixon, D. J., Johnston, K. P., Bodmeier, R. A., AIChE J., 39 (1993) 127.
[62] Kiran, E., in [33], page 541.
[63] Scholsky, K. M., J. Supercritical Fluids, 6 (1993) 103.
[64] McHugh, M. A., in [11], page 189.
[65] Buback, M., in [41], Vol. 3, page 93.
[67] Kleintjens, L. A., in [33], page 589.
[68] Newman, D. A., Hoefling, T. A., Beitle, R. R., Beckman, E. J., Enick, R. M., J. Supercritical Fluids, 6 (1993) 205.
[69] Hoefling , T., Stofesky, D., Reid, M., Beckman, E., Enick, R. M., J. Supercritical Fluids, 5 (1992) 237.

[70] DeSimone, J. M., Guan, Z., Elsbernd, C. S., Science, 257 (1992) 945.
[71] DeSimone, J. M., Maury, E. E., Guan, Z., Combes, J. R., McClain, J. B., Romack, T. J., Canelas, D. P., Betts, D. E., in [41], Vol. 3, page 277.
[72] Hoefing, T. A., Newman, D. A., Encik, R. M., Beckman, E. J., J. Supercritical Fluids, 6 (1993) 165.
[73] Modell, M., (Kolaczkowski, S. T., Crittenden, B. D. (eds.)), Management of hazardous and toxic waste in the process industries, 66, Elsevier Applied Science, Amsterdam, 1987.
[74] Groves Jr., F. R., Brady, B., Knopf, F. C., (Straub, C. P. (ed.)), Critical Reviews in Environmental Control, Vol. 15, 237, CRC Press, 1985.
[75] De Filippi, R. P., Chemistry and Industry, June 1982, 390.
[76] Andrews, A. T., Ahlert, R. C., Kosson, D. S., Environmental Progress, 9 (1990) 204.
[77] Dooley, K. M., Ghonasgi, D., Knopf, F. C., Environmental Progress, 9 (1990) 197.
[78] Erkey, C., Madras, G., Orejuela, M., Akgerman, A., Environ. Sci. Technol., 27 (1993) 1225.
[79] Hess, R. K., Erkey, C., Akgerman, A., J. Supercritical Fluids, 4 (1991) 47.
[80] Akgerman, A., (Sawyer, D. T., Martell, A. E. (eds.)), Industial Environmental Chemistry, 153, Plenum Press, New York, 1992.
[81] Lee, M. L., Markides, K. (eds.), Analytical Supercritical Fluid Chromatography and Extraction, Chromatography Conferences Inc., Provo, Utah, U.S.A., 1990.
[82] Wenclawiak, B. (ed.), Analysis with Supercritical Fluids, Springer-Verlag, Heidelberg, 1992.
[83] Dean, J. R. (ed.), Application of Supercritical Fluids in Industrial Analysis, Blackie Academic and Professional, Glasgow, 1993.
[84] Saito, M., Yamauchi, Y., Okuyama, T. (eds.), Fractionation by Packed-Column SFC and SFE, VCH Verlag, Weinheim, 1994.
[85] Luque de Castro, M. D., Valcarcel, M., Tena, M. T., Analytical Supercritical Fluid Extraction, Springer-Verlag, Heidelberg, 1994.
[86] Markides, K. (ed.), European Symposium on SFC and SFE, Proceedings 1st Conf.: Dec., 4–5, Wiesbaden, Germany, 1991; Riva del Garda, Sandra, P., Markides, K., (eds.), Proceedings 2nd Conf., May 27–28, 1993.
[87] Sielschott, W., KFA-Bericht no. 2624 (1992).
[88] Cassat, D., Perrut, M., Proceedings 2nd International Symposium on Supercritical Fluids, Nizza, Vol. 2, 771, 1988.
[89] Markowz, G., Subklev, G., in [41], Vol. 2, page 505.
[90] Schulten, H. R., Schnitzer, M., Soil Sci. Soc. Am. J., 55 (1991) 1603.
[91] Monin, J. C., Barth, D., Perrut, M., Espitalié, M., Durand, B., Org. Geochem., 13 (1988) 1079.
[92] Michel, S., Schulz, S., Chem.-Ing.-Tech., 64 (1992) 194.
[93] Lütge, C., Reiß, I., Schleußinger, A., Schulz, S., J. Supercritical Fluids, 7 (1994) 265.
[94] Kunert, D., Lütge, C., Schleußinger, A., Schulz, S., Chem.-Ing.-Tech., 66 (1994) 692.
[95] Lütge, C., Oswald, D., Schleußinger, A., Schulz, S., Terra Tech., 3 (1993) 80.
[96] Low, G. K. C., Duffy, G. J., Sharma, S. D., Chensee, M. D., Weir, S. W., Tibbett, A. R., in [41], Vol. 2, page 275.

7.4. REFERENCES

[97] Low, G. K. C., personal communication.

[98] Novak, K., Brunner, G., Proceedings 2nd International Symposium on Supercritical Fluids, Boston, 1991.

[99] Novak, K., Brunner, G., in [41], Vol. 2, page 223.

[100] Dahmen, N., Schön, J., Schwab, P., Wilde, H., Schmieder, H., KFK-Nachr., 26 (1994) 128.

[101] Schmitt, W. J., Reid, R. C., Chem. Eng. Comm., 64 (1988) 155.

[102] Kißler, H., "Praxis Forum" Oberflächentechnik, 1995, 113.

[103] De Werbier, P., Flandin-Rey, Y., patent, IMPHY STE Metallurgique, France, Patent-Nr.: 2686351 (1992).

[104] Elgin, J. C., Weinstock, J. J., J. Chem. Eng. Data, 4 (1959) 1.

[105] Bernad, L., Keller, A., Barth, D., J. Supercritical Fluids, 6 (1993) 9.

[106] Brunner, G., Kreim, K., Ger. Chem. Eng., 9 (1986) 246.

[107] Lim, J. S., Lee, Y.-W., Kim, J.D., Lee, Y. Y., J. Supercritical Fluids 8 (1995) 127.

[108] Horizoe, H., Tanimoto, T., Yamamoto, I., Kano, Y., Fluid Phase Equilibria 84 (1993) 297.

[109] Abboud, O. K., de Filippi, R. P., Goklen, K. E., Moses, J. M., Final Report OE/CS/40258-Tl, Vol. 1 & 2, 1983.

[110] Ehntholt, D. J., Eppig, C., Thrun, K. E., Project Report prepared for Health Effects Res. Lab. Office Res. and Dev., U.S. Env. Prot. Agency, NC, Nov 1984.

[111] Ehntholt, D. J., Thrun, K. E., Eppig, C., Ringhand, P., International J. Environ. Anal. Chem., 13 (1983) 219.

[112] Krukonis, V., Phasex Corp., Report for DOE, DOE/PETC/TR-86/7, 1986.

[113] Ghonasgi, D., Gupta, S., Dooley, K. M., Knopf, F. C., J. Supercritical Fluids, 4 (1991) 53, 181.

[114] Hedrick, J. L., Mulcahey, L. J., Taylor, L. T., (Bright, F. V., McNally, M. E. P. (eds.)), Supercritical Fluid Technology, ACS Symposium Series, 488 (1992) 206.

[115] Brewer, S. E., Kruus, P., J. Environ. Sci. Health, B28 (1993) 671.

[116] Brewer, S. E., Kruus, P., Burk, B., in [41], Vol. 2, page 269 .

[117] Ghonasgi, D., Gupta, S., Dooley, K. M., Knopf, F. C., AIChE J., 37 (1991) 944.

[118] Yeo, S.-D., Akgerman, A., AIChE J., 36 (1990) 1743.

[119] Sengupta, S., Gupta, S., Dooley, K. M., Knopf, F. C., J. Supercritical Fluids, 7 (1994) 201.

[120] Knez, Z., Golob, J., Posel, F., Krmelj, I., Proceedings 2nd International Symposium High Pressure Chemical Engineering, Sept. 24–26, Erlangen, Germany, 1990, 243.

[121] McGovern, W. E., Moses, J. M., Ann. Massachusetts Hazard. Waste Source Reduction Conf. Proceedings 3, Worcester, Mass., Oct. 23, 1986.

[122] Chum, H. L., Fildardo, G., US patent, Int. Appl. No. PCT [US89 [00494, 1989.

[123] Roop, R. K., Agkerman, A., Ind. Eng. Chem. Res., 28 (1989) 1542.

[124] Siskin, M., Katritzky, A. R., Science, 254 (1991) 231.

[125] Modell, M., Reid, R. C., Amin, S., Gasification Process, U.S. Patent 4, 113, 446, 1978.

[126] Yu, D., Aihara, M., Antal, M. J., Energy Fuels, 7 (1993) 5.
[127] Heger, K., Uematsu, M., Franck, E. U., Ber. Bunsenges. Phys. Chem., 84 (1980) 758.
[128] Uematsu, M., Franck, E. U., J. Phys. Chem. Ref. Data, 9 (1980) 129.
[129] Marshall, W. L., Franck, E. U., J. Phys. Chem. Ref. Data, 10 (1981) 295.
[130] Anderko, A., Pitzer, K. S., Geochim. Cosmochim. Acta, 57 (1993) 1657.
[131] Anderko, A., Pitzer, K. S., Geochim. Cosmochim. Acta, 57 (1993) 4885.
[132] Sterner, S. M., Chou, I. M., Downs, R. T., Pitzer, K. S., Geochim. Cosmochim. Acta, 56 (1992) 2295.
[133] Franck, E. U., (van Eldik, R., Jonas, J. (eds.)), High Pressure Chemistry and Biochemistry, NATO ASI Ser. C, Vol. 197, 93, Reidel Publishing Company, Dordrecht, 1987.
[134] Franck, E. U., J. Chem. Thermodynamics, 19 (1987) 225.
[135] Franck, E. U., (McHugh, M. A. (ed.)), Proceedings 2nd International Symposium on Supercritical Fluids, pp 91–96, Boston, 1991.
[136] Shaw, R. W., Brill, T. B., Clifford, A. A., Eckert, C. A., Franck, E. U., C & EN, 12 (1991) 2.
[137] Modell, M., Gaudet, G. G., Simson, M., Hong, G. T., Biemann, K., Solid Wastes Management, 25 (1982) 26.
[138] Thiel, R., Dietz, K. H., Rosenbaum, H. J., Steiner, S., U.S. Patent No. 4.141.829 (1979).
[139] Svensson, P., Chemical Technology Europe, Jan./Feb. (1995) 16.
[140] Gloyna, E. F., Li, L., Waste Management, 13 (1993) 379.
[141] Tester, J. W., Webley, P. A., Holgate, H. R., Ind. Eng. Chem. Res., 32 (1993) 236.
[142] Killilea, W. R., Swallow, K. C., Hong, G. T., J. Supercritical Fluids, 5 (1992) 72.
[143] Tester, J. W., Holgate, H. R., Armellini, F. J., Webley, P. A., Killilea, W. R., Hong, G. T., Barner, H. E., (Tedder, D. W., Pohland, F. G. (eds.)), Emerging technologies in hazardous waste managements III, ACS Symposium Series, 518 (1992) Chapter 3, 35.
[143a] Savage, P. E., Gopalan, S., Mizan, T. I., Martino, C. J., Brock, E. E., AIChE Journal, 41 (1995) 1723.
[144] Yang, H. H., Eckert, C. A., Ind. Eng. Chem. Res., 27 (1988) 2009.
[145] Helling, R. K., Tester, J. W., Environ. Sci. Technol., 22 (1988) 1319.
[146] Holgate, H. R., Tester, J. W., (McHugh, M. A. (ed.)), Proceedings 2nd International Symposium on Supercritical Fluids, pp 177–180, Boston, 1991.
[147] Holgate, H. R., Webley, P. A., Tester, J. W., Helling, R. K., Energy Fuels, 6 (1992) 586.
[148] Holgate, H. R., Tester, J. W., J. Phys. Chem., 98 (1994) 800.
[149] Webley, P. A., Tester, J. W., Holgate, H. R., Ind. Eng. Chem. Res., 30 (1991) 1745.
[150] Webley, P. A., Tester, J. W., Energy Fuels, 5 (1991) 411.
[151] Webley, P. A., Tester, J. W., Energy Fuels, 5 (1991) 617.
[152] Li, R., Savage, P. E., Szmukler, D., AIChE J., 39 (1993) 178.
[153] Thornton, T. D., Savage, P. E., (McHugh, M.A. (ed.)), Proceedings 2nd International Symposium on Supercritical Fluids, Boston, 1991.

7.4. REFERENCES

[154] Chang, K.-C., Li, L., Gloyna, E. F., J. Hazardous Materials, 33 (1993) 51.
[155] Lee, D. S., PhD Thesis, University of Texas, Austin, August 1990.
[156] Lee, D. S., Gloyna, E. F., Li, L., J. Supercritical Fluids, 3 (1990) 249.
[157] Rofer, C. K., Streit, G. E., Report, Los Alamos National Laboratory, CA, October 1988.
[158] Rofer, C. K., Streit, G. E., Report, Los Alamos National Laboratory, CA, September 1989.
[159] Hirth, T., Franck, E. U., Ber. Bunsenges. Phys. Chem., 97 (1993) 1091.
[160] Hirth, T., Franck., E. U., Chem. Ing. Techn., 66 (1994) 1355.
[161] Hovey, J. K., Pitzer, K. S., Tanger, J. C., Bischoff, J. L., Rosenbauer, R. J., J. Phys. Chem., 94 (1990) 1175.
[162] Li, Y., Pitzer, K. S., J. Chem. Ind. Eng. China, 1 (1986) 249.
[163] Rosasco, G. J., Proceedings 1st International Workshop on Supercritical Water Oxidation, WCM Forums, Amelia Island, Florida, February 1995.
[164] Armellini, F. J., Tester, J. W., J. Supercritical Fluids, 4 (1991) 254.
[165] Leistikow, S., Proceedings 1st International Workshop on Supercritical Water Oxidation, WCM Forums, Amelia Island, Florida, February 1995.
[166] Boukis, N., Landvatter, R., Habicht, W., Franz, G., Leistikow, S., Kraft, R., Jacobi, O., Proceedings 1st International Workshop on Supercritical Water Oxidation, WCM Forums, Amelia Island, Florida, February 1995.
[167] Hazlebeck, D. A., Downey, K. W., Elliott, J. P., Spritzer, M. H., Proceedings 1st International Workshop on Supercritical Water Oxidation, WCM Forums, Amelia Island, Florida, February 1995.
[168] Garcia, K. M., Mizia, R., Proceedings 1st International Workshop on Supercritical Water Oxidation, WCM Forums, Amelia Island, Florida, February 1995.
[169] Hong, G. T., Ordway, D. W., Zilberstein, V. A., Proceedings 1st International Workshop on Supercritical Water Oxidation, WCM Forums, Amelia Island, Florida, February 1995.
[170] Smith, J. M., (Freeman, H.M. (ed.)), Standard Handbook of Hazardous Waste Treatment and Disposal, 8.137–8.151, McGraw Hill, New York, 1989.
[171] VerTech – ein neues, umweltverträgliches und wirtschaftliches Verfahren zur Klärschlammaufbereitung, Mannesmann Anlagenbau, p. 20, 1995.
[172] Barner, H. E., Huang, C. Y., Johnson, T., Jacobs, G., Martch, M. A., Killilea, W. R., J. Hazardous Materials, 31 (1992) 1.
[173] Ahluwalia, K. S., Young, M. F., Haroldsen, B. L., Mills, B. E., Stoddard, M. C., Robinson, C. D., Proceedings 1st International Workshop on Supercritical Water Oxidation, WCM Forums, Amelia Island, Florida, February 1995.
[174] McGuinness, T. G., Proceedings 1st International Workshop on Supercritical Water Oxidation, WCM Forums, Amelia Island, Florida, February 1995.
[175] Goldacker, H., Personal communication, Aug. 1995.
[176] Fassbender, A. G., Robertus, R. J., Devermann, G. S., Proceedings 1st International Workshop on Supercritical Water Oxidation, WCM Forums, Amelia Island, Florida, February 1995.
[177] Schilling, W., Franck, E. U., Ber. Bunsenges. Phys. Chem., 92 (1988) 631.

[177a] Franck, E. U., Wiegand, G., Polish J. Chem., 70 (1996) 527.
[178] Saur, A. M., Behrendt, F., Franck, E. U., Ber. Bunsenges. Phys. Chem., 97 (1993) 900.
[179] Pohsner, G. M., Franck, E. U., Ber. Bunsenges. Phys. Chem., 98 (1994) 1082.
[180] Steeper, R. R., Brown, M. S., Johnston, S. C., J. Supercritical Fluids, 5 (1992) 262.
[181] LaRoche, H. L., Weber, M., Trepp, C., Proceedings 1st International Workshop on Supercritical Water Oxidation, WCM Forums, Amelia Island, Florida, February 1995.
[182] Wiegand, G., Bleyl, H. J., Goldacker, H., Petrich, G., Schmieder, H., Proceedings 1st International Workshop on Supercritical Water Oxidation, WCM Forums, Amelia Island, Florida, February 1995.
[183] Moore, S., Samdani, S., Ondrey, G., Parkinson, G., Chem. Engineering, 3 (1994) 32.
[184] Bleyl, H.-J., Abeln, J., Boukis, N., Goldacker, H., Kluth, M., Kruse, A., Petrich, G., Schmieder, H., Wiegand, G., Proceedings of the 9th Symposium on Separation Science and Technology for Energy Applications, Gatlinburg, Tennessee, October 22–26, 1995, to be published in Sep. Sci. Tech.
[185] Bunte, G., Eisenreich, N., Hirth, T., Krause, H., Energetic Materials – Insensitivity and Environmental Awareness, Vol. 24, International Annual Conference of ICT, pp 70.1–70.11, Fraunhofer Gesellschaft, Karlsruhe, June 29–July 2, 1993.
[186] Hirth, T., Eisenreich, N., Krause, H., Bunte, G., Energetic Materials – Insensitivity and Environmental Awareness, Vol. 24, International Annual Conference of ICT, pp 47.1–47.11, Fraunhofer Gesellschaft, Karlsruhe, June 29–July 2, 1993.
[187] Kemna, A. H., Kuharich, E., Proceedings 1st International Workshop on Supercritical Water Oxidation, WCM Forums, Amelia Island, Florida, February 1995.
[188] Tiltscher, H., Forster, M., Brandes, C., Fill, C., Stocker, S., BayFORREST Berichtsheft, 1 (1992) 193.
[189] Tiltscher, H., Forster, M., Brandes, C., Fill, C., Stocker, S., Kliemas, H., BayFORREST Berichtsheft, 2 (1994) 189.
[190] Welland, H., Reed, W., Valentich, D., Charlton, T., Proceedings 1st International Workshop on Supercritical Water Oxidation, WCM Forums, Amelia Island, Florida, February 1995.
[191] Hart, P. W., Proceedings 1st International Workshop on Supercritical Water Oxidation, WCM Forums, Amelia Island, Florida, February 1995.
[192] Hazlebeck, D. A., Downey, K. W., Jensen, D. D., Spritzer, M. H., Proceedings 12th International Conference on the Properties of Water and Steam, Orlando, Florida, September 1994.
[193] McBrayer, R. N., Proceedings 1st International Workshop on Supercritical Water Oxidation, WCM Forums, Amelia Island, Florida, February 1995.
[194] Suzuki, A., Proceedings 1st International Workshop on Supercritical Water Oxidation, WCM Forums, Amelia Island, Florida, February 1995.
[195] Webley, P. A., Holgate, H. R., Stevenson, D. M., Tester, J. W., Technical paper, Ser. No. 901333, Soc. of Automotive Engineers, 1990.
[196] Wightman, T. J., M.S. Thesis, Civil Engineering Department, University of California, Berkeley, 1981.
[197] Crain, N., Gloyna, E. F., Separation Research Program Fall Conference, University of Texas, Austin, 1992.

8

ULTRASOUND AS A NEW TOOL FOR SYNTHETIC CHEMISTS

T. J. Mason* and J. L. Luche

School of Natural and Environmental Sciences, Coventry University, Priory Street, Coventry CV1 5FB, England

8.1 INTRODUCTION
    8.1.1 The Sound Ranges Used by Chemists
    8.1.2 Sonochemisty Equipment
    8.1.3 The Generation of Power Ultrasound
    8.1.4 The Transmission of Ultrasonic Energy
8.2 THEORETICAL ASPECTS
    8.2.1 The Chemists View on Cavitation
    8.2.2 Cavitation Effects in Different Types of Systems
    8.2.3 An Attempt to Define the Laws of Sonochemisty
    8.2.4 Parameters of Importance in Sonochemical Reactions
8.3 SONOCHEMICAL REACTIONS IN SYNTHESIS
    8.3.1 Homogeneous Reactions
    8.3.2 Heterogeneous Reactions with Non-metals
    8.3.3 Organometallic Sonochemisty
    8.3.4 Sonochemically Assisted Catalysis
8.4 CONCLUSIONS
8.5 REFERENCES

At the beginning of laboratory research into sonochemistry, the use of ultrasonic waves in chemistry was viewed as merely a convenient technique for dissolving materials and for starting the more recalcitrant types of reactions. Its development in the past 15 years, however, has revealed both its broad applicability and the scientific challenge involved in understanding its underlying physical phenomenon – acoustic cavitation [1]. Fortunately an absolute understanding of the physical principles has not stopped an ever expanding number of applications in synthesis which has made sonochemistry attractive to many experimentalists. Its usage has spread beyond academic laboratories into industry and chemical engineering and there are now a number of texts available for study [2–8]. In recent years a major collaborative effort aimed at understanding the principles of sonochemistry has brought together chemists, physicists, engineers and mathematicians in an alliance which speaks wonders for the future of international cooperation in this domain.

8.1 INTRODUCTION

The term *sono*chemistry is used to describe a subject in which sound energy is used to affect chemical processes. This terminology is in keeping with that of the longer established methods which use light (*photo*chemistry) and electricity (*electro*chemistry) to achieve chemical activation.

The potential of sonochemistry was identified over sixty years ago in a wide ranging paper entitled "The Physical and Biological Effects of High Frequency Sound-Waves of Great Intensity" [9]. During the few years which followed this paper a great deal of pioneering work in sonochemistry was performed and, as a result of this, two reviews on the applications of ultrasound in polymer and chemical processes were published in the 1940's [10, 11]. Yet there are very few references to ultrasound in chemistry from about 1955 to 1970 when a major renaissance in the subject began to occur which has accelerated dramatically over the last few years. This revival is undoubtedly due to the more general availability of commercial ultrasonic equipment. At the time of writing world interest in the topic has seen the formation of national groups devoted to sonochemistry in the U.K, Germany, Romania, and Japan, together with a flourishing European Society of Sonochemistry and, of course, the formation of two COST D6 groups.

In this chapter the range of applications of sonochemistry in synthesis will be explored together with some attempts to relate its effects to the underlying principles.

8.1.1 The Sound Ranges Used by Chemists

Sonochemistry is normally associated with the use of sound which is beyond human hearing *i.e.*, ultrasound. The ultrasonic frequency range stretches from around 20 kHz up to many MHz (Figure 8.1) and all of it can be utilized by chemists. Two distinct frequency ranges have been identified in this figure and each is associated with different applications. In the higher (MHz) range, low power sound may be used as a distance measure using a pulse echo technique as in SONAR, a sophisticated descendant of which is medical scanning. At the very

8.1. INTRODUCTION

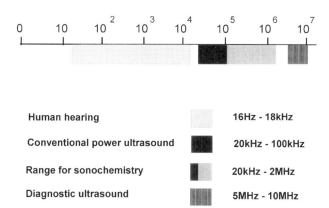

Figure 8.1 The frequency ranges of sound.

low energies and high frequencies used in scanning no chemical reactions are induced. However at lower frequencies, high powers can be generated leading to cavitation in the medium through which the sound wave passes and consequently leading to the potential to cause chemical changes.

The sound frequencies originally used by sonochemists were dictated by available equipment (commercial probe systems and baths) and were between 20 and 100 kHz. This is now considered to be somewhat restrictive since any sound frequency with sufficient energy (amplitude) to cause cavitation in a fluid can be used for sonochemistry. This extends sonochemistry into the low MHz range.

8.1.2 Sonochemistry Equipment

In the 1960's the ultrasonic cleaning bath began to make its appearance in metallurgical and chemical laboratories (Figure 8.2). It is perhaps not surprising that chemists, having seen the remarkable way in which these baths cleaned metal surfaces and soiled glassware, should have considered their potential use for the activation of chemical reactions. It was about this time that biology and biochemistry laboratories began using ultrasonic cell disruptors on a regular basis. These devices have now been adopted and developed by sonochemists and have become known as "probe systems" (Figure 8.3). Such instruments operate by the introduction of ultrasound directly into a solution by means of a vibrating metal probe, and offered the chance of irradiating chemical reactions with greatly increased ultrasonic power. Each piece of equipment has its own advantages (and disadvantages) which are outlined below (Table 8.1).

8.1.3 The Generation of Power Ultrasound

The essential device in the generation of power ultrasound for sonochemistry is the transducer. A transducer is any device which is capable of converting one form of energy into another. In the case of sound generation two simple

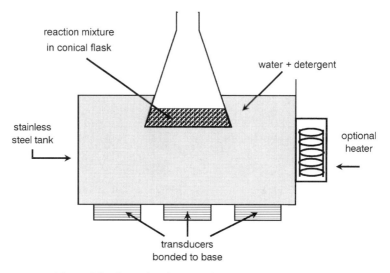

Figure 8.2 Sonochemistry equipment – cleaning bath.

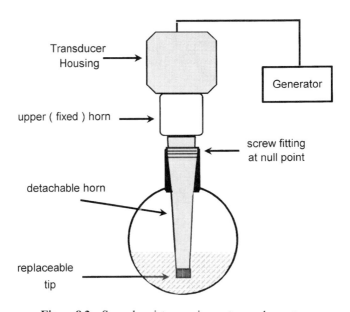

Figure 8.3 Sonochemistry equipment – probe system.

examples are a loudspeaker (which converts electrical to sound energy) and a whistle (gas movement into sound). There are three main types of ultrasonic transducers for sonochemistry: liquid driven, magnetostrictive and piezoelectric.

8.1. INTRODUCTION

Table 8.1 A comparison of cleaning bath and probe systems for sonochemistry.

System	Advantages	Disadvantages
Cleaning bath	The most widely available laboratory source of low power sonication.	Reduced power into vessel ($< 5 \text{ W cm}^{-2}$) through the reaction vessel walls.
	No special adaptation of reaction vessels required since they are simply dipped in the bath.	Fixed frequency (and different frequencies depending on type).
		Poor temperature control.
	Fairly even distribution of energy through reaction vessel walls.	Position of reaction vessel in bath affects intensity of sonication.
Probe	High power available (no losses due to transfer through vessel walls).	Fixed frequency (and different frequencies depending on type).
	Probes can be tuned to give optimum performance at different powers.	Temperature control requires jacketed reaction vessels.
		Tip erosion may occur leading to contamination by metallic particles.

Liquid-driven Transducers When the blade of a ship's propeller drives through water it produces pressure on one face and hence thrust. The trailing face however generates negative pressure as it cuts through the water and produces cavitation bubbles. These bubbles can produce enough energy to cause erosion damage to the blade surface. In essence a "liquid whistle" (Figure 8.4) generates cavitation by a similar mechanism. If a liquid is forced rapidly across a thin

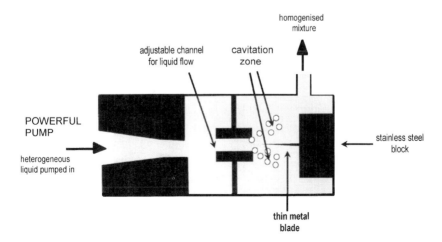

Figure 8.4 Sonochemistry equipment – liquid whistle.

metal blade anchored in a metal block the blade will begin to vibrate as a result of the liquid motion. The frequency of vibration is dependent on the flow rate and high flows produce ultrasonic vibrations. The liquid suffers cavitation as it passes across the face of the blade. When a mixture of immiscible liquids is forced across the blade the resulting cavitational mixing leads to extremely efficient homogenization. Such devices are particularly useful in industries where emulsification and homogenization is important.

Magnetostrictive Transducers Magnetostriction refers to an effect found in some ferromagnetic materials, *e.g.*, nickel or iron, which change in dimension on the application of a magnetic field. This type of transducer may be considered to be a form of solenoid with a magnetostrictive material as the core. The core is assembled as a laminate of many layers of thin nickel (or nickel alloy) plate and in its simplest form this comprises a closed square loop of core material with coils wound around two opposite sides (Figure 8.5). Applying a current to the coil produces a reduction in dimension of the core (magnetostriction) and consequently a reduction in the dimensions of the transducer. Switching off the current results in a return to the original dimensions. Repeated rapid switching on and off of the current generates the mechanical vibrations required. The dimensions of the transducer must be accurately designed so that the whole unit resonates at a precise sound frequency.

There are two disadvantages with this type of transducer. Firstly the frequency range is restricted to below 100 kHz and secondly the system is only about 60% electrically efficient, with losses occurring through heating. For this latter reason external cooling is often required. The major advantages are that the system is of an extremely robust construction and that very large driving forces can be obtained.

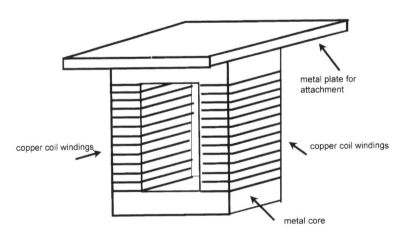

Figure 8.5 Construction of magnetostrictive transducer.

8.1. INTRODUCTION

Piezoelectric Transducers Currently the most common method employed for the generation of ultrasound utilizes the piezoelectric effect which relates to the instantaneous generation of an electric charge between opposite faces of certain materials when a sudden pressure is applied across them. Piezoelectric transducers use the inverse of this effect *i.e.*, if a charge is applied across such faces the material will respond by either expanding or contracting, depending on the polarity of the applied charges. Thus, applying charges alternating at 20 kHz or above leads to fluctuations in dimensions and the generation of ultrasonic vibrations. As with magnetostrictive transducers it is not possible to drive a given piece of piezoelectric crystal efficiently at every frequency. Optimum performance will only be obtained if the dimensions of the material correspond to the resonant length of the particular material.

Modern piezoelectric materials are made from ceramics containing dispersed piezoelectric compounds such as barium titanate ($BaTiO_3$), lead metaniobate ($PbNb_2O_6$) and the mixed crystal lead zirconate titanate. As ceramics they can be produced in a wide variety of shapes but the piezoceramic element commonly used in cleaners and for probe systems is produced in the form of a disk with metal coated contact faces and a central hole. A range of individual transducers can be constructed to provide highly efficient vibrations which span the whole ultrasonic frequency range.

It is normal practice to clamp piezoelectric elements between metal blocks which serve both to protect the delicate crystalline material and to prevent it from overheating by acting as a heat sink. Usually two elements are combined so that their overall mechanical motion is additive (Figure 8.6).

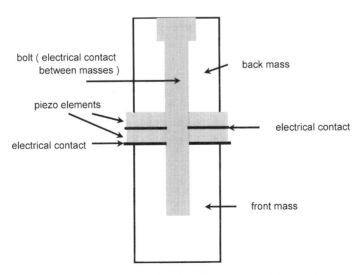

Figure 8.6 Construction of a piezoelectric sandwich transducer.

8.1.4 The Transmission of Ultrasonic Energy

In order to understand how sound energy can be transferred from transducer to a reaction it is important to recognize the way in which sound vibrations travel in a medium. Sound travels through a medium as a series of compression and rarefaction motions of molecules (Figure 8.7) and equal vibrational motion occurs at half wavelength distances of the sound. It is this distance – the half wavelength of sound in the medium – which governs the design of sonochemical equipment.

The transducer assembly is so constructed that the emitting face is at a maximum point of vibration. One or more of these can be directly bonded to the base of a stainless steel tank to drive an ultrasonic bath. In order to conduct higher vibrational energies from the transducer assembly to a laboratory reaction a probe system is generally used (Figure 8.3). A metal rod, known as a wave-guide, is used to transmit the vibrations and its length must be a multiple of half-wavelengths of the sound in the material from which the rod is constructed. The waveguide can be used to amplify the vibrational amplitude developed by the transducer by using a tapered guide (technically termed a horn). The type of horn dictates the amplification which easily can be ten-fold [12].

The Importance of Sound Wavelength   The wavelength of sound in a material is governed by the frequency and the velocity of sound in that material by the simple relationship:

$$sound\ velocity = wavelength \times frequency$$

The metal most commonly used as a waveguide in sonochemistry is titanium (or a titanium alloy) since it can transmit vibrations with a minimum of energy loss through heat and it is chemically unreactive. The velocity of sound in titanium is

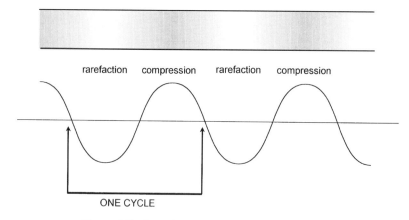

Figure 8.7 Sound transmission through a fluid.

5080 m s^{-1} so that at 20 kHz the waveguide must be either 12.7 cm in length (the half wavelength) or a multiple of this.

Transducers employing higher frequencies are much smaller and can use much shorter horns. Generally however, high frequency ultrasound, while quite capable of inducing cavitation, will provide far less power and penetration than those in the lower kHz range. As we will see later, however, higher frequencies seem to have a significant effect on free radical formation.

8.2 THEORETICAL ASPECTS

8.2.1 A Chemist's View on Cavitation

Acoustic energy is a mechanical vibration with frequencies many orders of magnitude below those of molecular vibrations. It is clear therefore, that ultrasonic irradiation cannot be directly absorbed by molecules as is the case with light. To trigger the remarkable sonochemical transformations which have been observed (see later) an indirect mechanism of activation must be involved. This activation process is acoustic cavitation. Like any sound wave, ultrasound is propagated *via* a series of compression and rarefaction waves induced in the molecules of the medium through which it passes. At sufficiently high power the rarefaction cycle may exceed the attractive forces of the molecules of the liquid and cavitation bubbles will form. It is the fate of these cavities that they collapse in succeeding compression cycles which generates the energy for chemical and mechanical effects.

The minimum acoustic power required to achieve cavitation in a liquid is known as the Blake Threshold and its value depends upon the conditions pertaining at the time of measurement. Essentially any process which aids the formation of cavitation will lower the Blake Threshold, *i.e.*, acoustic cavitation will be achieved at lower powers. Thus cavitation will be facilitated in systems containing dissolved (or entrained) gas, or when small particles which might act as nucleation sites, are present. In addition, any vapor pressure increase in the liquid induced by a rise in temperature, the presence of a more volatile component or a reduction in the overpressure on the system will enhance cavitation. As a rule-of-thumb it is reasonable to assume that the easier it is to produce acoustic cavitation the lower is the energy released on bubble collapse.

During the compression phase, the cavities can collapse violently, in very short times (10^{-6} s). Under these adiabatic conditions, their contents reach high temperatures (5000 K) and pressures (many hundreds of bars) [13]. This process has been named descriptively as the *hot spot* theory and its origins date back to 1956 and the work of Noltingk and Neppiras [14]. This is currently the most widely accepted interpretation of the collapsing acoustic bubble and permits a satisfactory interpretation of a large number of sonochemical results.

An alternative explanation has been offered based upon the production of an electric discharge during bubble collapse [15]. Various electrical theories have been developed, the earliest of which dates back to the work of Frenkel who suggested that electrical charges could be built up on opposite faces of a cavitation bubble as it is formed [16]. According to the more recent electrical theory, the bubble undergoes pulsations, then fragmentation, during which intense electrical fields, estimated to be at least $10^{11}\,\mathrm{V\,m^{-1}}$ are generated. More recently, it was envisaged that electrified sprays of the surrounding liquid are injected into the cavity [17]. Due to the intensity of the phenomenon, partial ionization of the bubble content should be produced and analogies with cold plasma chemistry were recorded. A schematic representation of these theories is shown (Figure 8.8).

It is of course quite possible that both interpretations are correct, but that the true situation is a hybrid which leans towards one or the other depending upon the precise reaction conditions. Whatever the correct theory might prove to be it is absolutely certain that highly energetic transient chemical species are indeed formed during sonolysis and that certain types of reactions are accelerated. It is to be hoped that current attempts to observe directly the activated species generated (*e.g.*, by ESR spectroscopy or sonoluminescence) will help to establish a general theoretical basis for sonochemistry.

8.2.2 Cavitation Effects in Different Types of Systems

In order to understand the way in which cavitational collapse can effect chemical processes, one must consider the possible effects of this collapse in different types of liquid systems.

In the case of a *homogeneous liquid phase*, as with any other liquid system, the cavity will almost certainly contain vapor from the liquid medium or dissolved volatile reagents (Figure 8.9). On collapse, these vapors will be subjected to extremely large increases in temperature and pressure resulting in molecular fragmentation with generation of highly reactive radical species. Such radicals might then react either within the collapsing bubble or after migration into the bulk liquid. Thus, for example, the sonication of water gives rise to radicals OH· and H·, and the subsequent formation of hydrogen peroxide. The sudden collapse of the bubble also results in an inrush of liquid to fill the void. So powerful is this inrush that it will produce shear forces in the surrounding bulk liquid capable of breaking chemical bonds in any polymeric materials which are dissolved in the fluid (see Chapter 9).

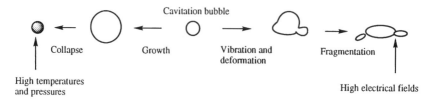

Figure 8.8 A schematic view of cavitational physical effects.

8.2. THEORETICAL ASPECTS

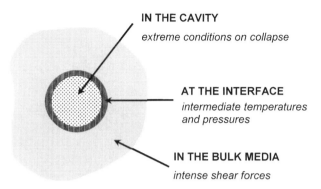

Figure 8.9 Cavitation in a homogeneous medium.

It is cavitation in a *heterogeneous* medium which is studied the most by sonochemists. When produced next to a phase interface, cavitation bubbles are strongly deformed. A liquid jet propagates across the bubble towards the interface at a velocity estimated to be hundreds of metres per second. At a *liquid-liquid interface*, the intense movement produces a mutual injection of droplets of one liquid into the other one, *i.e.*, an emulsion (Figure 8.10). Such emulsions, generated through sonication, are smaller in size and more stable than those obtained conventionally and often require little or no surfactant to maintain stability. It can be anticipated therefore that Phase Transfer Catalyzed (PTC) reactions will be improved by sonication. Examples are provided later in this chapter.

At a *solid-liquid interface*, the evolution of the cavitation bubbles has been observed with high speed cameras and the generation of a "jet" impinging upon the surface has been recorded (Figure 8.11) [18]. In this case, the shock on the surface produces erosion at and around the point of jet impact. Particles which may be ejected from the surface by this action react more efficiently than the

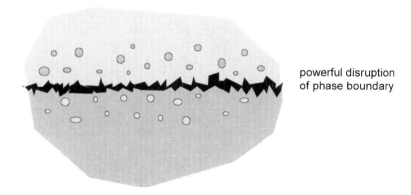

Figure 8.10 Cavitation in a biphasic liquid system.

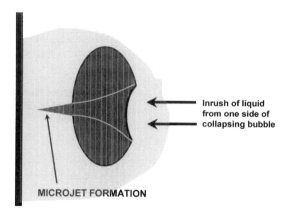

Figure 8.11 Cavitation in a solid-liquid system.

bulk solid surface due to an increased surface area. When a passivating coating is present on the solid surface the jetting will remove it, leaving a "clean" surface, and activation results [19]. At the same time, the surface is disorganized, local defects, dislocations and vacancies are produced and these generate a positive effect on reactivity. Associated with the collapse, and of particular importance in both catalysis and electrochemistry, is the increase in mass transfer to the surface by disruption of the interfacial boundary layers.

In the case of solid interfaces which are in the form of coarse powders, cavitation collapse can produce enough energy to cause fragmentation and activation through surface area increase. For very fine powders the particles are accelerated to high velocity by cavitational collapse and may collide to cause surface abrasion. For some metal powders these collisions generate sufficient heat to cause particle fusion (Figure 8.12).

The qualitative description above shows the complexity of the effects of cavitation on solids. The superficial modifications can be described by a

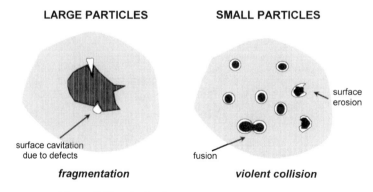

Figure 8.12 Cavitation in a powder-liquid system.

8.2. THEORETICAL ASPECTS

mathematical parameter, the *fractal number*. A value of 2 corresponds to a quasi planar geometry, 3 represents a three dimensional space. For silica, commonly used in solid supported reactions, the fractal number, originally 1.8–2.3, increases to 2.8 by sonication [20]. The higher reactivity of sonicated solids can then be explained by the transformation of a surface reaction to a process occurring in a volume. A more complex aspect is the tribochemical effect, according to which mechanical shocks are capable of ejecting high energy electrons from solids. As a result of such processes chemical reactions, *e.g.*, polymer formation or cleavage and radical generation from small molecules, can result [21]. It seems possible therefore that the tribochemical effect might well contribute to the higher reactivity of sonicated solids [22, 23].

8.2.3 An Attempt to Define the Laws of Sonochemistry

From a survey of past literature in sonochemistry it can be seen that ultrasonic irradiation in synthesis seems to have been developed on a practical rather than theoretical basis. Justification for enhanced reactivity was generally rationalized using the intuitively straightforward (at least in a qualitative sense) "hot spot" approach. Many advances were made by experimentalists, who proceeded essentially by "trial and error". Critics of the technique regarded sonochemistry merely as a kind of super-agitation. The first paper which provided evidence of the specific nature of sonochemistry was published by Ando *et al.* who described the first case of "sonochemical switching" [24]. The original system which led to this discovery (Figure 8.13) consisted of a suspension of benzyl bromide and alumina-supported potassium cyanide in toluene. The ***stirred*** reaction provided regioisomeric diphenylmethane products *via* a Friedel-Crafts reaction between the bromo compound and the solvent, catalyzed by the solid phase reagent. In contrast, ***sonication*** of the same constituents furnished only the substitution product, benzyl cyanide. According to the authors, a structural change on the catalytic sites of the solid support could be responsible for this effect.

Since this type of result shows that sonication is definitely not just another method of providing agitation of a medium, but exhibits its own peculiarities, it stands to reason that it should obey some rules of its own. An examination and classification of published material led to an empirical systemization of sonochemistry [23, 25]. This classification concentrates on the chemical effects in

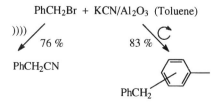

Figure 8.13 The original example of sonochemical switching.

sonochemistry but it should also be recognized that in some cases ultrasound does act in a mechanical sense achieving remarkable results through super agitation. Sometimes the mechanical and chemical effects occur together and rules may be invoked.

Rule 1 applies to homogeneous processes and states that those reactions which are sensitive to the sonochemical effect are those which proceed *via* radical or radical-ion intermediates. This statement means that sonication is able to effect reactions proceeding through radicals and that ionic reactions are not likely to be modified by such irradiation.

Rule 2 applies to heterogeneous systems where a more complex situation occurs and here reactions proceeding *via* ionic intermediates can be stimulated by the mechanical effects of cavitational agitation. This has been termed "false sonochemistry" although many industrialists would argue that the term "false" may not be correct, because if the result of ultrasonic irradiation assists a reaction it should still be considered to be assisted by sonication and thus "sonochemical". In fact the true test for "false sonochemistry" is that similar results should, in principle, be obtained using an efficient mixing system in place of sonication. Such a comparison is not always possible.

Rule 3 applies to heterogeneous reactions with mixed mechanisms, *i.e.*, radical and ionic. These will have their radical component enhanced by sonication although the general mechanical effect from Rule 2 may still apply. Two situations which may occur in heterogeneous systems involving two mechanistic paths are shown below (Figure 8.14). When the two mechanisms lead to the same product(s), which we will term a "convergent" process, only an overall rate increase results. If the radical and ionic mechanisms lead to different products, then sonochemical switching can take place by enhancing the radical pathway only. In such "divergent" processes, the nature of the reaction products is actually changed by sonication. Examples will be discussed later in this chapter.

8.2.4 Parameters of Importance in Sonochemical Reactions

At a simplistic level it might be considered that the only requirement to perform sonochemical reactions is cavitation and that once cavitation is achieved the chemistry follows. In fact, this is not always the case, and in order to optimize

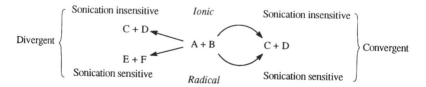

Figure 8.14 Two types of sonosensitive reactions.

8.2. THEORETICAL ASPECTS

the effect of sonication a number of parameters must be considered when conducting a sonochemical experiment.

It has been a fairly common misconception that the ***ultrasonic frequency*** used has no effect on the reaction despite the fact that in the past articles have appeared which contradict this [26, 27]. Studies of the role of frequency are still rare, since generators deliver only "monochromatic" vibrations. Changing the frequency can be accomplished only by changing the emitter; at the time of writing no continuously variable ultrasonic source is available. The frequencies generally employed in sonochemistry are in the narrow range between 20 to 100 kHz (power ultrasound). There is, however, a considerable amount of information on sonochemistry performed at frequencies much higher in the ultrasonic range (around 1 MHz) [28]. Even if we restrict our considerations to ultrasound in the kHz range, there are still some interesting frequency effects. In an important paper on this topic, Petrier has compared the effectiveness of 20 and 514 kHz irradiation for the generation of hydrogen peroxide in water at the same input power [29]. The rate of production of peroxide was about 12 times faster at the higher frequency. This result is ascribed to the fate of the OH· radical formed by the breakdown of water on cavitation bubble collapse. The OH· can be destroyed by reactions in the bubble or can migrate into the bulk and produce peroxide. At the higher frequency a shorter bubble lifetime allows more of the OH· to escape from the bubble into the bulk solution. A similar effect on OH· radical production has been shown over a narrower range of 20–60 kHz. In this case, the OH· radical, generated by sonolysis in aqueous sodium terephthalate, reacted with the dissolved substrate to form fluorescent hydroxyterephthalate. Its concentration could then be estimated spectroscopically [30]. The sonochemical efficiency for this reaction, after 30 minutes, was obtained by dividing OH· yield (directly proportional to fluorescence) by power input (obtained calorimetrically). The results clearly indicated that the efficiency for OH· production increases as the irradiation frequency is increased in the ratio 1 (20 kHz) : 2.7 (40 kHz) : 4.5 (60 kHz).

The ***ultrasonic power*** input must certainly have an effect on a sonochemical reaction. Once the power threshold to produce cavitation has been passed, there are several examples which show that an energy optimum exists beyond which the reaction efficiency decreases [31]. This decrease is an important point in sonochemistry because it signals a warning, often ignored, that maximum power is not always equated with maximum yield. The explanation lies in an effect in which the number of cavitation bubbles developed during sonication act as a barrier to the efficient transmission of acoustic energy from the vibrating source into the liquid medium.

In the case of ***reaction temperature*** there is again the possibility of an optimum value and this has been observed in a number of cases [31, 32]. This means that for a number of reactions there is a situation where a *rate increase* results from a *temperature decrease*.

The influence of ***overpressure*** has also been noted, but only in a limited number of cases. In one such case, a pressure lower than ambient was found to increase the reaction yield, possibly because of the higher efficiency for bubble resonance under these conditions [33].

The *solvent* used in sonochemistry is of great importance because the physical characteristics of the medium will dictate the amount of power which can be transmitted into the system *via* cavitational collapse. One obvious factor is vapor pressure (a high vapor pressure means that more vapor will enter the cavitation bubble and hence produce less violent collapse). This "cushioning effect" can be a limitation in sonochemistry, but can also be used with profit, for a "tuning" of the energy. Selectivity in sonochemical processes can thus be obtained [34]. Other factors to be considered include viscosity (which affects energy penetration and bulk heating) and, of course, chemical reactivity. It is this last factor which needs to be borne in mind before embarking on sonochemistry experiments. Due to the high energies produced in a liquid undergoing cavitation, it can be anticipated that every solvent can suffer some degree of degradation. Fortunately the majority of common solvents used in synthetic chemistry are relatively inert compared to the sonochemical reactivity of the dissolved or suspended substrate. Some classes of solvents, specifically halogenated materials, do undergo a degree of homolytic sonolysis with the formation of radicals, and these processes have been successfully used in a number of reactions (see later) [35, 36]. Somewhat surprisingly, it has been noted that a pH variation can also occur during halocarbon sonolysis. This is due to the formation of hydrogen halides and has been measured [37] and used for synthesis [38]. In non-volatile, viscous solutions, for instance in ethylene glycol, the total destruction of 10^{-2} M solutions of chloromethanes occurs in times as short as 10 min. From the foregoing it should be clear that the solvent effects considered in sonochemistry are somewhat different from those generally considered to be important in "classical" organic chemistry.

8.3 SONOCHEMICAL REACTIONS IN SYNTHESIS

A wide variety of sonochemical reactions has been studied by chemists over the last 15–20 years. For the sake of convenience, these will be subdivided into two types: homogeneous and heterogeneous. Of the two classes the former has received less attention and has not been reviewed in terms of synthetic possibilities. On the other hand heterogeneous systems have been extensively described in a number of articles and books [5–8, 39, 40]. For this reason, some emphasis will be placed on homogeneous sonochemistry in this chapter.

8.3.1 Homogeneous Reactions

Kinetic Studies The origins of sonochemistry lie in the study of homogeneous systems, and among the examples of early synthesis is the Curtius rearrangement which appeared in 1938 [41]. In this example benzazide when sonicated in benzene gives nitrogen and phenyl isocyanate (eq. 1), and the rate is increased in

8.3. SONOCHEMICAL REACTIONS IN SYNTHESIS

comparison to the normal thermal reaction. This reaction was not fully investigated at the time, and the observation that the reaction stopped after rapid initial steps was not explained.

$$PhCON_3 \xrightarrow{))))} Ph-N=C=O + N_2 \qquad (1)$$

Over the years since then until quite recently, homogeneous reactions were generally considered to be slow and low yielding and thus had no practical use in synthesis. This view has now altered and even if the cases are still not numerous, they do have important theoretical and applied interest.

Homogeneous solvolysis reactions have served to provide some detailed investigations of the kinetics of ultrasonic acceleration. In 1968 Fogler and Barnes [42] reported that ultrasonic irradiation (at 540 kHz) increased the rates of hydrolysis of methyl ethanoate and attributed the increase in reaction rate to the high temperatures reached within the cavitation bubbles (eq. 2). Studies on the effect of temperature on the sonochemical rate revealed a 28% enhancement at 20 °C with a smaller, 20% enhancement at 30 °C as expected from the effect of a change in vapor pressure (see Section 8.2.4).

$$CH_3COOCH_3 + H_2O \rightarrow CH_3COOH + CH_3OH \qquad (2)$$

In 1981 Kristol reported ultrasonically induced rate enhancements for the hydrolysis of the 4-nitrophenyl esters of a number of aliphatic carboxylic acids (**1**: R = Me, Et, i-Pr, t-Bu) (eq. 3) [43]. The rate enhancements at 35 °C were all in the range of 14–15% and were independent of the alkyl substituent (R) on the carboxylic acid. This was taken as evidence that the sonochemical effects were definitely not due to bulk heating of the reaction medium since this would be the same for all substrates and there are marked differences in the energy of activation (E_a) for the hydrolyses of each ester.

$$R-\overset{O}{\underset{\|}{C}}-O-\underset{}{\bigcirc}-NO_2 + H_2O \longrightarrow R-\overset{O}{\underset{\|}{C}}-OH + HO-\underset{}{\bigcirc}-NO_2 \qquad (3)$$

Ultrasound has been found to influence the homogeneous hydrolysis of 2-chloro-2-methylpropane in aqueous alcoholic media (eq. 4). This system has been the subject of numerous kinetic studies since it is a classic example of an S_N1 reaction.

$$(CH_3)_3C-Cl \xrightarrow{H_2O/CH_3CH_2OH} (CH_3)_3C-OH + HCl \qquad (4)$$

The homogeneous solvolysis of this substrate in aqueous ethanolic solvents can be monitored by the change in conductance as HCl is produced.

Detailed studies of the aqueous ethanol system led to the following main conclusions [44]:

(i) The effect of ultrasound increased with increased ethanol content and decreased temperature giving rate enhancements up to 20 fold at 10 °C in 60% w/w.
(ii) at ethanol concentrations of 50 and 60% w/w the actual rates of reaction under ultrasonic irradiation increased as the temperature was reduced from 20 °C to 10 °C (by factors of 1.4 and 2.1, respectively).
(iii) a maximum effect of ultrasound appeared to occur at a solvent composition of around 50% w/w, at 25 °C.

The effect of ultrasonic irradiation power on the chemical reactivity of this system revealed an optimum value for rate enhancement [45].

Sonolytic Processes Sonolysis of the solvent constitutes an important chemical process to be considered in any sonochemical synthesis. Normally, and fortunately for the synthetic chemist, such reactions comprise only a very minor proportion of the major reaction pathways involving homogeneous or heterogeneous reactants. Many studies have been made in the field of solvent breakdown with a significant proportion involving the study of water where the basic reaction is the homolytic cleavage of an O–H bond in the water molecule (eq. 5). This leads to the highly reactive hydroxyl radical [46], which can then react with many organic compounds.

$$H_2O \xrightarrow{))))} \cdot H + \cdot OH \qquad (5)$$

Although the first example of sonolysis in a non-aqueous solvent, the decolorization of the diphenylpicrylhydracyl (DPPH) radical in methanol, was reported in 1953 [47], it was some 20 years later when it was realized that cavitation could be supported successfully in organic solvents. Homogeneous non-aqueous sonochemistry is typified by the sonolysis of chloroform which has been studied using ultrasonic irradiation of frequency 300 kHz (I = 3.5 W cm^{-2}) to yield a large number of products amongst which are HCl, CCl_4 and C_2Cl_4 (Figure 8.15) [48]. Other chlorohydrocarbons have been studied and these also give complex mixtures on sonolysis [49]. The precise mechanisms involved in the decomposition are complex, but almost certainly involve the homolytic fission of chloroform to radicals and the formation of carbenoid intermediates.

Radicals form in sonicated pure organic solvents even under low energy sonication conditions, *i.e.*, in a cleaning bath (Figure 8.16). The radicals can be spin-trapped with suitable reagents, nitrones or aromatic nitroso compounds, and identified by ESR spectroscopy [35]. Sonolysis occurs, according to the authors *via* a pyrolytic mechanism, more efficiently with liquids of low volatility.

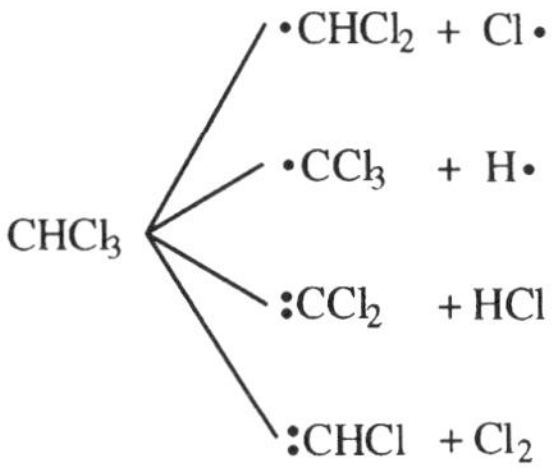

8.3. SONOCHEMICAL REACTIONS IN SYNTHESIS

$$CHCl_3 \begin{cases} \cdot CHCl_2 + Cl\cdot \\ \cdot CCl_3 + H\cdot \\ :CCl_2 + HCl \\ :CHCl + Cl_2 \end{cases}$$

Figure 8.15 Degradation of chloroform.

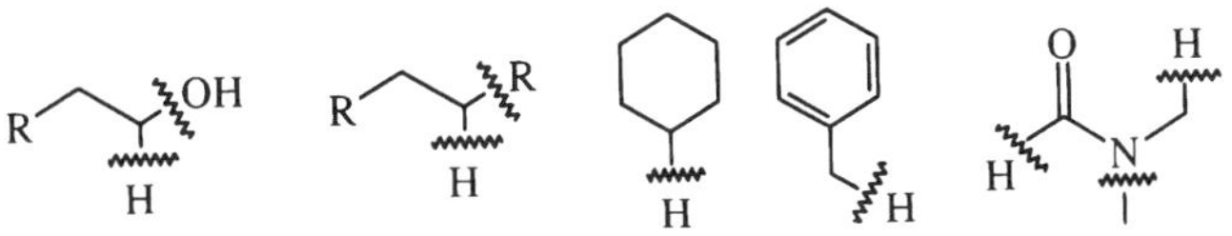

Figure 8.16 Identified cleavages in the sonolysis of a few solvents.

Such studies serve to emphasise that the use of certain solvents such as DMF in sonochemical syntheses can lead to difficulties because they may suffer decomposition.

Sonolysis of the carbon-halogen bond can also be achieved by sonication of a halocarbon in solution [37]. In a solvent acting as an electron acceptor, e.g., nitrobenzene, redox reactions take place [50], and a complex mixture is obtained. However, in a few instances, this sonolysis can lead to selective, high yielding reactions. One of these, reported over thirty years ago, involved the addition of sonochemically generated halogen atoms to maleic acid or its esters in a reversible manner [51]. The result is a *cis-trans* isomerization (Figure 8.17). The characteristics of this type of reaction have been studied, and it was shown that the mechanism seems to involve a chain reaction with a turnover number of ca. 7500 [52]. The influence of gases introduced into the medium has revealed unexpected results. According to the hot-spot theory, higher temperatures in the cavitating bubble should be obtained when the saturating gas has a high polytropic ratio and a low thermal conductivity [6]. Hence maleate to fumarate

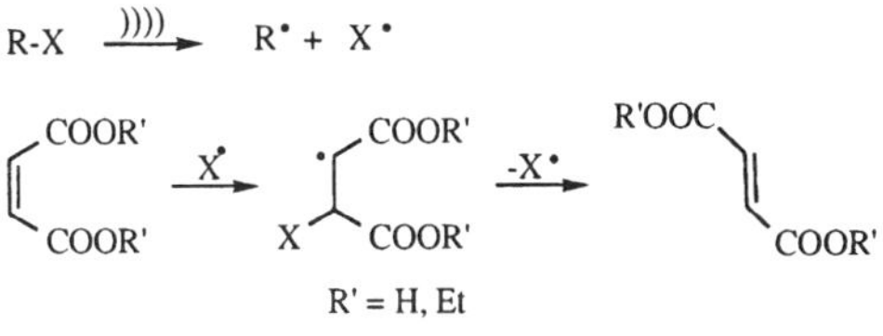

$$R\text{-}X \xrightarrow{))))} R^\bullet + X^\bullet$$

R' = H, Et

Figure 8.17 Sonoisomerization by halogen radicals.

isomerization proceeds more easily in the presence of argon, in comparison with other gases. In a notable exception to this it was found that carbon tetrafluoride was an efficient promoter [53]. This gas is also known to be an efficient energy carrier in plasma chemistry, an observation which provides some support for the electrical theory at the expenses of the hot-spot interpretation.

Other compounds known to undergo a (Z)-(E) isomerization are 1,2-dichloroethene and some vinyl sulfones. In the latter case, it was observed that in the presence of larger amounts of radical precursors addition of the halogen atom is no longer reversible but a second atom is trapped to give a dihalide product (eq. 6) [54].

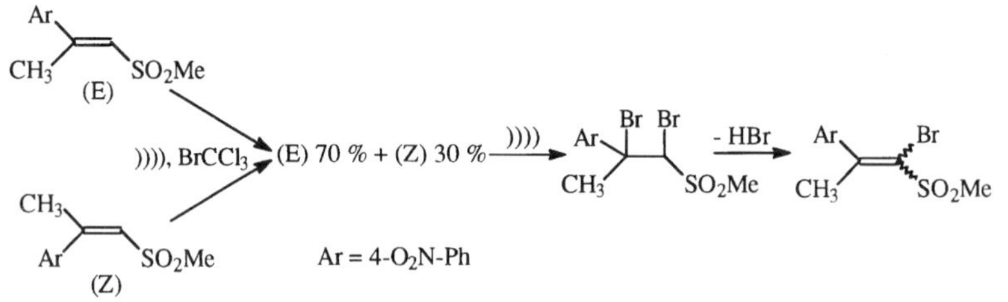

$$\text{(E) and (Z) isomers} \xrightarrow[\text{))))},\ \text{BrCCl}_3]{} (E)\ 70\% + (Z)\ 30\% \xrightarrow{))))} \underset{\text{CH}_3}{\overset{\text{Ar}}{\underset{|}{\text{C}}}}\!\!-\!\!\underset{\text{SO}_2\text{Me}}{\overset{\text{Br Br}}{\underset{|}{\text{C}}}} \xrightarrow{-\text{HBr}} \underset{\text{CH}_3}{\overset{\text{Ar}}{\text{C}}}\!\!=\!\!\underset{\text{SO}_2\text{Me}}{\overset{\text{Br}}{\text{C}}} \quad (6)$$

Ar = 4-O_2N-Ph

Dauben et al. found that the CCl_3 radical produced by sonolysis of carbon tetrachloride can be used in a decarboxylation-halogenation sequence (Figure 8.18) [55]. Sonication of the substrates at 33 °C for 10–50 min in carbon tetrachloride leads to the expected chlorides in high yields. In the presence of bromotrichloromethane or iodoform, bromides and iodides were formed in yields > 80%. The reaction which can be successfully applied to primary, secondary, or tertiary esters offers an interesting variant to the usual Hunsdiecker procedure.

Aromatic compounds undergo carbonization during sonication [56]. The reaction can occur either at the bubble interface or inside the cavity, according to the hydrophilicity of the substrate. It seems that apolar, hydrophobic

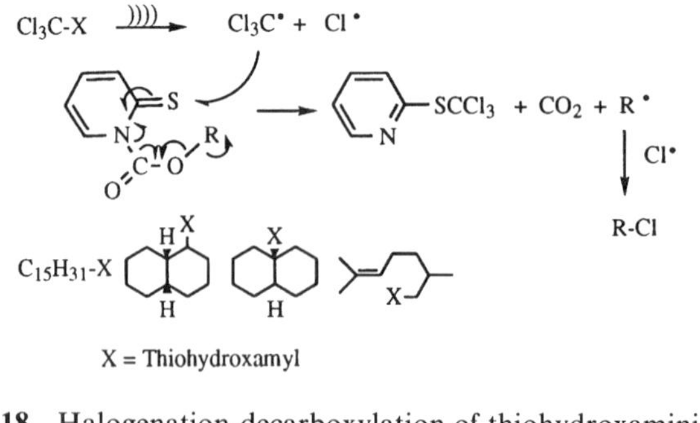

X = Thiohydroxamyl

Figure 8.18 Halogenation-decarboxylation of thiohydroxaminic esters.

8.3. SONOCHEMICAL REACTIONS IN SYNTHESIS

compounds, *e.g.*, benzene [57] and halocarbons [53, 58], are normally pyrolyzed inside the bubble (Figure 8.19).

Hydrophilic polar compounds, for instance phenols [59], carbohydrates [60] and nucleic acid constituents [61], unable to vaporize into the bubble, are more likely to be subjected to the electrical effects in the interfacial shell around the bubble, analogies for which are found with radiolytic degradations.

Applications of these degradative processes for the destruction of water pollutants are presently under active investigation. Decomposition is efficient even at low concentrations; for instance an initial 10^{-3} M concentration of phenol in solution can be reduced to 10^{-7} M without the necessity of other auxiliary treatment. However, the reaction times are relatively long, and diligent research efforts are being made to optimize conditions so that such treatments may become compatible with industrial exploitation. It is envisaged that sonication in combination with other technologies, *e.g.*, photocatalysis by inorganic semiconductors (TiO_2) [62] or with electrochemical methods, are likely to be the most successful methods of incorporating sonochemistry into environmental protection [63].

In 1983 Suslick reported the effects of high intensity (*ca.* 100 W cm^{-2}, 20 kHz) irradiation of alkanes at 25 °C under argon [64]. Under these extreme conditions the primary products were H_2, CH_4, C_2H_2 and shorter chain alk-1-enes. These products are not dissimilar to those produced by high temperature (1200 °C) alkane pyrolyses, in support of the hot-spot theory. The principal degradation process under ultrasonic irradiation was considered to be C–C bond fission with the production of radicals. By monitoring the decomposition of $Fe(CO)_5$ in different alkanes it was possible to demonstrate the inverse relationship between sonochemical effect and solvent vapor pressure [65].

Some hydrogen-heteroelement bonds undergo a sonolysis process similar to the labile carbon halogen bond. Significant examples of this occur in the case of tin hydrides, such as triphenyltin hydride [66]. The triphenyltin radical, which was also detected in the sonocleavage of hexa-alkyl or hexa-aryl ditin [67], adds to alkenes and alkynes in excellent yields, even at temperatures below 0 °C (Figure 8.20).

In contrast, the thermal reaction, which requires the presence of an initiator and heating at 50–100 °C, proceeds slower by a factor of at least 2 orders of

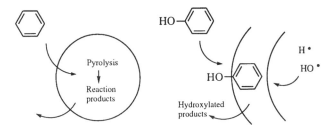

Figure 8.19 Influence of hydrophilicity on the reaction site.

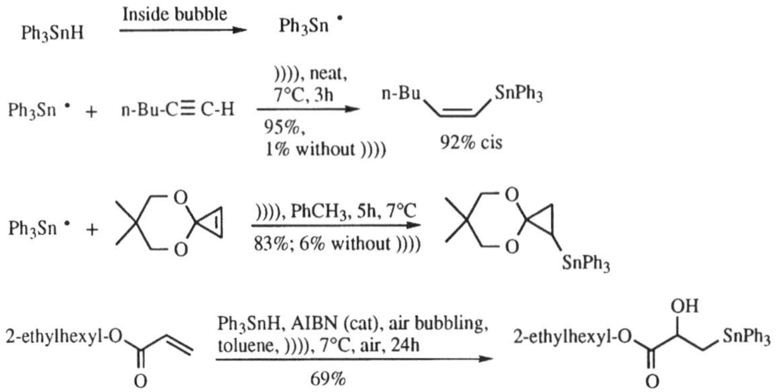

Figure 8.20 Sonochemical hydro- and hydroxy-stannations.

magnitude. This type of reaction is certainly of interest in synthesis, but the mechanistic aspects are also important. The authors indicate that the Sn–H bond cleavage takes place inside the cavitation bubble, and this seems quite possible for reaction in the absence of a solvent. However it does not seem to explain a similar cleavage for tributyltin hydride in solvents such as THF or toluene which possess reasonably high vapor pressures. When carried out in aerated solutions sonolysis leads to the previously unknown process of hydroxystannation (Figure 8.20).

The trialkyltin radical also reduces alkyl bromides and iodides even at low temperatures ($<0\,°C$) and, in the presence of oxygen (Figure 8.21), an OH group is introduced but is accompanied by a loss in configuration [68]. This stereochemical disadvantage is ameliorated by the fact that the milder conditions used in the sonochemical method preserves the configurational integrity of

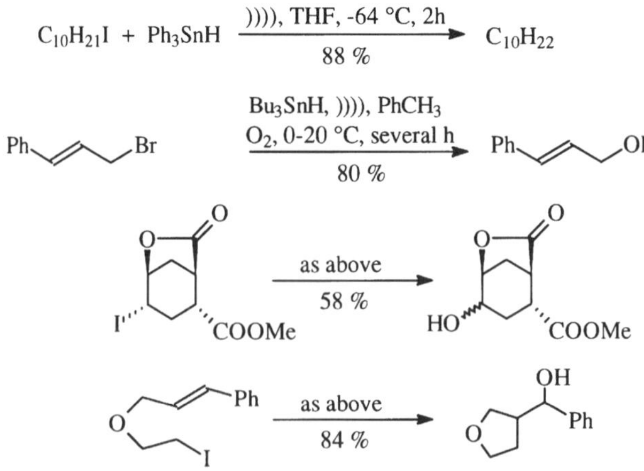

Figure 8.21 Sonochemical reductive processes induced by tin hydrides.

8.3. SONOCHEMICAL REACTIONS IN SYNTHESIS

the olefinic bond in allylic halides. Substrates with a 5-hexenyl structure underg%o cyclization before hydroxylation, a synthetic advantage over the usual reductive methods in which one functional group is lost. This reaction was used with success in the reductive deuteration of ribonucleosides [69].

Another example of heteroatom-H bond cleavage is found in the addition of phosphonates to imines (Figure 8.22) [70]. In contrast to the thermal reaction which has an induction period of about 1 h and produces a 50% yield after 2 h, sonication induces an immediate reaction which reaches > 90% yield in 1.5 h. The free radical nature of the mechanism was established by the use of inhibitors in the sonochemical reaction and promoters (AIBN = azobisisobutyronitrile) in the thermal reaction and was confirmed by trapping experiments. Unexpectedly in a related case ($R = C_{16}H_{31}$) there is a complete insensitivity to sonication. It is likely that this example involves polar rather than radical species since there is no ESR signal. From a fundamental viewpoint, the initial sonolysis of the P–H bond would probably not occur inside the bubble with such involatile compounds.

More extensive studies of reactions involving a sonolytic step have been in the field of transition metal compounds. One of the first examples was published by Suslick in 1981 [34]. In an appropriate solvent, metal carbonyl complexes can be cleaved by sonication to give transitory reactive compounds (Figure 8.23). With iron pentacarbonyl using relatively low level cavitational energy, a partial

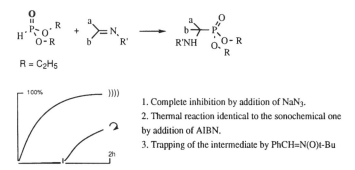

Figure 8.22 Addition of phosphonates to imines.

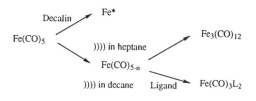

Figure 8.23 Sonolysis of iron carbonyl complexes.

decomposition occurs to an intermediate which yields a product not easily accessible by more traditional methods. This type of reaction was also observed with manganese, chromium, tungsten and molybdenum carbonyls. The partially decomposed intermediate species can be trapped by ligands added to the mixture. Under these sonochemical conditions catalytic amounts of iron pentacarbonyl can be used to cause isomerization of pent-1-ene to (E)- and (Z)-pent-2-ene. If, however, a low volatility medium is used to produce greater cavitational energy, the metal carbonyls can be decomposed totally to give a polydispersed metal [34, 71]. In the case of $Fe(CO)_5$ an amorphous iron powder which has important catalytic properties is produced.

Synthetic applications of the sonolysis of iron carbonyls which lead to useful ferrilactone synthons have been described (eq. 7). These are prepared easily and in good yields from vinyl epoxides and either iron pentacarbonyl or, for convenience and safety, diiron nonacarbonyl [72]. Somewhat surprisingly this reaction is efficient even in diethyl ether, a volatile solvent which delivers a low cavitation energy.

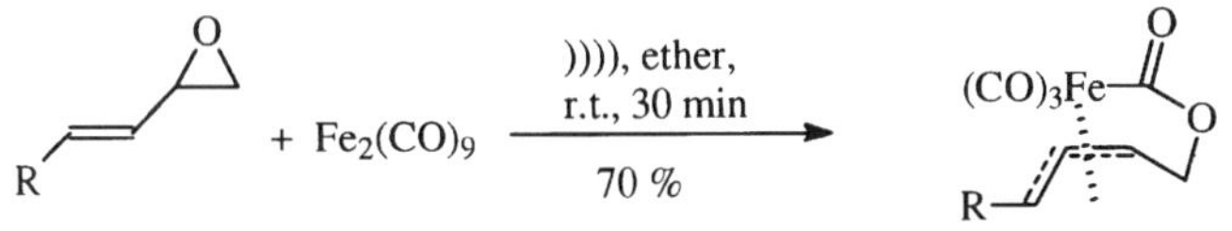

Sonochemical improvements have also been reported for the Heck palladium catalyzed addition-oxidation of halides to alkenes in its intramolecular version (eq. 8) [73], the Pauson Khand annulation using dicobalt hexacarbonyl-alkyne complexes (eq. 9) [74] and the reactions of metallocarbenes of chromium (eq. 10) [75].

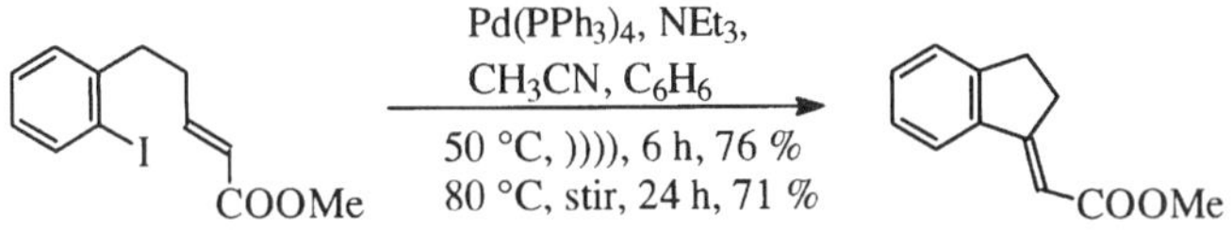

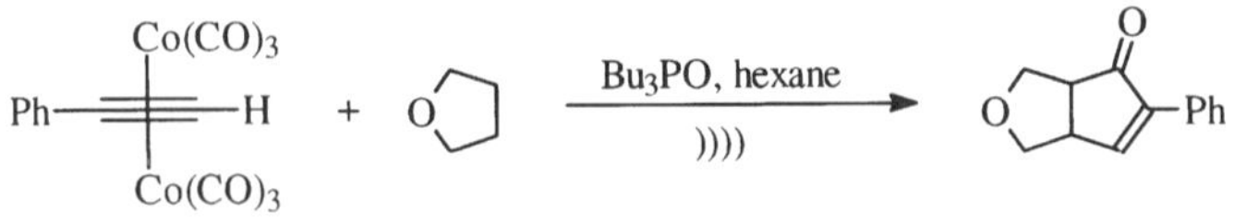

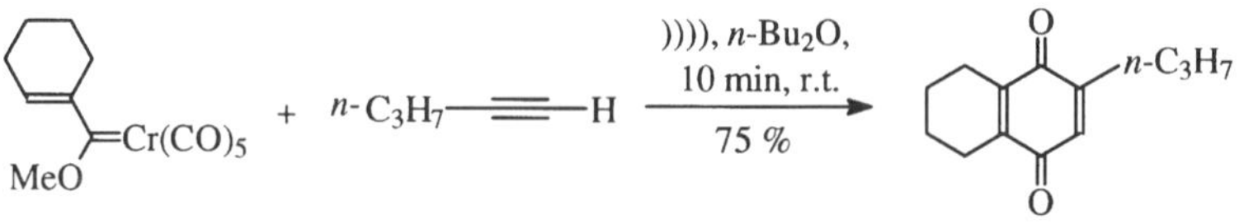

8.3. SONOCHEMICAL REACTIONS IN SYNTHESIS

Carbon-carbon Bond Formation Homogeneous sonochemical processes can be used in the formation of carbon-carbon bonds. This is exemplified by the generation of a carbon centred free radical ($\cdot CH_3$) from the sonolysis of lead tetraacetate [76]. In the presence of styrene it adds to the double bond to give 1-phenylpropyl acetate while *gem*- or *vic*-diacetates are formed *via* ionic mechanisms (Figure 8.24). In accord with the laws of sonochemistry it is the radical process which is favored under sonication. These reactions can be considered to be an example of sonochemical switching.

The Wittig-Horner reaction is substantially improved by sonication (eq. 11). Moreover, reactions involving enolizable aromatic ketones, which are normally useless for synthesis (yield <10%), reach 70–80% yields when performed under sonication [77]. The mechanism of the Wittig reaction, depending on the stabilized or semi-stabilized nature of the ylides, is a matter of debate [78]. Non-stabilized ylides should react *via* a single electron transfer (SET) process which would be sensitive to sonication. A stabilized reagent such as diethylmethylsulfonylmethyl phosphonate should follow a classical polar pathway and not be so affected by sonication. A more systematic study seems necessary to interpret the sonochemical data. The effects of sonication can be used to provide new evidence for the mechanistic pathways.

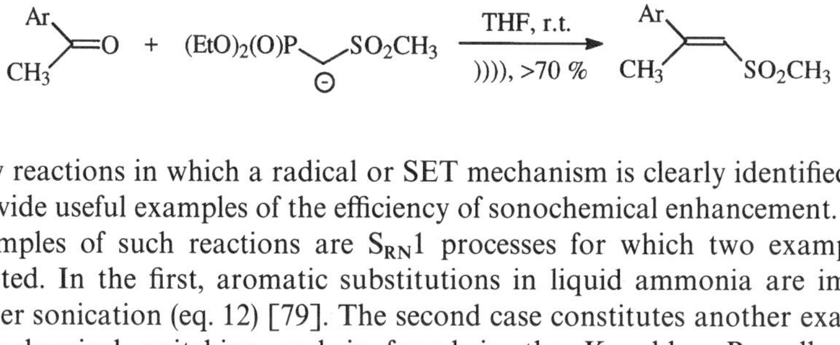

$$\underset{CH_3}{\overset{Ar}{>}}=O \;+\; (EtO)_2(O)P\underset{\ominus}{\frown}SO_2CH_3 \quad\xrightarrow[\text{)))), >70 \%}]{\text{THF, r.t.}}\quad \underset{CH_3}{\overset{Ar}{>}}=\!\!\!=\!\!\!\underset{SO_2CH_3}{} \qquad (11)$$

Any reactions in which a radical or SET mechanism is clearly identified would provide useful examples of the efficiency of sonochemical enhancement. Among examples of such reactions are $S_{RN}1$ processes for which two examples are quoted. In the first, aromatic substitutions in liquid ammonia are improved under sonication (eq. 12) [79]. The second case constitutes another example of sonochemical switching and is found in the Kornblum-Russell reaction (Figure 8.25). 4-Nitrobenzyl bromide was shown to react with 2-lithio-2-itro-propane *via* a predominantly polar mechanism to give, as a final product, 4-nitrobenzaldehyde. In contrast, the SET pathway leads to a dinitro

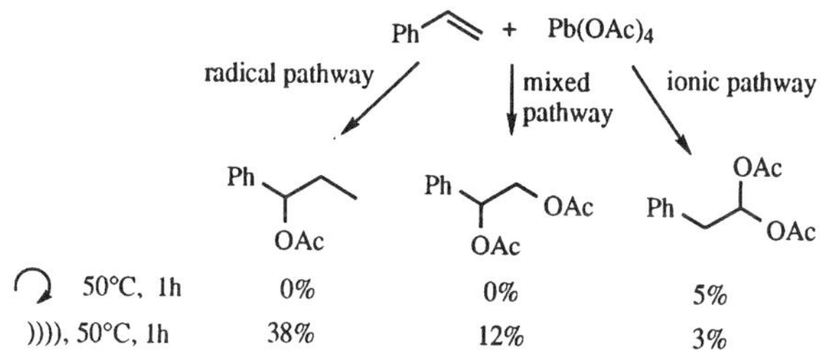

Figure 8.24 Competitive pathways in the reaction of lead tetraacetate with styrene.

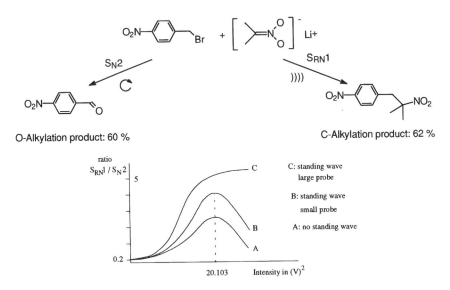

Figure 8.25 Sonochemical switching in the Kornblum-Russel reaction.

compound. Sonication changes the normal course of the reaction and gives preferentially the latter compound, in amounts depending on the irradiation conditions and the acoustic intensity [80].

$$\text{Naphthyl-X} + Ph_2P^- \xrightarrow[\text{2. [O]}]{\text{1.))))}, NH_3 \text{ liq.}, \text{r.t., 1h}} \text{Naphthyl-}P(O)Ph_2 \quad (12)$$

X = Cl,)))), 30 % (stirring 7 %)
Br,)))), 94 % (stirring 10 %)
I,)))), 70 % (stirring 45 %)

Other Examples In an analogy to the sonochemical $S_{RN}1$ reaction, discussed above, an aromatic chlorine atom can be substituted by azide (eq. 13) [81]. The author made a comparison between the sonochemical and photochemical substitution and used the sensitivity of the reaction to sonication to propose the SET mechanism which was confirmed in later studies.

$$\text{pyridyl-Cl} \xrightarrow{NaN_3,))))} \text{pyridyl-}N_3 \quad (13)$$

Another example of great synthetic interest involves the hydroboration reaction of alkenes [82]. In general, the addition of borane to alkenes proceeds stepwise,

8.3. SONOCHEMICAL REACTIONS IN SYNTHESIS

the final product being the trialkylborane. However, hindered alkenes react slowly, especially when the dialkylborane precipitates from the medium. It was found that trialkyl boranes could be obtained rapidly under sonication even with highly hindered substrates (Figure 8.26). Applications of this useful modification were published; among them was the reduction-hydroxylation of vinyl groups by 9-borabicyclononane (9-BBN) [83].

The sensitivity of this addition to sonication is quite intriguing in that the accepted mechanism does not seem to involve radical species.

8.3.2 Heterogeneous Reactions with Non-metals

In contrast to the foregoing examples of homogeneous reactions, heterogeneous systems appear to behave in a reasonably predictable manner when sonicated. Numerous synthetic applications have been successfully attempted [2–8, 24b], to such an extent that reactions with solid non-metal reagents or their aqueous solutions constitute an important part of modern synthetic chemistry. The efficiency of simple ultrasonically induced emulsification is a positive benefit in many cases, but the situation is more complex with respect to solid-liquid systems. The transfer of a reagent from a solid to a solution is a multistep process. For this transfer to occur, disturbance or destruction of the surface of the solid reagent is necessary. One classic way of achieving this is with the help of a quaternary ammonium salt as a Phase Transfer Catalyst (PTC). This type of activation can also be produced by sonication under appropriate conditions, depending on the properties of the solid (lattice energy, hardness), the wave

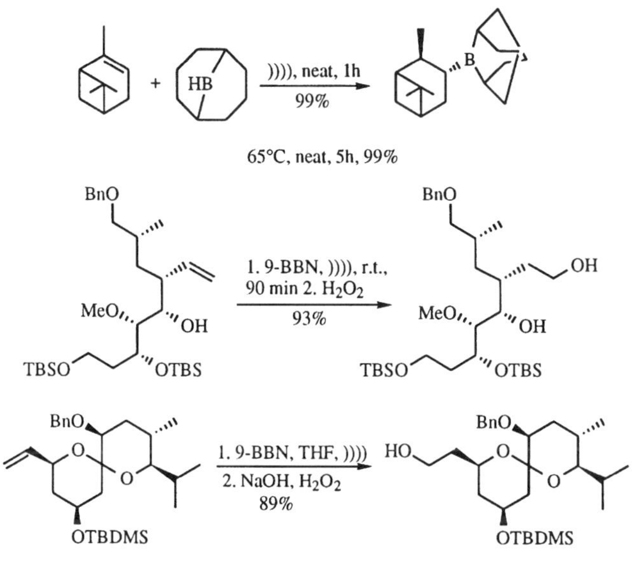

Figure 8.26 Sonochemical hydroborations.

(intensity, frequency) and the medium (cavitation energy). Mass transfer phenomena are enhanced by ultrasonic microstreaming and the overall accelerations are then a composite effect. In some situations the presence of a PTC may not be necessary, and some authors consider sonication as "a substitute to PTC" [24b]. In the following section a summary has been made of the abundant literature data [39].

Addition Reactions Radicals generated from carboxylic acids and manganese triacetate or ceric ammonium nitrate add to alkenes to give γ-lactones (Figure 8.27) [84]. Under sonication, this reaction becomes more efficient, and higher yields are obtained. It would appear that sonication induces a partial reoxidation to the trivalent salt which, in principle, could make the process catalytic with respect to manganese.

Perfluoroalkyl chains confer interesting properties on molecules (*e.g.*, as oxygen carriers), which is the reason why great efforts are being made to develop simple and cheap methods for their preparation. One of these consists of the addition of perfluoroalkyl iodides to alkenes or alkynes (eq. 14). The suggested mechanism involves a reduction of the reagent with sodium dithionite to a radical intermediate [85].

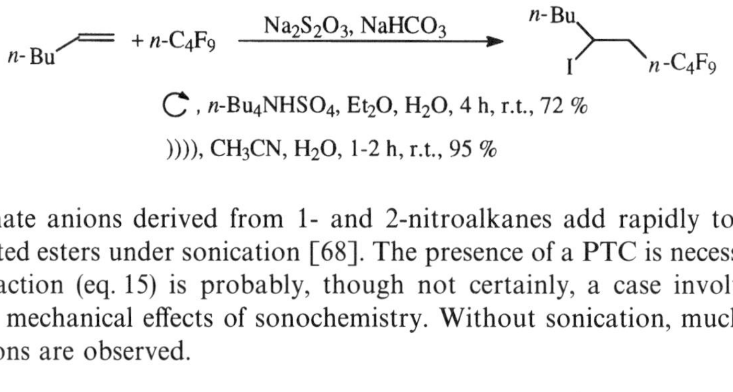

$$n\text{-Bu}\diagup\diagup \;+\; n\text{-C}_4\text{F}_9 \xrightarrow{\text{Na}_2\text{S}_2\text{O}_3,\; \text{NaHCO}_3} n\text{-Bu}\diagdown\text{CH(I)}\text{CH}(n\text{-C}_4\text{F}_9) \quad (14)$$

$\complement$, n-Bu$_4$NHSO$_4$, Et$_2$O, H$_2$O, 4 h, r.t., 72 %

)))), CH$_3$CN, H$_2$O, 1-2 h, r.t., 95 %

Nitronate anions derived from 1- and 2-nitroalkanes add rapidly to α,β-unsaturated esters under sonication [68]. The presence of a PTC is necessary and the reaction (eq. 15) is probably, though not certainly, a case involving the purely mechanical effects of sonochemistry. Without sonication, much slower additions are observed.

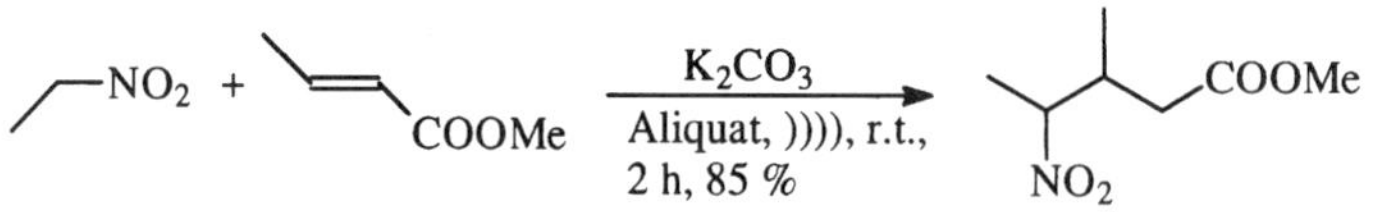

$$\diagup\text{NO}_2 \;+\; \diagdown\diagup\text{COOMe} \xrightarrow[\text{Aliquat,)))), r.t.,}]{\text{K}_2\text{CO}_3} \text{(NO}_2\text{)CH--CH}_2\text{--COOMe} \quad (15)$$

2 h, 85 %

cyclohexene + NC–CH$_2$–COOH $\xrightarrow[\text{AcOH,)))), 0-10°C, 20 min, 65%}]{\text{Mn(OAc)}_3,\; \text{KOAc}}$ bicyclic lactone (CN)

MeO–dihydropyran + MeOOC–CH$_2$–COOH $\xrightarrow[\text{CH}_3\text{CN,)))), r.t., 2h, 81\%}]{\text{CAN}}$ bicyclic lactone (COOMe)

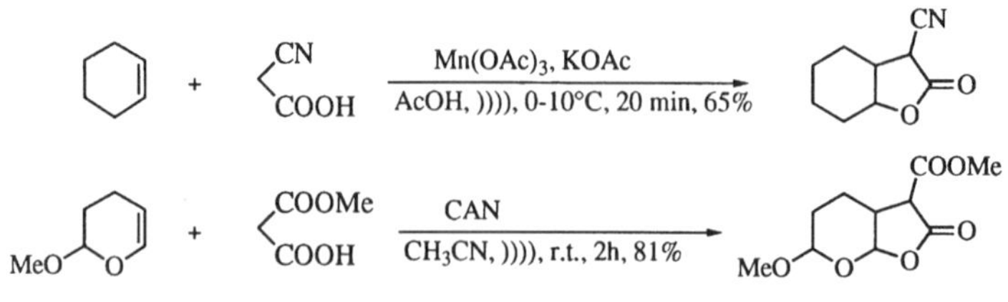

Figure 8.27 Lactonization of alkenes.

8.3. SONOCHEMICAL REACTIONS IN SYNTHESIS

Among the carbon-carbon forming reactions, the Wittig and related reactions have been studied and extended by sonochemists. Starting from allylic phosphoranes, it was shown that the initial deprotonation step could be effected with an increased efficiency using butyl lithium in THF or even in benzene (eq. 16) [87]. The insoluble phosphorane disappears after 5–15 min sonication. The mechanical effect may also be accompanied by a chemical effect [78]. This superposition of the two essential roles of sonication is reflected by the change in the stereoselectivity, larger proportions of the *trans*-diene being formed under irradiation.

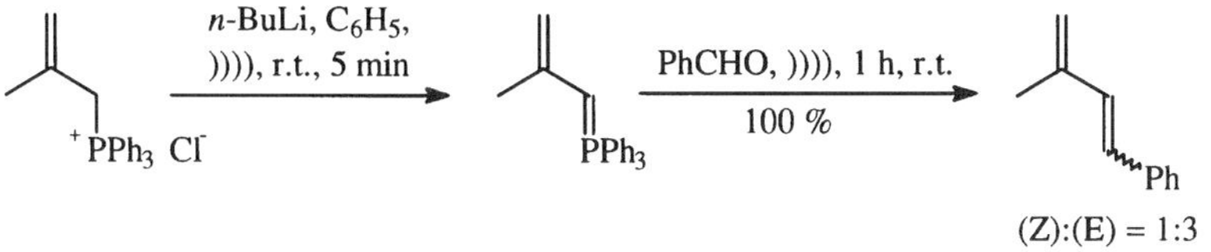

(16)

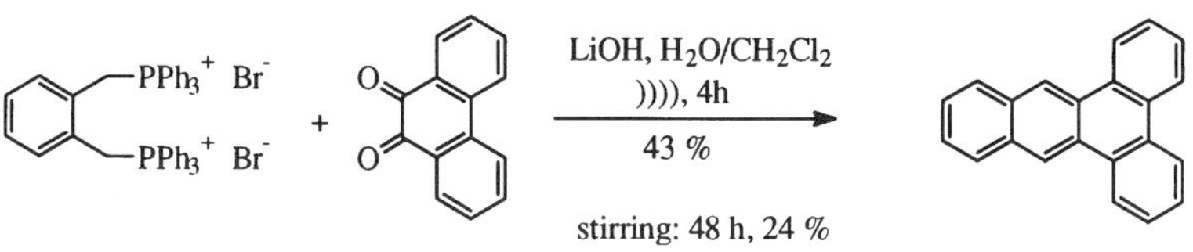

(17)

A rather difficult double Wittig reaction (eq. 17) has been effected with enhanced efficiency under sonication [88]. The process constitutes a novel type of annelation of an aromatic ring when applied to *o*-quinones. It is possible to simplify considerably experimental procedures with ultrasound which allow the use of bases which are insensitive to moisture or air. This is the case of barium hydroxide [89] or alkali metal carbonates [90]. In the former case (Figure 8.28), a mechanistic study has revealed that sonication has the double role of physical and chemical effects. The physical effect improves mass transport from the solution to the surface while the more complex chemical effects appear to favor an SET reaction between the phosphonate and the basic sites on the catalyst surface. Water molecules, necessary for the reaction to occur, undergo cleavage

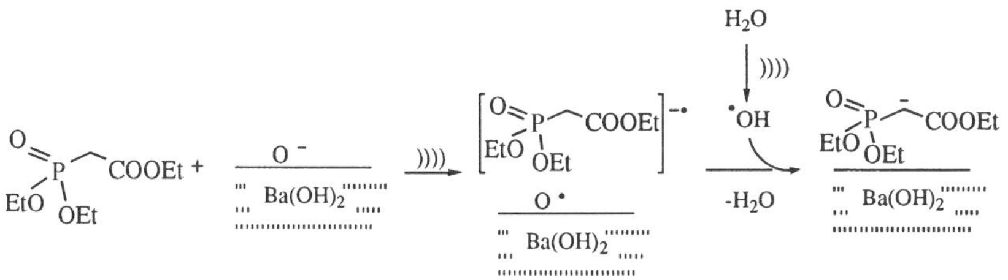

Figure 8.28 Sonochemical formation of a Wittig-Horner reagent on barium hydroxide.

to the hydroxyl radical, which oxidizes the radical anion to the expected anionic reagent.

Addition of cyanide ion to a carbonyl compound leads to a cyanohydrin, a process with many applications. One such is in the preparation of α-hydroxyacids or α-aminoacids (Figure 8.29). This has been significantly improved by sonication affording a rapid access to cyanohydrin esters from aldehydes through the trapping of the cyanohydrin anion with acyl chloride [24b, 91].

The formation and hydrolysis of esters, acetals and amides involve the addition of a base to the carbonyl group as the first step *via* a polar mechanism. It is therefore not surprising that it is only modestly accelerated by sonication in homogeneous solutions [92]. In contrast, when effected heterogeneously (eqs. 2, 18), a marked rate enhancement is observed as a result of the sonochemical emulsification [93]. The hydrolysis of an amide has also been reported in a solid-liquid system [94].

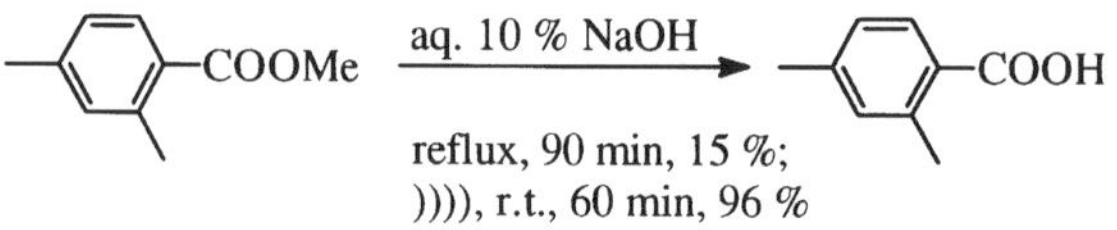

$$\text{Ar-COOMe} \xrightarrow[\substack{\text{reflux, 90 min, 15 \%;} \\ \text{))))), r.t., 60 min, 96 \%}}]{\text{aq. 10 \% NaOH}} \text{Ar-COOH} \quad (18)$$

Finally, in this selection of addition reactions it is interesting to note an original addition of the NO_2 radical to alkenes under mild conditions [95]. The process, while experimentally simple, has a complex chemistry which is not an unusual situation in sonochemistry (Figure 8.30).

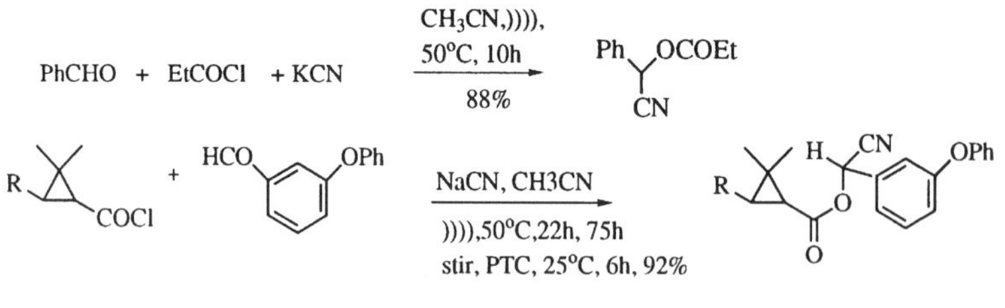

Figure 8.29 Addition of cyanide to carbonyl groups.

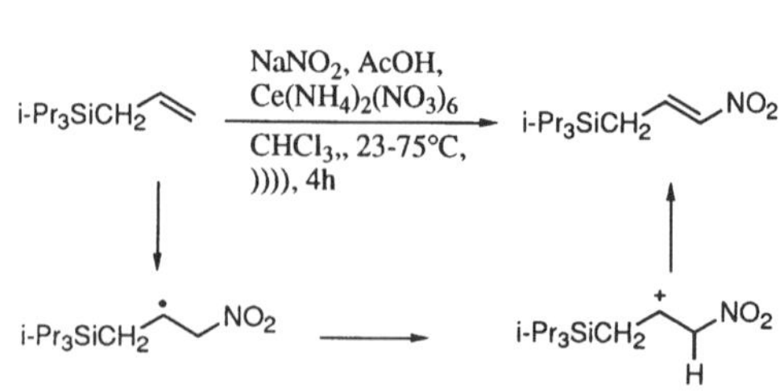

Figure 8.30 Synthesis of nitroolefines.

8.3. SONOCHEMICAL REACTIONS IN SYNTHESIS

The radical species is initially formed from sodium nitrite and acetic acid adds to the alkene. The resulting radical is then oxidized by a cerium(IV) salt which avoids the reverse elimination of NO_2. The last step, elimination of a proton, leads to the nitroalkene in excellent yields. In this way compounds inaccessible by conventional chemistry were prepared in high yields through sonochemistry.

Substitution Reactions A few examples of sonochemically induced substitution reactions are known, from which some cases were selected. In the synthesis of a modified adenosine compound, the required substitution of a tosylate group by cyanide can be effected successfully using an alkali metal cyanide, in the presence of 18-Crown-6 as the PTC [96] (Figure 8.31). The inconveniences involved in the usual procedure (long reaction times, high temperatures or the formation of undesired products) are avoided in the sonochemical reaction. 1,4-Dibromobut-2-yne can be transformed to cyclic trithiocarbonates with a propadiene side-chain [97]. 18-Crown-6 is required, but sonication speeds up the reaction by a factor of 10. The product was used for a simple preparation of butatriene.

Significant studies in this field have been reported by Ando *et al.*, who studied the effect of sonication on the substitution of halides in various compounds by acetate, thiocyanate and cyanide ions supported on alumina, silica gel or celite (eqs. 19, 20) [24b, 98]. In most cases, sonication improves the rate of substitution. One of the more important findings was obtained with cyanide, which led to the discovery of the sonochemical switching discussed previously under the laws of sonochemistry (Figure 8.13). Optimization of the conditions involved the combined use of the solid support, traces of water and sonication. The reaction can also be carried out in non-polar media resulting in small amounts of by-products (alcohols from hydrolysis) and yields higher than those obtained using conventional phase transfer catalysis.

$$PhCH_2Br + AcOK \xrightarrow[SiO_2,))))], 80\%]{C_6H_{12}, 50\ °C, 24\ h} PhCH_2OAc \qquad (19)$$

with stirring: 60%

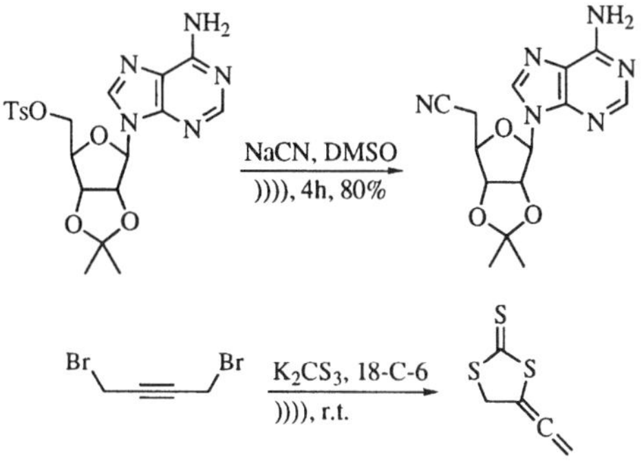

Figure 8.31 Two examples of substitution reactions.

$$1\text{-}C_8H_{17}\text{-}Br + KCN \xrightarrow[\text{)))), 50 °C, 48 h, 44\%}]{\text{PhCH}_3, \text{Al}_2\text{O}_3} 1\text{-}C_8H_{17}\text{-}CN \quad (20)$$

with stirring, 90 °C, 18-C-6: 23%

Formation of Anions Non-polar media cannot generally be used in the preparation of reactive anions, mainly because of the low solubility of the reagents. The use of PTC solves this problem; yields and rates are acceptable for synthetic purposes, and sonication in a number of instances exerts a synergistic effect with the PTC.

Deprotonation at a position which is vicinal or geminal to a leaving group results in elimination. Many examples of sonochemical carbene generation have been published since the first mention of such a process in 1982 [99]. The reaction can be performed in a cleaning bath and is therefore generally useable in laboratories. Additions to alkenes (eq. 21) occur in yields higher or at least equivalent to those of conventional methods, with the advantage of shorter reaction times [100]. From a mechanistic viewpoint, some evidence was obtained that the reaction may involve radical species (eq. 22) [101].

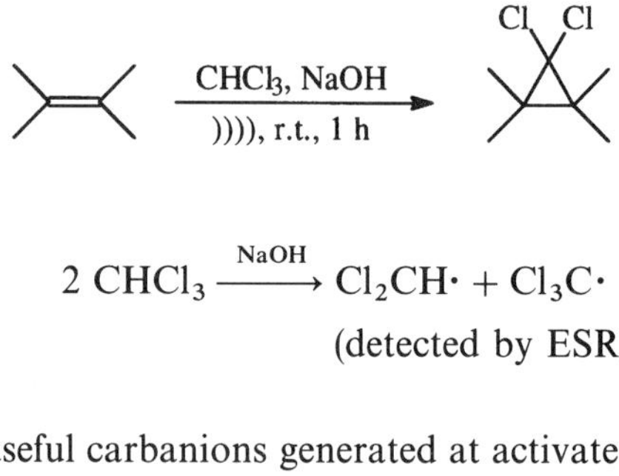

$$\text{Me}_2\text{C=CMe}_2 \xrightarrow[\text{)))), r.t., 1 h}]{\text{CHCl}_3, \text{NaOH}} \text{Me}_2\text{C(Cl)-C(Cl)Me}_2 \quad (21)$$

$$2\ CHCl_3 \xrightarrow{\text{NaOH}} Cl_2CH\cdot + Cl_3C\cdot \quad (22)$$

(detected by ESR)

The synthetically useful carbanions generated at activated methylene positions can be obtained easily under sonication, but the number of studies or uses quoted in the literature is surprisingly low. As an example, ethyl cyanoacetate and ethylene dibromide sonicated with potassium carbonate and polyethylene glycol in ethylene dichloride provide the expected cyclopropane in 85% yield (eq. 23) [102].

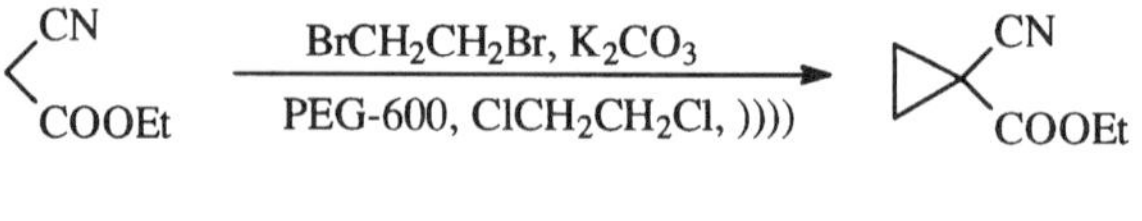

$$CH_2(CN)(COOEt) \xrightarrow[\text{PEG-600, ClCH}_2\text{CH}_2\text{Cl,))))}]{\text{BrCH}_2\text{CH}_2\text{Br, K}_2\text{CO}_3} \text{cyclopropane-C(CN)(COOEt)} \quad (23)$$

A second case, which has been somewhat overlooked, reports the formation of the highly useful methylsulfinyl anion, from DMSO and sodium hydride [103]. The usual preparations require heating of the reagents which must be controlled to within a specific temperature range, *ca.* 70 °C, in order to achieve a sufficiently rapid reaction without decomposing the thermally unstable reagent. Sonication

8.3. SONOCHEMICAL REACTIONS IN SYNTHESIS

at 800 kHz provides an easy and fast reaction (1 h) at 50 °C (eq. 24). It is also surprising to find that the reagent prepared by the sonochemical method is much more stable than those prepared conventionally and can be stored for 2 months.

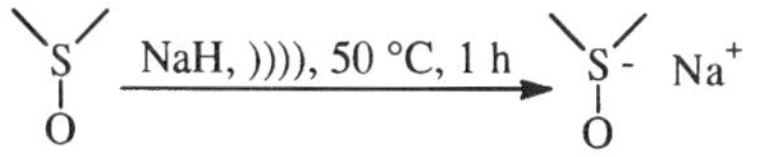

$$\underset{O}{\overset{}{S}}\overset{}{\underset{}{}} \xrightarrow{\text{NaH,)))), 50 °C, 1 h}} \underset{O}{\overset{}{S^-}} Na^+ \quad (24)$$

From the early days of sonochemistry there have been reports of alkyl group additions to anions such as alkoxides and amides. The O-alkylation of 5-hydroxychromones leads to potentially useful pharmaceutical compounds and is known to be difficult using conventional methodologies. This reaction has been studied comparatively using thermal, sonochemical, and PTC procedures in the absence of solvent (Figure 8.32) [104, 105]. Alkylation occurs with the same efficiency using sonication or PTC, both of these leading to improved results with respect to the thermal process. The nature of the PTC is important, and aliquat was efficient in the case of activated halides (benzyl, allyl), but useless with 1- and 2-butyl bromide. In this case, 18-Crown-6 must be used, making the PTC reaction the more efficient.

In the presence of polyethyleneglycol methyl ether or tetraalkylammonium chloride, phenol undergoes quantitative ethylation. Esterification also proceeds in high yields under similar conditions, with a particular advantage for fatty acids which react selectively to give esters without side-reactions [106].

Nitrogen alkylation occurs easily under sonochemical conditions [107]. Indole and carbazole undergo N-alkylation with increased rates and yields under sonication in the presence of polyethylene glycol methyl ether as PTC. No reaction occurs in the absence of the PTC. A few examples of O- and N-alkylation reactions of this type are shown (Figure 8.33).

A synthetically important N-acylation reaction is the formation of Boc protected amines [108]. Sonication allows this reaction to be effected in the absence of water using solid sodium hydrogen carbonate as the base in a

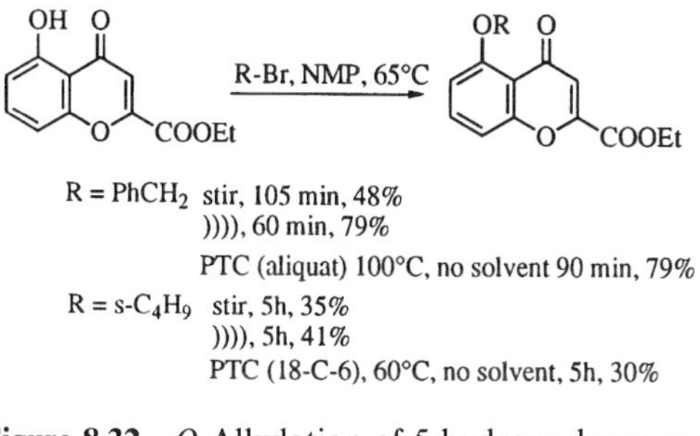

R = PhCH$_2$ stir, 105 min, 48%
 )))), 60 min, 79%
 PTC (aliquat) 100°C, no solvent 90 min, 79%
R = s-C$_4$H$_9$ stir, 5h, 35%
 )))), 5h, 41%
 PTC (18-C-6), 60°C, no solvent, 5h, 30%

Figure 8.32 O-Alkylation of 5-hydroxychromone.

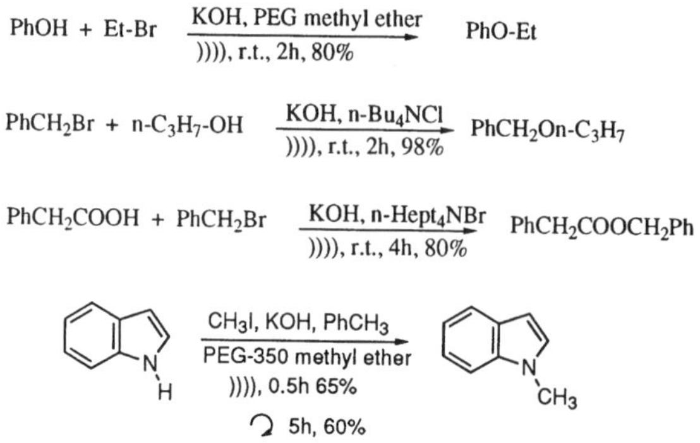

350   8. ULTRASOUND AS A NEW TOOL FOR SYNTHETIC CHEMISTS

$$\text{PhOH} + \text{Et-Br} \xrightarrow[\text{)))), r.t., 2h, 80\%}]{\text{KOH, PEG methyl ether}} \text{PhO-Et}$$

$$\text{PhCH}_2\text{Br} + n\text{-C}_3\text{H}_7\text{-OH} \xrightarrow[\text{)))), r.t., 2h, 98\%}]{\text{KOH, n-Bu}_4\text{NCl}} \text{PhCH}_2\text{On-C}_3\text{H}_7$$

$$\text{PhCH}_2\text{COOH} + \text{PhCH}_2\text{Br} \xrightarrow[\text{)))), r.t., 4h, 80\%}]{\text{KOH, n-Hept}_4\text{NBr}} \text{PhCH}_2\text{COOCH}_2\text{Ph}$$

indole + CH$_3$I, KOH, PhCH$_3$, PEG-350 methyl ether,)))), 0.5h 65%; ⟂ 5h, 60% → N-methylindole

Figure 8.33 Various *O*- and *N*-alkylations.

methanol suspension. An application is shown (eq. 25).

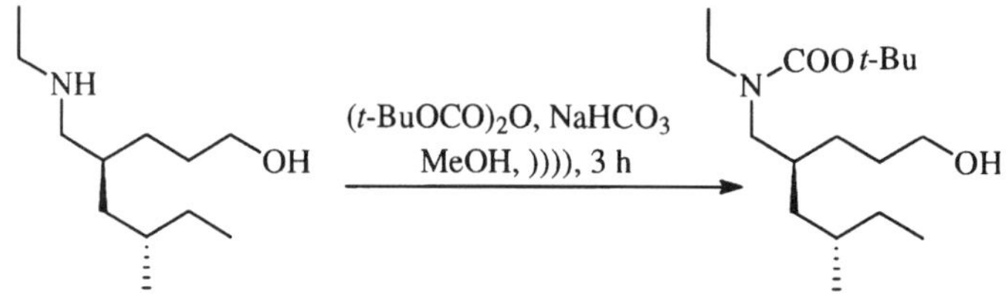

$$\text{R-NH-R'-OH} \xrightarrow[\text{MeOH,)))), 3 h}]{(t\text{-BuOCO})_2\text{O, NaHCO}_3} \text{R-N(COO}t\text{-Bu)-R'-OH} \quad (25)$$

Reduction and Oxidation Reactions A few examples of sonochemical reductions have been investigated. Thus the cleavage of an aromatic carbon-halogen bond by lithium aluminum hydride can be achieved under sonication even in difficult cases such as bromoanisole [109]. The reduction of heteroatom-halogen (or heteroatom-methoxy group) by a similar procedure is efficient even in apolar solvents [110]. Reduction of the aromatic nitro group to amine occurs in good yields using either a catalytic procedure with palladium and hydrogen [111], or less classically with a mixture of sulfur and hydrazine in the presence of activated carbon [112]. A last example is that of the reduction of aldehydes with tributyltin hydride in a biphasic system [113]. These reactions are summarized in Figure 8.34.

More references to oxidative processes than to reductions are found in the literature. Oxidation of C–H bonds, even non-activated ones, can be improved by sonication. An example of this is the oxidation of cyclohexane using molecular oxygen in the presence of perfluoroalkyl sulfonate salts. Although details are not readily available, it can be envisaged that a type of PTC reaction is developed in which the perfluorinated catalyst is able to help the transfer of oxygen to the liquid medium [114].

Oxidation at the benzylic position of indane by potassium permanganate (eq. 26) gives indanone in good yields and no PTC is necessary [33]. In

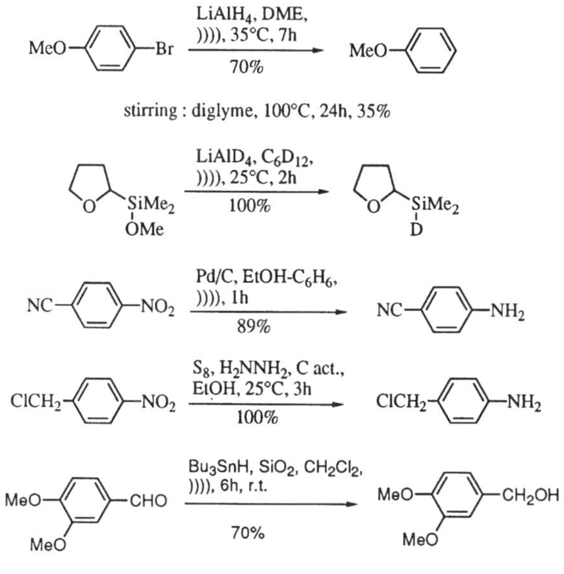

8.3. SONOCHEMICAL REACTIONS IN SYNTHESIS

Figure 8.34 Heterogeneous reductions.

a two-phase system consisting of an aqueous solution of $KMnO_4$ and indane in benzene an 80% yield can be obtained under a reduced pressure of *ca.* 450 Torr. The authors explain this effect by the size of the cavitation bubbles, which is dictated to some extent by the overpressure. An optimal energy transformation, from acoustic to chemical, can thus take place.

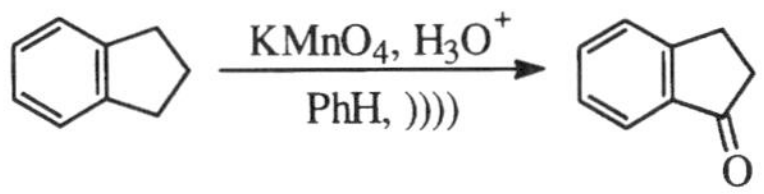

$$\text{indane} \xrightarrow[\text{PhH, })))]{KMnO_4, H_3O^+} \text{indanone} \quad (26)$$

Tosyloxylation vicinal to a keto group was reported with good yields using hydroxy-tosyloxy-iodobenzene [115]. The reaction is completed in a few minutes and exhibits a good selectivity (with no double α, α' functionalisation). The reagent is insoluble in the solvent (acetonitrile), and without sonication no product is formed.

The Etard reagents, *i.e.*, chromyl chloride and some derivatives, suffer from the problem that occasionally they can exhibit a lack of selectivity and low yields. They are useful in the selective oxidation of aromatic side-chains to a carbonyl group, aldehyde or ketone but in many instances, the formation of the initial complex is slow and yields are low because of difficulties in the work-up which lead to undesired over-reaction. Attempts have been made to solve these problems by the use of sonication [116]. A simple preparation of the liquid reagent was proposed, and the Etard reaction itself together with the hydrolytic step were conducted under sonication, with some success (Figure 8.35).

The bromination of aromatics under free radical conditions can also be considered as an oxidative process (eq. 27) [117]. Bromination using $CuBr_2$ requires forcing conditions, but Kodomari [118] reported that these reactions could be carried out under much milder conditions using $CuBr_2$ supported on alumina. It was found that the use of ultrasound during the impregnation of the Cu on the support followed by reaction under conventional conditions, gave the best results [119]. Studies of the effects of variation of the ultrasonic power used in the preparation of the $CuBr_2$/alumina reagent on reactivity in the bromination of naphthalene revealed that increasing the ultrasonic power up to an optimum level led to increasing surface disruption which was accompanied by increased reactivity. Further increases in power beyond the optimum led to agglomeration and decreased reactivity.

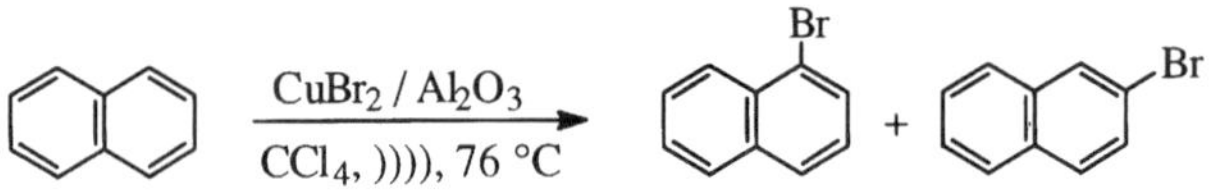

$$\text{naphthalene} \xrightarrow[CCl_4,)))), 76\,°C]{CuBr_2 / Al_2O_3} \text{1-Br-naphthalene} + \text{2-Br-naphthalene} \quad (27)$$

The oxidation of alcohols to carbonyl compounds has been studied by several authors; examples of such oxidations are illustrated (Figure 8.36) [120].

$$CrO_3 + TiCl_4 \xrightarrow{CH_2Cl_2,)))), \text{r.t.}, 2.5h} CrO_2Cl_2$$

$$CrO_2Cl_2 + ArCH_2R \xrightarrow{)))), 1h, \text{r.t.}} \underset{\text{Etard complex}}{ArCH_2RCrO_2Cl_2} \xrightarrow{Zn \text{ or } Na_2S_2O_3, H_2O} ArCOR$$

Figure 8.35 The Etard reaction.

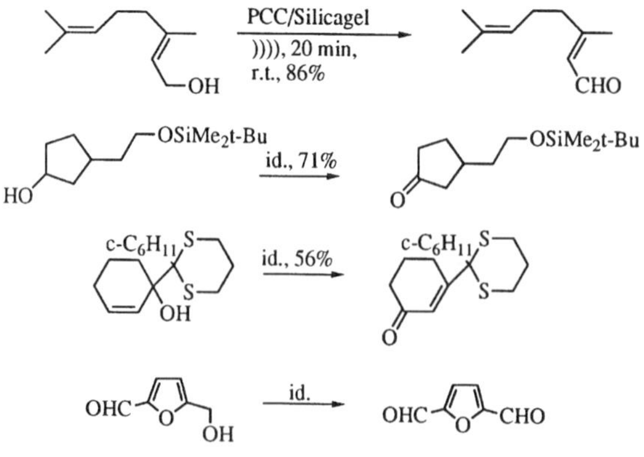

Figure 8.36 Oxidation of alcohols.

8.3. SONOCHEMICAL REACTIONS IN SYNTHESIS

Steroidal homoallylic alcohols can be converted to the corresponding 4-ene-3,6-diones using tetrapropylammonium perruthenate (TPAP) in catalytic amounts (eq. 28) [121]. In this case, the oxidizing agent is N-methyl morpholine N-oxide.

$$\text{HO-steroid} \xrightarrow[>90\%]{\text{NMO, CH}_2\text{Cl}_2\text{, Mol. Sieves,} \atop \text{TPAP,)))), 25°C, 90 min}} \text{O=steroid=O} \qquad (28)$$

8.3.3 Organometallic Sonochemistry

If one judges the significance of sonochemistry in different synthetic fields by the number of the papers published, it becomes clear that organometallic chemistry has seen the greatest impact. A recent review citing more than 230 references has appeared [40]. The erosion effects of cavitation on solid surfaces have been discussed already (see Section 8.2.2) [122]. In addition, the influence of ultrasonically enhanced corrosion is described [123]. The consequences for metal reactivity are important since passivating coatings which are frequently present on a metal surface (*e.g.*, oxides, carbonates and hydroxides) can be removed by the cavitational impacts. An illustration can be found with the activation of nickel powder and the determination of its surface composition by Auger spectroscopy (Figure 8.37) [124].

This "depassivation" effect, the importance of which seems to be strongly dependent on the particle size [125], is not the only factor responsible for the enhanced reactivity of metals. In some instances, the crystal itself is broken and a dispersion with a much larger surface area and increased reactivity is obtained [126]. Some other hypotheses should be considered to account for the chemical effects of sonication to supplement the purely mechanical effects. One such influence is the tribochemical effect (mechanochemistry) [22], in which electrons are ejected from the surface of a metal which is treated mechanically. Since electron transfer is known to constitute a crucial step in the formation of organometallic compounds [127], it is expected that sonochemistry, through a tribochemical effect, should significantly help in their preparation.

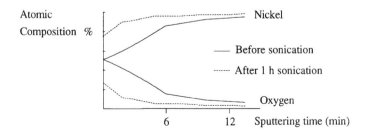

Figure 8.37 Auger effect spectroscopy of a nickel surface.

Preparation of Active Metals from Metals, their Salts and Complexes The preparation of an activated metal for use in an organometallic synthesis can be effected in a separate sonication step. Thus, sodium or potassium metal in aromatic solvents rapidly transform to a silvery blue "colloidal" suspension [126]. A blue-purple emulsion can be prepared from molten sodium in an inert solvent [128], an activation in which a frequency effect has been observed [129]. Molten zinc gives active powders, less than 100 μm in size, under sonication [130]. Dispersed palladium, platinum and rhodium prepared by sonochemical reduction of aqueous salts also exhibit a higher reactivity, assigned to a surface area increase [131].

The metallic couples of zinc with copper, nickel or cobalt form readily by sonication of zinc dust with salts, mainly halides, of the second metal in aqueous organic solvents. This method avoids the usual tedious preparative methods [132–135].

Reduction of metal salts to a finely divided active free metal, known as Rieke's method, has been explored using a series of metal chlorides [135, 136]. Sonication allows the reduction of copper salts to be achieved by lithium in THF at room temperature instead of potassium in the solvent under reflux. Sonication of titanium trichloride and lithium at 30 °C provides the so-called "low valent" titanium, whereas the silent reaction requires reflux conditions [136].

The reduction of transition metal halides by sodium sand under a carbon monoxide atmosphere with a low overpressure of 1–4 bar generates the corresponding metal carbonyl [137]. The usual method requires pressures of the order of 200 atmospheres and elevated temperatures. Significant examples of this type of activation are shown (Figure 8.38).

In situ Sonochemical Activation of Metals In many syntheses activation is not effected by sonochemical preparation of the metal alone, but rather by sonication of a mixture of the metal and an organic reagent(s). The first example was published almost half a century ago by Renaud, who reported the beneficial role of sonication in the preparation of organolithium, -magnesium, and -mercury compounds [138]. For many years, these important findings were not followed

$$TiCl_3 + Li \xrightarrow[)))))]{THF, 30°C} \text{"Ti"}$$

$$WCl_6 \xrightarrow[10°C,)))), 40\%]{Na, CO\ (1\ atm.),\ THF} W_2(CO)_{10}^{2-}$$

$$VCl_3(THF) \xrightarrow[THF,\ 10°C,\ 35\%]{Na,\ CO\ (4.4\ atm.)} V(CO)_6^-$$

Silent reaction: 160 °C, 200 atm.

Figure 8.38 Sonochemical reduction of metallic salts.

8.3. SONOCHEMICAL REACTIONS IN SYNTHESIS

up, possibly because of the lack of simple, cheap ultrasonic equipment. More recently, however, ultrasonic methods of this type have been widely adopted.

Alkali Metals and Magnesium

(a) Reduction of Conjugated Systems The use of sonication in the formation of arene-metal compounds, first described in 1957 as applied to benzoquinoline-sodium [139], has been extended to the preparations of naphthalene-, anthracene-, biphenyl-sodium (and lithium) [140]. Among these, lithium 4,4'-di(*t*-butyl)biphenyl (LiDBB) has frequently been used by synthetic sonochemists. This radical anion, used as an electron carrier, promotes several reactions including Barbier-type reactions and reductions of aromatic ketones, carboxylic esters or acids (Figure 8.39) [141].

The complex between magnesium and anthracene is generated and decomposed under sonication (eq. 29). During this cycle pieces of metal are converted into an activated magnesium powder which can be exploited in the preparation of Grignard reagents [142].

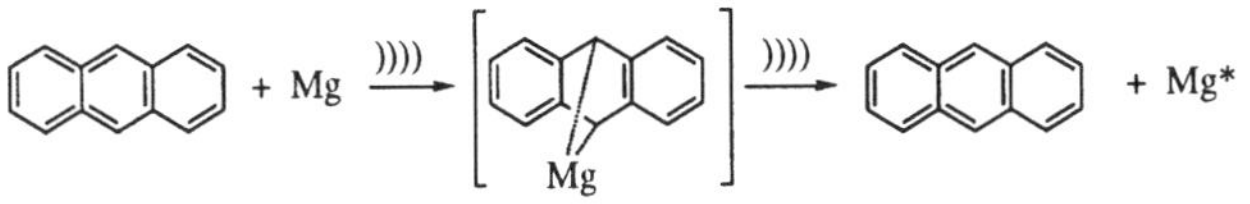

$$\text{anthracene} + \text{Mg} \xrightarrow{))))} [\text{anthracene-Mg complex}] \xrightarrow{))))} \text{anthracene} + \text{Mg}^* \quad (29)$$

Radical anions derived from conjugated dienes are also accessible through sonication. They behave as bases able to abstract a proton from an amine and can be used to prepare reagents such as lithium diisopropylamide (eq. 30) [143]. This preparation can be accomplished in one pot containing the substrate to be deprotonated (eq. 31).

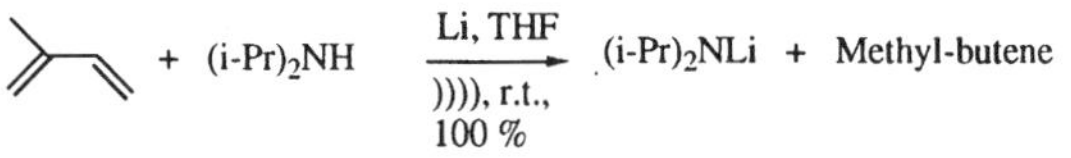

$$\text{isoprene} + (i\text{-Pr})_2\text{NH} \xrightarrow[\substack{))))\text{, r.t.,} \\ 100\%}]{\text{Li, THF}} (i\text{-Pr})_2\text{NLi} + \text{Methyl-butene} \quad (30)$$

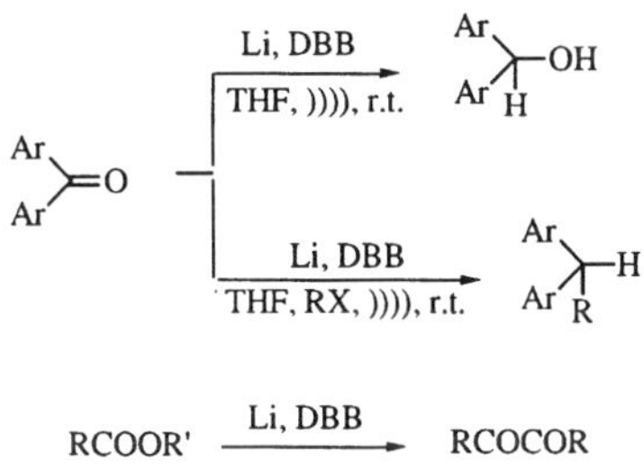

$$\underset{\text{Ar}}{\overset{\text{Ar}}{>}}\!\!=\!\!\text{O} \quad \begin{cases} \xrightarrow[\text{THF,)))), r.t.}]{\text{Li, DBB}} & \underset{\text{Ar}}{\overset{\text{Ar}}{>}}\!\!\!\underset{\text{H}}{\overset{\text{OH}}{\,}} \\ \xrightarrow[\text{THF, RX,)))), r.t.}]{\text{Li, DBB}} & \underset{\text{Ar}}{\overset{\text{Ar}}{>}}\!\!\!\underset{\text{R}}{\overset{\text{H}}{\,}} \end{cases}$$

$$\text{RCOOR'} \xrightarrow{\text{Li, DBB}} \text{RCOCOR}$$

Figure 8.39 Reductions of carbonyl groups in the presence of LiDBB.

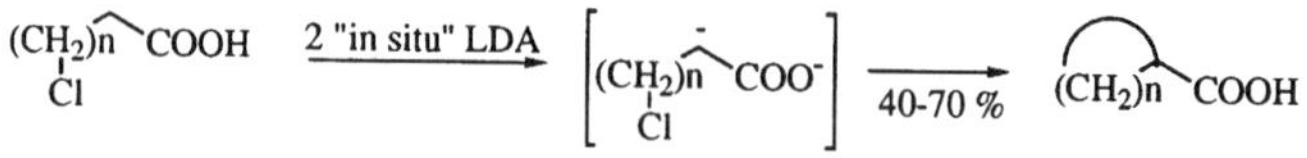

$$(CH_2)_n\underset{Cl}{\overset{}{\frown}} COOH \xrightarrow{2 \text{ "in situ" LDA}} \left[(CH_2)_n\underset{Cl}{\overset{-}{\frown}} COO^-\right] \xrightarrow{40\text{-}70\%} (CH_2)_n\overset{}{\frown} COOH \quad (31)$$

(b) Reduction of Carbonyl Groups Alkali metals and magnesium react with carbonyl groups to give radical anions (ketyls) widely used in synthesis. The reduction of C=O bonds by the "dissolving metal" process is known to follow a complex mechanism [144]. The sonochemical reduction of camphor by alkali metals in THF gave a mixture of borneol and isoborneol in the same ratio as in the absence of ultrasound [145]. This prompted the authors to exclude a dianionic mechanism for the process. If the initial ketyl radical anion is generated next to a functionality able to undergo addition or coupling, cyclization can occur. One such reaction is the acyloin condensation of diesters [146] performed smoothly by sonication instead of in refluxing solvents (eq. 32).

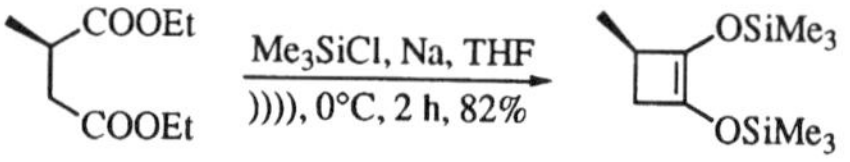

$$\underset{COOEt}{\overset{COOEt}{\diagup\!\!\!\diagdown}} \xrightarrow[\text{))))}, 0°C, 2\text{ h}, 82\%]{Me_3SiCl, Na, THF} \underset{OSiMe_3}{\overset{OSiMe_3}{\square}} \quad (32)$$

A sonochemical switching was observed in the formation of the indanone nucleus from *o*-allyl benzamides (Figure 8.40) [147]. The ketyl radical anion cyclizes to 2-methyl indanone and liberates an amide ion which deprotonates the allyl moiety. The resulting carbanion then undergoes cyclization to α-naphthol. Under sonication the first step of the process is accelerated and the ketyl is generated much more rapidly so that only the cyclization to 2-methyl indanone occurs.

(c) Reduction of Single Bonds A number of examples of the sonochemical cleavage of bonds involving heteroatoms in synthesis have been reported. Carbon-sulfur bonds in an allylic position in sulfones are reduced by ultrasonically dispersed potassium (eqs. 33, 34) [148]. The intermediate radical-anion can be trapped by electrophiles in a number of instances, and when the reaction is performed on a cyclic substrate, a new open chain sulfone is obtained. Sulfones

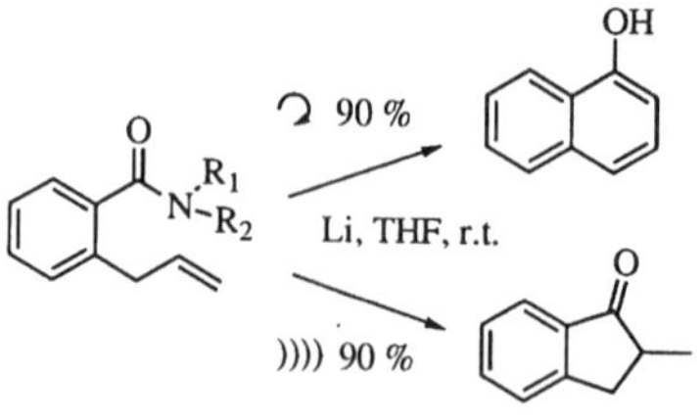

Figure 8.40 Reductive cyclizations of *o*-allyl-benzamide.

8.3. SONOCHEMICAL REACTIONS IN SYNTHESIS

unreactive under these conditions can, however, be activated when water or phenol is added to the reaction medium (eq. 35) [149]. Cleavages have also been reported which involve sulfur or selenium compounds. Thus, the Se–Se bond in diphenyldiselenide is reduced by sodium under irradiation at 50 kHz to give sodium phenylselenide (eq. 36) [150]. Catalytic amounts of benzophenone accelerate the reaction. The reagent prepared by this method exhibits improved nucleophilic properties.

$$\text{(sultine, R)} \xrightarrow[\text{)))), 1 min, >80 \%}]{\text{K, PhMe}} \text{R-CH=CH-R} \quad (33)$$

$$\text{(sultine, R,R')} \xrightarrow[\text{)))), 1 min, 90 \%}]{\text{K, PhMe}} \text{R-CH=CH-R} \quad (34)$$

$$\text{(bicyclic sultine)} \xrightarrow[\text{2. CH}_3\text{I, 50 \%}]{\text{K, PhMe-H}_2\text{O,))))}} \text{bicyclic-SO}_2\text{Me} \quad (35)$$

(0 % without water)

$$\text{PhSeSePh} + 2\,\text{Na} \xrightarrow[\text{)))), r.t., 5 min}]{\text{THF, Ph}_2\text{CO (cat)}} 2\,\text{PhSeNa} \quad (36)$$

A particularly important case of bond reduction is the formation of organometallic reagents from organic halides including widely used synthetic Grignard and organolithium reagents. The applications are extremely broad, and so great interest is shown in the reaction parameters for these processes together with the kinetics and especially the factors which influence initiation. Frequently, there is an induction period which can make these reactions somewhat hazardous when performed on an industrial scale. Sonication usually reduces the induction period or even completely suppresses it [151]. Organolithium compounds are obtained in good yields from primary alkyl bromides, although secondary and tertiary alkyl bromides require longer reaction times (>1 h) (eq. 37). Magnesium activation is easily accomplished by sonication [151, 152], even in wet solvents, and kinetic studies show the importance of surface activation and transport phenomena [153]. Functionalized reagents and others, inaccessible by conventional methods, can be prepared smoothly (eq. 38) [154].

$$\text{R-X} + 2\,\text{Li} \xrightarrow[\text{))))}]{\text{THF, r.t.}} \text{R-Li} + \text{LiX} \quad (37)$$

R = Et, *n*-Bu, *i*-Pr, 2-propenyl, Ph, *c*-pentyl, ...

$$\underset{\text{Br}}{\diagup\!\!\!\diagup\!\!\!\diagup\!\!\!\diagdown}\text{SiMe}_3 \xrightarrow[\text{))), 1 h, 50°C}]{\text{Mg, THF}} \underset{\text{MgBr}}{\diagup\!\!\!\diagup\!\!\!\diagup\!\!\!\diagdown}\text{SiMe}_3 \quad (38)$$

The Barbier reaction of an organic halide and a carbonyl compound in the presence of a metal can be considered to be essentially a one-pot Grignard synthesis. It allows the chemist to avoid the sometimes difficult or tedious initial preparation of the organometallic reagent [155]. Sonication has been shown to broaden substantially the applicability of the Barbier reaction [151]. In a study of the kinetics of such a reaction under sonication an inverse relationship between temperature and reaction rate was found. When the rates were determined between -50 to $+25\,°C$, an optimum reaction rate was found at *ca.* $0\,°C$ [31]. This result suggested that sonication was promoting electron transfer from the metal surface and this was confirmed by a study of the mechanism [156].

Since there are intrinsic advantages of sonochemical methodology, many synthetic applications are to be found in the literature. Highly successful syntheses involve the use of benzylic halides [157], and of allylic phosphates as a source of allyl anions [158]. Magnesium provides good yields in the preparation of perfluoroalkyl carbinols [159]. A particular case of an intramolecular Barbier reaction is shown (eq. 39) [160]. An extension of this procedure to the synthesis of formamides, known as the Bouveault process, leads to aldehydes (eq. 40) [161].

[indole alkaloid structure with I and COOMe] $\xrightarrow[\text{)))), r.t.}]{\text{Na, THF}}$ [cyclized product with COOMe] (39)

$$\text{Ph-Br} + \text{HOC-N}\diagup\!\!\!\diagdown\text{N}- \xrightarrow[\substack{\text{)))) 50 kHz, r.t., 1h, 0\%,} \\ \text{)))), 500 kHz, 45 min, 77\%}]{\text{Li, Et}_2\text{O}} \text{Ph-CHO} \quad (40)$$

Another extension of the Barbier reaction, in this case to carboxylate salts [162], affords a simple access to furanyl ketones [163]. By sonication of a mixture of a lithium carboxylate, an alkyl chloride and lithium in THF at room temperature, the intermediate organo-lithium reagent forms rapidly, then generates the 2-furanyl lithium which adds to the carboxylate group in high yields. The method constitutes an example of a reaction "cascade", in which several intermediates are generated sequentially (Figure 8.41).

(d) Coupling Reactions Many studies have been devoted to Wurtz-type coupling reactions under sonochemical conditions. Alkyl, aryl, and heteroaryl halides give homocoupling products with lithium in THF, when sonicated at room

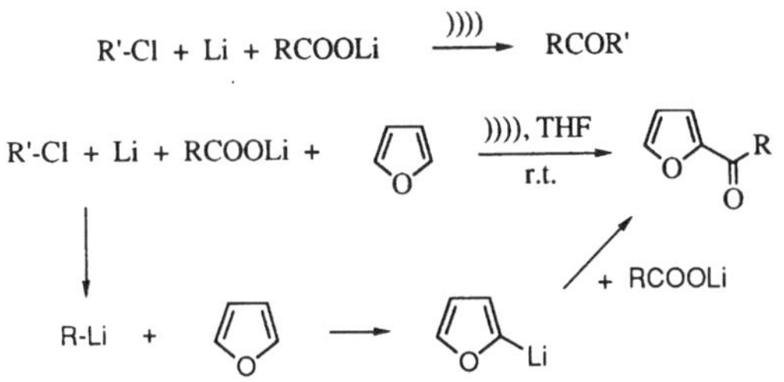

8.3. SONOCHEMICAL REACTIONS IN SYNTHESIS

Figure 8.41 Preparation of furanylketones *via* a reaction cascade.

temperature (eqs. 41, 42) [164]. Yields are usually moderate to good, but little or no reaction takes place in the absence of ultrasound. The coupling of bromotoluenes and bromo-pyridines provides a mixture of isomeric products which implicates a radical intermediate [165]. This was confirmed by the inhibition of the reaction in the presence of radical scavengers.

$$n\text{-}C_3H_7\text{-}Cl \xrightarrow[\text{)))), r.t., 17 h, 72\%}]{\text{Li, THF}} n\text{-}C_6H_{14} \tag{41}$$

$$\text{Ar-X} \xrightarrow[\text{)))), r.t., 10-20 h, 36-70\%}]{\text{Li, THF}} \text{Ar-Ar} \tag{42}$$

The source of the majority of research in homo or heterocoupling reactions has been in organosilicon chemistry aimed at finding new routes to polymers [166]. Homocoupling reactions (*e.g.*, eq. 43) were first described by Boudjouk *et al.* who noted an improved yield of the reaction with less reactive substrates could be achieved in the presence of anthracene [167].

$$R_3\text{SiCl} \xrightarrow[\substack{R = \text{Et:)))), r.t., 60 h, 15\%} \\ \text{with anthracene: 2 h, 58\%}}]{\text{Li, THF}} R_3\text{SiSiR}_3 \tag{43}$$

The case of dichlorosilanes is much more complex since, in these cases, polymerization is accompanied by the formation of cyclic oligomers (eq. 44) [166, 168, 169]. The relative importance of cyclic oligomers *vs* linear polymers depends on the solvent and the metal, *e.g.*, sodium in toluene gives mainly a linear product, but lithium in THF gives 5 or 6 membered rings. It is interesting to note that the intriguing phenomenon of chemical oscillation occurs in these reactions. The equilibrium between the cyclic and linear compounds is obtained only after a certain time which depends on the reaction conditions (600 min in the presence of 50% excess of sodium).

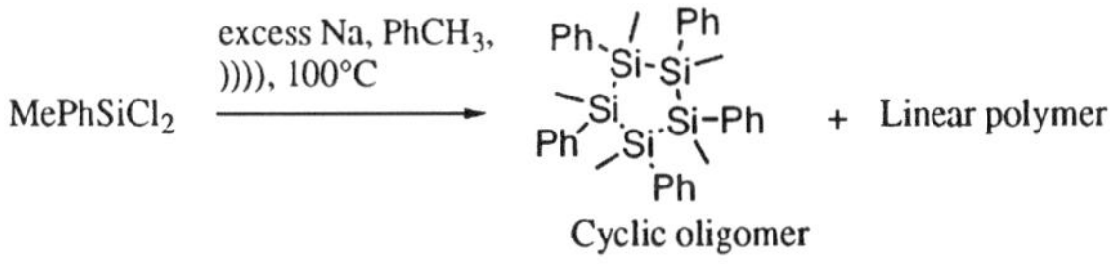

$$\text{MePhSiCl}_2 \xrightarrow[\text{)))), 100°C}]{\text{excess Na, PhCH}_3} \text{Cyclic oligomer} + \text{Linear polymer} \tag{44}$$

In these reactions, after the first metallation has occurred, it is possible that an α-elimination can occur which is equivalent to the process leading to carbenes. With silicon as the atom undergoing the reaction, a "silylene" would be involved. This possibility has been investigated, and two supporting pieces of evidence have been produced. The dimerization of a highly hindered compound gave a disilene compound [170], and trapping by isomeric alkenes gave isomeric silacyclopropanes [171]. These reactions only occur under the influence of ultrasound (eq. 45).

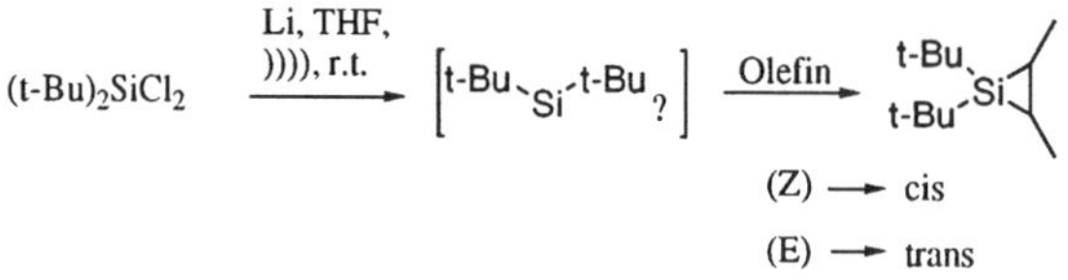

$$(\text{t-Bu})_2\text{SiCl}_2 \xrightarrow{\text{Li, THF,)))), r.t.}} [\text{t-Bu-Si-t-Bu}_?] \xrightarrow{\text{Olefin}} \begin{array}{c} \text{t-Bu} \\ \text{t-Bu} \end{array}\!\!\!\text{Si}\!\!\!\!\!\!\begin{array}{c} \\ \\ \end{array} \qquad (45)$$

(Z) → cis
(E) → trans

The formation of α-disulfones by a coupling reaction has been described (eq. 46) [172]. These compounds, useful as polymerization photosensitizers, are formed in much greater yields when ultrasound is used.

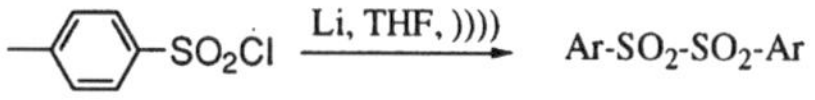

$$\text{Ar-SO}_2\text{Cl} \xrightarrow{\text{Li, THF,))))}} \text{Ar-SO}_2\text{-SO}_2\text{-Ar} \qquad (46)$$

(e) Other Uses of in situ Generated Organometallics The *in situ* generated organoalkali metal reagents can be used to effect deprotonation of dithiane, terminal alkynes, aromatic compounds and, importantly, secondary amines (Figure 8.42) [173, 174]. A cascade reaction giving the isobutyric acid dianion was achieved *via* the initial formation of butyl lithium from its precursors. Subsequent deprotonation of diisopropylamine to LDA and then double deprotonation of the acid by LDA led to the reactive intermediate which, on addition of an aldehyde, generates the hydroxyacid in 78% yield.

Preparation of Organometallic Reagents by Transmetallation Metal exchange from lithium or magnesium reagents and a metallic salt is a procedure frequently used to prepare organometallics not easily accessible by direct reaction of the

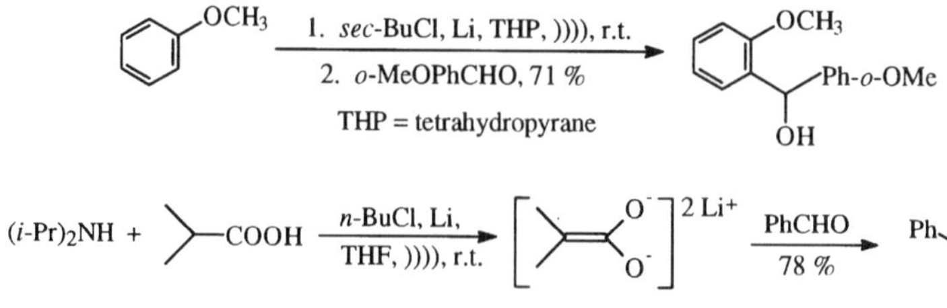

PhOCH₃ →(1. sec-BuCl, Li, THP,)))), r.t.; 2. o-MeOPhCHO, 71%)→ Ph(OCH₃)(CH(OH)Ph-o-OMe)
THP = tetrahydropyrane

$(i\text{-Pr})_2\text{NH} + \text{(i-Pr)COOH} \xrightarrow{n\text{-BuCl, Li, THF,)))), r.t.}} [\text{dianion}]^{2-}\,2\text{Li}^+ \xrightarrow{\text{PhCHO}, 78\%} \text{Ph-CH(OH)-C(Me)}_2\text{-COOH}$

Figure 8.42 Deprotonations with *in situ* generated butyllithium.

8.3. SONOCHEMICAL REACTIONS IN SYNTHESIS

metal. Sonication has made these transmetallations possible in rapid one-pot reactions and the preparation of organozinc compounds illustrates this methodology. Alkyl, vinyl or aryl reagents are obtained in short times and excellent yields from lithium, alkyl halides and a zinc salt (Figure 8.43) [175]. Di-organozinc compounds undergo conjugate addition to α-enones, even when they are sterically congested. Copper reagents are ineffective in such cases [176]. The coupling of dialkylzinc compounds with a cyclopropyl bromide under sonication has also been described [177].

Triorganoboranes are obtained *via* the *in situ* formation of Grignard reagents and metal exchange with boron trifluoride etherate (eq. 47) [178]. The interest in this method lies in the possibility of preparing organoboranes which were previously inaccessible through the usual hydroboration method.

$$PhCH_2Br \xrightarrow[\text{))))}, 30 \text{ min}, 99\,\%]{Mg,\ BF_3\ OEt_2} (PhCH_2)_3B \qquad (47)$$

$$\circlearrowright\ 24\text{ h}, 99\,\%$$

Reactions with Zinc Metal Historically, zinc was the first metal to be used in organic chemistry [179]. Since its synthetic potential is extremely diverse and it is in common usage [180], it is not surprising that sonochemists have turned their attention to its extensive use in synthesis.

(a) Reduction of Functional Groups by Zinc The zinc-nickel couple will reduce C=C bonds in aqueous alcohol dispersions [132, 181]. The chemical role of nickel seems to be threefold, activation of zinc, easier reduction of water to hydrogen, then catalysis of the hydrogenation step. The selectivity of the system can be modified by an adjustment of the conditions (Figure 8.44).

Zinc transfers electrons to carbonyl groups, with a complex formation of the ketyl intermediates, as in the case of the Clemmensen reduction. To avoid the usual strongly acidic conditions, sonochemical improvements have been made

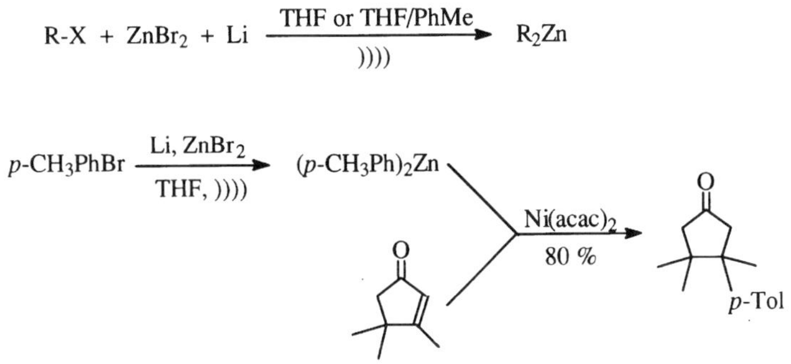

$$R\text{-}X + ZnBr_2 + Li \xrightarrow[\text{))))}]{THF\text{ or }THF/PhMe} R_2Zn$$

$$p\text{-}CH_3PhBr \xrightarrow[THF,\ ))))]{Li,\ ZnBr_2} (p\text{-}CH_3Ph)_2Zn \xrightarrow[80\,\%]{Ni(acac)_2}$$

Figure 8.43 Preparation and reaction of dialkylzinc reagents.

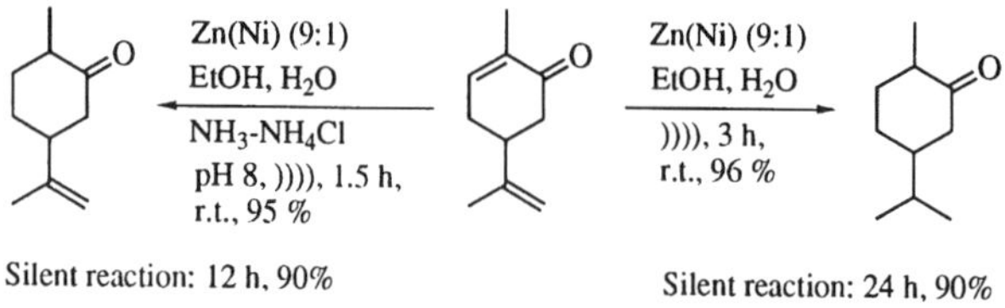

Figure 8.44 Selective reductions in the presence of zinc-nickel couple.

as in the mild reduction of 3-keto and 3-keto-Δ^4 steroids in acetic acid (eq. 48) to the expected deoxygenated products [182].

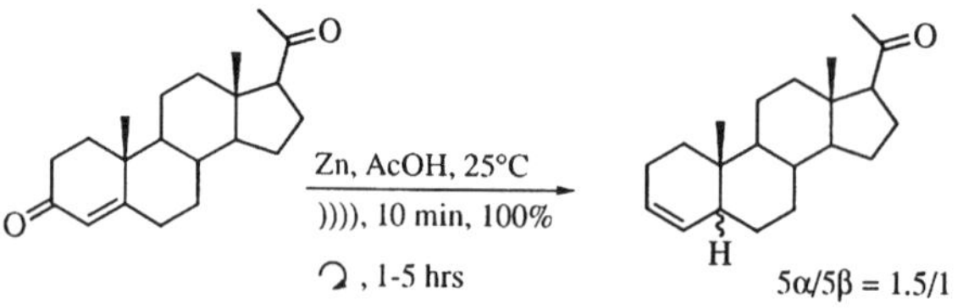

$$5\alpha/5\beta = 1.5/1 \tag{48}$$

The ketyl radical anions undergo a different reaction sequence in neutral media where a pinacolization reaction occurs. This is often referred to as a facile reaction but in fact it presents many difficulties when used in synthesis. Some sonochemical improvements have been made, for instance by performing the reaction in the presence of trimethylchlorosilane (which traps the dialkoxide) [183]. When dimerization is not possible, as in the case of quinones or α-diketones, the result is a disilyl ether (Figure 8.45) [184].

(b) Preparation and Reactions of Organozinc Reagents The preparation of organozinc reagents from organic halides can be achieved under sonication. An example is the formation of aromatic derivatives from the corresponding iodides [185]. In most examples, however, reaction occurs in the preparation medium, *i.e.*, a one-pot method, as in a "Barbier" procedure. Coupling with vinylic and allylic halides in the presence of a palladium catalyst has been described [186].

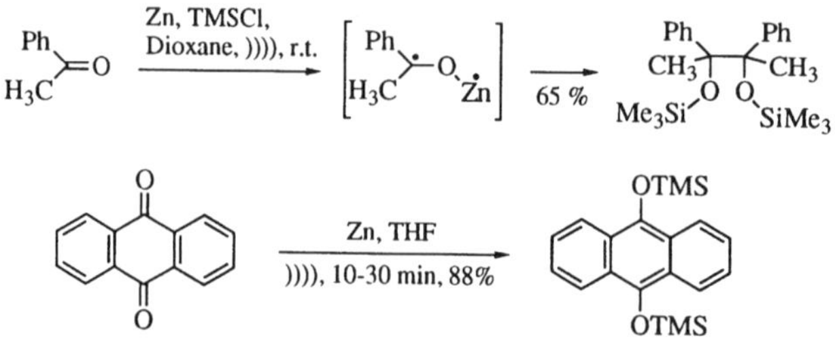

Figure 8.45 Carbonyl reductions by zinc/TMSCl.

8.3. SONOCHEMICAL REACTIONS IN SYNTHESIS

Terminal alkynes and isoprene undergo hydroperfluoroalkylation with perfluoroalkyl iodides and zinc in the presence of copper, palladium or titanium catalysts. They also add to unsaturated bonds to effect carbometallation reactions [187]. A selection of this zinc sonochemistry is shown (Figure 8.46).

Carbenoid reagents add to alkenes in the Simmons-Smith cyclopropanation and this was one of the first organometallic reactions studied under sonication [188]. The unpredictable induction period is suppressed, and good, reproducible yields, significantly higher than could be achieved using conventional methodology, are obtained using various types of zinc (eq. 49).

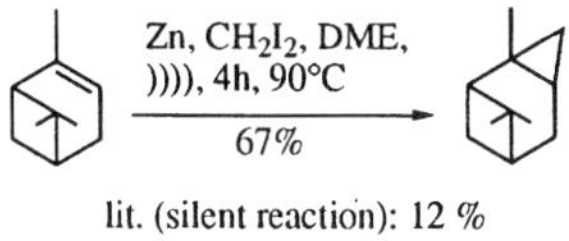

$$\text{norbornene} \xrightarrow[67\%]{\text{Zn, CH}_2\text{I}_2, \text{DME},))), 4\text{h}, 90°\text{C}} \text{cyclopropanated product} \tag{49}$$

lit. (silent reaction): 12 %

When a β-keto alkene is used as substrate the carbenoid does not add directly to the double bond, instead it induces cyclization to a furan (eq. 50) [189]. The corresponding homoallylic alcohol undergoes the expected Simmons-Smith reaction, although in rather moderate yield (31%).

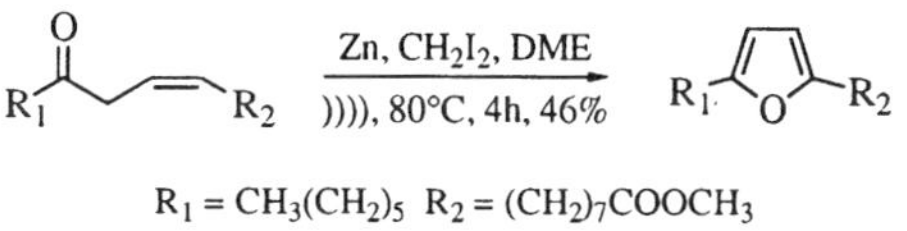

$$R_1\text{-CO-CH}_2\text{-CH=CH-}R_2 \xrightarrow[))), 80°\text{C}, 4\text{h}, 46\%]{\text{Zn, CH}_2\text{I}_2, \text{DME}} R_1\text{-furan-}R_2 \tag{50}$$

$R_1 = CH_3(CH_2)_5$ $R_2 = (CH_2)_7COOCH_3$

Ph−Br + CF$_3$I $\xrightarrow[))),\ \text{r.t.},\ 1\text{h},\ 65\%]{\text{Zn, THF, Pd(PPh}_3)_4}$ Ph−CF$_3$

H$_3$C−Br + n-C$_4$F$_9$-I $\xrightarrow[))),\ \text{r.t.},\ 1\text{h},\ 51\text{-}78\%]{\text{Zn, THF, Pd(OAc)}_2}$ CH$_3$−CH=CH−n-C$_4$F$_9$

CF$_3$I + HO−C≡CH $\xrightarrow[))),\ \text{r.t.},\ 2\text{h},\ 61\%]{\text{Zn, CuI, THF}}$ F$_3$C−CH=CH−OH

CF$_3$I + isoprene $\xrightarrow[))),\ 56\%]{\text{Zn, THF, Cp}_2\text{TiCl}_2\ (\text{cat})}$ F$_3$C−CH=C(CH$_3$)−CH=CH$_2$

ROCO−CHBr−CH$_3$ + ≡−R $\xrightarrow[))),45\text{-}50°\text{C}\ 48\text{-}81\%]{\text{Zn, THF}}$ ROCO−C(=CH$_2$)−R

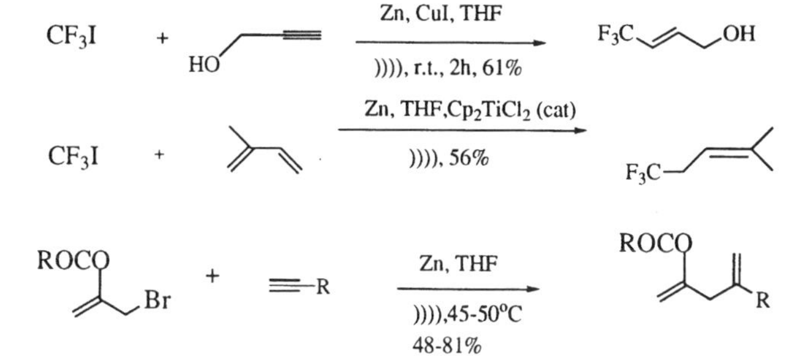

Figure 8.46 Coupling and addition reactions of organozinc reagents.

(c) Zinc in Barbier Type Processes Zinc can be used with success in Barbier type additions [190], and the selectivity thus obtained can be quite different from reactions involving lithium or magnesium. Thus, in the reaction of 4-bromo-2-sulfolene with various carbonyl compounds in the presence of the zinc-silver couple addition occurs at carbon 4 (eq. 51). When the reaction is performed with magnesium, however, the alcohol moiety becomes attached to carbon 2 [191].

$$\text{4-bromo-2-sulfolene} \xrightarrow[\text{)))), r.t., 5 h, 97 \%}]{\text{Zn(Ag), THF, CH}_3\text{CHO}} \text{product} \qquad (51)$$

The reaction of α,α'-dibromo-*o*-xylene, an activated alkene, and zinc gives rise to cyclic adducts (eq. 52) [192].

$$\text{dibromoxylene} + \text{alkene} \xrightarrow[\text{)))), 67-89\%}]{\text{Zn, dioxane}} \text{cyclic adduct} \qquad (52)$$

Z = electron withdrawing group, Z' = idem or H

No reaction without))))

The sonochemical Barbier procedure was used to perform a number of diverse perfluoroalkylations of aldehydes and ketones in DMF. In an illustrative example, tricarbonylchromium benzaldehyde is readily transformed to trifluoromethyl carbinol (eq. 53) [193]. Under these conditions, the cleavage of the ligand-metal bond which is a possible side-reaction [34] did not take place.

$$(CO)_3Cr\text{-ArCHO} \xrightarrow[\text{)))), 20-30 °C, 1 h, 85 \%}]{C_2F_5I, Zn, DMF} (CO)_3Cr\text{-Ar-CH(C}_2F_5)\text{OH} \qquad (53)$$

(d) The Reformatsky Reaction The Reformatsky reaction [194] has been used in many synthetic applications under sonochemical conditions with various substrates, *viz.* aldehydes, ketones, nitriles, imines and some less common functional groups [195]. In model cases, Boudjouk, followed by Suslick, observed significant rate and yield improvements [196]. The origin of the higher reactivity under sonication was ascribed to morphological and composition changes of the metal. The reaction proceeds well even with uncommon bromoesters such as ethyl bromo difluoroacetate [197]. In another case, an intramolecular Blaise reaction (Reformatsky applied to nitriles) is accomplished using the unusual zinc-silver couple [198] to give the stable enamine. A few examples of typical sonochemical Reformatsky reactions are given (Figure 8.47).

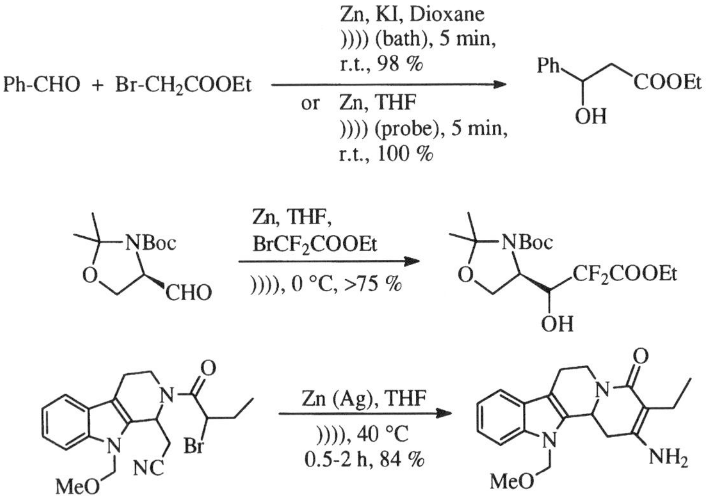

8.3. SONOCHEMICAL REACTIONS IN SYNTHESIS

Figure 8.47 Reformatsky reactions.

Instead of being prepared *in situ*, some Reformatsky reagents can be obtained in a separate step under sonication using a zinc-copper couple in benzene-dimethylacetamide solution [199]. They react with acyl chlorides or aryl iodides in the presence of palladium to give acyl- or arylaminoacid derivatives (Figure 8.48).

(e) Zinc in "Organometallic" Reactions in Aqueous Solvents One of the most surprising aspects of sonochemistry is that some reactions which are known to require anhydrous conditions under conventional methodology can be run in aqueous solvents using sonochemistry. This is the case for allylation of carbonyl compounds in the presence of zinc [200]. The first examples were performed in aqueous THF in the presence of zinc, but it was found later that the reaction could be performed without sonication in the presence of ammonium chloride.

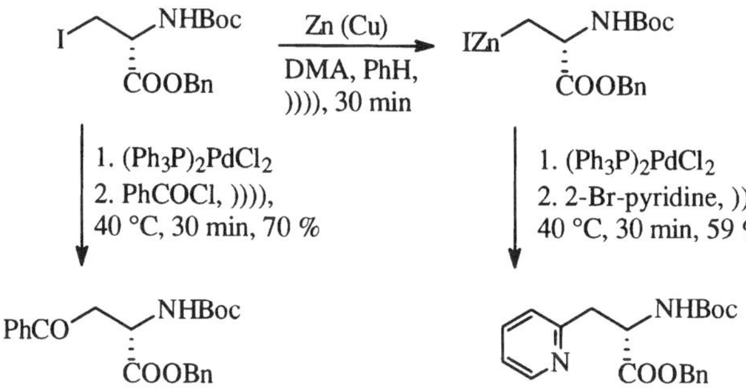

Figure 8.48 Synthesis of β-acyl- or arylaminoacids.

Since organozinc compounds are known to react with water, it seems likely that the aqueous reaction proceeds *via* radicals. In some instances, zinc can be replaced by tin, which provides higher yields. At the same time the selectivity in favour of aldehydes is increased, and the reaction is feasible in the presence of unprotected ketones. Later, further improvements were obtained by using indium as the metal (Figure 8.49) [201].

In the above reactions only the unsaturated halides react and the saturated analogues remain unchanged. However, saturated halides undergo clean conjugate additions to electron deficient alkenes in the presence of the zinc-copper couple [202]. Secondary and tertiary alkyl bromides and iodides give excellent yields, but iodides are preferred for primary derivatives. Because chlorides are inert and a hydroxyl group does not inhibit the reaction, functionalized groups can be introduced without any complications through side reactions and without the need for any side-group protection (Figure 8.50). Epoxyalkyl halides add without affecting the oxirane ring [203]. The activated alkene can be an aldehyde, ketone, nitrile, or amide. Yields better than those obtained with the usual tributyltin method are observed, but no stereochemical change is produced by these unusual reaction conditions when compared with classical ones [204]. This methodology has been applied to the synthesis of vitamin D_3 analogs [205].

Other Metals The Ullmann coupling of aryl halides by copper, under a series of conditions, was studied by Lindley [206]. The best results, exhibiting a substantial rate increase, were obtained with copper flakes in four-fold excess in DMF under probe sonication (eq. 54). An observed decrease in the metal particle size proved to be insufficient to explain all of the improvement, and contributions from a breakdown of intermediates and/or the desorption of the products from the metal surface were suggested.

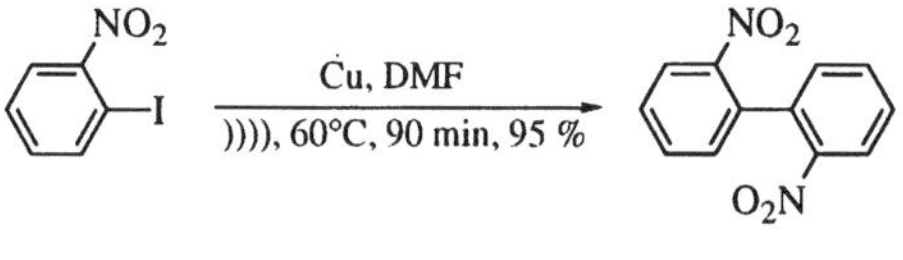

$$(54)$$

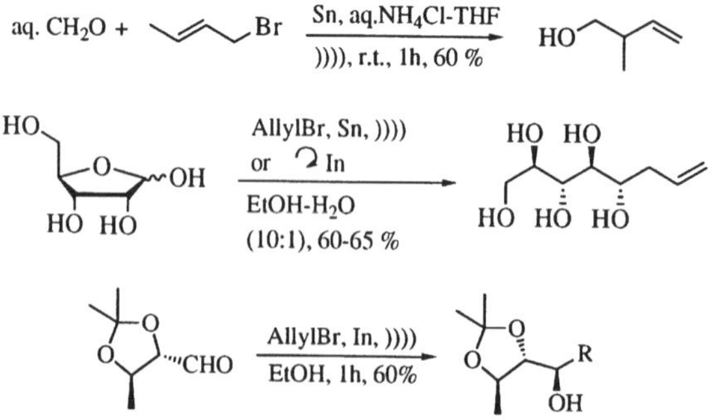

Figure 8.49 Aqueous allylation of carbonyl groups.

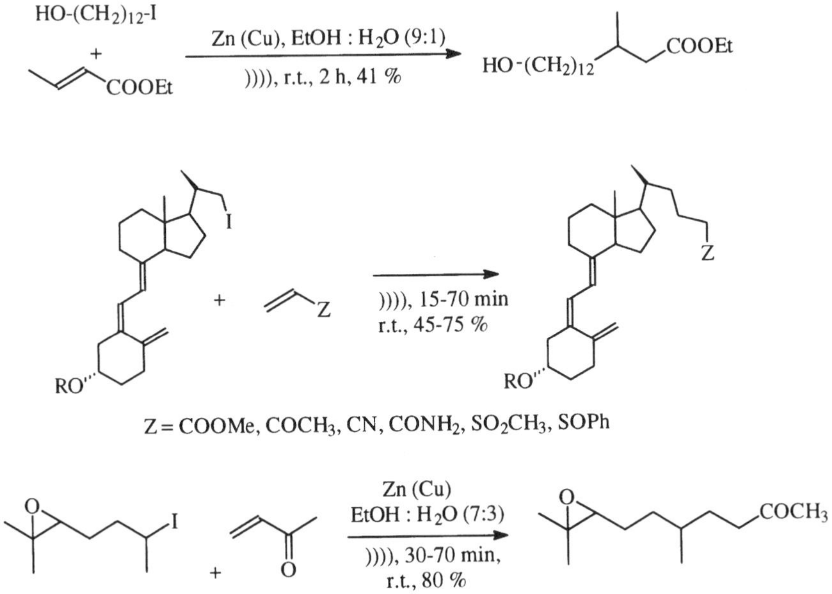

8.3. SONOCHEMICAL REACTIONS IN SYNTHESIS

Figure 8.50 Conjugate additions in aqueous media.

Picryl bromide can be coupled at or below room temperatures by copper powder used in only 10% excess to give coupling or reduction products depending on the solvent and stoichiometric conditions (eq. 55) [207]. Sonications were performed with a pulsed wave, a technique not frequently used.

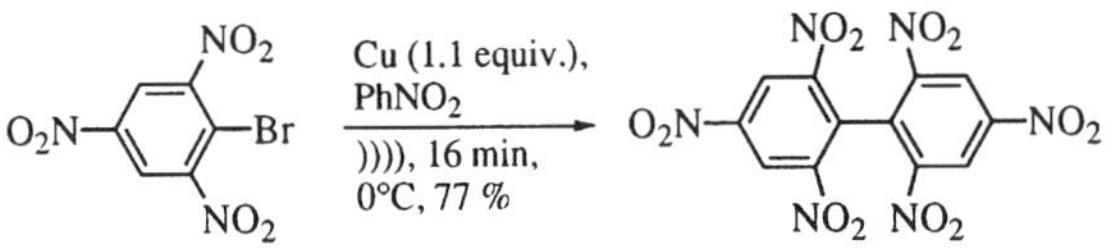

(55)

Aluminum is easily activated by sonication. The organic sonochemistry of this metal is, however, limited to only a few papers dealing with the preparation of its sesqui- or trialkyl compounds [208], or the use of aluminium amalgam in the reduction of epoxides (eq. 56) [209], nitro groups [210], or phthalimides [211]. In each case sonication induced rate and yield increases.

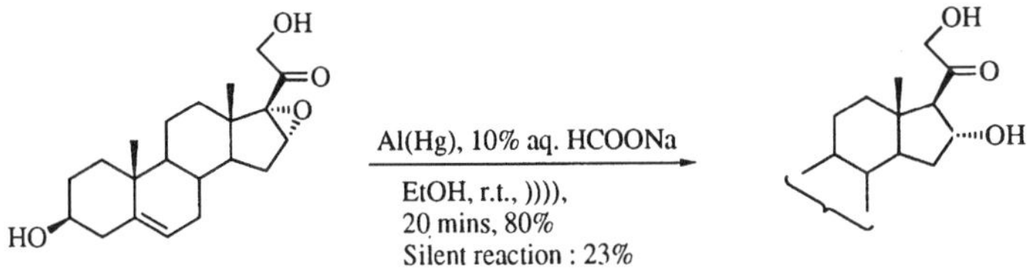

(56)

8.3.4 Sonochemically Assisted Catalysis

Sonochemically activated catalysis is currently an active area of interest, and there are many examples in the literature. Unfortunately, much of this is not easily accessible in the West [131]. Irradiation with ultrasound can be expected to have important uses at two stages in catalytic processes: either in the preparation of the catalyst itself or during the catalytic reaction [212].

Catalyst Preparation Metal powders such as nickel are generally poor catalysts, but can be activated to perform as good catalysts by sonication before use. Thus 3 micron nickel which is a poor catalyst in hydrogenation can be made quite efficient by sonication in ethanol before being used in the hydrogenation of oct-1-ene [213]. The reason for this increase in activity is undoubtedly the removal of a passivating outer oxide layer [214]. The Auger effect analysis of the surface of nickel after sonication clearly reveals the effect of cavitation upon the change in the composition of the surface [124]. Other analyses (XPS, Ar^+-ion sputtering, area measurement) confirmed that the surface area increase is negligible, but reactive sites are created and the elemental distribution is modified [215].

Totally new catalyst types can be produced, as in the generation of amorphous metals, *e.g.*, iron from the sonolysis of $Fe(CO)_5$ (0.4 M) in decane under argon [124]. In development of this process nanosized particles have been produced with ultrasound, *e.g.*, Mo_2C, with the high surface area of 188 $m^2 g^{-1}$ is generated on sonication of molybdenum carbonyl in hexadecane at 90 °C under argon [216]. In preliminary trials, this material has been shown to be 100% selective in the dehydrogenation of cyclohexane to benzene.

New supported catalysts have been produced through the sonochemical impregnation of catalytic materials into supports, *e.g.*, the production of ruthenium 1% (w/w) on alumina [217]. Sonically prepared material differs markedly from that prepared conventionally in that the metal is impregnated as a zone below the alumina surface rather than at the surface. The sonochemical impregnation of $CuBr_2$ on alumina as a medium for catalytic bromination of naphthalene has been explored [218]. Initial results in this area showed a markedly improved efficiency in the preparation of the supported catalyst *via* sonochemical methodology.

Recent studies on the effect of sonication on suspended powders have shown that the particles can be forced into such violent collision that, in the case of metals, fusion can occur [219]. In some cases, the colliding powders can undergo chemical reaction: thus, a mixture of powdered copper and sulfur sonicated in hexane generates CuS [220]. Particle collision is implicated in the sono-catalytic decomposition of aqueous NaOCl in the presence of two metal oxide powders: CuO and MnO_2 [221]. Certainly one of the results of the sonication is the reduction in particle size [10]. Both MnO_2 and CuO were reduced from 21 μm and 32 μm (vmd), respectively, to approximately 5 μm in under 5 minutes. If a mixture of these materials is pre-sonicated in water prior to

8.3. SONOCHEMICAL REACTIONS IN SYNTHESIS

normal thermal reaction with NaOCl, the efficiency was improved compared with unsonicated material, whereas if the metal oxides were pre-sonicated separately the reaction was not effected. This suggests that sonication has produced a more effective catalytic material.

Sonication During Catalysis Nickel, either the Raney form or the sonochemically activated metal powder, seems to have received more attention than the other catalysts. Important results were obtained by Saracco and Arzano in a study of the role of sonication in the hydrogenation of unsaturated fatty acid esters with Raney nickel in cyclohexane [222]. As previously discussed, an optimal rate results from the adjustment of the acoustic intensity. Studies reporting frequency effects are not numerous, and in this case, a clear influence of this parameter was shown (Figure 8.51).

Hydrogenolysis of hydrazines using a Raney nickel catalyst is performed with hydrogen at normal pressure and room temperature, with advantages in terms of functional selectivity (benzylic C–N bonds are not cleaved) and stereoselectivity (no racemization at chiral centers) (eq. 57) [223].

$$\text{(57)}$$

Raney nickel modified by tartaric acid becomes a highly enantiodifferentiating catalyst in the hydrogenation of β-dicarbonyl compounds [224]. The authors determined that the enantioselective catalytic sites correspond to pure crystalline nickel domains. Domains containing aluminium have a disordered structure and are not enantioselective. The role of sonication during the activation step consists of dissolving the disordered zones. Enantioselectivities higher than 85% are obtained with this catalyst (eq. 58). The reaction could be conducted with 10 kg of methyl acetylacetonate and 190 g of catalyst, to give

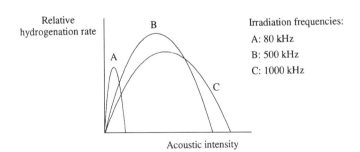

Figure 8.51 Hydrogenation rates as a function of acoustic intensity and frequency.

3-hydroxybutanoate in 81% e.e..

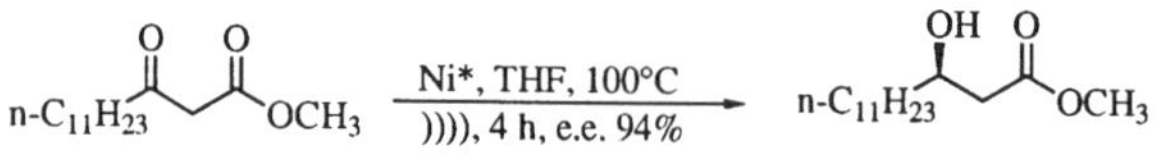

$$\text{n-C}_{11}\text{H}_{23}\text{COCH}_2\text{COOCH}_3 \xrightarrow[\text{))))}, 4\text{ h, e.e. 94\%}]{\text{Ni*, THF, 100°C}} \text{n-C}_{11}\text{H}_{23}\text{CH(OH)CH}_2\text{COOCH}_3 \quad (58)$$

The use of palladium for the hydrogenolysis of a benzyloxy group has been mentioned (without experimental details of the sonolysis) for the synthesis of 3-aminonocardicinic acid (eq. 59) [225]. The same catalyst gives excellent results for the hydrogenation of difluoro alkenes (eq. 60) but fails in the absence of sonication [226]. Alkenes undergo hydrogenation in the presence of Pd/C with formic acid as the source of hydrogen [227]. Sonication provides considerable acceleration of a reaction which occurs without an induction period.

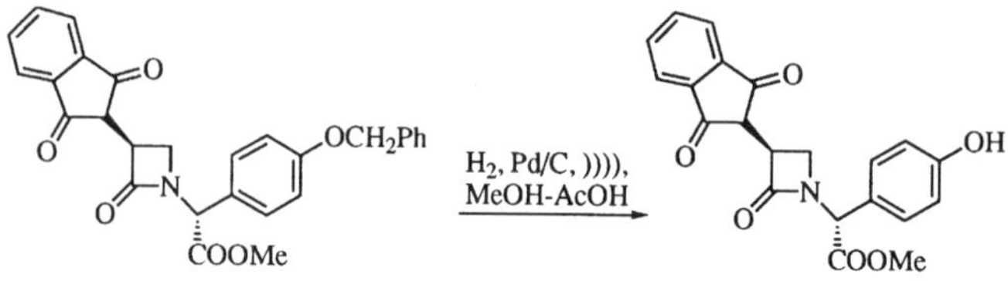

$$\text{(azetidinone-CH(Ar-OCH}_2\text{Ph)-COOMe)} \xrightarrow[\text{MeOH-AcOH}]{\text{H}_2, \text{Pd/C,))))}} \text{(azetidinone-CH(Ar-OH)-COOMe)} \quad (59)$$

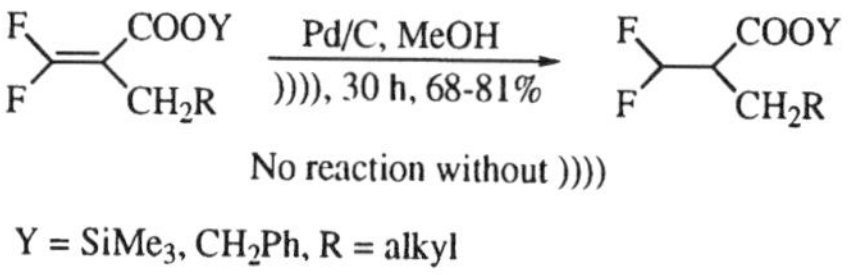

$$\underset{F}{\overset{F}{>}}\!\!=\!\!\underset{CH_2R}{\overset{COOY}{<}} \xrightarrow[\text{))))}, 30\text{ h, 68-81\%}]{\text{Pd/C, MeOH}} \underset{F}{\overset{F}{>}}\text{CH-CH(COOY)(CH}_2\text{R)} \quad (60)$$

No reaction without))))

Y = SiMe$_3$, CH$_2$Ph, R = alkyl

Catalytic hydrosilylation reactions have been investigated under sonication conditions. Sonication of hydrosilanes and alkenes with a Pt/C catalyst at 30 °C and atmospheric pressure gives hydrosilylation products in high yields (70–96%) in 1–2 h [228]. Addition of trichlorosilane to 1-hexene occurs quantitatively in 40 min with a specially prepared activated nickel catalyst (Figure 8.52) [229]. The interest of this result lies in the possibility of its use as an inexpensive replacement for platinum compounds.

Sonication during the catalytic dehydrogenation of tetrahydronaphthalene to naphthalene using 3% Pd/C in digol has been reported [230]. Under

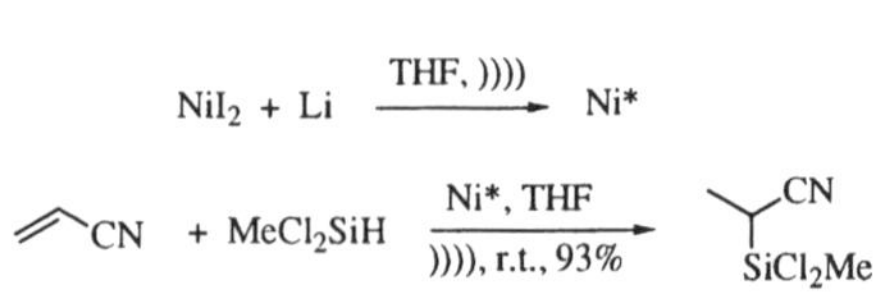

$$\text{NiI}_2 + \text{Li} \xrightarrow{\text{THF,))))}} \text{Ni*}$$

$$\text{CH}_2=\text{CHCN} + \text{MeCl}_2\text{SiH} \xrightarrow[\text{))))}, \text{r.t.}, 93\%]{\text{Ni*, THF}} \text{MeCl}_2\text{Si-CH(CN)-CH}_3$$

Figure 8.52 Catalytic hydrosilylation reactions.

continuous sonication the reaction is accelerated, and there is no sign of the reduction in the catalyst activity normally observed. Sonication improved the overall reaction rate with no reduction in activity even at a reduced bulk temperature. Most significantly, sonication proved equally effective in a pulse mode operating at 50% cycle (0.5 second bursts of ultrasound during each second).

8.4 CONCLUSIONS

The field of sonochemistry has progressed dramatically from its beginnings in the 1940's and 50's, and through its renaissance in the 1970's to where it stands today as a major new technology available to chemists and chemical technologists. It is an expanding field of study which continues to thrive on outstanding laboratory results which have even more significance now that scale-up systems are available. In this chapter we have concentrated on the theme of chemical synthesis and sonocatalysis, but the field is much broader than this. Particular areas of current interest include the beneficial combinations of sonochemistry and electrochemistry, biotechnology, and photochemistry. The future prospects are likely to encompass an even wider range of applications, as frequency effects are exploited and more studies of sonochemical reaction mechanisms are begun. Interest in sonochemistry has been accompanied by a surge in the development of new equipment for the generation of ultrasound. This is particularly evident in new development work for the scale-up of sonochemistry reactions and other processing.

Sonochemical applications can be found in all types of chemical systems. Even in homogeneous reactions, which began as the main subject for theoretical studies, laboratory uses have been found. It is, however, in heterogeneous reactions that sonochemical syntheses are most widely developed, and very few chemical laboratories would feel complete without a small cleaning bath for the promotion of organometallic type reactions.

Sonochemistry is now well accepted on a practical level, and it is in the interests of all involved to try to understand the underlying principles. Empirical understanding has taken sonochemists a long way, but the time has come for more theoretical interpretations. Several groups are addressing the exact meaning of cavitational energy and how it should be harnessed (see Chapter 10). What actually occurs in the collapsing bubble and its immediate environment is a major key to the prediction of sonochemical reactivity. Is there a hot spot, an electrical discharge, a plasma of some sort – or is it a composite of all of these? In heterogeneous reactions should we be looking for a mechanochemical contribution to the undoubted existence of a high speed jet?

Whatever the underlying laws of sonochemistry might be, it is clear that sonochemistry has arrived, that sonochemistry is expanding and that chemists from all disciplines will find within the subject plenty that will be of interest to them.

ACKNOWLEDGEMENTS

The authors thank their colleagues members of the COST Networks, Profs. J. Berlan[†] (Toulouse, France), A. Campos-Neves (Coimbra, Portugal), H. Delmas (Toulouse, France), G. Descotes (Lyon, France), H. Fillion (Lyon, France), H. Heusinger (München, Germany), J. Jurczak (Warsaw, Poland), T. Lepoint (Brussels, Belgium), R. Miethchen (Rostock, Germany), C. Petrier (Chambery, France), J. Reisse (Brussels, Belgium), S. Toma (Bratislava, Slovakia), J.P. Lorimer, J. Lindley and N. Baker (Coventry, U.K.). Financial support by CNRS (France) and EPSRC (U.K.).

8.5 REFERENCES

[1] Roy, R. A., Ultrasonics Sonochemistry, 1 (1994) S5.
[2] Mason, T. J., Lorimer, J. P., Sonochemistry, Theory, Applications and Uses of Ultrasound in Chemistry, Ellis Horwood, Chichester, 1989.
[3] Ley, S. V., Low, C. M. R., Ultrasound in Synthesis, Springer Verlag, Berlin, 1989.
[4] Mason, T. J., Practical Sonochemistry, A Users Guide to Applications in Chemistry and Chemical Engineering, Ellis Horwood, Chichester, 1991.
[5] Mason, T. J. (ed.), Advances in Sonochemistry, JAI Press, London; a) (1990) Vol. 1; b) (1991) Vol. 2; c) (1993) Vol. 3.
[6] Suslick, K. S. (ed.), Ultrasound, its Physical, Biological and Chemical Effects, VCH, Weinheim, 1988.
[7] Price, G. J. (ed.), Current Trends in Sonochemistry, Royal Society of Chemistry, Oxford, 1993.
[8] a) Einhorn, C., Einhorn, J., Luche, J. L., Synthesis, (1989) 787; b) Luche, J. L., Ultrasonics Sonochemistry, 1 (1994) S111.
[9] Wood, R. W., Loomis, A. L., Phil. Mag., 4 (1927) 414.
[10] Mark, H., J. Acoust. Soc. Amer., E16 (1945) 183.
[11] Weissler, A., J. Chem. Ed., 28 (1948) 28.
[12] Perkins, J. P., (Mason, T. J. (ed.)), Sonochemistry: the Uses of Ultrasound in Chemistry, Chap. 4, pp 47–59, Royal Society of Chemistry, London, 1990.
[13] Jeffries, J. B., Copeland, R. A., Suslick, K. S., Flint, E. B., Science, 256 (1992) 248.
[14] Noltingt, B. E., Neppiras, E. A., Proc. Phys. Soc. (London), 63B (1950) 674.
[15] Margulis, M. A., Ultrasonics, 23 (1985) 157.
[16] Frenkel, Y. I., Russ. J. Phys. Chem., 14 (1940) 305.
[17] Lepoint, T., Mullie, F., Ultrasonics Sonochemistry, 1 (1994) S13.
[18] See for example Lauterborn, W., Hentschell, W., Ultrasonics, 23 (1985) 260.
[19] See for example a) Pugin, B., Turner, A. T., in ref. 5a) Vol. 1 (1990) 81; b) ref. 8b) and c) Suslick, K. S., Docktycz, D., ref. 2a) (1990) 197.
[20] Starchev, K., Stoilov, S., Phys. Rev. B, Condens. Matter., 47 (1993) 11725.

[†]Jacques Berlan died on April 1st, 1995.

8.5. REFERENCES

[21] Kashiwagi, H., Enomoto, S., Inoue, M., Chem. Pharm. Bull., 37 (1989) 3181.
[22] Boldyrev, V. V., J. Chim. Phys., 83 (1986) 821.
[23] Luche, J. L., ref. 5c) (1993) 85.
[24] a) Ando, T., Sumi, S., Kawate, T., Ichihara, J., Hanafusa, T., J. Chem. Soc., Chem. Commun., (1984) 439; b) Ando, T., Kimura, T., ref. 5b) 211.
[25] Luche, J. L., Einhorn, C., Einhorn, J., Sinisterra-Gago, J. V., Tetrahedron Lett., 31 (1990) 4125.
[26] Boucher, R. M. J., British Chemical Engineering, 15 (1970) 363.
[27] Cum, G., Galli, G., Gallo, R., Spadaro, A., Ultrasonics, 30 (1992) 267.
[28] Henglein, A., (Mason, T. J. (ed.)), Advances in Sonochemistry, JAI Press, 3 (1992) 1.
[29] Petrier, C., Jeunet, A., Luche, J. L., Reverdy, G., J. Am. Chem. Soc., 114 (1992) 3148.
[30] For methodologies used in these estimations see ref. 4, pp 36–46.
[31] de Souza Barboza, J. C., Petrier, C., Luche, J. L., J. Org. Chem., 53 (1988) 1212.
[32] Mason, T. J., Lorimer, J. P., Mistry, B. P., Tetrahedron Lett., 24 (1983) 4371.
[33] Cum, G., Galli, G., Gallo, R., Spadaro, A., J. Chem. Soc., Perkin Trans. II, (1988) 375.
[34] Suslick, K. S., Schubert, P. F., Goodale, J. W., J. Am. Chem. Soc., 103 (1981) 7342; Suslick, MK. S., Goodale, J. W., Schubert, P. F., Wang, H. H., J. Am. Chem. Soc., 105 (1983) 5781.
[35] Misik, V., Riesz, P., J. Phys. Chem., 98 (1994) 1634.
[36] Suslick, K. S., Gawienowsky, J. J., Schubert, P. F., Wang, H. H., Ultrasonics, (1984) 33.
[37] Petrier, C., Reyman, D., Luche, J. L., Ultrasonics Sonochemistry, 1 (1994) S103.
[38] Eshuis, J. J. W., Tetrahedron Lett., 35 (1994) 7833.
[39] Loupy, A., Luche, J. L., (Sasson, Y. (ed.)), Phase Transfer Catalysed Reactions, Chapman & Hall, in press.
[40] Luche, J. L., Cintas P., (Fürstner, A. (ed.)), Activated Metals, VCH, Weinheim, in press.
[41] Porter, C. W., Young, L., J. Am. Chem. Soc., 60 (1938) 1497.
[42] Folger, S., Barnes, D., Ind. Eng. Chem. Fundam., 7 (1968) 222.
[43] Kristol, D. S., Klotz, H., Parker, R. C., Tetrahedron Lett., 22 (1981) 907.
[44] Mason, T. J., Lorimer, J. P., Mistry, B. P., Tetrahedron, 41 (1985) 5201.
[45] Mason, T. J., Lorimer, J. P., Mistry, B. P., Ultrasonics International '85, Proceedings, Butterworth, U.K. (1985) 839.
[46] Henglein, A., in Ref. 5b).
[47] Schultz, R., Henglein, A., Z. Naturforsch., 8 (1953) 160.
[48] Henglein, A., Fischer, C. H., Ber. Bunsenges. Phys. Chem., 88 (1984) 1196.
[49] Prakash, S., Pandey, J. D., Tetrahedron, 21 (1965) 903.
[50] Vinatoru, M., Iancu, A., Bartha, E., Petride, A., Badescu, V., Niculescu-Duvaz, D., Badea, F., Ultrasonics Sonochemistry, 1 (1994) S27.
[51] Elpiner, I. E., Sokolskaia, A. V., Margulis, M. A., Nature, 208 (1965) 945.
[52] Reisse, J., J. Org. Chem., 60 (1995), in press.

[53] Reisse, J., Danhui Yang Maeck, M., van der Cammen, J., van der Donckt, E., Ultrasonics, 30 (1992) 397.
[54] Peters, D., Pautet, F., El Fakih, H., Luche, J. L., Fillion, H., J. Prakt. Chem., in press.
[55] Dauben, W. G., Bridon, D. P., Kowalczyk, V., J. Org. Chem., 54 (1989) 6101.
[56] Kotronarou, A., Mills, G., Hoffmann, M. R., J. Phys. Chem., 95 (1991) 3630.
[57] Lamy, M. F., Petrier, C., Reverdy, G., 3rd Meeting, European Society of Sonochemistry, (1993).
[58] Cheung, H. M., Kurup, S., Environ. Sci. Technol., 28 (1994) 1619.
[59] Petrier, C., Micolle, M., Merlin, G., Luche, J. L., Reverdy, G., Environ. Sci. Technol., 26 (1992) 1639.
[60] Fuchs, E., Heusinger, H., Z. Lebensm. Unters. Forsch., 198 (1994) 486.
[61] Alegria, A. E., Lion, Y., Kondo, T., Riesz, P., J. Phys. Chem., 93 (1989) 4908.
[62] Serpone, N., Terzian, R., Hidaka, H., Pelizzetti, E., J. Phys. Chem., 98 (1994) 2634.
[63] a) Petrie, C., Lamy, M. F., Francony, A., Benahcene, A., David, B., Renaudin, V., Gondrexon, N., J. Phys. Chem., 98 (1994) 10515; b) Benahcene, A., Labbé, P., Petrier, C., Reverdy, G., New J. Chem., (1995) in press.
[64] Suslick, K. S., Gawlenowski, J. J., Schubert, P. F., Wang, H. H., J. Phys. Chem., 87 (1983) 2299.
[65] Suslick, K. S., Hammerton, D. A., Cline Jr., R. F., J. Am. Chem. Soc., 108 (1986) 5641.
[66] Nakamura, E., Inubushi, T., Aoki, S., Machii, D., J. Am. Chem. Soc., 113 (1991) 8990; Nakamura, E., Imanishi, Y., Machii, D., J. Org. Chem., 59 (1994) 8178.
[67] Rehorek, D., Janzen, E. G., J. Organomet. Chem., 268 (1984) 135.
[68] Nakamura, E., Inubushi, T., Aoki, S., Machii, D., J. Am. Chem. Soc., 113 (1991) 8980.
[69] Kawashima, E., Aoyama, Y., Sekine, T., Nakamura, E., Kainosho, M., Kyogoku, Y., Ishido, Y., Tetrahedron Lett., 34 (1993) 1317.
[70] Hubert, C., Munoz, A., Garrigues, B., Luche J. L., J. Org. Chem., 60 (1995) 1488.
[71] Grimstaff, M. W., Cichowlas, A. A., Choe S.-B., Suslick, K. S., Ultrasonics, 30 (1992) 168.
[72] Horton, A. M., Hollinshead, D. M., Ley S. V., Tetrahedron, 40 (1984) 1737.
[73] Cheng, J., Luo, F., Bull. Inst. Chem. Acad. Sin., 36 (1989) 9.
[74] Billington, D. C., Helps, I. M., Pauson, P. L., Thomson, W., Willison, D., J. Organomet. Chem., 354 (1988) 233.
[75] Harrity, J. P. A., Kerr, W. J., Middlemass, D., Tetrahedron, 49 (1993) 5565.
[76] Ando, T., Bauchat, P., Foucaud, A., Fujita, M., Kimura, T., Sohmiya, H., Tetrahedron Lett., 32 (1991) 6379.
[77] El Fakih, H., Pautet, F., Fillion, H., Luche, J. L., Tetrahedron Lett., 33 (1992) 4909.
[78] Maryanoff, B. E., Reitz, A. B., Chem. Rev., 89 (1989) 863.
[79] Manzo, P. G., Palacios, S. M., Alonso, R. A., Tetrahedron Lett., 35 (1994) 677.
[80] Dickens, M. J., Luche, J. L., Tetrahedron Lett., 32 (1991) 4709.
[81] Creary, X., Acc. Chem. Res., 25 (1992) 31.

8.5. REFERENCES

[82] Brown, H. C., Racherla, U. S., Tetrahedron Lett., 26 (1985) 2187.
[83] Crimmins, M. T., O'Mahony, R., Tetrahedron Lett., 30 (1989) 5993; Keck, G. E., Palani, A., McHardy, S. F., J. Org. Chem., 59 (1994) 3113.
[84] a) Bosman, C., D'Annibale, A., Resta, S., Trogolo, C., Tetrahedron, 50 (1993) 10705; b) D'Annibale, A., Trogolo, C., Tetrahedron Lett., 35 (1994) 2083.
[85] Rong, G., Keese, R., Tetrahedron Lett., 31 (1990) 5615.
[86] Jouglet, B., Blanco, L., Rousseau, G., Synlett, (1991) 907.
[87] Low, C. M. R., Synlett, (1991) 123.
[88] Yang, C., Yang, D. T. C., Harvey, R. G., Synlett, (1992) 799.
[89] Sinisterra, J. V., Fuentes, A., Marinas, J. M., J. Org. Chem., 52 (1987) 3875.
[90] Bankova, V. S., J. Nat. Products, 53 (1990) 821.
[91] Yamaguchi, S., Miyamoto, K., Hanafusa, T., Bull. Chem. Soc. Jpn., 62 (1989) 3036.
[92] For a discussion on kinetic effects see ref. 2, chapter 5.
[93] Moon, S., Duchin, L., Cooney, J. V., Tetrahedron Lett., 20 (1979) 3917.
[94] Sainsbury, M., Williams, C., Naylor, A., Scopes, D. I., Tetrahedron Lett., 31 (1990) 2763.
[95] Hwu, J. R., Chen, K., Ananthan, S., J. Chem. Soc., Chem. Commun., (1994) 1425.
[96] Singh, A. K., Synth. Commun., 20 (1990) 3547.
[97] Herges, R., Hoock, C., Synthesis, 12 (1991) 1151.
[98] Ando, T., Kawate, T., Ishihara, J., Hanafusa, T., Chem. Lett., (1984) 725.
[99] Regen, S. L., Singh, A., J. Org. Chem., 47 (1982) 1587.
[100] a) Xu, L., Smith, W. B., Brinker, U. H., J. Am. Chem. Soc., 114 (1992) 783; b) Lukevics, E., Gevorgyan, V., Goldberg, Y., Gaukhman, A., Gavars, M., Popelis, J., Shimanska, M., J. Organomet. Chem., 265 (1984) 237.
[101] Tao, F., Huang, H., Sun, M., Xu, L., Yingyong Huaxue, 5 (1988) 91; Chem. Abstr., 111 (1989) 96594f.
[102] Lin, Q., Zhang, Y., Zhang, C., Song, W., Qiu, Q., Chin. Chem. Lett., 2 (1991) 517; Chem. Abstr., 116 (1992) 193753t.
[103] Sjoberg, K., Tetrahedron Lett., 7 (1966) 6383.
[104] Mason, T. J., Lorimer, J. P., Turner, A. T., Harris, A. R., J. Chem. Res. (S), (1988) 80.
[105] Mason, T. J., Lorimer, J. P., Paniwnik, L., Harris, A. R., Wright, P. W., Bram, G., Loupy, A., Ferradou, G., Sansoulet, J., Synth. Commun., 20 (1990) 3411.
[106] Davidson, R. S., Safdar, A., Spencer, J. D., Robinson, B., Ultrasonics, 25 (1987) 35.
[107] Davidson, R. S., Patel, A. M., Safdar, A., Thornthwaite, D., Tetrahedron Lett., 24 (1983) 5907.
[108] Einhorn, J., Einhorn, C., Luche, J. L., Synlett, (1991) 37; Enders, D., Tiebes, J., De Kimpe, N., Keppens, M., Stevens, C., Smagghe, G., Betz, O., J. Org. Chem., 58 (1993) 4881; Josien, H., Martin, A., Chassaing, G., Tetrahedron Lett., 32 (1991) 6547.
[109] Han, B. H., Boudjouk, P., Tetrahedron Lett., 23 (1982) 1643.
[110] Lukevics, E., Gevorgyan, V. N., Goldberg, Y. S., Tetrahedron Lett., 25 (1984) 1415.

[111] Han, B. H., Shin, D. H., Lee, H. R., Jang, D. G., J. Korean Chem. Soc., 33 (1989) 436.
[112] Jang, D. G., Han, B. H., J. Korean Chem. Soc., 35 (1991) 179.
[113] Figadere, B., Chaboche, C., Franck, X., Peyrat, J. F., Cavé, A., J. Org. Chem., 59 (1994) 7138.
[114] Mokryi, E. N., Ludin, A. N., Reutskii, V. V., Dopov. Akad. Nauk Ukr., 7 (1993) 99; Chem. Abstr., 121 (1994) 107751q.
[115] Tuncay, A., Dustman, J. A., Fisher, G., Tuncay, C. I., Suslick, K. S., Tetrahedron Lett., 33 (1992) 7647.
[116] Luzzio, F. A., Moore, W. J., J. Org. Chem., 58 (1993) 512.
[117] Zhao, Y., Mason, T. J., Lindley, J., Youji Huaxue, 14 (1994) 435; Chem. Abstr., 121 (1994) 255381j.
[118] Kodomari, M., Satoh, H., Yoshitomi, S., J. Org. Chem., 53 (1988) 2093.
[119] Lindley, J., Ultrasonics, 30 (1992) 163.
[120] a) Adams, L. L., Luzzio, F. A., J. Org. Chem., 54 (1989) 5387; b) Luzzio, F. A., Moore, W. J., J. Org. Chem., 58 (1993) 2966; c) Han, B. H., Shin, D. H., Jang, D. G., Kim, S. N., Bull. Korean Chem. Soc., 11 (1990) 157; d) Cottier, L., Descotes, G., Lewkowski, J., Skowronski, R., Pol. J. Chem., 68 (1994) 693.
[121] Miranda Moreno, M. J. S., Sa e Melo, M. L., Campos Neves, A. S., Tetrahedron Lett., 32 (1991) 3201.
[122] Pugin, B., Turner, A., in ref. 5a), 81.
[123] Tomlinson, W. J., in ref. 5a), 173.
[124] Suslick, K. S., Doktycz, S. J., in ref. 5a), 187.
[125] Teoh, C. C. A., Goh, N. K., Chia, L. S., J. Chem. Soc., Chem. Commun., (1995) 201.
[126] Luche, J. L., Petrier, C., Dupuy, C., Tetrahedron Lett., 25 (1984) 753.
[127] Garst, J. F., Ungvary, F., Batlaw, R., Lawrence, K. E., J. Am. Chem. Soc., 113 (1991) 5392; Hamdouchi, C., Topolski, M., Goedken, V., Walborski, H. M., J. Org. Chem., 58 (1993) 3148.
[128] Pratt, M. W. T., Helsby, R., Nature, 184 (1959) 1694.
[129] Margulis, M. A., Los, G. P., Bashkatova, A. A., Beilin, A. G., Skorokhodov, I. I., Zinovev, O. I., Russ. J. Phys. Chem., 65 (1991) 1618.
[130] Kruus, P., Ultrasonics, 26 (1988) 216.
[131] Maltsev, A. N., Russ. J. Phys. Chem., 50 (1976) 995.
[132] Petrier, C., Dupuy, C., Luche, J. L., Tetrahedron Lett., 27 (1986) 3149.
[133] Petrier, C., Luche, J. L., Tetrahedron Lett., 28 (1987) 2347.
[134] Kruus, P., Robertson, D. A., McMillen, L. A., Ultrasonics, 29 (1991) 370.
[135] Burkhardt, E. R., Rieke, R. D., J. Org. Chem., 50 (1985) 416; Parker, W. L., Boudjouk, P., Rajkumar, A. B., J. Am. Chem. Soc., 113 (1991) 2785.
[136] Nayak, S. K., Banerji, A., J. Org. Chem., 56 (1991) 1940.
[137] Suslick, K. S., Johnson, R. E., J. Am. Chem. Soc., 106 (1984) 6856.
[138] Renaud, P., Bull. Soc. Chim. Fr., (1950) 1044.
[139] Slough, W., Ubbelohde, A. R., J. Chem. Soc., (1957) 918.

8.5. REFERENCES

[140] Boudjouk, P., Sooriyakumaran, R., Han, B. H., J. Org. Chem., 51 (1986) 2818.
[141] Badejo, I. T., Karaman, R., Lee, N. W. I., Lutz, E. C., Mamanta, M. T., Fry, J. L., J. Chem. Soc., Chem. Commun., (1989) 566; Badejo, I. T., Karaman, R., Fry, J. L., J. Org. Chem., 54 (1989) 4591; Karaman, R., Kohlman, D. T., Fry, J. L., Tetrahedron Lett., 31 (1990) 6155; Karaman, R., Fry, J. L., Tetrahedron Lett., 30 (1989) 4931, 4935 and 6267.
[142] Bönnemann, H., Bogdanovic, B., Brinkman, R., He, D. W., Spliethoff, B., Angew. Chem. Int. Ed. Engl., 22 (1983) 728.
[143] De Nicola, A., Einhorn, J., Einhorn, C., Luche, J. L., J. Chem. Soc., Chem. Commun., (1994) 879.
[144] Huffman, J. W., (Trost, B. M., Fleming, I. (eds.)), Comprehensive Organic Synthesis, Pergamon Press, New York, 1993, Vol. 8, 107.
[145] Huffman, J. W., Liao, W. P., Wallace, R. H., Tetrahedron Lett., 28 (1987) 3315.
[146] Fadel, A., Canet, J. L., Salaun, J., Synlett, (1990) 89.
[147] Einhorn, J., Einhorn, C., Luche, J. L., Tetrahedron Lett., 29 (1988) 2183.
[148] Chou, T. S., You, M. L., J. Org. Chem., 52 (1987) 2224.
[149] Chou, T. S., Hung, S. H., Peng, M. L., Lee, S. J., Tetrahedron Lett., 32 (1991) 3551.
[150] Ley, S. V., O'Neil, Y. A., Low, C. M. R., Tetrahedron, 42 (1986) 5363.
[151] Luche, J. L., Damiano, J. C., J. Am. Chem. Soc., 102 (1980) 7927.
[152] Sprich, J. D., Lewandos, G. S., Inorg. Chim. Acta, 76 (1983) L241.
[153] Tuulmets, A., Karelson, M., Proc. Estonian Acad. Sci. Chem., 43 (1994) 51.
[154] Yamaguchi, R., Kawasaki, H., Kawanisi, M., Synth. Commun., 12 (1982) 1027.
[155] Blomberg, C., The Barbier Reaction and Related One Step Processes, Springer Verlag, Berlin, 1993.
[156] Moyano, A., Pericas, M. A., Riera, A., Luche, J. L., Tetrahedron Lett., 31 (1990) 7619.
[157] Burkow, I. C., Sydnes, R. K., Ubeda, D. C. N., Acta Chem. Scand., B41 (1987) 235; Singh, S. B., Pettit, G. R., Synth. Commun., 17 (1987) 877.
[158] Araki, S., Butsugan, Y., Chem. Lett., (1988) 457.
[159] Rong, G., Keese, R., Tetrahedron Lett., 31 (1990) 5617.
[160] Hugel, G., Cartier, D., Levy, J., Tetrahedron Lett., 30 (1989) 4513.
[161] Petrier, C., Gemal, A. L., Luche, J. L., Tetrahedron Lett., 23 (1982) 3361; Einhorn, J., Luche, J. L., Tetrahedron Lett., 27 (1986) 1791.
[162] Yang, D. H., Einhorn, J., Einhorn, C., Aurell, M. J., Luche, J. L., J. Chem. Soc., Chem. Commun., (1994) 1815.
[163] Aurell Piquer, M. J., Einhorn, C., Einhorn, J., Luche, J. L., J. Org. Chem., 60 (1995) 8.
[164] Han, B. H., Boudjouk, P., Tetrahedron Lett., 22 (1981) 2757.
[165] Price, G. J., Clifton, A. A., Tetrahedron Lett., 32 (1991) 7133; Osborne, A. G., Glass, K. J., Staley, M. L., Tetrahedron Lett., 30 (1989) 3567.
[166] Zinov'iev, O. I., Margulis, M. A., in ref. 5c), 165.
[167] Boudjouk, P., Han, B. H., Tetrahedron Lett., 22 (1981) 3813.
[168] Margulis, M. A., Los, G. P., Zinov'iev, O. I., Russ. J. Phys. Chem., 65 (1991) 1614.

[169] Price, G. J., J. Chem. Soc., Chem. Commun., (1992) 1209.
[170] Boudjouk, P., Han, B. H., Anderson, K. R., J. Am. Chem. Soc., 104 (1982) 4992.
[171] Boudjouk, P., Black, E., Kumarathasan, R., Organometallics, 10 (1991) 2095.
[172] Prokes, I., Toma, S., Luche, J. L., Tetrahedron Lett., 36 (1995) 3849.
[173] Einhorn, J., Luche, J. L., J. Org. Chem., 52 (1987) 4124.
[174] Banerji, A., Nayak, S. K., Current Science, 58 (1989) 249.
[175] Petrier, C., De Souza Barboza, J. C., Dupuy, C., Luche, J. L., J. Org. Chem., 50 (1985) 5761.
[176] Luche, J. L., Petrier, C., Lansard, J. P., Greene, A. E., J. Org. Chem., 48 (1983) 3837.
[177] Dehmlow, E. V., Büker, S., Chem. Ber., 126 (1993) 2759.
[178] Brown, H. C., Racherla, U. S., J. Org. Chem., 51 (1986) 427.
[179] Frankland, E., J. Chem. Soc., 2 (1849) 263.
[180] Erdik, E., Tetrahedron, 43 (1987) 2203.
[181] Petrier, C., Luche, J. L., Lavaitte, S., Morat, C., J. Org. Chem., 54 (1989) 5313.
[182] Salvador, J. A. R., Sa e Melo, M. L., Campos Neves, A. S., Tetrahedron Lett., 34 (1993) 357, 361.
[183] So, J. H., Park, M. K., Boudjouk, P., J. Org. Chem., 53 (1988) 5871.
[184] Boudjouk, P., So, J. H., Synth. Commun., 16 (1986) 775.
[185] Takagi, K., Chem. Letters, (1993) 469.
[186] Kitazume, T., Ishikawa, N., J. Am. Chem. Soc., 107 (1985) 5186.
[187] Knochel, P., Singer, R. D., Chem. Rev., 93 (1993) 2117.
[188] Repic, O., Vogt, S., Tetrahedron Lett., 23 (1982) 2729.
[189] Jie, M. L. K., Lam, W. L. K., J. Chem. Soc., Chem. Commun., (1987) 1460.
[190] Kitazume, T., Ultrasonics, 28 (1990) 322.
[191] Tso, H. H., Chou, T. S., Hung, S. C., J. Chem. Soc., Chem. Commun., (1987) 1552.
[192] Han, B. H., Boudjouk, P., J. Org. Chem., 47 (1982) 751.
[193] Solladie-Cavallo, A., Farkhani, D., Fritz, S., Lazrak, T., Suffert, J., Tetrahedron Lett., 25 (1984) 4117.
[194] Rathke, M. W., Weipert, P., (Trost, B. M., Fleming, I., Heathcock, C. H. (eds.)), Comprehensive Organic Synthesis, Pergamon Press, Oxford, U.K., 1991, Vol. 2, Chapter 1.8.
[195] Lim, C. L., Han, B. H., J. Korean Chem. Soc., 35 (1991) 762; Chem. Abstr., 116 (1992) 128122x; Kashima, C., Huang, X. C., Harada, Y., Hosomi, A., J. Org. Chem., 58 (1993) 793; Bose, A. K., Gupta, K., Manhas, M. S., J. Chem. Soc., Chem. Commun., (1984) 86; Oguni, N., Tomago, T., Nagata, N., Chem. Express, 1 (1986) 495.
[196] Han, B. H., Boudjouk, P., J. Org. Chem., 47 (1982) 5030; Suslick, K. S., Doktycz, S. J., J. Am. Chem. Soc., 111 (1989) 2342.
[197] Kim, K. S., Qian, L., Tetrahedron Lett., 34 (1993) 7195.
[198] Beard, R. L., Meyers, A. I., J. Org. Chem., 56 (1991) 2091.
[199] Jackson, R. F. W., Wishart, N., Wood, A., James, K., Wythes, M. J., J. Org. Chem., 57 (1992) 3397.

8.5. REFERENCES

[200] Petrier, C., Luche, J. L., J. Org. Chem., 50 (1985) 910; Einhorn, C., Luche, J. L., J. Organomet. Chem., 322 (1987) 177; Petrier, C., Einhorn, J., Luche, J. L., Tetrahedron Lett., 26 (1985) 1449.

[201] Kim, E., Gordon, D. M., Schmid, W., Whitesides, G. M., J. Org. Chem., 58 (1993) 5500; Binder, W. H., Prenner, R. H., Schmid, W., Tetrahedron, 50 (1994) 749.

[202] Dupuy, C., Petrier, C., Sarandeses, L., Luche, J. L., Synth. Commun., 21 (1991) 643.

[203] Sarandeses, L. A., Mourino, A., Luche, J. L., J. Chem. Soc., Chem. Commun., (1991) 818; Sarandeses, L. A., Mourino, A., Luche, J. L., J. Chem. Soc., Chem. Commun., (1992) 798.

[204] Ohno, M., Ishizaki, K., Eguchi, S., J. Org. Chem., 53 (1988) 1285; Giese, B., Damm, W., Roth, M., Zehnder, M., Synlett, (1992) 441.

[205] Pérez-Sestelo, J., Mascarenas, J. L., Castedo, L., Mourino, A., J. Org. Chem., 58 (1993) 118; Pérez-Sestelo, J., Mascarenas, J. L., Castedo, L., Mourino, A., Tetrahedron Lett., 35 (1994) 275.

[206] Lindley, J., Lorimer, J. P., Mason, T. J., Ultrasonics, 24 (1986) 292.

[207] Nelson, K. A., Adolph, H. G., Synth. Commun., 21 (1991) 293.

[208] Liou, K., Yang, P., Lin, Y., J. Organomet. Chem., 294 (1985) 145; Yang, P. H., Liou, K. F., Lin, Y. T., J. Organomet. Chem., 307 (1986) 273; Lin, Y. T., J. Organomet. Chem., 317 (1986) 277.

[209] Miranda Moreno, M. J. S., Sa e Melo, M. L., Campos Neves, A. S., Tetrahedron Lett., 34 (1993) 353.

[210] Fitch, R. W., Luzzio, F. A., Tetrahedron Lett., 35 (1994) 6013.

[211] Luzzio, F. A., O'Hara, L. C., Synth. Commun., 20 (1990) 3223.

[212] Lindley, J., in ref. 7, 123.

[213] Cains, P. W., McCausland, L. J., Bates, D. M., Mason, T. J, Ultrasonics Sonochemistry, 1 (1994) S45.

[214] Suslick, K. S., Casadonte, D. J., J. Am. Chem. Soc., 109 3459 (1987).

[215] Cioffi, E. A., Willis, W. S., Suib, S. L., Langmuir, 4 (1988) 697.

[216] Suslick, K. S., Hyeon, T., Fang, M., Cichowlas, A. A., (Gonsalves, M. R. S. (ed.)), Molecularly Designed Nanostructured Materials, Pittsburgh, 1994.

[217] Bianchi, C. L., Carli, R., Lanzani, S., Lorenzetti, D., Vergani, G., Ragaini, V., Ultrasonics Sonochemistry, 1 (1994) S47.

[218] Lindley, J., (Price, G. J. (ed.)), Current Trends in Sonochemistry, Royal Society of Chemistry, Special pub. 116, 1992.

[219] Casadonte, D. J., Sweet, J. D., Vedamuthu, M. S., Ultrasonics, 32 (1994) 477.

[220] Goh, N. K., Teah, A., Chia, L. S., Ultrasonics Sonochemistry, 1 (1994) S43.

[221] Mason, T. J., Newman, A., Lorimer, J. P., Lindley, J., Hutt, K., Ultrasonics Sonochemistry, 2 (1995) in press.

[222] Saracco, G., Arzano, F., Chimica e Industria (Milano), 50 (1968) 314.

[223] Alexakis, A., Lensen, N., Mangeney, P., Synlett, (1991) 625; Enders, D., Klatt, M., Funk, R., Synlett, (1993) 226

[224] Tai, A., Kikukawa, T., Sugimura, T., Inoue, Y., Ozawa, T., Fujii, S., J. Chem. Soc., Chem. Commun., (1991) 795.

[225] Townsend, C. A., Nguyen, L. T., J. Am. Chem. Soc., 103 (1981) 4582.

[226] Kitazume, T., Ohnogi, T., Miyaushi, H., Yamazaki, T., Watanabe, S., J. Org. Chem., 54 (1989) 5630.
[227] Boudjouk, P., Han, B. H., J. Catal., 79 (1983) 489.
[228] Boudjouk, P., in ref. 7, 110.
[229] Boudjouk, P., Han, B. H., Jacobsen, J. R., Hauck, B. H., J. Chem. Soc., Chem. Commun., (1991) 1424.
[230] Mason, T. J., Lorimer, J. P., Paniwnyk, L., Wright, P. W., Harris, A. R., J. Catalysis, 147 (1994) 1.

9

APPLICATIONS OF HIGH INTENSITY ULTRASOUND IN POLYMER CHEMISTRY

Gareth J. Price
School of Chemistry, University of Bath, Bath, Claverton Down, BA2 7AY, U.K.

9.1 INTRODUCTION
    9.1.1 Why Study Polymers?
    9.1.2 Ultrasound, Cavitation and Sonochemistry
    9.1.3 Factors Affecting Cavitation
    9.1.4 Experimental Techniques
    9.1.5 Ultrasound and Polymers
9.2 ULTRASOUND AND POLYMER PROCESSING
    9.2.1 Ultrasound as a Diagnostic Tool
    9.2.2 Ultrasonic Welding of Thermoplastics
    9.2.3 Ultrasound Assisted Dispersal and Encapsulation
    9.2.4 Sonochemical Modification of Polymer Surfaces
9.3 SONOCHEMICAL SYNTHESIS OF VINYL POLYMERS
    9.3.1 Sonochemical Initiation by Radical Generation
    9.3.2 The Initiation Process
    9.3.3 The Polymerization Process
    9.3.4 Polymerization by Ionic Mechanisms
    9.3.5 Emulsion and Suspension Polymerization
9.4 SONOCHEMICAL POLYMERIZATION OF NON-VINYL MONOMERS
    9.4.1 Ring Opening Polymerizations
    9.4.2 Condensation Polymerization
    9.4.3 Electrochemically Promoted Polymerization Reactions
    9.4.4 Ziegler-Natta Polymerizations
    9.4.5 Grignard Coupling of Polyphenylenes
    9.4.6 Wurtz Coupling of Poly(organosilanes)

9.5 ULTRASONIC DEGRADATION OF POLYMERS IN SOLUTION
    9.5.1 Kinetics and Mechanism of Degradation
    9.5.2 Applications of Ultrasonic Degradation
    9.5.3 Synthesis versus Degradation
9.6 CONCLUSIONS AND FUTURE PROSPECTS
9.7 REFERENCES

9.1 INTRODUCTION

As outlined in the previous chapter, the application of ultrasound to chemical reactions has become a widespread, almost routine, technique. This started in the early 1970's with the work of Boudjouk, Suslick, Luche and others on Grignard, Barbier, organolithium and similar reactions dealt within the previous chapter [1]. The application of ultrasound to polymer systems in fact considerably predates these other chemical applications and originates from the 1920's and 1930's when the reduction in the viscosity of solutions of natural polymers such as agar, starch and gelatin on sonication was noted by several workers [2–4]. In view of this, the relatively small number of workers currently researching in polymer sonochemistry is perhaps rather surprising. However, the interest has steadily increased over the past few years [5, 6] and this chapter will review this work and attempt to predict some of the areas which will become important over the next few years.

9.1.1 Why Study Polymers?

The impact of polymers and plastics has been immense over the past half-century. In considering polymer systems, most people immediately think of the commodity plastics which are encountered in everyday life. The most common examples, accounting for between 50–60% of all polymer production include the various grades of polyethylene, polypropylene, polystyrene, polyvinyl chloride (PVC), and poly(ethylene terephthalate) (polyester). Together, these have an enormous volume of production. For example, it is estimated [7] that in 1994 34 million tons of these five plastics were produced in the United States alone. Added to these must be the nylons which have a range of uses, textile fibers such as acrylics, engineering thermoplastics such as polycarbonates, poly(ether ketone)s as well as polyurethanes and a number of others. More recently, conducting polymers, other photo- and electro-active materials as well as inorganic systems such as the silicones have risen in importance.

Polymers are produced in a wide variety of chemical reactions [8] involving a range of intermediates. The use of ultrasound in many of these has been investigated and will be described in the following sections. The distinguishing

9.1. INTRODUCTION

factor between polymers and low molar mass species is that they consist of a large number of repeat units organised in a chain. Clearly, in common with other branches of chemistry, the rate and yield of polymerization reactions are important areas of study as is the product stereochemistry. However, in this case, the stereochemistry, or *tacticity* of the polymer, refers to the arrangement of substituents in relation to the chain backbone rather than to their absolute position (see Section 9.3.3). The statistical nature of polymerization reactions leads to a particular sample containing chains of varying lengths so that only an average chain length or molecular weight can be measured. This property plays a major and fundamental role in determining the material properties and is perhaps the most important characteristic of the polymer. However, a distribution of chain lengths will be present and this has important consequences and can lead to two samples with the same average molecular weight having very different physical and material properties. This distribution is characterised by deriving from the statistical moments of the distribution, the *polydispersity*, γ, which takes a value of 1.0 if all the polymer chains are the same length (synthetically impossible at present) and increases with the breadth of the distribution. Molecular weights can be measured by a number of methods, but one of the most useful since it also gives the polydispersity is Gel Permeation Chromatography [9]. This is a form of liquid chromatography which utilises a porous solid to separate the polymer chains according to their size to give a chromatogram of the form shown in Figure 9.1. Since the size of the chain directly depends on its molecular weight, comparison of the chromatogram with those from standards with known molecular weight allows calculation of the various molecular weights as well as the breadth of the distribution.

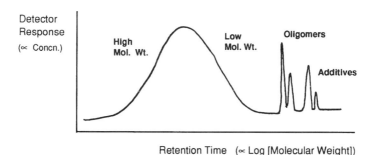

Figure 9.1 Schematic Gel Permeation Chromatogram showing the distribution of molecular weights.

9.1.2 Ultrasound, Cavitation and Sonochemistry

Uses of ultrasound are commonly divided into two categories. Ultrasound of the type used in medical diagnostics or non-destructive testing can be used to determine structure and conformational changes in polymers. This normally uses high frequency ultrasound, >1 MHz, to obtain good resolution and relatively low powers (<1 W) so as not to change the material. Some typical applications in polymer chemistry are briefly described in Section 9.2. This type of sound has limited application in influencing chemical processes or reactions. For sonochemistry, *power ultrasound* is used, usually in the frequency range 20–500 kHz, and at powers of up to several hundred Watts. Such high power can lead to the formation of microbubbles in a liquid and it is the growth and rapid collapse of these bubbles, a process known as *cavitation*, which provides the source of energy for chemical reactions (see Section 8.2).

The origin of cavitation and resulting production of chemically active species has been described in the previous chapter of this book [1]. While there are conflicting theories to explain the effects, what is certain is that highly reactive species such as free radicals and other dissociated, vibrationally excited moieties can be formed. Thus, in any system, three zones can be identified as shown schematically in Figure 9.2.

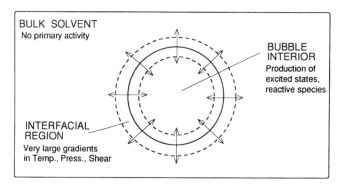

Figure 9.2 Schematic representation of bubble motion during cavitation.

The primary sonochemical activity such as the production of radicals takes place inside the bubble. Secondly, the bulk liquid has no *primary* activity although subsequent reaction with sonochemically generated species may occur. Finally, the interfacial region around the bubbles has very large temperature, pressure and electric field gradients and also very large shear gradients caused by rapid movement of solvent molecules. This last factor has a special importance in polymer sonochemistry as will be described in Section 9.5. The molecules of the liquid can be set into very rapid motion causing mixing and stirring, a process known as acoustic streaming. There is an extra effect in heterogeneous systems where cavitation can occur in the region of an interface. In immiscible liquid mixtures, the interface is disrupted and efficient mixing and dispersion occurs. Bubble collapse near a solid surface is asymmetric and can result in jets of liquid impinging on the surface at high speed. These effects are largely responsible for the sonochemical enhancement of reactions over metal surfaces or other catalysts [10] by, for example, removing passivating oxide layers and

9.1. INTRODUCTION

other poisoning species. Additionally, mass transport of reactants to and from the surface can be enhanced.

9.1.3 Factors Affecting Cavitation

Since virtually all the chemical effects can be traced back to cavitation, it is important to understand which factors affect the threshold and intensity of cavitation in a liquid. These factors have been described in detail [11, 12] including in the previous chapter [1] but it is worthwhile summarizing those of most importance to account for the behavior described in the following sections.

Perhaps the least studied factor, due to the scarcity of suitable apparatus, is the *ultrasound frequency*. It was thought that the amount of cavitation produced for the same intensity would fall as the frequency rose. However, there are some recent studies [13, 14] which show that there is an optimum value between 100 kHz and 1 MHz which depends on the system under investigation. There are few studies of the effect of changing frequency in polymer systems.

The most important ultrasound parameter is the *ultrasound intensity* which defines the amount of energy entering the reaction or process. As might intuitively be expected, the amount of cavitational activity rises with increasing intensity but there is an optimum value above which activity falls [15, 16].

Since a liquid is necessary for the transmission of sound, *solvent properties* such as density and viscosity which affect the mobility of the molecules are important. However, of more influence on chemical activity is the *vapor pressure* of the solvent. Volatile solvents will evaporate into the bubble as it grows and "cushion" its collapse, lowering the final temperatures and pressures achieved. This type of effect also explains the *temperature* dependence of sonochemical reactions. At lower temperatures, less vapor enters the bubbles so that collapse is more intense and chemical effects are enhanced. Thus, an **acceleration at lower temperatures** or "inverse-Arrhenius" behaviour is often observed.

Thus, complete characterization of a sonochemical system is complicated by the large number of parameters that can affect cavitation [11, 17]. However, the role of most of them is at least qualitatively understood and their manipulation offers the possibility of controlling a reaction with a good deal of precision.

9.1.4 Experimental Techniques

The apparatus for performing polymer sonochemistry in the laboratory is essentially the same as that used in organic or organometallic procedures [18, 19] described in the preceding chapter [1]. Some useful reactions have been carried out in cleaning baths but most have utilized the higher powers and more controllable intensities and temperature conditions allowed by a horn system. The type of equipment used in our laboratories is shown schematically in Figure 9.3 and consists of a jacketed cell through which thermostatted water is flowed to provide temperature control. There are various ports to allow

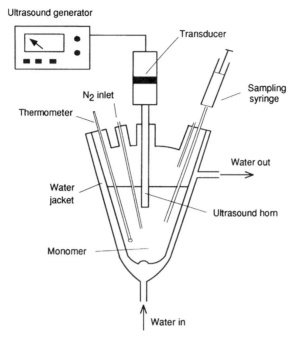

Figure 9.3 Schematic diagram of "horn" apparatus for sonochemical reactions. (Shown as used for inert atmosphere and kinetic work.)

reaction under inert gases and to remove samples for analysis. A number of other designs are available.

A different apparatus which has been used in polymer work is the "whistle reactor", the principle of which is that a liquid is pumped over a vibrating blade which essentially acts as a low intensity mixer or emulsifier. However, this has not been as widely used as the bath or horn systems. A wide variety of "home-built" equipment has also been described.

9.1.5 Ultrasound and Polymers

So, what are the possible benefits for polymer chemists in using ultrasound? Firstly, any process which results in radical production can be used as a method of initiation for vinyl monomers. A range of other reactions such as ring opening can also be promoted by excited species. A number of polymers are produced by heterogeneous reactions, and increases in yield and rate might be expected under sonication. Among the most common uses of ultrasound in synthesis have been in organometallic procedures such as Grignard or organolithium reactions [20–22], a range of which have recently been used to synthesize novel polymers. Finally, very large shear gradients can be set up around cavitation bubbles and polymers often undergo modification under these conditions. Hence, there is

9.2. ULTRASOUND AND POLYMER PROCESSING

a range of sonochemical effects for the polymer chemist to exploit. This chapter will deal only with those aspects of sonochemistry that are appropriate for the preparation of polymeric materials, either by synthesis from low molecular weight monomers or by modification of preexisting polymers. Polymer sonochemistry is a particularly rich area since, as will be seen, the opportunity exists to utilize all of the various effects caused by ultrasound, and a survey of these applications will be given. Polymerizations occurring by a variety of mechanisms will be discussed and the advantages offered by ultrasound to each will be indicated.

9.2 ULTRASOUND AND POLYMER PROCESSING

Before commencing on a discussion of ultrasound in polymer synthesis, it is appropriate to distinguish between distinct areas of application. The use of high frequency ultrasound at relatively low power has already been alluded to. Here no chemical changes are induced in a material and brief comments about applications in polymer chemistry will be made below. However, this book is mainly concerned with *chemistry* under extreme conditions and this involves the use of power ultrasound. Even here, two distinct areas can be identified. Irradiation with ultrasound can cause solely physical changes such as rapid stirring and mixing or heating. Cavitation is not always necessary for these effects but they almost always occur in a cavitating system. Examples include the formation of emulsions and melting plastics in welding systems. These can be referred to as *sonoprocessing*. In contrast, sonication in other applications can result in chemical changes, almost invariably as a result of cavitation, and this has led to a branch of chemistry often known as *sonochemistry*.

9.2.1 Ultrasound as a Diagnostic Tool

In general, to achieve high spacial resolution when using ultrasound as a diagnostic tool, high frequencies are used. For the applications briefly described here, typical frequencies could be in the range of 500 kHz–5 MHz. When sound passes through a medium, some of its energy is lost by *attenuation* due to viscous drag and coupling with other processes such as molecular vibrations. Measurement of the sound velocity in the medium along with this attenuation and its frequency dependence allows a range of properties of systems to be probed. A complete treatment of the propagation of sound through solids and fluids is outside the scope of this book and the reader is referred to more specialist texts [23–25].

During a polymerization, the viscosity and hence acoustic properties such as the compressibility will change throughout the reaction. Measurement of these gives a useful method for following the extent and rate of reaction. It has been applied to a number of systems [26, 27] involving addition polymerizations including polystyrene and polyvinyl acetate as well as condensation processes

[28] and the curing of epoxy resins [29]. Composite materials, consisting of an inorganic filler surrounded by a polymer matrix (*e.g.*, glass-fiber) are gaining popularity as construction materials. The polymer-filler interface and quality of the dispersion can also be investigated using this method [30].

The interaction of sound with the reaction mixture can cause isomerizations or motion of particular segments of a polymer chain, the relaxation of which yields information on the dynamics, energy and steric requirements of conformational changes. For example, rotational isomerization of the phenyl ring in polystyrene [31] and the methyl groups in poly(dimethyl-1,4-phenylene oxide) have been observed [32]. The attenuation in solid polymers can also be used to calculate elastic and mechanical properties such as moduli [33]. Finally, measurements on solutions over a wide range of concentrations have yielded [23] information on how the solvent influences the polymer conformation and hence on the thermodynamic interactions between polymer and solvent. Perhaps the most complex systems studied to date have been polyelectrolytes where the processes such as side-chain and backbone motion which occur in normal polymers are accompanied by ion-pair, solvation and polar interactions [34].

9.2.2 Ultrasonic Welding of Thermoplastics

This process represents perhaps the major application of ultrasonics in the polymer industry and has now become a routine joining technique in a wide range of areas. A particular example is the assembly of automotive components. Ultrasonic welding works well for amorphous plastics that are relatively hard and rigid such as polystyrene and polycarbonate. Ultrasound at very high powers of 1–2 kW, usually at around 20 kHz, is applied to the parts to be joined [35, 36] and the alternating stress causes intense local heating and hence plastic flow. The two polymers can then mix to give a strong joint between the materials. The local heating minimizes distortion to other parts of the component and, since the heat is generated inside the component rather than being conducted from outside, less degradation occurs. Typical joints can be made in ~ 1 s and the procedure can easily be automated.

9.2.3 Ultrasound Assisted Dispersal and Encapsulation

It is well known that particles of solids suspended in liquids are set into motion due to acoustic streaming and the shear forces around cavitation bubbles. This can lead to inter-particle collisions, and the shock waves caused by cavitation can also impinge on the particles, both factors leading to modification of the particle size. For example in metallic powders, particles can either be fused together or fragmented depending on the physical properties of the metal [37, 38].

With inorganic solids, two processes can be identified. Firstly, large (>10 μm) particles can be fragmented. Alternatively, loosely bound aggregates of small particles (<0.5–1 μm) can be broken up to give a better dispersion of the

9.2. ULTRASOUND AND POLYMER PROCESSING

original material. This has particular application in the production of polymers containing fillers where the properties are largely determined by the quality of the dispersion. Particular target compounds are TiO_2 which is used as a filler and white pigment in many paints and carbon black, used in many inks. Organic systems such as anti-static agents, anti-oxidants, dyes and pigments are also of interest.

Stoffer and Fahim [39] studied some typical paint formulations containing TiO_2 and found that using ultrasound for the dispersal and mixing phases of the process led to smaller, more even particle sizes in shorter times than conventional methods such as ball-milling or high-shear homogenizers. They also estimated a large saving in energy costs. Similar results have been reported by Utsimi et al. for the production of toner inks for copying applications [40].

This type of technology has also been applied on a larger, pilot-plant scale, for example to prepare TiO_2 encapsulated in polymer [41,42]. Inclusion of TiO_2 in an emulsion polymerization (see Section 9.3.5) of PVC led to encapsulated particles in which the coating was much more uniform when carried out in the presence of ultrasound. Scale-up studies were also performed and the process was carried out on a pilot plant to produce 200 kg of coated material. A process utilizing an ultrasonic mixer to incorporate pyrogenic silica into polyester resin has been described [43]. The process operated at rates approaching $12,000 \, L \, h^{-1}$ and gave a much better dispersion than other methods.

Recent work [44] has investigated the effect of sonication on polymer particles dispersed in water. Different materials behaved differently in that some underwent particle fusion leading to an increase in the average particle size while others fragmented to increase the number of small particles, as shown in Figure 9.4. The behavior correlated with the physical properties of the polymer in that soft, easily deformable materials tended to fragment while collisions between harder, glassy polymers led to some fragmentation but also some fusion between particles.

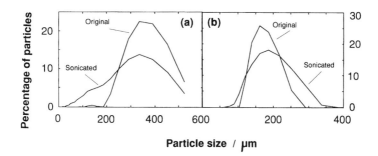

Figure 9.4 Change of particle size after 1 hr sonication in water. (a) polypropylene; (b) PVC.

9.2.4 Sonochemical Modification of Polymer Surfaces

The cleaning action of ultrasound to remove surface contaminants is one of the most well established sonochemical processes. This takes advantage of the asymmetric collapse of a cavitation bubble near a surface which "jets" streams of liquid against the surface. Recently, this effect has been utilized in the modification of the surfaces of polymer properties.

One of the earliest reports was by Urban and Salazar-Rojas [45] in which HF was removed from the surface of poly(vinylidene difluoride), PVF_2 to give a surface consisting of conjugated C=C double bonds (Scheme 9.1). PVF_2 is a piezoelectric material but is an insulator whereas a conjugated surface would confer conductive properties. A more conventional approach would be to reflux powdered polymer in a solution of a strong base such as sodium hydroxide together with a phase transfer catalyst, *e.g.*, tetrabutyl ammonium bromide, TBAB, to enhance wetting of the surface.

Little reaction was found in the absence of TBAB but the use of ultrasound accelerated the dehydrofluorination, particularly at low temperatures. However, the maximum thickness of the surface layer was thinner when ultrasound was used, which was attributed to ultrasonic erosion of the modified surface.

In addition to conferring conductivity, the unsaturation gives sites which can be further modified by subsequent chemical reaction. Examples [46, 47] of this have been seen by the grafting of silicon and germanium centred phthalocyanines (Pc) onto the surface of PVF_2. Lithium was used with $(Pc)SiCl_2$ or its germanium analogue for the reaction which was shown to anchor the (Pc) to the surface *via* a SiC or GeC bond.

In our work at Bath [48], we have used a similar approach to modify the surface of PVC, an inexpensive, commodity polymer. If the surface could be functionalized it would be an inexpensive method of producing materials with the properties of more expensive polymers. We have shown that it is possible to graft a number of compounds including metal ions, dyes and hydrophilic monomers. A more difficult example is polyethylene, PE, which having a very unreactive surface needs to be activated by *e.g.*, chromic acid or other strong oxidizing agents prior to adhesion or printing. Using ultrasound, we were able to modify rapidly the character of PE using milder, more environmentally friendly oxidizing agents.

9.3 SONOCHEMICAL SYNTHESIS OF VINYL POLYMERS

With an appreciation of the basic effects of ultrasound and the type of processes which it can induce, we can now proceed to a discussion of its applications in

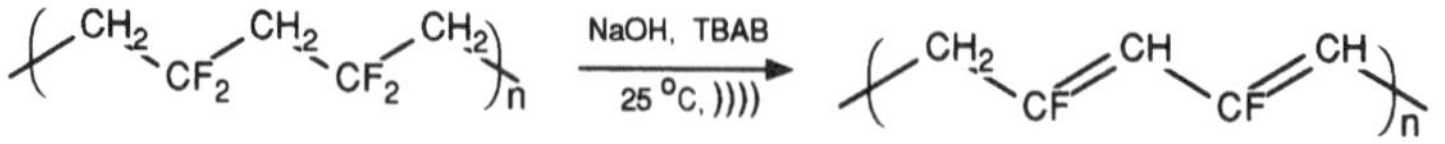

Scheme 9.1

polymer synthesis. Vinyl monomers are those such as styrene, vinyl chloride, acrylonitrile and (meth)acrylates which have a C=C double bond activated by one or more polar or aromatic groups. Polymers derived from these monomers are in very widespread use for a range of applications. Their polymerization depends on the addition of some functionality such as a radical or an ion across the active double bond.

9.3.1 Sonochemical Initiation by Radical Generation

The most common method for vinyl polymerization is that using radical initiation [49], either by simple thermal decomposition of the monomer or by the addition of a labile initiator. A major problem with this reaction, caused by the high reactivity of the radical center, is control over the stereochemistry of the resulting polymer. Another major topic of current research is the control over the molecular weights of polymers produced by radical initiation, usually achieved by adding compounds known as chain transfer agents which limit the growth of the growing chain. As already noted, sonication can produce high concentrations of radicals as a result of cavitation. Hence, application to vinyl monomers would provide an alternative method of initiation with the possibility of a great deal of control over the process. This process is perhaps the most studied of sonochemical polymerizations and considerable progress has been made in some systems toward a complete understanding and predictive model.

Water is particularly susceptible to cavitation [50–52] and the radicals H· and OH· generated were used by Lindstrom and Lamm [53] and Henglein [54] to prepare poly(acrylonitrile) in aqueous solution. It is also now known that radicals can be produced as a result of cavitation in virtually all organic liquids [55, 56]. However, depending on the physical properties of the particular liquid, it is often less efficient than in water and it was long believed that polymerization would not take place in organic media. For example, El'Piner [57] stated that "... polymerization of monomers in an ultrasonic field does not occur if these are thoroughly dried and do not contain substances in the polymerized state." Several other workers also suggested that no reaction would take place unless it was carried out with an initiator or was "seeded" with preformed polymer which would degrade to yield the initiating radicals [58, 59]. However, this was contradicted by the work of Melville [60] who polymerized styrene, methyl methacrylate and vinyl acetate both in the presence and absence of poly(methyl methacrylate). More recently, Miyata and Nakashio [61] investigated the azobisisobutyronitrile (AIBN) initiated polymerization of styrene and found that higher molecular weight polymers were formed under sonication and the polymerization rate decreased linearly with the ultrasonic intensity. Also, O'Driscoll and Sridharan [58] found that sonication caused a rapid increase in the conversion of a thermally initiated polymerization of methyl methacrylate.

In other recent work, Stoffer and co-workers [62, 63] investigated the sonochemical polymerization of methyl methacrylate (MMA) and acrylamide using very high ultrasound intensities from a large 'horn' apparatus. No polymerization was detected in the absence of an initiator, which in this case was dodecane thiol which reportedly produced RS· and H· radicals. The effect of changing the ultrasound intensity and initiator concentrations were reported and were in accord with the usual treatment of radical initiated polymerization kinetics. Copolymers of styrene and maleic anhydride were also produced but the reports did not indicate whether there was any difference between these polymers and those produced thermally.

In other sonochemically induced polymerizations, Lorimer et al. [64] found that in addition to the rate of polymerization, there was an increase in molecular weight during the ultrasonic polymerization of N-vinyl carbazole, but that there was an optimum intensity for the rate of reaction. The same workers [65] also investigated the polymerization of N-vinyl pyrrolidone, a system that is unusual since it does not follow the usual kinetic dependence on monomer concentration due to formation of hydrogen bonded complexes with the solvent (water). On the basis that ultrasound was known to disrupt this type of bonding, an increased polymerization rate was expected but, experimentally, the opposite was found. Fujiwara and Goto [66] polymerized diallyl terephthalate essentially using radicals formed by breakdown of the aqueous solvent and the mechanism of initiation and the kinetics of the polymerization were deduced.

Most sonochemical polymerizations have used high intensities of ultrasound but polymers can be produced using much lower powers. Orszulik [67] showed that while polymerization did not occur in the absence of an initiator, the decomposition of AIBN could be greatly accelerated in a laboratory cleaning bath and that moderately high yields of a range of acrylic polymers and copolymers were produced after 17 h sonication.

Although perhaps not strictly used as a synthetic method, Heusinger and co-workers have shown that polysaccharides can be produced by sonication of aqueous sugar solutions. The results of irradiation with ultrasound were shown to be very similar to those using γ-irradiation from a ^{60}Co source and to be a result of radical formation in the solution [68]. Analysis of sonicated dilute solutions of D-glucose by gas chromatography-mass spectrometry showed a number of products including derivatives of gluconic and hexulosonic acids together with a relatively small proportion of polymeric components. Similar results were found with solutions of ascorbic acid [69].

Thus, despite the rather conflicting early reports, it has become clear over the past few years that vinyl monomers can be polymerized solely by irradiation with ultrasound and this has led to more detailed study of the various parts of the reaction.

A series of papers describing a detailed study of the mechanism of polymerization using sonochemically generated radicals has been published by Kruus and co-workers [70–74]. Early work showed that under some conditions pyrolysis of the monomer occurred inside cavitation bubbles causing the formation of significant amounts of insoluble chars in addition to linear polymers. However, it was also shown that, as long as the monomers were properly purified and deoxygenated, soluble high molecular weight polymers of methyl methacrylate

9.3. SONOCHEMICAL SYNTHESIS OF VINYL POLYMERS

and styrene could be produced. It was found that reasonable rates of conversion could be achieved over a range of temperatures but, significantly, the reaction stopped when the ultrasound was switched off.

Our work in this area at Bath [75, 76] has also focused on model studies using similar monomers. Figure 9.5 shows some of the features of sonochemical polymerization. High molecular weight polymer is formed at very early stages of the reaction but at longer times the molecular weights are lower. This is not the same trend as found with conventionally initiated radical polymerization [49] and the reasons for this will be discussed below (see Section 9.5). As shown in Figure 9.6, a conversion of $\sim 2-3\%$ h^{-1} was achieved with MMA, at 25 °C.

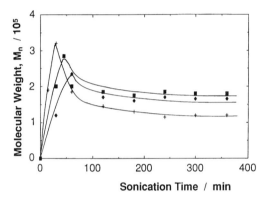

Figure 9.5 Variation in molecular weight during sonochemical polymerization (Redrawn from results in [76] and [81]). + styrene; ■ methyl methacrylate; ◆ n-butyl methacrylate.

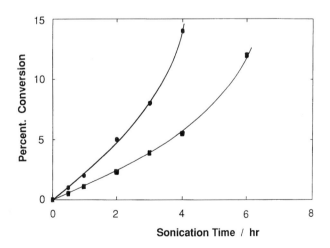

Figure 9.6 Ultrasonic polymerization of methyl methacrylate at 25 °C (reproduced from [76]). ■ bulk MMA; ● MMA + 0.1% AIBN.

However, above a particular conversion, cavitation in the solution essentially stopped and no further conversion to polymer occurred. This is probably due to the increased viscosity of the solution restricting movement of the solvent molecules and suppressing cavitation, hence preventing formation of radicals.

In its simplest form, radical initiated polymerization can be treated as a three stage reaction [49]. *Initiation* of the reaction occurs as a result of the formation of radicals, either by breakdown of the monomer or by the decomposition of an added initiator. Common initiators in organic systems are aroyl (*e.g.* benzoyl) peroxides or azo-compounds such as AIBN, while persulfates or metal ion catalyzed redox decomposition products of hydrogen peroxide are often used in aqueous systems. The growth of the polymer chain by addition of monomer to its reactive end is the *propagation phase* and the final step of the process is *termination* where the radicals are consumed by such processes as combination or disproportionation. To appreciate fully the potential usefulness of ultrasound in polymerization chemistry, it is necessary to understand the effect on each phase of the reaction as well as other processes which can occur in solution such as chain transfer and abstraction from solvent. Detailed kinetic studies have only recently been carried out and progress towards a quantitative model has been made.

9.3.1 The Initiation Process

Sonochemical initiation could take place by two routes. Pure monomers could produce initiating radicals through decomposition, or ultrasound could accelerate the decomposition of added initiators. Both of these mechanisms have been studied recently.

The rate of initiation during the sonochemical polymerization of N-vinyl carbazole has been studied by Lorimer *et al.* [64] by monitoring the decomposition of the initiator (AIBN) by following its loss at 60 °C using HPLC. The rate of initiator breakdown increased with rising ultrasound intensity, reached a maximum value and fell at high intensities to a rate lower than that in the absence of ultrasound. The same workers also showed that a sonochemical acceleration of ~40% in the rate of initiator breakdown occurred in aqueous solutions of potassium persulfate [77].

In our work with MMA, we used an alternative method involving trapping the radicals [76] using an excess concentration of 2,2′-diphenyl-1-picryl hydrazyl, DPPH. Doubts have been cast on the accuracy of trapping methods. For example, there will be a degree of radical recombination and other side reactions, but since the radical must escape the vicinity of the cavitation bubble whether to initiate polymerization or to be trapped by DPPH, we feel that the rate of trapping will closely mimic the rate of initiation. To avoid reactions with growing polymer chains, a model solvent was used. A solvent was needed that behaved in an identical manner to MMA when sonicated, *i.e.*, had closely matched physical properties. That most closely matching MMA is methyl butyrate. Thus, the application of ultrasound to these two "cavitationally

9.3. SONOCHEMICAL SYNTHESIS OF VINYL POLYMERS

similar" liquids would be expected to produce similar numbers of radicals on sonication. By this method we were able to measure the rate constants for radical production under a range of conditions and the results are shown in Table 9.1. Our work showed that the results agreed well with a first order reaction although the fit was somewhat better to a zero order process, as might be expected if the sonochemical step was the rate limiting process. To compare with thermal measurements, the rate constants considered here are those calculated for a first order reaction.

The rate constant for purely thermal production of radicals, estimated by extrapolation from higher temperatures [78], is negligible compared with that of the sonochemical reaction. The sonochemical rates are also much higher than the corresponding values for the thermal decomposition of AIBN in MMA. The results show that simply by using ultrasound alone, we can achieve at 25 °C similar rates of initiation as are usually found in thermally initiated polymerizations. To further compare with more conventional experiments, we also sonicated a solution of 0.1%wt AIBN in MeOBu, the rate constant also being shown in Table 9.1. This is some three orders of magnitude higher than that expected for the thermal decomposition AIBN at this temperature. It is evident that the sonication process is clearly able to accelerate the decomposition of AIBN in solution as well as to produce radicals directly by decomposition of the solvent.

Control of the sonochemical reaction demands an understanding of the effect of conditions on the process. The effect of changing the temperature is shown in Table 9.1 and indicates that the rate of radical production is faster at lower temperatures. As noted in Section 9.1.3, this "inverse Arrhenius" behaviour is not unusual in sonochemical systems. The usual "Arrhenius" treatment gave an apparent activation energy of $-18.9 \, \text{kJ} \, \text{mol}^{-1}$. Kruus and co-workers [70], from inhibition measurements, reported this value to be $-15 \, \text{kJ} \, \text{mol}^{-1}$ for styrene although, considering their polymerization data, they suggested that the initiation was insensitive to temperature for MMA. These values may be compared with that of $+159.8 \, \text{kJ} \, \text{mol}^{-1}$ for the decomposition of AIBN in the

Table 9.1 **Rate constants for DPPH radical trapping in methyl butyrate (from [76]).**

System	Temperature (°C)	Rate constant/s^{-1} Ultrasound	Conventional
MeOBu	-10	6.35×10^{-5}	
MeOBu	25	2.21×10^{-5}	$\sim 1 \times 10^{-15}$
MeOBu	60	1.03×10^{-5}	
MeOBu + 0.1% AIBN	25	9.13×10^{-5}	$\sim 2 \times 10^{-8}$
MeOBu + 0.1% AIBN	70		3.1×10^{-5}

absence of ultrasound [78] and $+92$ kJ mol^{-1} for solely thermal initiation in MMA at high temperatures. The other parameter of importance is the ultrasound intensity, I_{us}, and it was found that below an intensity of approximately 12–13 W cm^{-2} there was no production of radicals; this presumably corresponded to the cavitation threshold in this system. Above this minimum value, there was a linear dependence on the rate of DPPH consumption.

These considerations led us [76] to account for the rate of the initiation step, R_i, by an equation of the form:

$$R_i = (k_d f)[CS] = (k_d f) A I_{us} = k_{us} I_{us}$$

where [CS] is the concentration of active cavitation bubbles in the system, k_d represents the decomposition rate of the initiator (or monomer for a bulk reaction) in the presence of ultrasound and f is a factor to account for the efficiency of a radical in initiating polymerization. Since the number of cavitation bubbles is determined primarily by the ultrasound intensity, this can be expressed by introducing the proportionality constant, A. Incorporation of all the constant terms in a "composite" rate constant, k_{us}, leads to a prediction of the observed linear dependence of the rate on intensity as long as I_{us} is defined as the ultrasonic intensity in excess of the threshold value. It should be noted that this linear dependence will only hold for a limited range of I_{us} before passing through a maximum.

Thus, by suitable manipulation of the conditions such as temperature and ultrasound intensity, we can predict and control the rate of initiation. The effect of sonication on the polymerization process was then considered.

9.3.3 The Polymerization Process

To develop a model for polymerization, it is necessary to understand the effect of each of the factors on the propagation and termination reactions as well as initiation since these will determine the structure of the resulting polymer.

The effect of experimental conditions on the rate of polymerization and on the properties of the final polymers has been measured by Kruus et al. as well as by our group in the work referred to earlier. For example, the rate was found to be proportional to the monomer concentration, [M], and to depend on the square root of I_{us}. The final molecular weight varied inversely with [M] and scaled with $I_{us}^{-1/2}$. These findings are totally in accord with the assumption that initiation depends on the intensity, as described above. In contrast to the initiation, higher temperatures led to an increase in the rate of polymerization. The termination steps, being bimolecular radical reactions, do not depend strongly on temperature. Hence, the main effect will be on the propagation. However, these reactions are rapid compared with the initiation rate. For example, the rate constants [8] at 50 °C for propagation and termination in styrene are 209 and 1.15×10^9 dm^3 mol^{-1} s^{-1}, respectively, and the corresponding values for methyl methacrylate 410 and 2.4×10^8 dm^3 mol^{-1} s^{-1}, respectively.

9.3. SONOCHEMICAL SYNTHESIS OF VINYL POLYMERS

Some of the properties of PMMA polymers produced sonochemically are given in Table 9.2 which shows in particular the effect of the temperature and sound intensity. While the rate of polymerization increases with rising temperature, there is little variation in the final conversion achieved. Again, this supports the suggestion that polymerization is limited by the suppression of cavitation as the solution viscosity rises. The molecular weight is larger at higher temperatures, again indicating that propagation is more efficient. The final molecular weight decreases with rising ultrasound intensity. The reasons for this will be discussed in Section 9.5.

From the kinetic results, it can be concluded that the ultrasound has relatively little effect (perhaps a two or three fold acceleration) on the propagation or termination reactions and is insignificant compared with the acceleration of the initiation. This conclusion is also supported by the small amount of published work on copolymerization where two or more different monomers can be incorporated into the same polymer chain. If ultrasound were affecting the propagation, differences in the sequences of the two monomers along the chain would be expected. No significant differences were found by Miyata and Nakashio [79] for styrene-acrylonitrile copolymerizations or by Price et al. [80] for mixtures of styrene with methyl or butyl methacrylates.

The final factor which has been investigated for these systems is the stereochemical arrangement of the monomer units along the chain. In polymer chemistry, this is described by the *tacticity* of the polymer. Chains which have all of the substituents arranged on the same side of the chain are referred to as isotactic while those which have an alternating arrangement are syndiotactic. This is shown for the case of PMMA in Figure 9.7. Polymers with a random arrangement are termed atactic or heterotactic. The tacticity in PMMA is readily determined using NMR spectroscopy [81] and Table 9.3 shows the relative degrees of tacticity in sonochemically produced PMMA [76] together with literature results for more conventional polymerizations.

Table 9.2 Properties of PMMA produced by 6 h sonication of pure monomer (from [78]).

Temperature (°C)	Intensity (W cm^{-2})	Molecular weight	Polydispersity	Conversion (%)
−10	20	107,000	1.8	11
0	20	119,000	1.7	10
25	20	147,000	1.9	11
40	20	241,000	1.6	11
60	20	356,000	1.4	13
25	13	144,000	2.3	2
25	29	131,000	1.9	10
25	35	97,000	1.8	14
25	58	35,000	1.5	12

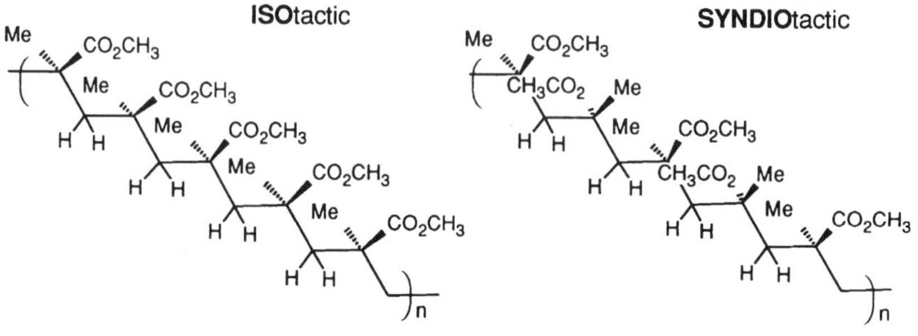

Figure 9.7 Stereochemical arrangement of substituents in PMMA (an atactic polymer would have a random arrangement of the substituents).

Table 9.3 Stereochemical tacticity ratios for radically initiated PMMA (from [78]). i=*iso*tactic; a=*a*tactic; s=*syndio*tactic.

Polymerization conditions		Ratio		
		i	a	s
Monomer[a]	100 °C	8.9	37.5	53.9
Solution[a]	50 °C	6.3	37.6	56.0
Ultrasound	60 °C	4.3	41.0	54.7
Ultrasound	40 °C	4.0	40.4	56.4
Ultrasound	25 °C	2.8	34.3	62.9
Ultrasound	0 °C	1.7	33.8	64.6
Ultrasound	−10 °C	0.8	25.6	73.6

[a] Using benzoyl peroxide as initiator.

It is clear that conventional initiation using a peroxide initiator leads to predominantly syndiotactic polymers although there are significant sequences of atactic and isotactic groups. At high temperatures sonication and thermal initiation both produce polymers with similar stereochemistry again suggesting that sonication has little or no effect on the propagation reaction. Polymerizations occurring by an anionic mechanism using, for example, n-butyl lithium as the initiator give 60–70% isotactic polymer, because the rate of propagation is slow compared with the initiation rate. In sonochemically promoted reactions at low temperatures, the initiation is accelerated while the propagation rate is decelerated compared to the thermal polymerization, and hence the proportion of syndiotacticity along the chain is raised. Steric hindrance between the bulky ester groups makes syndiotactic addition thermodynamically more favorable, although the energy difference is small. As the temperature is lowered, the propagation rate is slowed and there is more chance of the favored addition taking place.

9.3. SONOCHEMICAL SYNTHESIS OF VINYL POLYMERS

By analogy with the usual treatment [49] of radical polymerization kinetics, the following mechanism was proposed for the sonochemical polymerization. Although a somewhat simplified treatment, it can account for concurrent polymerization and degradation reactions and at least qualitatively explain all the observations described above. In the following, "CS" represents an active cavitation site and "P" a polymer chain.

$$M + CS \xrightarrow{k_1} 2R\cdot \quad (A)$$

$$R\cdot + M \xrightarrow{k_2} RM\cdot \quad (B)$$

$$RM_n\cdot + M \xrightarrow{k_3} RM_{n+1}\cdot \quad (C)$$

$$RM_n\cdot + RM_n\cdot \xrightarrow{k_4} P \quad (D)$$

$$P + CS \xrightarrow{k_5} 2RM_n\cdot \quad (E)$$

The process in (E) accounts for the effect of ultrasound on polymer chains once they are formed in solution; this is described in detail in Section 9.5. Using this mechanism, equations can be derived to describe the time dependence of the rate of polymerization, the molecular weight and the conversion to polymer [69, 74] which are in good agreement with the experimental observations.

Thus, we are now in a position where the mechanism and kinetics of the polymerization are understood in sufficient detail to model the process. However, there are still problems to be overcome, the main one being the relatively low final conversions achieved thus far, before the method will enjoy more widespread acceptance.

9.3.4 Polymerization by Ionic Mechanisms

One method which has been adopted to overcome the deficiencies of radical polymerization in vinyl monomers is to initiate the reaction using ionic species. Cationic initiators such as a number of Lewis acids can be used, although anionic mechanisms utilizing, for example, organolithium species are perhaps more common. These reactions yield polymers with very narrow molecular weight distributions and polydispersities as low as 1.03. However, they require the use of very pure reagents and dry, oxygen-free conditions.

It is the general understanding [21] that sonication is most effective in single electron transfer reactions and in mechanisms involving radicals. Indeed, there are a number of reports on the synthesis of low molar mass compounds where ionic mechanisms are completely suppressed in favor of radical processes. Thus, there has been only a small amount of work in this area. In a very recent report [82], Schultz et al. described how sonication had little effect on the anionic polymerization (addition of styrene to an ion pair at the end of a growing chain) but allowed the initiators (such as butyl lithium) to be prepared at faster rates than under conventional conditions. In addition, the need for dry conditions was removed as successful initiators were prepared in undried solvents.

9.3.5 Suspension and Emulsion Polymerizations

Alternative routes by which vinyl monomers can be polymerized are emulsion and/or suspension polymerization [83] where the organic phase is dispersed, usually in an aqueous system. In addition to water and the monomer(s), a number of additives such as stabilizers, dispersants and initiators are usually added. The efficient mixing and dispersion of liquids using ultrasound is a common commercial process, for example in the food industry. The large degree of motion induced by acoustic streaming provides this efficient mixing while the very high shear forces around cavitation bubbles act to break up droplets of liquid and maintain a small and even distribution of droplet sizes.

There are various aspects of these polymerizations where ultrasound could be beneficial to this type of polymerization. The dispersion could be formed and maintained during polymerization by using ultrasound. The production of radicals in the aqueous phase could also be used as a method of initiation. There is thus scope for the reduction or elimination of some of the additives such as emulsifiers and initiators used in the polymerization. In addition, since the polymer is in a dispersed phase, the viscosity does not increase to the same extent as in bulk polymerizations so that the suppression of cavitation will not be as much of a problem.

Ultrasound was first applied to this type of reaction in 1950 by Ostroski [84] who found that sonication increased the dispersion of styrene and other components and also accelerated significantly the rate of polymerization. Hatate *et al.* [85, 86] also studied the suspension polymerization of styrene and found that ultrasound prevented the droplets from agglomerating or sticking to the walls of the container, so minimizing the build up of heat in the reactor. At the intensities used, little effect on the kinetics of the polymerization or on the molecular weight was seen but the particle size could be influenced by varying the ultrasonic frequency. More recently, Lorimer and co-workers have also compared the emulsion polymerization of styrene carried out in the presence and absence of ultrasound from a high intensity horn [77].

The decomposition of potassium persulfate initiator in aqueous solution was accelerated under ultrasound and this, together with the production of more stable emulsions, increased the rate of polymerization as well as eliminated the induction period. An additional effect was that lower amounts of emulsifier (sodium dodecyl sulfate) were needed to maintain the reaction but that lower molecular weights were obtained than in the conventional reaction.

The properties of **PMMA** materials produced by a sonochemical emulsion polymerization have been described in a brief report by Stoffer *et al.* [87] who produced very high molecular weight ($2.5-3.5 \times 10^6$) polymers. Their tacticity was found to depend on the intensity used, an effect attributed to the lower intensities leading to lower final temperatures and hence thermodynamically favoured propagation, as described above.

Davidson and co-workers [88–90] have taken an alternative approach to sonochemical emulsion polymerization. They were interested in obtaining uniform emulsions for use as pressure sensitive adhesives of the type used in adhesive tapes. However, they needed to avoid the initiation and degradation processes which can occur. Therefore, a source of low intensity ultrasound, the so-called "whistle reactor" was employed. The use of this equipment is well established in the food industry. Here, it was used to prepare and maintain the emulsion while employing conventional thermal or photochemical initiation methods. Using this system with both oil and water soluble initiators gave latexes of poly(2-ethylhexyl acrylate) which were stable for several months even when prepared in the absence of surfactants. A smaller and more even size distribution was also obtained leading to adhesives with superior properties to those from a conventionally prepared emulsion.

In the most recent example of this type, Grieser and co-workers [91] have used a horn system to produce latexes of polystyrene, poly(butyl acrylate) and poly(vinyl acetate) with low amounts of or even no surfactant and smaller particle sizes than in conventional processes. The rates of reaction were also accelerated but, significantly, this depended on the vapor pressure of the monomer. In the case of vinyl acetate, which has a high vapor pressure, the polymerization was relatively slow. This was attributed to monomer vapor entering the cavitation bubbles (primarily in the aqueous phase) and so reducing the intensity of cavitational collapse and hence the number of radicals formed. In contrast to the bulk or solution polymerizations described above, conversions close to 100% were routinely achieved.

9.4 SONOCHEMICAL POLYMERIZATION OF NON-VINYL MONOMERS

While vinyl polymers are an important class, there is a wide variety of other monomers which are polymerized by other mechanisms. Many of these are described below but there is one which does not conveniently fit into any classification and merely relies on the heating effect produced by ultrasound as it passes through a material.

Rubber articles such as tubes and tyres are made from rubber (natural or synthetic) which is lightly crosslinked. The crosslinking joins all the chains together so that while the product retains some elasticity, the chains cannot be separated to flow or dissolve. Crosslinking is often carried out by heating with sulfur compounds, a process known as *vulcanization*. This is very energy intensive since rubber is a poor conductor of heat. In the sonochemical process [92], use is made of the fact that the heat is generated inside the article due to attenuation (see Section 9.2.1) so that it is relatively rapid and less degradation of the rubber occurs. In some cases, both the treatment time and the energy requirements were lowered by up to a half. The same ultrasonic heating method can be applied to other systems such as epoxy resins [93].

9.4.1 Ring Opening Polymerizations

A number of commercially important polymers are produced by a ring opening mechanism of a cyclic monomer [94]. A range of polyesters can be produced from cyclic lactones but probably the most commercially significant in terms of amount of polymer manufactured is the reaction of ε-caprolactam to give Nylon-6 (Scheme 9.2).

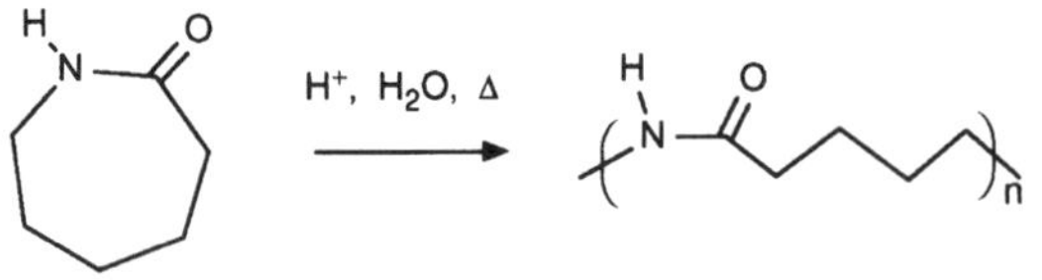

Scheme 9.2

Nylon-6 is conventionally produced in a two stage process. The initial ring opening is catalyzed by a small amount (~1%) of water and this is followed by polymerization to high molecular weight under vacuum. This sonochemical reaction has been studied by Ragaini et al. [95, 96] in Milan. Their work showed that ultrasound enhanced the ring opening phase allowing a single step polymerization without the need to add water to start the reaction. High molecular weight materials with narrower distributions were formed in shorter reaction times than when using the conventional process. An additional advantage allowed by sonication is that the polymerization could be operated at significantly lower temperatures than the conventional process.

In work at Bath, we have recently commenced study of lactone polymerization. In other work, the ring opening reaction of octamethylcyclotetrasiloxane, catalyzed by sulfuric acid, to poly(dimethyl siloxane), PDMS [97] has been used as a model reaction (Scheme 9.3). PDMS is the base material of a large number of silicone materials.

The first study of this reaction was by Kogan and Smirnov [98] who demonstrated dramatic rate enhancements under ultrasound. In our work [99], we found that conventional and sonochemical polymerizations at room temperature gave similar yields although the rate of the latter was faster. However, the

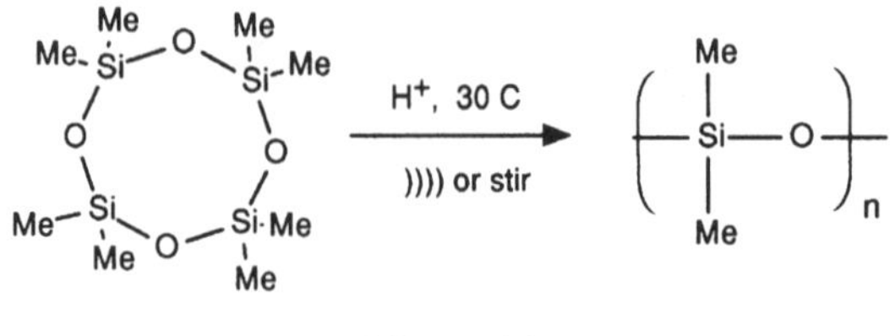

Scheme 9.3

9.4. SONOCHEMICAL POLYMERIZATION OF NON-VINYL MONOMERS

molecular weight distributions showed considerable differences, as shown by the GPC traces in Figure 9.8. Clearly, the ultrasound is accelerating the ring opening reaction as well as yielding higher molecular weight materials. The distribution of molecular weights is also narrower. However, it was clear from other results that the amount of catalyst was the main factor in determining the extent of reaction and, in particular, no polymerization occurred in the absence of added acid even at the high intensities produced by a horn system. It seems probable that the acceleration of the polymerization is due to a much more efficient dispersion of the acid catalyst throughout the reacting system, leading to a more homogeneous reaction which therefore leads to a lower distribution of chain lengths as well as a faster rate of reaction.

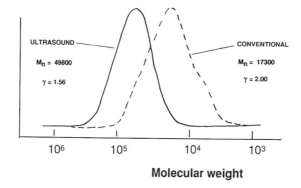

ULTRASOUND
$M_n = 49800$
$\gamma = 1.56$

CONVENTIONAL
$M_n = 17300$
$\gamma = 2.00$

$10^6 \quad 10^5 \quad 10^4 \quad 10^3$

Molecular weight

Figure 9.8 GPC chromatograms of PDMS before and after sonication in toluene.

As a final example of this class of reaction, Stoessel [100] has recently reported the use of ultrasound in the ring opening of small polycarbonate rings. A mixture of these cyclic substances, such as shown, can be used as a low viscosity "pre-polymer" for ease of processing before adding a nucleophilic initiator such as a metal carbonate at around 250 °C to obtain the final, high molecular weight material (Scheme 9.4).

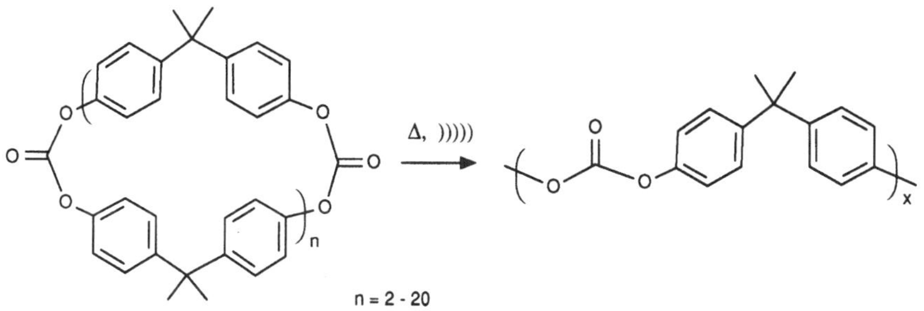

$n = 2 - 20$

Scheme 9.4

The use of ultrasound at very high intensities from a large horn system was found to obviate the need for an added initiator in some cases and to allow greater control over the process, including its operation at lower temperatures to minimize polymer degradation. While the precise mechanism of the ring opening was not deduced, it was suggested that the results were inconsistent with a radical intermediate and that polymerization resulted either from the high local temperatures produced by cavitation or from the sonochemical promotion of impurities such as sodium hydroxide or carbonate inducing the ring cleavage.

9.4.2 Condensation Polymerizations

A number of industrially very important polymers and plastics such as polyesters, polyurethanes and nylons are prepared *via* condensation reactions. In view of this, there have been surprisingly few reports of the application of ultrasound in this area. Watanabe et al. [101] applied ultrasound from a cleaning bath to the preparation of aromatic polyformals from, for example, bisphenol-A and methylene bromide. This is a two phase system using solid potassium hydroxide and normally requires the use of a phase transfer catalyst such as TBAB (tetrabutyl ammonium bromide) to promote the surface mediated reaction (Scheme 9.5).

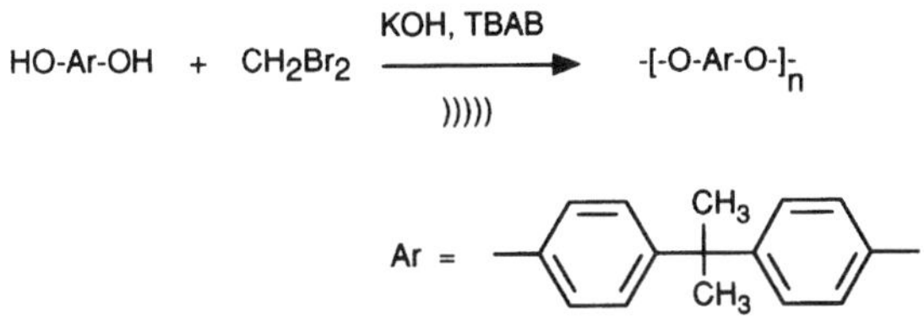

$$\text{HO-Ar-OH} + \text{CH}_2\text{Br}_2 \xrightarrow[\text{)))))}]{\text{KOH, TBAB}} \text{-[-O-Ar-O-]}_n$$

$$\text{Ar} = -\text{C}_6\text{H}_4-\text{C}(\text{CH}_3)_2-\text{C}_6\text{H}_4-$$

Scheme 9.5

The sonicated reactions gave considerably higher yields of polymers with much higher inherent viscosities (and hence larger molecular weights) as shown in Figure 9.9. Although no reaction occurred in the absence of the phase transfer catalyst, ultrasound clearly assisted its effect and promoted the transport of reactants to the surface of the powdered base. This is typical of sonochemical acceleration of heterogeneous processes, a number of phase transfer examples having been reported.

Amongst the very few other published studies of condensation processes is the work of Long who, in a wide-ranging patent [102] described various reactors which incorporated ultrasonically vibrating walls. These could be used for the precise control of both when and where polymerization took place for several polyurethane systems. The sonochemical set-up was especially useful in producing foams.

9.4. SONOCHEMICAL POLYMERIZATION OF NON-VINYL MONOMERS

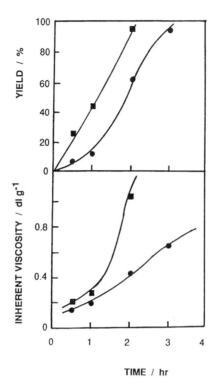

Figure 9.9 The effect of ultrasound on the yield and inherent viscosity of polyformals (redrawn from [101]). ■ with ultrasound; ● without ultrasound.

9.4.3 Electrochemically Promoted Polymerizations

The potential beneficial consequences of the combination of electrochemistry with ultrasound have stimulated considerable recent interest [103–106]. Beneficial effects arise from cleaning and regeneration of the electrode surface as well as from enhanced mass transport and prevention of ion depletion from the diffusion layer at the electrodes.

Several types of polymers including polypyrroles, polythiophenes and polyanilines as well as those of the more usual vinyl monomers can be prepared by electrochemical methods. In a preliminary study of the electrosynthesis of copolymers containing isoprene and α-methyl styrene, Topare and co-workers [107, 108] reported that higher yields were produced under sonication and that the reactivity ratios (or the preference for a monomer to react with a different monomer rather than itself) of both components were raised from those in the conventional reaction. The differences were attributed to the ultrasonic cleaning action allowing the continual renewal of fresh electrode surface and the prevention of irreversible adsorption of impurities and fouling of the electrode. These workers have also applied ultrasound to the electroinitiated polymerization of butadiene sulfone [109]. The polymer obtained had the same structure as that from radical initiated bulk or solution polymerization. As shown in Figure 9.10, the rate of polymerization was much faster under sonication, an effect again attributed to enhanced mass transfer effects near the electrode.

Ito and co-workers [110] have studied extensively the electrochemical polymerization of thiophene under the influence of ultrasound from a cleaning bath operating at 45 kHz. As shown in Table 9.4, sonication gave dramatically improved yields, particularly at higher current densities. In addition, the polymer films produced using ultrasound were more homogeneous and had higher conductivities and superior mechanical properties. These workers suggested that the effects were due to the efficient mixing and agitation provided by cavitation near the surface which would disrupt any diffusion layer, and enhance mass transfer of reactants to the surface and products away into solution.

A related example involved the deposition of polypyrrole onto microporous membranes to form conducting films [111]. Sonication allowed better

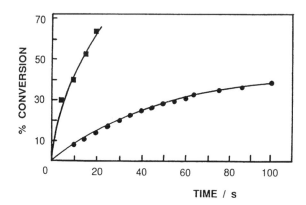

Figure 9.10 Conversion versus time plots for the electroinitiated polymerization of butadiene sulfone (data taken from [109]). ■ with ultrasound; ● without ultrasound.

Table 9.4 Properties of ultrasonically promoted electro-polymerized polythiophene[a] at various current densities.

Current density (mA cm^{-2})	Polymer yield (%)		Conductivity (S cm^{-1})	
	NU	US	NU	US
0.7	84	96	15	25
3.0	63	95	40	80
5.0	52	93	60	90

[a] Polymerizations at 5 °C and 0.3 mol dm^{-3} monomer concentration. NU denotes no ultrasound, US denotes the use of ultrasound. (Results taken from Ito et al. [110], Table 1.)

9.4. SONOCHEMICAL POLYMERIZATION OF NON-VINYL MONOMERS

impregnation and a more even film to form giving enhanced material properties such as electromagnetic shielding.

9.4.4 Ziegler-Natta Polymerizations

The discovery of Ziegler-Natta catalysts in the late 1940's revolutionized polymer science in providing a method to polymerize monomers such as ethylene and propylene which are not susceptible to radical reactions. This process and more modern variants are now exploited commercially on a massive scale [112] for the production of polyolefins. The catalysts usually involve a mixed metal species prepared by reacting, for example, a trialkyl aluminum compound with titanium tri- or tetrachloride yielding, in the most simple view, a complex with a vacant coordination site onto which the alkene is attached. A repeated alkene insertion reaction then results in polymer formation (Scheme 9.6).

Although polyolefins can be produced by other methods, the stereochemical selectivity of the coordination site allows a large degree of control over the structure of the polymer and it is the favored method for the production of *iso*- and *syndio*tactic polyolefins. However, molecular weight control is difficult due to the complexity of the reaction system.

Ultrasound was first applied to coordination polymerization as long ago as 1957 by Mertes [113] who patented a sonochemical procedure which yielded more uniform polyethylenes resulting from better dispersion of the catalyst and prevention of catalyst deactivation. Several forms of catalysts including $TiCl_3/Al(C_2H_5)_3$ or CrO_3 impregnated onto alumina, were found to have higher activities and longer lifetimes when the polymerizations were carried out under 200–400 kHz ultrasound at low intensities of $<10\,W\,cm^{-2}$. In our work at Bath, we have made a preliminary study of the effect of ultrasound on the heterogeneous, Ziegler-Natta polymerization of styrene [114] using a $TiCl_4/Al(C_2H_5)_3$ catalyst to determine whether ultrasound could influence the rate and yield of the polymerization and/or have an effect on the molecular weight distribution and microstructure of the polymer.

Simply preparing the catalyst using ultrasound and allowing the polymerization to proceed under "silent" conditions yielded polymers with identical properties to those without any sonication. However, maintaining the sonication throughout the polymerization increased the rate of reaction with no change in tacticity suggesting that ultrasound has no significant effect on the

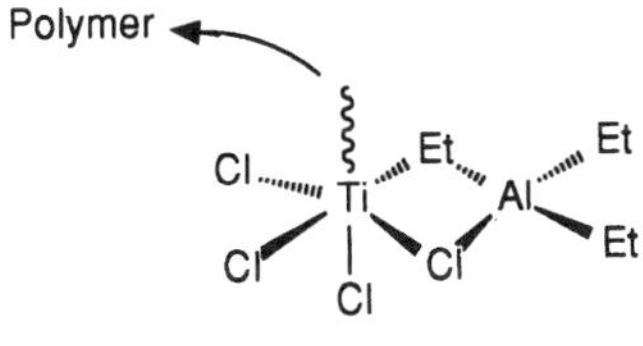

Scheme 9.6

catalyst structure. Examples of GPC chromatograms from the polymers are shown in Figure 9.11. Differences in molecular weight caused by sonication are obvious. The reasons for the increased yields and rates of reaction are not totally clear, but are probably related to sonication causing very efficient mixing and faster mass transfer of monomer to the reactive site on the surface of the catalyst. There may possibly also be a reduction in the catalyst particle size with consequent increase in active area of the catalyst. The effect of ultrasound on the polymer in solution, outlined in Section 9.5, must also be considered.

9.4.5 Grignard Coupling of Polyphenylenes

These polymers, consisting exclusively of linked aromatic rings, attracted considerable attention some years ago as prototypes for conducting materials. However, their potential has never been fully exploited due to problems in the synthetic methods available [115] as well as processing problems caused by their very high melting points and very low solubilities. These have recently been overcome by preparing materials substituted with, for example, alkyl chains to confer solubility or by employing complex precursor routes. Interestingly though, a number of the original synthetic methods are of the type to which ultrasound has been successfully applied in organic synthesis [1, 20–22] so that we felt it was worthwhile to explore some of these methods as part of our program.

The most straightforward method for preparing polyphenylene is the oxidative coupling of benzene using a Lewis acid catalyst (Scheme 9.7), the mechanism involving a one electron transfer step [116], precisely the type of reaction that would be expected to be influenced most by ultrasound.

Conventionally, the reaction is carried out in a paste of benzene and the oxidizing agents. Not surprisingly, this was not affected by ultrasound [5] since

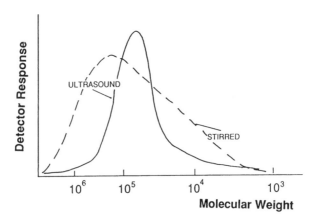

Figure 9.11 GPC of polystyrenes produced by Ziegler-Natta polymerization at 30 °C. ---- Stirred; ———— Ultrasound.

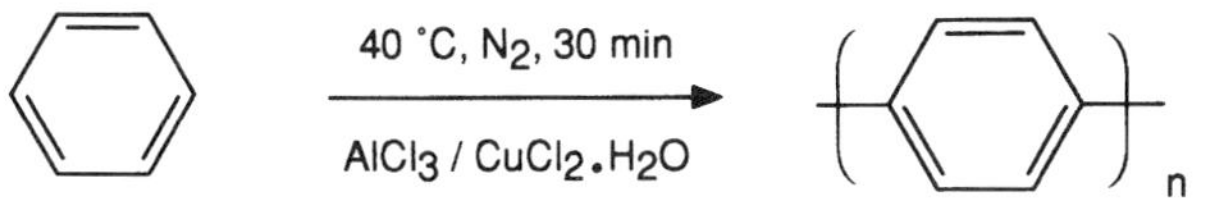

9.4. SONOCHEMICAL POLYMERIZATION OF NON-VINYL MONOMERS

$$\text{benzene} \xrightarrow[\text{AlCl}_3 / \text{CuCl}_2 \cdot \text{H}_2\text{O}]{40\,°\text{C, N}_2,\ 30\text{ min}} \text{poly(phenylene)}_n$$

Scheme 9.7

a liquid is needed for cavitation. The reaction was then repeated in an excess amount of benzene to give a suspension so that a liquid phase was present to enable the transmission of the sound. Operation in a cleaning bath doubled the amount of polymer recovered.

Other methods for preparing polyphenylenes are based on coupling reactions of dihaloaromatics but they often give low yields and sometimes react only as far as the dimer or trimer. One of the more successful schemes is due to Rehann *et al.* using a nickel complex catalyzed Grignard type reaction [117] (Scheme 9.8).

Ultrasound was applied to this scheme under various conditions and the results, which show a clear ultrasonic acceleration, are given in Table 9.5.

Table 9.5 Effect of ultrasound on Grignard coupling of dibromobenzene.

Conditions	Ultrasound	Polymer yield (%)
Reflux, 2 h	No	20
Reflux, 5 h	No	35
20 °C, 24 h	No	40
60 °C, 2 h	Bath	48
20 °C, 2 h	Horn	33
60 °C, 2 h	Horn	45

In each case, sonication gives better yields than the equivalent "silent" conditions as well as allowing the reaction to proceed at lower temperatures. There are routes to related polymers which involve a number of organometallic reactions including Wurtz and Ullmann couplings.

9.4.6 Wurtz Coupling of Poly(organosilanes)

These polymers have a backbone of silicon atoms substituted with a variety of organic groups and are currently attracting considerable interest due to their

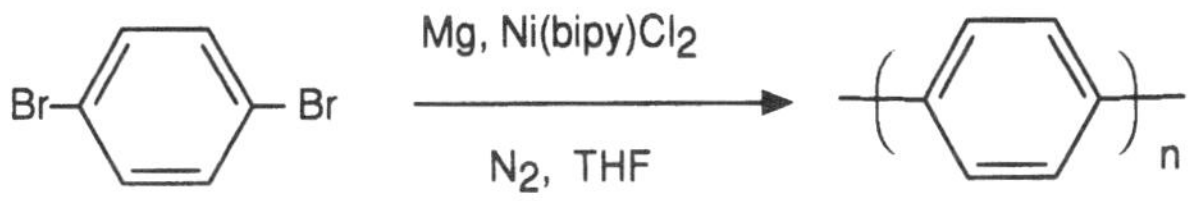

$$\text{Br-C}_6\text{H}_4\text{-Br} \xrightarrow[\text{N}_2,\ \text{THF}]{\text{Mg, Ni(bipy)Cl}_2} \text{poly(phenylene)}_n$$

Scheme 9.8

range of potential applications [118, 119] as photoactive and photoconductive materials including photoresists and photoinitiators and also as precursors to ceramic materials. Their more widespread use has been prevented by problems in their synthesis. The usual synthetic method [118–121] is a Wurtz type coupling of dichlorodiorganosilanes using molten sodium in refluxing toluene (Scheme 9.9). However, the reactions do not proceed reproducibly, the yields are rather low and the polymers often have a very wide, usually bi- or tri-modal, molecular weight distribution.

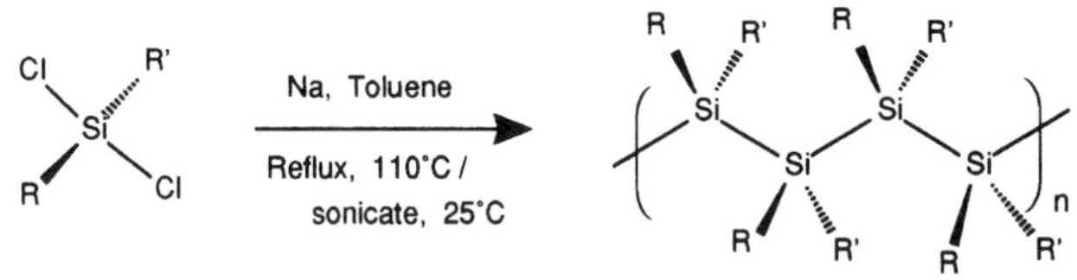

Scheme 9.9

To achieve commercial use, polymers with a controlled structure and, preferably, monomodal distribution are needed. A synthesis under more environmentally acceptable conditions is also desirable. While polymers with reasonable distributions have been produced by carrying out the reaction in the presence of additives such as crown ethers [122, 123], a synthetic method to produce them directly remains a significant target.

The possibility of applying ultrasound to this type of reaction comes from the use in the early 1980's by Han and Boudjouk of lithium to couple organochlorosilanes to give R_3SiSiR_3 [124]. The work has been extended by using R_2SiCl_2 to give the polymeric materials. While polymers with a range of substituents have been prepared, most work has been done with poly(methyl phenyl silane) as it yields soluble, easily characterizable polymers. The dimethyl- and diphenyl- polymers are crystalline and insoluble so that molecular weight distributions cannot be measured.

The first reports of ultrasound being applied to the synthesis of this poly(organosilane) were by Matyjaszewski and co-workers [125]. Their extensive studies [126] produced materials with monomodal molecular weight distributions and polydispersities as low as 1.2, albeit in rather low yield (11–15%), using ultrasound at 60 °C in toluene. Conversely, Miller et al. [127] reported somewhat conflicting results in that the sonication method did not yield polymers with a monomodal distribution unless diglyme or 15-crown-5 were added to the solvent. In the absence of such additives, bimodal distributions were obtained in which the higher molecular weight fraction ($\sim 146{,}000$) comprised about 65% of the polymer, the remainder being of relatively low molecular weight (~ 9700).

We have studied a number of silanes under a range of conditions [99]. Reaction of the dimethyl or diphenyl compounds showed enhanced yields of polymer when the sodium was dispersed using ultrasound from a cleaning bath,

9.4. SONOCHEMICAL POLYMERIZATION OF NON-VINYL MONOMERS

but to allow ready characterization, the majority of our work has involved the methyl phenyl silane which yields soluble materials.

Again, enhanced yields and reaction rates were measured under sonochemical conditions. However, perhaps of more significance are the changes in the molecular weights and distributions of the polymers. The GPC chromatograms, with molecular weights given relative to polystyrene standards, of some of these polymers are shown in Figure 9.12. The conventional, reflux method gives a polymer with a very wide, bimodal distribution. The polydispersity is vastly reduced under the influence of ultrasound and use of the high intensity probe system gave a monomodal, though broad, distribution.

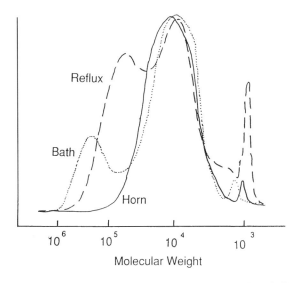

Figure 9.12 GPC chromatograms of poly(methyl phenyl silanes).

Differences in the distributions given by the bath or horn apparatus gave a clue to a possible explanation for the differences in the published results of Matyjaszewski and Miller. Four polymerizations were performed using the horn system under identical conditions except that the intensity of the ultrasound was varied [128]. The resulting molecular weight distributions are shown in Figure 9.13 and clearly show the importance the ultrasound intensity has in determining the structure of the polymers. This effect of intensity seems the most likely explanation to account for the differences in literature results.

The use of higher intensities resulted in narrower molecular weight distributions. Many sonochemical reactions have been interpreted in terms of preferential promotion of radical and single electron transfer processes over those involving ionic intermediates. However, there is no firm evidence of radical intermediates in this polymerization and the explanation of the effects probably

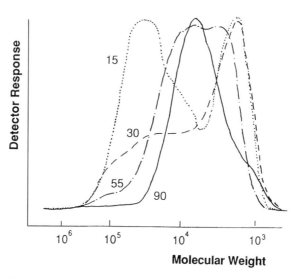

Figure 9.13 Effect of ultrasound intensity on GPC chromatograms of poly(methyl phenyl silane) (from [128]). Values indicate ultrasound intensity in W cm^{-2}.

lies in the physical rather than chemical effects of sonication. Sonochemical acceleration of heterogeneous reactions is usually attributed to increased mass transfer and the continual sweeping of the surface leading to a greater number and faster regeneration of reactive sites. This would, in this case, give a more homogeneous chain growth and hence a narrower distribution of chain lengths. In addition, as described in Section 9.5, high molecular weight material formed early in the reaction will be degraded by the ultrasound, a process known to be more efficient at high intensity.

A modification of the sonochemical synthesis of linear polymers was used by Bianconi and co-workers [129]. It involved sonochemical activation of Na/K alloy in the preparation of polysilynes, $(RSi)_n$.

$$RSiCl_3 \xrightarrow{Na/K} (RSi)_n \quad R = n\text{-}C_4H_9, \ n\text{-}C_6H_{13}$$

Again, the use of ultrasound removed the very high molecular weight fractions and hence allowed the synthesis of polymers with molecular weights in the range 10,000–100,000 and monomodal distributions.

In a final example of this type of coupling reaction, Hohol and Urban [130] prepared polymers consisting of chains of phthalocyanine units with germanium in their centers (Scheme 9.10). The use of ultrasound allowed a one-step, room temperature synthesis which gave higher yields than conventional methodology.

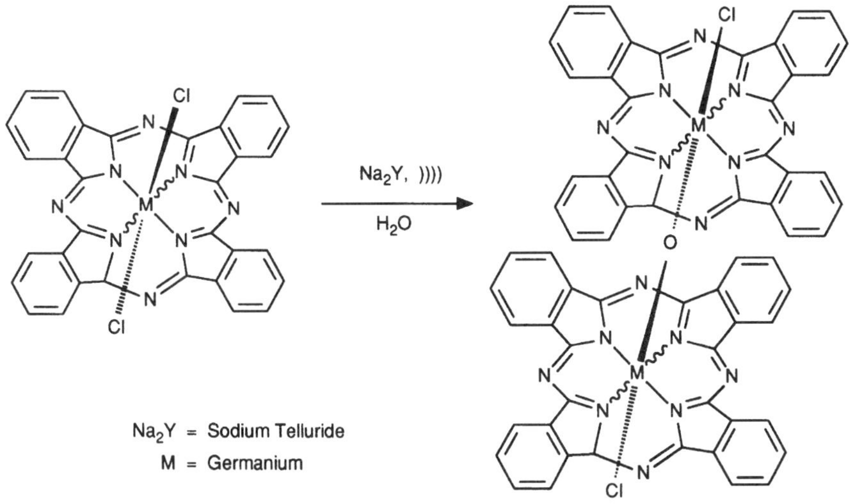

9.5. ULTRASONIC DEGRADATION OF POLYMERS IN SOLUTION

Na₂Y = Sodium Telluride
M = Germanium

Scheme 9.10

9.5 ULTRASONIC DEGRADATION OF POLYMERS IN SOLUTION

While the use of ultrasound in polymer synthesis is relatively recent, the effect on dissolved polymers in solution has been known for much longer. It should be stressed that in this regard, "degradation" is taken simply to mean an irreversible lowering of the molecular weight due to chain breakage. It is important to understand this effect as it will occur on polymer chains during synthetic procedures. Indeed, it has already been demonstrated that sonochemical polymerizations often lead to very different molecular weights than conventional procedures, and this can usually be associated with the ultrasonic degradation.

The first effects of ultrasound on polymer solutions were noted in the 1920's. After considerable discussion as to whether changes in solution viscosity were permanent or due to thixotropic effects, it was realised that the process was one of breakage of the polymer chains leading to lower molecular weights. This led to a large amount of work over the succeeding two decades to characterize the process in terms of the rate of bond cleavage for a wide range of polymers and the effect of the solution and ultrasound parameters. Further work has been carried out more recently with the advent of better apparatus for sonicating chemical reactions and, in particular, better methods of polymer analysis. The process is now understood with sufficient detail to make it commercially applicable in a number of areas [131, 132].

The basic effects [133] of irradiating a polymer solution with power ultrasound are shown in Figure 9.14 using polystyrene in toluene as an example. The degradation proceeds more rapidly at higher molecular weights and approaches

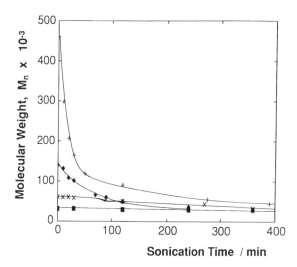

Figure 9.14 Ultrasonic degradation of 1% solutions of narrow polydispersity polystyrenes with different starting molecular weights in toluene.

a limiting value, M_{lim}, below which no further degradation takes place, in this case ca. 30,000. Polymers with this, or lower values, are unaffected by ultrasound under these conditions. Since the rate of degradation is molecular weight dependent, longer chains are removed from the sample and the polydispersity of the polymer is changed. These effects appear to be universal in that they have been seen for a wide range of organic polymers in organic solvents [131, 132], for inorganic polymers such as poly (organosiloxanes) [99, 134] and poly-organosilanes [135] and for aqueous systems such as polyethylene oxide [136, 137], cellulose [138], polypeptides, proteins [139] and DNA [140]. This common behavior suggests that a physical process is taking place that is more dependent on the size of the macromolecule in solution than on its chemical nature.

Most studies have shown that the degradation is relatively insensitive to the nature of the polymer. Schoon and Rieber [141, 142] found little difference in M_{lim} for polystyrene, polyisobutylene, polychloroprene, polybutadiene, poly(dimethyl siloxane), poly(vinyl acetate) or natural rubber sonicated under the same conditions. Melville and Murray [143] degraded benzene solutions of PMMA and two poly(methyl methacrylate-acrylonitrile) copolymers with MMA:AN ratios of 40:1 and 411:1. Within experimental error, all three polymers gave the same degradation rate and limiting molecular weight, contrasting with other work [144] which showed that under purely thermal conditions, the chains broke at the somewhat weaker acrylonitrile linkages. Evidence that sonochemical degradation can occur at "weak spots" in the chain was provided by Encina et al. [145] who found that the degradation of poly(vinyl pyrrolidone) was speeded up by tenfold when the polymer was prepared with

~0.3% peroxide linkages in the backbone. They suggested that chain cleavage occurred 5000 times faster at O–O bonds than at C–C. However, it is clear that for this effect to be noticed, there must be a substantial difference in the relative bond energies.

9.5.1 Kinetics and Mechanism of Degradation

A large number of studies have been performed to characterize the rate of degradation in order to develop quantitative models of the process. These have led to a number of equations being proposed to describe the degradation. One of the earliest equations suggested, which is still in widespread use, is that of Ovenall [146] who, following earlier work by Schmid, proposed that the rate of bond cleavage was first order, but depended on the chain length in excess of M_{lim}. Hence, if x is the number of chain breaks,

$$\frac{dx}{dt} = 0 \text{ for } M \sim M_{lim} \text{ and } \frac{dx}{dt} = k(M - M_{lim}) \text{ for } M > M_{lim}.$$

This leads to the following expression for the molecular weight, M_t, of the polymer at time, t, during the degradation

$$\ln\left(\frac{1}{M_{lim}} - \frac{1}{M_t}\right) = \ln\left(\frac{1}{M_{lim}} - \frac{1}{M_i}\right) - k\left(\frac{M_{lim}}{c\,m_o}\right)t$$

so that if the equation describes the degradiation a plot of $\ln(1/M_{lim} - 1/M_t)$ versus t produces a linear relationship, the gradient of which, knowing the solution concentration, c, and the monomer molecular weight, m_o, yields the rate constant k. This model has given a good fit to the majority of results obtained, particularly in the early stages of the degradation. Other workers have suggested a number of alternative derivations and equations, most of which can only be applied to specific systems. These have been reviewed by Basedow and Ebert [131] and by Price [132].

The degradation can therefore be characterized in two ways; the limiting value of the molecular weight reached after long sonication times, M_{lim}, and also in terms of the rate constant for the degradation, k. To be able to predict behavior under a wide range of conditions, the effect of varying all possible experimental conditions on the degradation must be understood. This has been documented in detail, but for present purposes, it is sufficient to summarize the main trends of the results.

In summary, the degradation proceeds faster and to lower molecular weights at lower temperatures, in more dilute solutions and in solvents with low volatility as would be expected from the discussion of the effect of these parameters on cavitation in Section 9.1.3. Other factors which have been quantified are the ultrasound intensity and the nature of dissolved gases. Hence, by suitable manipulation of the experimental conditions, we can exert a great deal of control over the process, exploitation of which allows the modification of existing polymers into new materials.

While there is still some debate about the precise origins of the degradation, it has been shown to be a direct consequence of cavitation. Under conditions which suppress cavitation, no degradation has been found. The mechanism can briefly be best described as the polymer chain being caught in the rapid flow of solvent molecules caused by the collapse of cavitation bubbles. A second cause of solvent movement is the shock waves generated after the implosion of the bubbles. The chains are thus subjected to extremely large shear forces resulting in the stretching of the chain and, if the force is sufficiently large breakage of a bond in the chain as schematically shown in Figure 9.15 (note that this is not drawn to scale; a polymer chain will be <1 μm in length compared with the 50–200 μm of the bubble). Some workers have also interpreted the effect in terms of frictional forces between the solvent molecules and polymer chains.

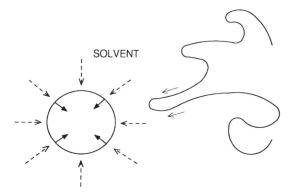

Figure 9.15 Schematic representation of polymer degradation due to cavitation.

There is no evidence that the extreme conditions of temperature found in cavitation bubbles contribute to the degradation. Even after very long sonication times for polystyrene dissolved in a number of solvents, none of the major product of thermal degradation, styrene monomer, was detected [147]. The degradation kinetics were also different from those of the thermal process. There are some results which indicate that other processes can take place as a result of radicals generated by sonolysis of the solvent, but these are generally at a much lower level than the chain breakage. Additionally, thermal degradation produces cleavage at random points along the chain in many polymers while ultrasonic degradation is much more specific. A number of workers have shown that cleavage occurs preferentially near the middle of the chain [148, 149]. The most persuasive evidence for this comes from the work of van der Hoff *et al.* [150–152] who investigated the degradation of polystyrene in THF. They found that the degradation could be best expressed as the product of two probabilities, one accounting for the chain length dependence and the other for the position

9.5. ULTRASONIC DEGRADATION OF POLYMERS IN SOLUTION

along the chain where breakage occurred. The best fit to the data was given when the breakage was distributed in a Gaussian manner within $\pm 15\%$ of the center of the chain. Their work clearly showed that neither a random model nor one where breakage occurred exclusively at the chain center fitted the experimental results as well as one derived from the gaussian distribution. The same model was recently applied by Koda et al. [137] to the degradation of several polymers in aqueous solutions and a similar gaussian probability function was found to best describe the results.

This "center cleavage" model is also consistent with the stretching and breakage mechanism outlined above. Degradation of polymer solutions in shear fields formed by extensional flow in narrow capillaries [153–156], in other flow systems [157, 158] as well as in high shear stirrers [159] also results in preferential breakage at the chain centres.

Since we have realistic rate models for the degradation, as well as being able to model the changes in molecular weight distribution with time, and the variation with changes in the experimental conditions, there is a good deal of control over the process.

9.5.2 Applications of Ultrasonic Degradation

At its most straightforward level, the degradation can be used as an additional processing parameter to control the molecular weight distribution. For example, Figure 9.16 shows GPC chromatograms of a polymer solution undergoing sonication [147]. The degradation of the higher molecular weight species narrows the distribution markedly with consequent modification of the physical

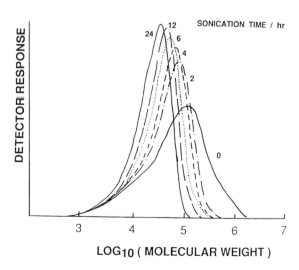

Figure 9.16 GPC chromatograms of a polyalkane sonicated in toluene (the values indicate the sonication time in hours).

properties of the polymer. A potential application would be to have a sonochemical step as the final processing stage during manufacture so as to "fine-tune" the polydispersity and molecular weight distribution to give the desired material properties of the polymer. There are several examples where this has been applied to commercial processes on at least pilot-plant scale.

In all the carbon backbone polymers studied to date, the primary product of the degradation is a radical species arising from homolytic bond breakage along the chain. The first evidence for the macromolecular radicals was first obtained by Henglein [52, 160] who trapped the radicals with the stable free radical DPPH, and monitored the process by the color change. More recently, Tabata and co-workers [161–163] have used electron spin resonance spectroscopy to investigate the radicals produced on sonication. For example, the radicals produced on sonicating benzene solutions of polystyrene, PMMA or poly(vinylacetate) were trapped with pentamethyl nitrosobenzene and the ESR spectra showed that they arose from chain cleavage rather than from secondary reactions with the solvent. Similar results in aqueous solutions were found by Taranukha et al. [164]. The only major exception to the production of radicals has been poly(dimethyl siloxane), PDMS. Thomas and de Vries [134] sonicated PDMS in various solvents and observed no discoloration of DPPH even though the usual lowering of molecular weight took place. More recent work [147] has shown no ESR signal with this polymer.

A second application of the degradation utilizes these macromolecular radicals as initiating species in the preparation of copolymers and end-capped materials. A large number of workers have sonicated mixtures of two polymers dissolved in a common solvent, resulting in a cross reaction between the two types of radicals to form a block copolymer. This approach was used by Melville and Murray [58] to show the presence of radical intermediates in the degradation. Henglein, as long ago as 1954 [52, 159], prepared copolymers consisting of blocks of acrylonitrile and acrylamide in aqueous solution, and a large number of similar reactions have since been carried out [132] on a wide range of materials including styrene with alkyl methacrylates, substituted styrenes with poly(methyl vinyl ketone) or poly(methyl vinyl ether), poly(ethylene oxide) with hydroxyethyl cellulose, as well as a range of vinyl acetate copolymers. However, the drawback here arises from the difficulty in recovering the products by selective precipitation and also in controlling their structure [165]. The approach which we have adopted in Bath is to sonicate a polymer dissolved in a solution containing the second monomer such as polystyrene and methyl methacrylate as shown in Figure 9.17.

Using the results of degradation studies, the structure and block length of the first polymer can be controlled quite precisely. By changing the concentration of monomer in solution the block size of the second polymer can also be varied, allowing a large degree of control over the resulting material structure. A related approach is to sonicate the polymer in the presence of a species labile to radical attack where "end capped" polymers are formed. We have used this to prepare, for example, polystyrenes and poly(alkanes) bearing fluorescent groups.

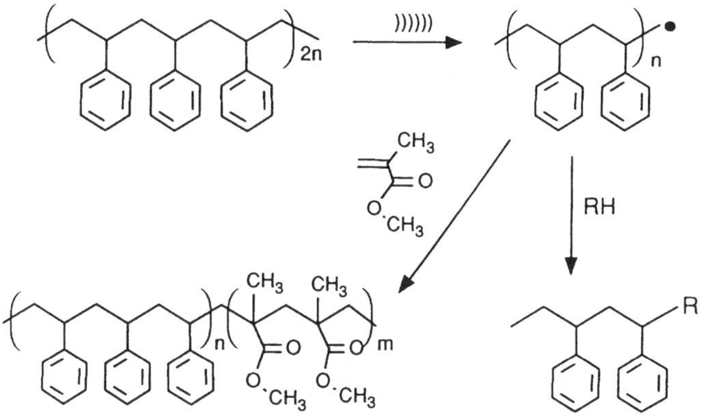

9.6. CONCLUSIONS AND FUTURE PROSPECTS

Figure 9.17 Sonochemical production of end-capped polymers and block copolymers.

9.5.3 Synthesis versus Degradation

The degradation process will operate whenever dissolved polymers are subjected to sonication, including during their synthesis. Indeed, often the degradation is significant at sound intensities below those needed to form significant amounts of radicals. This factor can be used to explain, at least in part, most of the molecular weight effects described in earlier sections of this chapter.

The rapid growth of molecular weight followed by a gradual reduction during radical polymerization is caused by this process. Long chains formed in the initial stages of the reaction are then subjected to shear and cleaved to give the rather unusual plots seen in Figure 9.5. The situation in the other polymerizations is not so clear since other factors are also important. For example, in the ring opening synthesis of PDMS, higher molecular weight polymer is produced under sonication. This is primarily due to ultrasound giving a much better dispersion of the catalyst through the system. The degradation does play a part though in cleaving the very long chains so that the polydispersity is lower than in the conventional reactions. Similar explanations in terms of better mass transfer but fractionation of very high molecular weight material apply in the heterogeneous Ziegler and Wurtz polymerizations described in Section 9.4. Thus, ultrasound has its part to play in accelerating and improving the yield of polymerization reactions and mirrors its effects in the synthesis of low molar mass compounds. However, in polymers it has the extra advantage of giving a convenient method to control the structure, and hence properties, of the resulting materials.

9.6 CONCLUSIONS AND FUTURE PROSPECTS

To date, there has been a relatively small amount of work related to sonochemical polymerization. However, the benefits offered in a range of reaction types warrant further studies in the area, in terms of acceleration of rate and increased yields but particularly in the control of molecular weight during polymerization.

Perhaps the major factor which has made industrial acceptance of sonochemistry slow is that it is regarded by many as a small-scale, laboratory technique. However, in addition to laboratory apparatus, equipment for operating at much larger, commercially realistic scales is being developed [166, 167]. Extremely large cleaning baths are available [168] (with capacities of ~ 8000 L, designed for cleaning aircraft engines) but it is unrealistic to consider these for chemical applications. One or more large horns have been incorporated into flow systems and variations on this, where the reactants flow through or between arrays of transducers in a variety of arrangements, are commercially available to process volumes up to 150 L min^{-1} depending on the required intensity. This type of arrangement seems to be favored for the relatively few scale-up studies which have been performed to date. However, should there be sufficient value in sonochemical polymerizations, the equipment exists to exploit the effects on at least pilot-plant and possibly low-volume production scale. Perhaps the most obvious application would be in an "add-on" to existing processes where the ultrasonic degradation can be utilized to control the final molecular weight of polymers. When any polymerization process in solution is subject to sonication, the degradation will always occur concurrently with chain growth processes. However, it seems unlikely that a sonochemical process will replace current methodology unless it allows *significant* improvement of material properties or saving in production or energy costs. Thus, sonochemical manufacture of polystyrene, polyethylene and similar materials on a large industrial scale seems improbable. However, ultrasound has the possibility in a number of areas of producing polymers with no initiators or emulsifiers being needed. The presence of additives and impurities in biomedical materials and food products is undesirable, but the added value of the products in their absence would support the (at present) extra costs of introducing sonochemical technology.

As far as the author is aware, there have been no examples to date of commercial processes for ultrasonically producing vinyl polymers. Indeed, the possibility of using the methodology may be very limited. While there are situations where low temperature, site-specific initiation can be useful, this can often be achieved photochemically, particularly in thin films. It seems unlikely that sonochemically initiated polymerization will be used on a large scale, but there may be small-scale, specialized cases where the absence of initiator fragments may be useful. One example is the initiation in PMMA pastes containing inorganic fillers used as dental fillings [169]. The ability to work at low temperatures may find use with monomers such as α-methyl styrene which have ceiling temperatures below those which allow the convenient use of thermal initiators. Finally, the ability to control the molecular weight without adding chain transfer agents, as well as switching off the initiation during the polymerization to prevent heat build-up in highly exothermic reactions, may be advantages that can be exploited in particular cases.

In similar fashion, it is difficult to envisage high tonnage production of polyolefins by Ziegler polymerization under ultrasound, but there may well be some cases where relatively small amounts of high value materials needing specific molecular weight distributions are required in which sonochemistry can play a part. Electrochemically initiated polymerization is, by its nature, a relatively small scale and expensive process so that incorporation of sonochemical technology may be more straightforward in this case. Perhaps the most promising (and so-far least studied) examples are those which can be classified as "organometallic" polymerizations. These, and other heterogeneous reactions can be used in the production of a range of highly functionalized materials such as conducting and/or NLO polymers, side-chain liquid crystals or dendrimers. These are low-volume, high-value products and it is in this general area where ultrasound may have a major role in the near future.

The purpose of this chapter was to show that there are a number of areas where ultrasound may be of great benefit in the preparation of polymeric materials. It remains to completely characterize and develop models for each of the reactions, particularly for heterogeneous processes. However, the advantages offered may well outweigh the problems associated with larger scale sonochemistry in a number of cases. The huge amount of polymer production around the world provides a spur for interest in sonochemistry but also provides a problem in its acceptance. It is still regarded as a small-scale process so that its place in the production of polymers may be limited, albeit that larger scale operation is now possible, as described above. Commodity plastics are relatively low-cost materials so that the extra cost associated with the introduction of new processes cannot be borne unless sonochemical methods offer a considerable advantage over existing technology. Thus, there is more chance in the near future of a breakthrough for sonochemistry in the industrial synthesis of low-volume, high-cost materials such as pharmaceuticals or complex intermediates. However, there are now some materials, such as photo- and electro-active polymers which fall into this category and it is to be hoped that ultrasound will find its role in their production. The outlook is encouraging and it appears that sonochemical polymerization reactions are well placed to become commercially viable in the near future.

9.7 REFERENCES

[1] Mason, T. J., Luche, J. L., Previous chapter of this book.
[2] Flosdorf, E. W., Chambers, L. A., J. Am. Chem. Soc., 55 (1933) 3051.
[3] Szalay, A. S., Phys. Chem. A., 164 (1933) 234.
[4] Gyorgi, A. S., Nature, 131 (1933) 278.
[5] Price, G. J., (Price, G. J. (ed.)), Current Trends in Sonochemistry, R.S.C. Special Publication 116, R.S.C, Cambridge, 1992, p 87.
[6] Price, G. J., (Ebdon, J. R., Eastmond, G. C. (eds.)), New methods of polymer synthesis, Vol. II, Blackie, Glasgow, in press.

[7] Reisch, M. S., Chem. Eng. News, 22 (1995) 30.
[8] Billmeyer, F. W., Textbook of Polymer Science, John Wiley & Sons, 3rd ed., New York, 1984.
[9] Hunt, B. J., Holding, S. R., Size Exclusion Chromatography, Blackie, London, 1989.
[10] Boudjouk, P., (Suslick, K. S. (ed.)), Ultrasound: Its chemical, physical and biological effects, VCH Publishers, New York, 1990, p 165.
[11] Leighton, T. G., The Acoustic Bubble, Academic Press, London, 1994.
[12] Hengein, A., Herberger, D., Guiterrez, M., J. Phys. Chem., 96 (1992) 1126.
[13] Petrier, C., Jeunet, A., Luche, J. L., Reverdy, G., J. Am. Chem. Soc., 114 (1992) 3148.
[14] Cum, G., Galli, G., Gallo, R., Spadaro, A., Ultrasonics, 30 (1992) 267.
[15] Sata, N., Nakashima, K., Bull. Chem. Soc. Japan, 18 (1943) 220.
[16] Price, G. J., Smith, P. F., Polymer, 34 (1993) 4111.
[17] Young, F. R., Cavitation, McGraw-Hill, London, 1990.
[18] Mason, T. J., Practical Sonochemistry, Ellis Horwood, Chichester, 1991.
[19] Suslick, K. S., Flint, E. B., (Wayda, A., Darensbourg, M. B. (eds.)), Experimental Organometallic Chemistry, A.C.S., Washington, 1987, p 195.
[20] Ley, S. V., Low, C. R., Ultrasound in Chemistry, Springer Verlag, London, 1989.
[21] Luche, J. L., (Price, G. J. (ed.)), Current Trends in Sonochemistry, R.S.C. Special Publication 116, R.S.C, Cambridge, 1992, p 34.
[22] Boudjouk, P., (Price, G. J. (ed.)), Current Trends in Sonochemistry, R.S.C. Special Publication 116, R.S.C, Cambridge, 1992, p 110.
[23] Pethrick, R. A., Advances in Sonochemistry, 2 (1991) 65.
[24] Matheson, A. J., Molecular Acoustics, Wiley Interscience, London, 1971.
[25] North, A. M., Pethrick, R. A., (Allen, G., Buckingham, A. D. (eds.)), International Review of Science, Physical Chemistry, Series 1, Vol 2, MTP Press, Dordecht, 1972.
[26] Sladky, P., Pelant, L., Parman, L., Ultrasonics, 16 (1979) 32.
[27] Hauptman, P., Dinger, F., Sauberlich, R., Polymer, 26 (1985) 1741.
[28] Emery, J. R., Durand, D., Tabbelout, M., Pethrick, R. A., Polymer, 28 (1987) 1435.
[29] Sofer, G. A., Diesz, G. H., Hauser, E. A., Ind. Eng. Chem., 45 (1975) 2743.
[30] Reynolds, W. N., Scudder, L. P., Pressman, H., Polymer Testing, (1986) 325.
[31] Frolich, B., Jasse, B., Noel, C., Monnerie, L., J. Chem. Soc., Faraday Trans. II, 74 (1978) 445.
[32] Evman, A., North, A. M., Pethrick, R. A., Wandelt, B., J. Chem. Soc., Faraday Trans. II, 72 (1976) 1957.
[33] Pethrick, R.A., (Bevington, J.C., Allen, G. (eds.)), Comprehensive Polymer Science Vol. 3, Ch. 17, Pergamon Press, Oxford, 1989.
[34] Kato, S., Yamauchi, H., Nomura, H., Mirahara, Y., Macromolecules, 18 (1985) 1496.
[35] Shoh, A., Ultrasonics, 14 (1976) 209.
[36] Rawson, F., Brit. Plastics and Rubber, May 1994, p 22.

9.7. REFERENCES

[37] Suslick, K. S., Casadonte, D. J., Green, M. L. H., Thompson, M. E., Ultrasonics, 25 (1987) 56.

[38] Doktycz, S., Suslick, K. S., Science, 247 (1990) 1067.

[39] Stoffer, J. O, Fahim, M., J. Coating Technol., 63 (1991) 61.

[40] Utsimi, H., Shinzo, K., Kuriyama, K., Suguwara, R., Fukida, M., Jiraishi, S., Eur. Patent Appl., 87111252.0, 1987.

[41] Mason, T. J., Lorimer, J. P., Sonochemistry: theory, applications and uses of ultrasound in chemistry, Ellis Horwood, Chichester, 1988.

[42] Templeton-Knight, R., Chemistry and Industry, (1990) 513.

[43] Lorimer, J. P., Mason, T. J., Kershaw, D., Livsey, I., Templeton-Knight, R., Coll. Polym. Sci., 29 (1991) 392.

[44] Price, G. J., Clifton, A. A., Polymer, 36 (1995) 4919.

[45] Urban, M. W., Salazar-Rojas, E. M., Macromolecules, 21 (1988) 372.

[46] Exsted, B. J., Urban, M. W., J. Inorg. Organomet. Polym., 3 (1993) 105.

[47] Exsted, B. J., Urban, M. W., Polymer, 35 (1994) 5560.

[48] Price, G. J., Clifton, A. A., manuscripts in preparation.

[49] Bevington, J. C., (Bevington, J. C., Allen, G. (eds.)), Comprehensive Polymer Science, Vol. 3, Ch. 6, Pergamon Press, Oxford, 1989.

[50] Hart, E., Henglein, A., J. Phys. Chem., 90 (1986) 5992.

[51] Riesz, P., Berdahland, D., Christmoer, C., Environ. Health. Perspect., 64 (1985) 233.

[52] Riesz, P., Advances in Sonochemistry, 2 (1992) 23.

[53] Lindstrom, O., Lamm, O., J. Phys. Colloid Chem., 55 (1951) 1139.

[54] Henglein, A., Makromol. Chem., 14 (1954) 15.

[55] Hart, E., Henglein, A., J. Phys. Chem., 90 (1986) 5889.

[56] Suslick, K. S., Ultrasound: Its chemical, physical and biological effects, VCH Publishers, New York, 1990, Chapter 4.

[57] El'Piner, I. E., Ultrasound: Physical, chemical and biological effects, Consultants Bureau, New York, 1964.

[58] O'Driscoll, K. F., Shridhan, A. U., J. Polym. Sci. Polym. Chem., 11 (1973) 1111.

[59] Fujiwara, H., Kakiuchi, H., Kanmaki, K., Goto, K., Kobunshi Ronbunshu (Engl. Edn.), 5 (1976) 256.

[60] Melville, H. W., Murray, A., Trans. Faraday Soc., 46 (1950) 996.

[61] Miyata, T., Nakashio, F., J. Chem. Eng. Japan, 8 (1975) 463.

[62] Stoffer, J. O., Sitton, O. C., Kao, H. L., Polym. Mater. Sci. Eng. Prepr., 65 (1991) 42.

[63] Stoffer, J. O., Sitton, O. C., Kim, Y. H., Polym. Mater. Sci. Eng. Prepr., 67 (1992) 242.

[64] Lorimer, J. P., Mason, T. J., Kershaw, D., J. Chem. Soc., Chem. Commun., (1991) 1217.

[65] Lorimer, J. P., Mason, T. J., Ultrasonics International Conference Proceedings, 1987, 762.

[66] Fujiwara, H., Goto, K., Polymer Bulletin, 25 (1991) 571.

[67] Orszulik, S. T., Polymer, 34 (1993) 1320.
[68] Heusinger, H., Carbohydrate Research, 209 (1991) 109.
[69] Portenlanger, G., Heusinger, H., Carbohydrate Research, 232 (1992) 291.
[70] Kruus, P., Ultrasonics, 21 (1983) 193.
[71] Kruus, P., Patraboy, T. J., J. Phys. Chem., 89 (1985) 3379.
[72] Kruus, P., Lawrie, J., O'Neill, M. L., Ultrasonics, 26 (1988) 352.
[73] Kruus, P., O'Neill, M. L., Robertson, D., Ultrasonics, 28 (1990) 304.
[74] Kruus, P., Advances in Sonochemistry, 2 (1991) 1.
[75] Price, G. J., Smith, P. F., West, P. J., Ultrasonics, 29 (1991) 166.
[76] Price, G. J., Norris, D. J., West, P. J., Macromolecules, 25 (1992) 6447.
[77] Lorimer, J. P., Mason, T. J., Fiddy, K., Kershaw, D., Groves, T., Dodgson, D., Ultrasonics International Conference Proceedings, (1989) 1283.
[78] Brandrup, J., Immergut, E. H. (eds.), Polymer Handbook 3rd ed., Section II, p 2, Wiley, New York, 1990.
[79] Miyata, T., Nakashio, F., J. Chem. Eng. Japan, 8 (1975) 469.
[80] Price, G. J., Daw, M. R., Newcombe, N. J., Smith, P. F., Br. Polym. J., 23 (1990) 63.
[81] Bovey, F. A., (Bevington, J. C., Allen, G. (eds.)), Comprehensive Polymer Science, Vol. 1, Ch. 17, Pergamon Press, Oxford, 1989.
[82] Schultz, D. N., Sissano, J. A., Costello, C. A., Polym. Prepr., 35 (1994) 514.
[83] Bassett, D. R., Hamielec, A. E., Emulsion polymers and emulsion polymerization, A.C.S. Symposium Series, 165 (1981).
[84] Ostroski, A. S., Stanbaugh, R. B., J. Appl. Phys., 21 (1950) 478.
[85] Hatate, Y., Ikeura, T., Shinonome, M., Kondo, K., Nakashio, F., J. Chem. Eng. Japan, 14 (1981) 38.
[86] Hatate, Y., Ikari, A., Kondo, K., Nakashio, F., Chem. Eng. Commun., 34 (1985) 325.
[87] Chou, H. C., Lin, W. Y., Stoffer, J. O., Polym. Mater. Sci. Eng., 72 (1995) 363.
[88] Allen, K. W., Davidson, R. S., Zhang, H. S., British Patent Appl., 90177544 (1990).
[89] Allen, K. W., Davidson, R. S., Zhang, H. S., Proceedings of "Radtech Europe" Conference, Edinburgh, 1991.
[90] Davidson, R. S., Private communication.
[91] Cooper, G., Grieser, F., Biggs, S., Chem. Austral., Feb. (1995) 22.
[92] Senapati, N., Mangaraj, D., U.S. Patent, 45489771 (1985).
[93] Salensa, T. K., Babbar, N. K., Ultrasonics, 13 (1971) 21.
[94] Bailey, W. J., (Bevington, J. C., Allen, G. (eds.)), Comprehensive Polymer Science Vol. 3, Ch. 22, p 283, Pergamon Press, Oxford, 1989.
[95] Ragaini, V., Italian Patent Appl., 20478-A/90.
[96] Carli, R., Bianchi, C. L., Gariboldi, P., Ragaini, V., Proc. 3rd Europ. Sonochem. Soc., Coimbra, Portugal, 1993.
[97] Kendrick, T. C., Parbhoo, B. M., White, J. W., (Bevington, J. C., Allen, G. (eds.)), Comprehensive Polymer Science, Vol. 4, Ch. 25, Pergamon Press, Oxford, 1989.

9.7. REFERENCES

[98] Kogan, E. V., Smirnov, N. I., Zh. Priklad Khim., 35 (1962) 1382.
[99] Price, G. J., Wallace, E. N., Patel, A. M., (Jones, R. G. (ed.)), Silicon Containing Polymers, R.S.C., Cambridge, in press.
[100] Stoessel, S. J., J. Appl. Poly. Sci., 48 (1993) 505.
[101] Watanabe, S., Matsubara, I., Kakimoto, M., Imai, Y., Polym. J., 25 (1993) 989.
[102] Long, G. B., U.S. Patent, 3346472 (1967).
[103] Mason, T. J., Lorimer, J. P., Walton, D. J., Ultrasonics, 28 (1990) 251.
[104] Walton, D. J., Chyla, A., Lorimer, J. P., Mason, T. J., Synthetic Commun., 20 (1990) 1843.
[105] Compton, R. G., Eklund, J. C., Page, S. D., Sanders, G. H. W., Booth, J., J. Phys. Chem., 98 (1994) 12410.
[106] Walker, M. R., Chem. in Brit., 26 (1991) 251.
[107] Topare, L., Eren, S., Akbulut, U., Polymer Commun., 28 (1987) 36.
[108] Akbulut, U., Topare, L., Yurttas, B., Polymer, 27 (1986) 803.
[109] Aybar, S. P., Hacioglu, B., Akbulut, U., Toppare, L., J. Polym. Sci. Polym. Chem., 29 (1991) 1971.
[110] Osawa, S., Ito, M., Tanake, K., Kuwano, J., J. Polym. Sci. Polym. Phys., 30 (1992) 19.
[111] Kathirgamanathan, P., Souter, A. M., Baulch, D., J. Appl. Electrochem., 24 (1994) 283.
[112] Tait, P. J. T., (Bevington, J. C., Allen, G. (eds.)), Comprehensive Polymer Science Vol. 4, Ch. 1, p 1, Pergamon Press, Oxford, 1989.
[113] Mertes, T. S., U.S. Patent, 2899414 (1960).
[114] Price, G. J., Patel, A. M., Polymer Commun., 33 (1992) 4435.
[115] Jones, M. B., Kovacic, P., (Bevington, J. C., Allen, G. (eds.)), Comprehensive Polymer Science, Vol. 5, p 465, Pergamon Press, Oxford, 1989.
[116] Kovacic, P., Kyriakis, A., J. Am. Chem. Soc., 85 (1963) 454.
[117] Rehann, M., Schluter, A., Feast, W. J., Synthesis, (1988) 386.
[118] Miller, R. D., Michl, J., Chem. Rev., 89 (1989) 1359.
[119] West, R., J. Organomet. Chem., 300 (1986) 327.
[120] Devaux, J., Sledz, J., Schue, F., Giral, L., Naarmann, H., Eur. Polym. J., 25 (1989) 263.
[121] Gauthier, S., Worsfold, D. J., Macromolecules, 22 (1989) 2213.
[122] Cragg, R. H., Jones, R. G., Swain, A. C., Webb, S. J., J. Chem. Soc., Chem. Commun., (1990) 1143.
[123] Fujino, M., Isaka, H., J. Chem. Soc., Chem. Commun., (1989) 466.
[124] Han, B. H., Boudjouk, P., Tetrahedron Lett., 22 (1981) 3813.
[125] Kim, H. K., Matyjaszewski, K., J. Am. Chem. Soc., 110 (1989) 3321.
[126] Kim, H. K., Uchida, H., Matyjaszewski, K., Macromolecules, 28 (1995) 59.
[127] Miller, R. D., Thompson, D., Sooriyakumaran, R., Fickes, G. N., J. Polym. Sci. Polym. Chem., 29 (1991) 813.
[128] Price, G. J., J. Chem. Soc., Chem. Commun., (1992) 1209.

[129] Weidman, T. W., Bianconi, P. A., Kwock, E. W., Ultrasonics, 28 (1990) 310.
[130] Hohol, M. D., Urban, M. W., Polymer, 34 (1993) 1995.
[131] Price, G. J., Advances in Sonochemistry, 1 (1990) 231.
[132] Basedow, A. M., Ebert, K., Adv. Polym. Sci., 22 (1977) 83.
[133] Price, G. J., Smith, P. F., Eur. Polym. J., 29 (1993) 419.
[134] Thomas, J. R., de Vries, D. L., J. Phys. Chem., 63 (1959) 254.
[135] Irie, S., Irie, M., Radiat. Phys. Chem., 40 (1992) 107.
[136] Chen, K., Shen, Y., Li, H., Xu, X., Gaofenzi Tongxun, 6 (1985) 401.
[137] Koda, S., Mori, H., Matsumoto, K., Nomura, H., Polymer, 35 (1994) 30.
[138] Sato, T., Nalepa, D. E., J. Coating Technol., 49 (1977) 45.
[139] Bradbury, J. H., O'Shea, J., Aust. J. Biol. Sci., 26 (1973) 583.
[140] Davison, P. F., Freifelder, D., Biophys. J., 2 (1962) 235.
[141] Schoon, T. G., Rieber, T., Angew. Makromol. Chem., 23 (1972) 43.
[142] Schoon, T. G., Rieber, T., Angew. Makromol. Chem., 15 (1971) 263.
[143] Melville, H. W., Murray, A., Trans. Faraday Soc., 46 (1950) 996.
[144] Grassie, N., Melville, H. W., Proc. Roy. Soc. A., 199 (1949) 39.
[145] Encina, M. V., Lissi, E., Sarasusa, M., Gargallo, L., Radic, D., J. Polym. Sci. Polym. Lett., 18 (1980) 757.
[146] Ovenall, D. W., Hastings, G. W., Allen, P. E. M., J. Polym. Sci., 33 (1958) 207.
[147] Price, G. J., *et al.*, Unpublished results – manuscripts in preparation.
[148] van der Hoff, B. M. E., Glynn, P. A. R., J. Macromol. Sci. Macromol. Chem., A8 (1974) 429.
[149] Smith, W. B., Temple, H. W., J. Phys. Chem., 72 (1968) 4613.
[150] Glynn, P. A. R., van der Hoff, B. M. E., Reilly, P. M., J. Macromol. Sci., A6 (1972) 1653.
[151] Glynn, P. A. R., van der Hoff, B. M. E., Reilly, P. M., J. Macromol. Sci. A7 (1973) 1695.
[152] van der Hoff, B. M. E., Gall, C. E., J. Macromol. Sci., A11 (1977) 1739.
[153] Odell, J. A., Keller, A., J. Polym. Sci. Polym. Phys., 24 (1986) 1889.
[154] Muller, A. J., Odell, J. A., Keller, A., Polymer Commun., 30 (1989) 298.
[155] Muller, A. J., Odell, J. A., Keller, A., Macromolecules, 23 (1990) 3090.
[156] Moan, M., Omari, A., Polymer Degrad. and Stability, 35 (1991) 277.
[157] Nguyen, T. Q., Kausch, H. H., J. Non-Newt. Fluid Mechan., 30 (1988) 125.
[158] Nguyen, T. Q., Kausch, H. H., Adv. Polym. Sci., 100 (1992) 73.
[159] Watanabe, O., Tabata, M., Kuedo, T., Sohma, J., Ogiwara, T., Prog. Polym. Phys. Japan, 28 (1985) 285.
[160] Henglein, A., Makromol. Chem., 15 (1955) 188.
[161] Tabata, M., Miyawaza, T., Sohma, J., Proc. 3rd Yamada Conf. on Free Radicals Osaka, Japan, 1979, p 243.
[162] Tabata, M., Miyawaza, T., Sohma, J., Kobayashi, O., Chem. Phys. Lett., 73 (1980) 178.
[163] Tabata, M., Sohma, J., Eur. Polym. J., 16 (1980) 589.

9.7. REFERENCES

[164] Taranukah, O., Logvinenko, P. N., Dmitrieva, T. V., Dopov. Akad. Nauk. Ukr. R.S.R., 7 (1985) 47.
[165] Shen, Y., Chen, K., Wang, Q., Xu, H., Xu, X., J. Macromol. Sci., A23 (1986) 141.
[166] Martin, P., (Price, G. J. (ed.)), Current Trends in Sonochemistry, R.S.C. Special Publication 116, R.S.C, Cambridge, 1992, p 158.
[167] Berlan, J., Mason, T. J., Ultrasonics, 30 (1992) 203.
[168] Hunicke, R. L., Ultrasonics, 28 (1990) 291.
[169] Limin, G., Jian, L., Gang, W., Zhen, H., Biomat. Art. Cells Immobili. Biotechnol., 20 (1992) 125.

10

CHEMISTRY UNDER EXTREME CONDITIONS IN WATER INDUCED ELECTROHYDRAULIC CAVITATION AND PULSED-PLASMA DISCHARGES

Michael R. Hoffmann*, Inez Hua, Ralf Höchemer,
Dean Willberg, Patrick Lang and Axel Kratel

W.M. Keck Laboratories, California Institute of Technology, Pasadena, California 91125

10.1 ULTRASONIC IRRADIATION OF WATER AND THE CHEMICAL EFFECTS OF ELECTROHYDRAULIC CAVITATION
    10.1.1 Background
    10.1.2 Physical Principles
    10.1.3 Chemical Consequences of Extreme Transient Conditions
10.2 PULSED-PLASMA DISCHARGES IN WATER
    10.2.1 Background
    10.2.2 Physical Principles
    10.2.3 Chemical Consequences of Transient Plasma Formation
10.3 SONOLYTIC REACTIONS IN WATER: KINETICS AND MECHANISMS
    10.3.1 Degradation of *p*-Nitrophenol
    10.3.2 Hydrolysis and Degradation of *p*-Nitrophenyl Acetate
    10.3.3 Degradation of Carbon Tetrachloride
    10.3.4 Degradation of TNT
    10.3.5 Sonochemical Autoxidation of Iodide
    10.3.6 Other Sonochemical Reactions of Interest in Water
10.4 CHEMICAL REACTIONS INDUCED BY EHD PLASMAS
10.5 REACTOR DESIGN CONSIDERATIONS
10.6 CONCLUSIONS
10.7 REFERENCES

10.1 ULTRASONIC IRRADIATION OF WATER AND THE CHEMICAL EFFECTS OF ELECTROHYDRAULIC CAVITATION

10.1.1 Background

The introduction of high power ultrasound (*i.e.*, sound energy with frequencies in the range 15 kHz to 1 MHz) into liquid reaction mixtures is known to cause a variety of chemical transformations [1–47]. The chemical effects of ultrasound on chemical reactions were first reported by Richards and Loomis [48] in 1927. This early report was followed by a detailed investigation of the catalytic effect of ultrasonic irradiation on the autoxidation of the iodide ion [49]. Since then, the application of ultrasound as a catalyst in chemical synthesis has become an important field of research as described by Mason in Chapter 8.

In recent years, due to the growing need to eliminate undesirable chemical compounds, the utilization of high energy ultrasound for hazardous waste treatment has been explored with great interest [31, 50–63].

10.1.2 Physical Principles

Ultrasonic irradiation of liquid reaction mixtures induces electrohydraulic cavitation, which is a process during which the radii of preexisting gas cavities in the liquid oscillate in a periodically changing pressure field created by the ultrasonic waves. These oscillations eventually become unstable, forcing the violent implosion of the gas bubbles. The rapid implosion of a gaseous cavity is accompanied by adiabatic heating of the vapor phase of the bubble, yielding localized and transient high temperatures and pressures. Temperatures on the order of 4200 K and pressures of 975 bar have been estimated [7]. Noltingk and Neppiras [64, 65] estimate even higher temperature and pressure values that range up to 10,000 K and 10,000 bar. Experimental values of $p = 313$ atm and $T = 3360$ K have been reported [66] for aqueous systems, while temperatures in excess of 5000 K have been reported [16, 67] for cavitation in organic and polymeric liquids. Recent experimental results on the phenomenon of sonoluminescence [67–84] suggest that even more extreme temperatures and pressures are obtained during cavitational bubble collapse [85]. Thus, the apparent chemical effects in liquid reaction media are either direct or indirect consequences of these extreme conditions.

In a recent study, Riesz and co-workers [86] used the semiclassical model of the temperature dependence of the kinetic isotope effect for H· and D· atom formation to estimate the effective temperature of the hot cavitation regions in which H· and D· atoms are formed by ultrasound-induced pyrolysis of water molecules. The H· and D· atoms were formed in argon-saturated H_2O and D_2O mixtures (1:1) exposed to 50 kHz ultrasound and were detected by spin trapping with nitrone spin traps detected by and quantified by ESR. From these results, they estimated average cavitational collapse temperatures in the range of 2000 to 4000 K.

10.1. ULTRASONIC IRRADIATION

Equations describing the inception and dynamics of a single cavitation bubble have been developed [64, 65, 87, 88]. Based on approximate solutions to the Rayleigh-Plesset equation, Noltingk and Nepprias [64, 65] gave the following equation to predict the temperature at the center of a collapsed cavitation bubble:

$$T_{max} = T_0 \left\{ \frac{p_m(K-1)}{p} \right\} = T_0 \left(\frac{R_0}{R_{min}} \right)^{3(K-1)} \quad (1)$$

where T_0 = temperature of the bulk solution, $K = C_p/C_v$ = polytropic index of the cavity medium, p = pressure in the bubble at its maximum size, p_m = pressure in the bubble at the moment of transient collapse. Thus, the relative temperature of bubble collapse can be adjusted by saturating the solution with gases characterized by substantially different specific heats, thermal conductivities, and solubilities. From eq. (1) we see that an important factor controlling the collapse temperature is the polytropic constant, K, of the saturating gas. The value of K is associated with the amount of heat released from the gas inside the bubble during compression. As K increases, the heat released upon bubble collapse also increases. Additional physicochemical properties that may influence the temperature attained during bubble collapse include thermal conductivity (of both the liquid and gas), λ, and gas solubility. A low thermal conductivity favors high collapse temperatures, because the heat of collapse will dissipate less quickly from the cavitation site. Highly soluble gases should result in the formation of a larger number of cavitation nuclei and more extensive bubble collapse, since gases with higher solubilities are more readily forced back into the aqueous phase. Thus, a gas with both a low thermal conductivity and high water solubility should yield the highest temperature upon cavitational bubble collapse. Based on known physical properties, krypton should reach the highest temperatures upon cavitational bubble collapse in aqueous solution. A list of background gases and their physicochemical properties is given in Table 10.1.

Even though the basic physical and chemical consequences of cavitation are fairly well understood, many fundamental questions about the cavitation site in aqueous solution remain unanswered. In particular, the dynamic temperature

Table 10.1 Physical properties of selected gases used during sonolysis[a]

Gas	Polytropic index, κ	Thermal conductivity, λ[b] (mW/m K)	Gas solubility in water (m_g^3/m_{aq}^2)
Kr	1.66	17.1	0.0594
Ar	1.66	30.6	0.032
He	1.63	252.4	0.0086

[a] Data at 298 K unless otherwise indicated.
[b] Data at 600 K.

and pressure changes at the bubble interface and their effects on chemical reactions need further exploration. Since this region is likely to have transient temperatures and pressures in excess of 647 K and 221 bar for periods of microseconds to milliseconds, we [89] have proposed that supercritical water (SCW) provides an additional phase for chemical reactions during ultrasonic irradiation in water. Supercritical water exists above the critical temperature, T_c, of 647 K and the critical pressure, p_c, of 221 bar and has physical characteristics intermediate between those of a gas and a liquid [89, 90]. The physicochemical properties of water such as viscosity, ion-activity product, density, and heat capacity change dramatically in the supercritical region. These changes favor substantial increases for rates of most chemical reactions. SCW has been used in industrial applications such as extraction [91], enhanced hydrolysis [92, 93] and for the oxidation of hydrocarbons [94] and phenols [95, 96] (see Chapter 7).

The transient temperatures and pressures within a collapsing cavitation bubble clearly exceed the critical point of water for some finite period of time. Based on previously estimated temperatures within a collapsed bubble and a smaller layer of surrounding liquid, one can describe the spatial and temporal temperature distribution around a bubble just after its collapse. Several severe and somewhat unrealistic assumptions are needed in order to obtain a simple first-order approximation of the heat transport from the interior of a hot bubble to the surrounding bulk liquid. First, one can assume that the collapsed bubble is an instantaneous point source of heat embedded in an infinite matrix at ambient temperatures, then one can assume that conduction provides the only means of heat transfer, and thus ignore heat transport by convection and radiation. Thirdly, one assumes that the bubble retains its spherical shape after collapse and that the heat capacity, thermal conductivity and density of the collapsed bubble are the same as of the surrounding water at room temperature. Finally, one assumes that a uniform temperature is attained within the bubble immediately after collapse.

The following values for the physical properties of liquid water at 300 K are used in the calculation (e.g., $C_p = 4178.4$ J/kg K, the thermal conductivity, $\lambda = 0.6154$ J/s m K, $\rho = 995.65$ kg/m^3). The initial radius, $a = 150$ μm and the initial temperature of the collapsed bubble, $T_0 = 5000$ K have previously been estimated [67, 97]. The temperature of the water surrounding the collapsed bubble is $T_{med} = 300$ K.

Based on these assumptions, the simple heat transport equation is written as eq. (2)

$$\frac{\partial T}{\partial t} = k \nabla^2 T \tag{2}$$

where the thermal diffusivity, k, in units of [m^2/s] is given as $\lambda/(\rho\, C_p)$. The Laplacian, written in spherical polar coordinates is as follows:

$$\nabla^2 = \frac{1}{r^2}\frac{\partial}{\partial r}\left[r^2\frac{\partial}{\partial r}\right] + \frac{1}{r^2 \sin\theta}\frac{\partial}{\partial \theta}\left[\sin\theta\frac{\partial}{\partial \theta}\right] + \frac{1}{r^2 \sin^2\theta}\frac{\partial^2}{\partial \phi^2} \tag{3}$$

10.1. ULTRASONIC IRRADIATION

Eq. (3) can be simplified, since we have assumed that the collapsed bubble retains its spherical shape and that it has a uniform initial temperature T_0. The resulting temperature distribution, $T(r)$, is, therefore, only a function of the distance from the bubble center and is independent of the angles θ and ϕ. With these assumptions we can write:

$$\frac{1}{k}\frac{\partial T}{\partial t} = \frac{\partial^2 T}{\partial r^2} + \frac{2}{r}\frac{\partial T}{\partial r} \quad (4)$$

For the initial conditions, $T = T_0$, for $t = 0$, $0 < r < a$ and $T = T_{med}$, for $t = 0$, $r > a$ and the boundary condition T = finite, for $r = 0$, the solution to eq. (4) is given by eq. (5) [89]:

$$T_{red} = \frac{1}{2}\left[\text{erf}\left(\frac{\frac{r}{a}+1}{2\sqrt{k\frac{t}{a^2}}}\right) - \text{erf}\left(\frac{\frac{r}{a}-1}{2\sqrt{k\frac{t}{a^2}}}\right)\right]\dots$$

$$\dots - \frac{a}{r\sqrt{\pi}}\sqrt{k\frac{t}{a^2}}\left[\exp\left(-\left(\frac{\frac{r}{a}-1}{2\sqrt{k\frac{t}{a^2}}}\right)\right) - \exp\left(-\left(\frac{\frac{r}{a}+1}{2\sqrt{k\frac{t}{a^2}}}\right)\right)\right] \quad (5)$$

where the error function is given by eq. (6):

$$\text{erf}(x) = \frac{2}{\sqrt{\pi}}\int_0^x \exp[-y^2]\,dy \quad (6)$$

while the reduced temperature is defined as eq. (7):

$$T_{red} = \frac{T - T_{med}}{T_0 - T_{med}} \quad (7)$$

where T_0 is the initial temperature of the collapsed bubble, T_{med} is the initial temperature of the surrounding water, T_{red} is the reduced temperature as a function of the distance from the center of the bubble, r, at an elapsed time, t. The actual temperature can be calculated from the definition of the reduced temperature.

The numerically-calculated times and corresponding reduced distances (i.e., extent of liquid shell under supercritical conditions) are given in Table 10.2. The estimated lifetime and spatial extent of the supercritical phase at a single cavitation site are on the order of milliseconds and microns, respectively. After

Table 10.2 Reduced radii and volumes of the hot layer around a collapsed cavity and corresponding elapsed times.

No, i	Time after collapse (msec)	Reduced distance $T > T_c$ $(r(t_i)/a)$	$V_{shell}(t_i)$ (pL)
0	0	1.000	0
1	1	1.147	7.2
2	2	1.200	10.3
3	5	1.307	17.4
4	10	1.400	24.7
5	15	1.467	30.5
6	30	1.573	40.9
7	50	1.613	45.2
8	70	1.573	40.9
9	90	1.467	30.5
10	110	1.307	17.4
11	120	1.200	10.3
12	130	1.067	3.04

10 ms the radius of the supercritical region around a collapsed bubble extends about 40% farther into the bulk solution than the original cavity. The radius of the supercritical shell expands up to 160% of the original bubble radius at 50 ms after collapse.

The fraction of the total volume of a sonified aqueous solution that is actually in the supercritical state can be readily estimated based on the above calculations. The volume $V_{shell}(t_i)$ of the hot layer, where $T > T_c$, around a collapsed cavity can now be calculated as a function of time.

$$V_{shell}(t_i) = \frac{4}{3}\pi a^3 \left[\left(\frac{r(t_i)}{a}\right)^3 - 1\right] \qquad (8)$$

If N is the number of cavities that collapse per unit volume per unit time, the fraction of the aqueous solution that is in the supercritical state, x_{scw}, is given as

$$\frac{V_{scw}}{V_{total}} \equiv x_{scw} = N \sum_i V_{shell}(t_i)[t_i - t_{i-1}] \qquad (9)$$

The number density of nuclei and their size distribution has been measured by Katz and Acosta [98]. However there are no experimental data for the number density of nuclei or actual cavitation bubbles in water during ultrasonic irradiation. Suslick and Hammerton [99] give an estimate for the number of collapsing bubbles per time and volume. Based on an estimate of $N = 4 \times 10^8 \text{ s}^{-1} \text{ m}^{-3}$,

10.1. ULTRASONIC IRRADIATION

the dynamic fraction of water in the supercritical domain during sonolysis is $x_{scw} = 0.0015$. Depending on the extent to which supercritical water accelerates chemical reactions, this fraction may represent a substantial contribution to reaction rate enhancements that have been reported previously for aqueous-phase chemical reactions in the presence of ultrasound.

The output power in a sonochemical reactor can be determined readily by following the rise in temperature, T, of a known volume of water with time, t, in the absence of cooling. A simplified heat transfer analysis as described by Kotronarou et al. [53] gives the rate of energy input, Q, to the system as follows:

$$Q = M \cdot C_p \cdot \frac{dT}{dt} + U \cdot A_w (T_0 - T_{amb}) \qquad (10)$$

where T_0 and T_{amb} are the initial temperature of the liquid and ambient temperature, respectively, A_w is the wetted area, M is the mass of water, C_p is the specific heat, and U is the bulk heat loss coefficient. The solution to this equation gives the temperature variation with time in a non-thermostatted sonochemical reactor as follows:

$$T = (T_0 - T_{amb} - Q/U \cdot A_w)\exp(-U \cdot A_w t/M \cdot C_p) + T_{amb} + Q(U \cdot A_w) \qquad (11)$$

10.1.3 Chemical Consequences of Extreme Transient Conditions

Two distinct sites for chemical reaction exist during a single cavitation event [42]. They are the gas phase in the center of a collapsing cavitation bubble and a thin shell of superheated liquid surrounding the vapor phase. The volume of the gaseous region is estimated to be larger than that of the thin liquid shell by a factor of $\sim 2 \times 10^4$ [97, 100].

During cavitational-bubble collapse, which occurs within 100 ns, H_2O undergoes thermal dissociation within the vapor phase to give hydroxy radicals and hydrogen atoms as in eq. (12):

$$H_2O \xrightarrow{\Delta} H\cdot + \cdot OH \qquad (12)$$

The concentration of $\cdot OH$ at a bubble interface in water has been estimated to be 4×10^{-3} M [44]. Many of the chemical effects of ultrasonically induced cavitation have been attributed to the secondary effects of $\cdot OH$ and $\cdot H$ production [55, 56, 58, 87, 101–106].

Based on work in our laboratory [89], we believe that sonochemical reactions in water are characterized by the simultaneous occurrence of supercritical water reactions, direct pyrolyses and radical reactions especially at high solute concentrations. Volatile solutes such as carbon tetrachloride [107] and hydrogen sulfide [53] will undergo direct pyrolysis reactions within the gas phase of the collapsing bubbles or within the hot interfacial region as shown below eqs. (13)–(15):

$$CCl_4 \xrightarrow{\Delta} \cdot CCl_3 + Cl \cdot \qquad (13)$$

$$\cdot CCl_3 \xrightarrow{\Delta} :CCl_2 + Cl \cdot \qquad (14)$$

$$H_2S \xrightarrow{\Delta} HS \cdot + H \cdot \qquad (15)$$

while low-volatility solutes such as thiophosphoric acid esters [48] and phenylate esters [89] can react in transient supercritical phases generated within a collapsing bubble as in eq. (16):

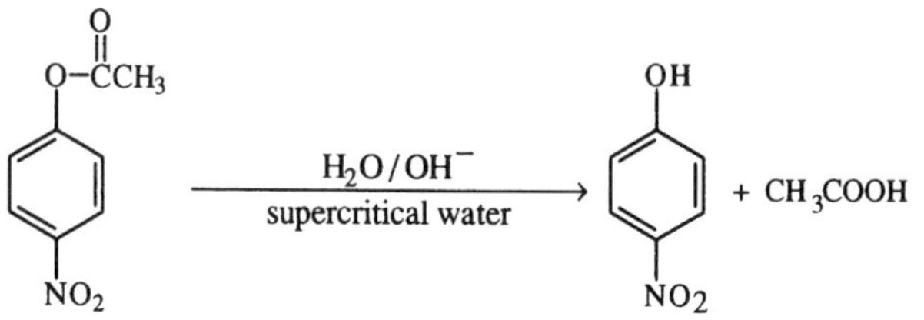

$$\underset{NO_2}{\underset{|}{C_6H_4}}-O-\overset{O}{\overset{\|}{C}}CH_3 \xrightarrow[\text{supercritical water}]{H_2O/OH^-} \underset{NO_2}{\underset{|}{C_6H_4}}-OH + CH_3COOH \qquad (16)$$

In the case of ester hydrolysis, reaction rates are accelerated 10^2 to 10^4 times the corresponding rates under controlled kinetic conditions (*i.e.*, same pH, ionic strength, and controlled overall temperature). This effect can best be illustrated by the catalytic effect of ultrasonic irradiation on the rate of hydrolysis of parathion in water at pH 7, eq. (17).

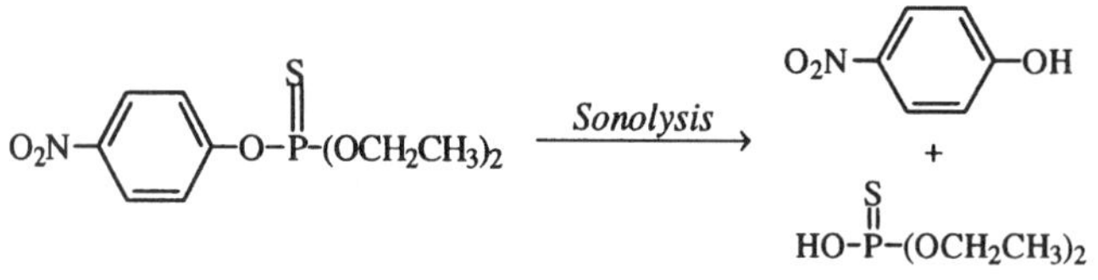

$$O_2N-C_6H_4-O-\overset{S}{\overset{\|}{P}}-(OCH_2CH_3)_2 \xrightarrow{\text{Sonolysis}} O_2N-C_6H_4-OH + HO-\overset{S}{\overset{\|}{P}}-(OCH_2CH_3)_2 \qquad (17)$$

The half-life for parathion hydrolysis at pH 7.4, in the absence of ultrasound at 25 °C, is 108 days. However, in the presence of ultrasound the half-life is reduced to 20 minutes [52].

Pyrolysis (*i.e.*, combustion) and supercritical water reactions in the interfacial region are predominant at high solute concentrations, while at low solute concentrations free radical reactions are likely to predominate. Depending on its physical properties, a molecule can simultaneously or sequentially react in both the gas and interfacial liquid regions.

10.1. ULTRASONIC IRRADIATION

In the specific case of hydrogen sulfide gas dissolved in water, both pyrolysis in the vapor phase of the collapsing bubbles and hydroxyl radical attack in the quasi liquid interfacial region occur simultaneously as in eqs. (18)–(19):

$$H_2S \xrightarrow{\Delta} HS\cdot + H\cdot \qquad (18)$$

$$H_2S + \cdot OH \longrightarrow HS\cdot + H_2O \qquad (19)$$

The kinetics of reactions occurring during ultrasonic irradiation follow classical rate expressions. For example, Kotronarou et al. [53] established that the experimentally-determined rate law for the autoxidation of H_2S in the presence of ultrasonic irradiation consisted of two terms, a first-order term, representing degradation initiated by the pyrolytic decomposition of hydrogen sulfide in the hot vapor phase and a zero-order term representing the combined effects of hydroxyl radical attack in both the hot vapor phase (H_2S) and in the hot interfacial liquid domain (HS^-). The rate law expressed in terms of total sulfide ($[H_2S]_T = [H_2S] + [HS^-]$) where $pK_{a1} = 7.0$) is

$$-\frac{d[H_2S]}{dt} = k_0 + k_1[H_2S]_T \qquad (20)$$

The integrated solution to eq. (20) is given in eq. (21) where k_0 is the rate constant for hydroxyl radical attack, k_1 is the rate constant for pyrolysis, $[H_2S]_{T,t}$ is the concentration of total sulfide at time $= t$ and $[H_2S]_{T,t}$ is the concentration of total sulfide at $t = 0$.

$$[H_2S]_{T,t} = \left([H_2S]_{T,0} + \frac{k_0}{k_1}\right)e^{-k_1 t} + \frac{k_0}{k_1} \qquad (21)$$

In addition to an apparent traditional kinetic dependence on concentration, pH, and ionic strength, a linear relationship between the applied power at a fixed frequency and the observed rate of loss of sulfide over the range from 10 to 120 W (10 to 120 W cm^{-2}) was reported [53].

Kotronarou and Hoffmann [108] have presented a detailed kinetic model for the sonolytic oxidation of HS^- by $\cdot OH$ at high pH (in a pH domain where pyrolytic decomposition is negligible). The oxidation of S^{2-} is initiated by reaction with hydroxyl radical but it is further propagated by a free-radical chain sequence involving O_2. Their mechanism can adequately model the observed oxidation of S^{2-} in air-saturated aqueous solutions sonicated at 20 kHz and 75 W cm^{-2} at pH ≥ 10, assuming a continuous and uniform hydroxyl radical input into solution from the imploding cavitation bubbles. These results suggest that the use of simplified approaches for modeling the liquid-phase sonochemistry of a well-mixed solution may be justified when $\cdot OH$ radical reactions predominate. For the probe-type reactor utilized in their study, a uniform release rate of hydroxyl radical into the bulk solution was observed (3.5 µM min^{-1}) with a corresponding steady-state concentration ≤ 0.1 µM.

The most obvious evidence for the role of pyrolysis during ultrasonically-induced transient cavitation is the production of NO, NO_2, NO_2^-, NO_3^- and H_2O_2 during the sonolysis of water in the presence of N_2 and O_2 alone as background gases. The following mechanism (eqs. (22)–(30)) [51], in which the first steps are analogous to those occurring during high-temperature combustion, appears to be operative:

$$O_2 \xrightarrow{\Delta} 2O \cdot \tag{22}$$

$$N_2 + O^{\cdot} \longrightarrow N_2O^{\cdot} \tag{23}$$

$$N_2O^{\cdot} + O^{\cdot} \longrightarrow 2NO \tag{24}$$

$$NO + O^{\cdot} \longrightarrow NO_2^{\cdot} \tag{25}$$

$$NO_2^{\cdot} + O \cdot \longrightarrow NO_3 \tag{26}$$

$$NO_2 + NO_3 \longrightarrow N_2O_5 \tag{27}$$

$$N_2O_5 + H_2O \longrightarrow 2HNO_3 \tag{28}$$

$$NO + HO \cdot \longrightarrow HONO \tag{29}$$

$$2NO_{2\,(aq)} + H_2O \longrightarrow HNO_3 + HNO_2 \tag{30}$$

Comparative rate thermometry, as described below, employing p-nitrophenol (PNP) as a probe molecule can be used in order to estimate the average effective temperatures achieved during bubble collapse. PNP is a suitable probe molecule since its kinetics and mechanism of degradation are well understood [50, 89]. The primary step during the sonolytic degradation of PNP has been shown to be as in eq. (31) [50, 63]:

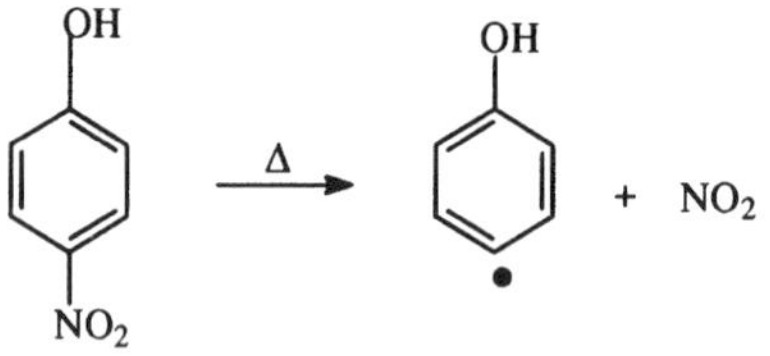

$$\text{HO-C}_6\text{H}_4\text{-NO}_2 \xrightarrow{\Delta} \text{HO-C}_6\text{H}_4 \cdot + NO_2 \tag{31}$$

The activation energy for the pyrolytic carbon-nitrogen bond cleavage, can be estimated from shock tube studies of nitrobenzene decomposition [109]. The reaction rate constant for carbon-nitrogen bond-cleavage in nitrobenzene at high temperatures follows a classical Arrhenius relationship (eq. (32)):

$$k = A \exp\left\{\frac{-E_a}{RT}\right\} \tag{32}$$

where $A = 1.9 \times 10^{15}\ \text{s}^{-1}$ and $E_a/R = 33{,}026\ \text{K}$ for the following elementary unimolecular reaction, eq. (33):

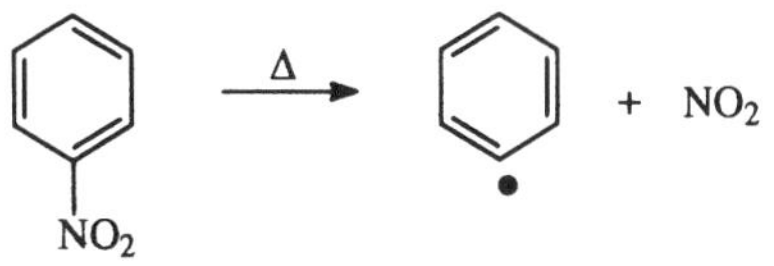

$$\text{C}_6\text{H}_5\text{NO}_2 \xrightarrow{\Delta} \text{C}_6\text{H}_5{}^\bullet + \text{NO}_2 \tag{33}$$

Using these values for A and E_a/R, the effective temperature during electro-hydraulic cavitation can be estimated as shown in eq. (34):

$$T_{\text{eff}} = \frac{-E_a/R}{\ln(k/A)} = \frac{-33{,}026}{\ln(k/1.9 \times 10^{15})}\ [\text{K}] \tag{34}$$

The relative temperature of bubble collapse can be adjusted by saturating the solution with gases characterized by substantially different specific heats, thermal conductivities, and solubilities. As mentioned above, an important factor controlling bubble collapse temperature is the polytropic constant, K, of the saturating gas. From a knowledge of K we can estimate the maximum temperature obtained during bubble collapse from eq. (1). Based on physical properties (Table 10.1), we predict that krypton will yield the highest rate of PNP degradation, while helium should yield the lowest relative rate.

The sonolytic degradation of PNP is found to be a first-order reaction for all gases as shown in Figure 10.1. Using the rate constants obtained from these data and eq. (34) we estimate the effective average temperatures at the interface of the collapsing bubbles for each gas (Table 10.3). The highest effective temperature is achieved in a Kr-saturated solution, whereas a He-saturated solution results in the lowest effective temperature, as predicted above. Although the value of K is similar for all three gases, helium has an unusually high thermal conductivity and a relatively low solubility in comparison to Ar and Kr. Thus, the difference in the resulting effective temperature is also larger than that between Ar and Kr.

10.2 PULSED-PLASMA DISCHARGES IN WATER

10.2.1 Background

Pulsed-power plasma discharges into water are an electrohydraulic phenomenon characterized by a periodic rapid release of accumulated electrical energy across a submerged electrode gap (1–2 cm). The power source is a bank of charged capacitors capable of delivering a high voltage, high amperage electrical current to the submerged electrodes at a moderate frequency. Each electrical discharge produces a short (20 μs) burst of electrical energy at a high power density within the electrode gap (i.e., 25 kJ). This highly ionized and pressurized

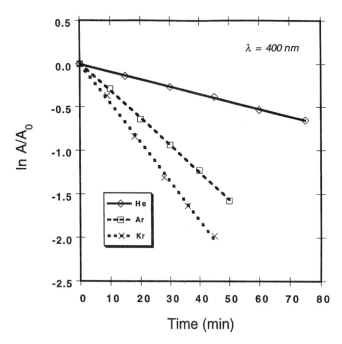

Figure 10.1 First-order plot of PNP degradation at 25 °C in a sonicated solution with different saturating gases. Rate constants in units of s^{-1}: a) Kr: 7.5×10^{-4}, b) Ar: 5.2×10^{-4}, c) He: 1.4×10^{-4}. A = absorbance at 400 nm at time t after start of sonication, and A_i = initial absorbance at 400 nm.

Table 10.3 Calculated effective temperature from comparative rate thermometry with PNP as reacting system.

Dissolved gas	Rate constant (s^{-1})	Calculated temperature (K)
Kr	7.5×10^{-4}	779
Ar	4.6×10^{-4}	772
He	1.2×10^{-5}	750

plasma has the ability to transfer energy to wastewater *via* dissociation, excitation, and ionization (*e.g.*, H_2O^+, e_{aq}^-, $\cdot OH$, $H\cdot$, etc.) with a corresponding increase in temperature. The rapidly-expanding plasma produces a very high pressure shock wave ($> 14{,}000$ atm). If the propagating shock wave is reflected back from either a free surface or from a material with a different acoustic impedance, intense cavitation occurs with the associated chemical changes as described above for the chemical effects of ultrasound. Thus, a pulsed-power discharge in water provides a unique method for promoting pyrolytic, $\cdot OH$

10.2. PULSED-PLASMA DISCHARGES IN WATER

radical, aquated electron and UV-promoted reactions. Additional reaction pathways result from the direct reactions of the rapidly expanding plasma gases with the chemical substrates of interest and from indirect production of ·OH radicals due to the release of soft X-rays and high energy UV radiation from the energized plasma.

The pulsed-plasma discharge process or electrohydraulic discharge process (EHD) is a non-thermal technique to inject energy directly into an aqueous solution through a plasma channel formed by a high-current/high-voltage electrical discharge between two submersed electrodes [110–113]. A typical EHD system, as shown in Figure 10.2, consists of two major components, which are the pulsed-power electrical discharge circuit and the reaction chamber. Electrical energy is stored in a large capacitance (135 µF) pulsed-power circuit. This energy is released as a pulsed electrical discharge using fast ignitron switches. Each pulse can have energies ranging from 5 to 25 kJ with a duration of 20 to 100 µs. Peak powers are in the megawatt to gigawatt range. The reactor vessel (Figure 10.2a) contains an electrode assembly and the aqueous solution to be treated. The reactor is designed to withstand intense shockwaves generated by the electrohydraulic discharges and to accommodate the high electrical currents and voltages required by the process.

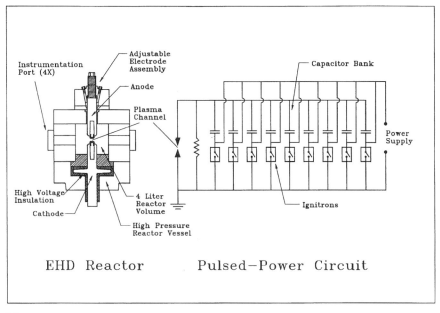

(a)

Figure 10.2 a) Schematic diagram of the EHD reactor and the corresponding electronic circuit. b) Current and voltage transients for the three different EHD: A) $V_T = 1.0$ L, $E_{CB} = 4.0$ kJ. B) $V_T = 1.0$ L, $E_{CB} = 7.0$ kJ. C) $V_T = 3.5$ L, $E_{CB} = 7.0$ kJ. The solid and dashed lines are for the current and voltage transients, respectively.

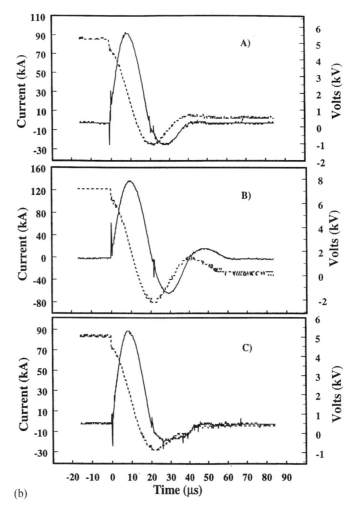

Figure 10.2 Continued

The plasma channel formed during an electrohydraulic discharge reaches temperatures of 20,000–50,000 K [113] and thus functions as a blackbody radiation source with a maximum emittance in the vacuum ultraviolet (VUV) region of the spectrum. The VUV emitted from the hot plasma is absorbed by the water layer immediately surrounding the plasma channel [114], although UV light with $\lambda > 185$ nm penetrates into the bulk aqueous solution. During the formation of the plasma channel (1–2 μs) an intense 5 to 20 kbar shockwave is generated, due to the rapidly expanding plasma [111]. The resulting shockwave can induce pyrolytic and free radical reactions indirectly *via* cavitation as described above. As the plasma channel cools over 1–3 ms, thermal energy is transferred to the surrounding water resulting in the formation of a steam

10.2. PULSED-PLASMA DISCHARGES IN WATER

bubble [115]. Within a steam bubble, the temperatures and pressures are high enough for the formation of transient supercritical water [116].

10.2.2 Physical Principles

The underwater discharge can be separated into three distinct temporal phases. When the switch of the LRC circuit is closed with the subsequent buildup of a voltage across the underwater electrodes, a discharge initiation phase takes place during which a plasma channel is formed. During a second "energy transfer phase", the bulk of the energy stored in the capacitor is dissipated into the plasma channel. The plasma channel in turn does work against the surrounding fluid as it expands, and in the process, it generates a shock wave. Some of the energy discharged to the plasma channel is released as blackbody radiation with channel temperatures $\geq 20{,}000$ K. During the energy transfer phase, energy losses due to heat conduction are negligible, since the initial-event time scales are too fast for heat transfer to occur. The discharge enters its final phase once heat transfer is initiated and the plasma channel is converted into a spherical steam bubble, which behaves as a cavitating bubble, until all of the remaining energy is dissipated *via* acoustic radiation and viscous friction.

The general circuit illustrated in Figure 10.2 can be expressed by the following differential equation (35):

$$L\frac{d^2 q(t)}{dt^2} + R(t)\frac{dq(t)}{dt} + \frac{q(t)}{C} = 0 \qquad (35)$$

where L is the inductance of the circuit, C is the capacitance, and q is the stored charge. $R(t)$ is the resistance of the circuit and is dominated by the time-dependent resistance of the plasma channel.

The energy transferred from the circuit to the plasma channel is partitioned into three processes; they are the mechanical work on the surrounding fluid as the plasma channel expands, the internal energy of the plasma itself, and electromagnetic radiation as the channel heats up. That energy process is described by the following continuity equation (36):

$$\frac{dE_d(t)}{dt} = \frac{d}{dt}(W(t)V(t)) + p(t)\frac{dV(t)}{dt} + (1-f)\sigma_B T(t)^4 S \qquad (36)$$

The term on the left side of eq. (36) represents the rate of energy transferred to the plasma channel from the capacitor bank. The first term on the right describes the energy transferred to change the internal energy state of the plasma channel where $W(t)$ denotes the internal energy density of the plasma, $V(t)$ the volume and $S(t)$ the surface area of the plasma channel. The second term on the right represents the work done by the channel on the surrounding fluid where $p(t)$ is the pressure. The last term on the right represents the radiative transfer where σ_B is the Stephan-Boltzmann constant. Some of the black-body radiation fails to escape the plasma channel and this is factored into account by the term $(1-f)$ where the fraction of the spectrum denoted by f represents radiation energy which is absorbed at the boundaries to ionize particles. The energy lost to these particles is recovered since these particles are inducted into the plasma channel. However, the resulting particle flux is non-negligible and contributes to the mass balance of the plasma channel. Since the plasma continually inducts particles from the channel walls into the main body of the plasma, the number of particles $N(t)$ in the plasma channel is not constant, and has the following time-dependence (eq. (37))

$$\frac{dN(t)}{dt} = \frac{S(t)f\sigma_B T^4}{\varepsilon_{ionization}} \tag{37}$$

where $S(t)$ is the surface area of the channel, $N(t)$ is the total number of particles within the channel, $\varepsilon_{ionization}$ is the energy required to dissociate water, and f is the fraction of the radiation spectrum absorbed in the fluid layers immediately surrounding the plasma channel.

The plasma channel can be viewed as an expanding cylindrical piston, in which the pressure on the surface of the channel is related to the expansion velocity using the Rankine-Hugoniot relationship for an expanding shock front (eq. (38)) [111].

$$(p+\beta)^{3/7} = \left(\frac{3\sqrt{\rho_0}}{\sqrt{7\alpha^{1/14}}}\right)\frac{da(t)}{dt} + \beta^{3/7} \tag{38}$$

where p = pressure, α = 3001 bar, β = 3000 bar and ρ_0 = 1000 kg m^{-3}. The plasma channel can be modeled as a resistor with a time varying resistance $R(t)$ undergoing resistive heating. From Kirchoff's Law, the potential across the LRC circuit is given by eq. (39):

$$L\frac{dI(t)}{dt} + [R_c + R(t)]I(t) + \frac{q(t)}{C} = 0 \tag{39}$$

where

$$\frac{dq(t)}{dt} = I \tag{40}$$

and where L is the inductance of the circuit, C is the capacitance of the energy storage bank. $R(t)$ is the resistance of the plasma channel given by eq. (41):

$$R(t) = \frac{l}{\sigma_{cond} \pi a^2} \tag{41}$$

10.2. PULSED-PLASMA DISCHARGES IN WATER

where σ_{cond}, the conductivity of the plasma channel, is given by eq. (42) [110]

$$\sigma_{\text{cond}} = c_2\, T^{3/2} \exp(-5000/T) \tag{42}$$

where c_2 is a constant that is $\approx 10^{-3}\,\text{W}^{-1}\,\text{K}^{-1.5}$. R_c, which is the equivalent resistance of the circuit, results in some dissipation of energy, that would otherwise be transferred to the underwater gap. Thus, not all of the energy stored in the capacitor bank (eq. (43))

$$E = 0.5 \cdot C U_0^2 \tag{43}$$

is actually released in the underwater discharge gap.

The internal energy of the plasma channel, W, is directly proportional to the external pressure on the surface of the channel and independent of the temperature of the channel over the temperature ranges of interest. Therefore, W can be written as in eq. (44):

$$W = \frac{1}{\gamma - 1} p \tag{44}$$

where $\gamma = 1.22 \equiv$ isentropic index of the plasma. Eq. (44) follows from the ideal gas law, $p = nkT$, and the Saha ionization equation, neglecting inter-particle coulombic interactions and the "pinch effect."

The volume and surface area of the plasma channel are given by eq. (45):

$$V = \pi a^2 l \qquad S = 2\pi a l \tag{45}$$

The above equations describing the temporal behavior of the plasma channel can be simplified to give:

$$\frac{I^2 l}{\sigma_{\text{cond}}\, \pi a(t)^2} = \frac{d}{dt}(W(t) \cdot V(t)) + p(t)\frac{dV(t)}{dt} + (1 - f)\sigma_B\, T(t)^4\, S(t) \tag{46}$$

where $\sigma_B = 5.67051 \times 10^{-8}\,\text{W/m}^2\,\text{K}^4$. These 4 coupled differential equations involving 4 unknowns (i.e., $I(t)$, $T(t)$, $a(t)$, and $N(t)$) are solved numerically to give the microsecond time history of a single plasma discharge in terms of current, energy, temperature and pressure. Agreement between theory and experiment is quite satifactory.

From simulations based on numerical classical approaches, the energy dissipated into various modes can be calculated. The energy transferred into the shock wave is given by eq. (47):

$$E_{\text{shockwave}} = \int p(t)\, dV(t) \tag{47}$$

while the energy dissipated through black-body radiation is given by eq. (48):

$$E_{radiation} = (1-f) \int \sigma_B \, T(t)^4 \, dt \qquad (48)$$

The total energy transferred from the capacitor bank into the plasma channel is given by eq. (49):

$$E_{discharge} = \int R(t) I(t)^2 \, dt \qquad (49)$$

This energy, which is dissipated by the net resistance in the circuit, is not the same as the total energy stored in the capacitor bank due to internal circuit losses during discharge *via* the circuit resistance, R_c.

The energy not dissipated by means of shock waves or radiation contributes to a post-discharge steam bubble. The maximum radius can be estimated using the following continuity relationship (eq. (50)):

$$E_{discharge} - E_{shock} - E_{radiation} = \tfrac{4}{3} \pi p_0 a_{max}^3 \qquad (50)$$

10.2.3 Chemical Consequences of Transient Plasma Formation

In the EHD process, chemical degradation occurs within the plasma channel directly due to pyrolysis and free radical reactions. However, the small volume of the plasma channel (1–3 mL) limits the amount of solution that can be exposed directly by high temperature processes in a large chamber reactor. On the other hand, experiments with exploding wires have shown that electrohydraulic discharges induce extreme electromagnetic and mechanical conditions in the bulk solutions outside of the plasma channel region [115]. These experiments have established that there are three primary physical processes (*i.e.*, ultraviolet (UV) photolysis, electrohydraulic cavitation, and supercritical water oxidation) with the potential to induce significant oxidative chemistry in the bulk solution.

Since intense black body radiation is emitted directly from the plasma channel into the secondary reaction volume of the reactor, we expect that direct photolysis will play a very important role in compound degradation. In the case of 0.5 L solution of 3,4-chloroaniline exposed to a series of 120 pulses in which each pulse had a total energy of 5.0 kJ, 74% of 3,4-dichloroaniline was degraded primarily by direct photolysis in 2 ms of power utilization according to the stoichiometry in eq. (51):

$$\text{3,4-dichloroaniline} \xrightarrow[8\,O_2]{h\nu} 2\,HCl + 6\,CO_2 + H_2O + HNO_3 \qquad (51)$$

10.2. PULSED-PLASMA DISCHARGES IN WATER

Direct photolysis reactions are quantified by the following rate equation:

$$-\frac{dC}{dt} = \Phi(\lambda) I_0(\lambda)(1 - \exp(-2.3 \varepsilon l C)) \quad (52)$$

where C is the concentration of the chromophoric reactant, $\Phi(\lambda)$ is the photolysis quantum yield, I_0 is the intensity of the UV source, ε is the extinction coefficient of the primary chromophore, and l is the pathlength of radiation. When $I_0(l, t)$ is constant and in the case of long pathlengths, high concentrations, and/or large extinction coefficients eq. (52) reduces to the zero-order expression given in eq. (53):

$$-\frac{dC}{dt} = \Phi(\lambda) I_0(\lambda) \quad (53)$$

Since the UV source in the EHD reactor is pulsed, $I_0(l, t)$ varies significantly over the lifetime of the pulse. However, the total energy per discharge is a well defined quantity. Thus, one can assume that the total UV flux per discharge is relatively constant as long as the discharge conditions remain constant. Under these conditions t is replaced with N to obtain eq. (54)

$$-\frac{dC}{dN} = \Phi(\lambda) Q(\lambda)(1 - \exp(-2.3 \varepsilon l C)) \quad (54)$$

which for large values of $\varepsilon\, l\, C$ gives eq. (55):

$$-\frac{dC}{dN} = \Phi(\lambda) Q(\lambda) = k_0 \quad (55)$$

where $Q(l)$ is the radiant energy per discharge. Since N is dimensionless, the rate constants will have non-conventional units. A zero-order rate constant, k_0, will have dimensions of μM, and a first-order rate constant, k_1, will be dimensionless. N-based kinetic constants can be readily transformed into intrinsic rate constants based on time.

If photochemical degradation is the only oxidative mechanism operative in the reactor, the process should exhibit zero-order behavior at sufficiently high substrate concentrations. However, under high concentrations strict zero-order behavior breaks down and the kinetics switch to apparent first-order behavior. To account for this transition, we consider the EHD system to be a two-compartment reactor. The first reactor compartment is the bulk solution which is exposed to the extended effects (UV and shockwave radiation) generated by the EHD discharge. Discounting any cavitational effects generated by the shockwave, degradation in this region is due to purely photochemical effects and therefore can be treated as a zero-order process as discussed above.

The second reactor compartment is the small volume contained within and immediately surrounding the plasma channel. The solution in this region is exposed to the localized chemical effects. The substrate is oxidized within the high temperature plasma channel and by hydroxyl radicals generated by the VUV photolysis of water. Since the conditions in and around the plasma channel are so extreme, it is assumed that any substrate within this region will be completely oxidized. These combined processes will appear to be first-order in kinetic behavior, when they are considered over the scale of the entire reactor (*vide infra*). If the solution in the reactor chamber is subsequently well-mixed after one discharge the measured concentration of the substrate will be:

$$C = \left(\frac{V_T - V_R}{V_T}\right) C_i - k_0 \tag{56}$$

where C_i is the initial concentration, V_T is the total solution volume, V_R is the plasma channel volume, and k_0 is the zero-order rate constant due to direct photolysis. After N discharges, it can be shown that the measured concentration will be:

$$C = a^N C_i - k_0 \sum_{n=0}^{N-1} a^n = a^N C_i - k_0 \left(\frac{1 - a^N}{1 - a}\right) \tag{57}$$

where,

$$a = \left(\frac{V_T - V_R}{V_T}\right) \tag{58}$$

Since N can be treated as a continuous variable for the kinetics analysis, the derivative of eq. (57) with respect to N yields:

$$\frac{dC}{dN} = e^{\ln(a)N} C_i + k_0 \frac{\ln(a) e^{\ln(a)N}}{1 - a} \tag{59}$$

Since $a \approx 1$ and $N < 40$, we can assume that $\ln(a) = (a - 1)$ and $\exp(\ln(a)N) = (1 + \ln(a)N)$ to obtain

$$\frac{dC}{dN} = \ln(a) C_i - k_0 - k_0 \ln(a) N + \ln^2(a) C_i \tag{60}$$

The last two terms of eq. (60) can be neglected since they are much smaller in magnitude than the first two terms and are much smaller than the experimental error. Therefore, the final differential equation that describes the initial rate of degradation in an EHD reactor can be described as the sum of zero and first-order terms as follows:

$$\frac{dC}{dN} = -k_1 C_i - k_0 = -k_A \tag{61}$$

The observed first-order rate constant will be $k_1 = -\ln(a)$. The value of k_1 is a function of the ratio of V_R to V_T. In our reactor $V_R \ll V_T$, and therefore k_1 has a small ($\approx 9 \times 10^{-4}$ discharge^{-1}) but significant effect.

10.3 SONOLYTIC REACTIONS IN WATER: KINETICS AND MECHANISMS

10.3.1 Degradation of p-Nitrophenol

The application of ultrasonic irradiation for the controlled degradation of chemical contaminants in water has been investigated using several model compounds [50, 52–58, 63, 89, 107]. For example, p-nitrophenol (PNP) is degraded completely by sonolysis to yield short-chain carboxylic acids, CO_2, NO_3^- and NO_2^- [52, 54, 89]. In the presence of ultrasound (20 kHz, 84 W), p-nitrophenol degradation occurs primarily by denitration to yield NO_2^-, NO_3^-, benzoquinone, hydroquinone, 4-nitrocatechol, formate, and oxalate. These reaction products and the kinetic observations are consistent with a model involving high-temperature reactions of p-nitrophenol in the interfacial region of cavitation bubbles. The main reaction pathway appears to be carbon-nitrogen bond cleavage. Reaction with hydroxyl radical provides a secondary reaction channel. The average effective temperature of the interfacial region surrounding the cavitation bubbles was estimated to be near 900 K.

Kotronarou et al. [50] proposed the following reaction mechanism (eqs. (62)–(65)) that was found to be consistent with their kinetic observations:

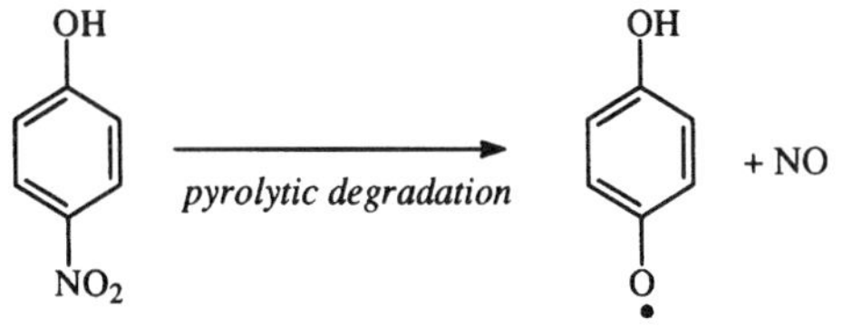

$$\text{PNP} \xrightarrow{\text{pyrolytic degradation}} \text{phenoxyl radical} + NO \tag{62}$$

$$\text{PNP} \xrightarrow{\text{pyrolytic degradation}} \text{phenoxyl radical} + NO_2 \tag{63}$$

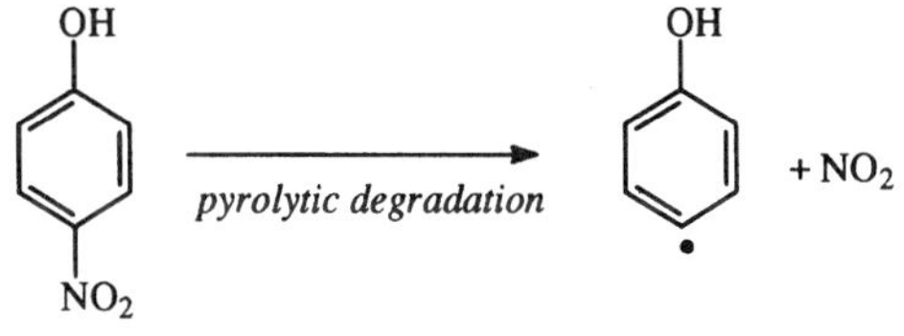

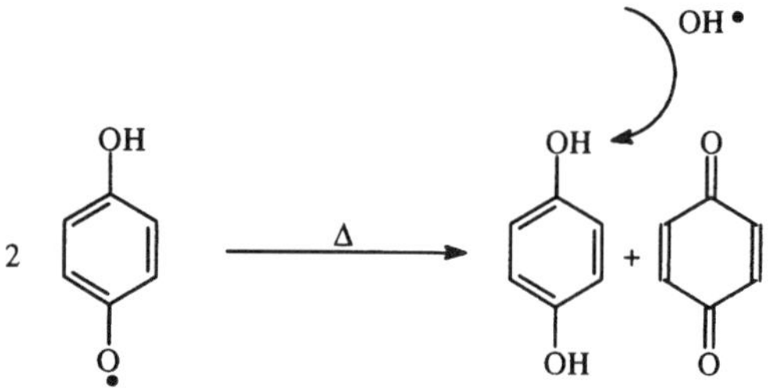

$$2 \; \text{HO-C}_6\text{H}_4\text{-O}^{\bullet} \xrightarrow{\Delta} \text{HO-C}_6\text{H}_4\text{-OH} + \text{O=C}_6\text{H}_4\text{=O} \xleftarrow{\text{OH}^\bullet} \quad (64)$$

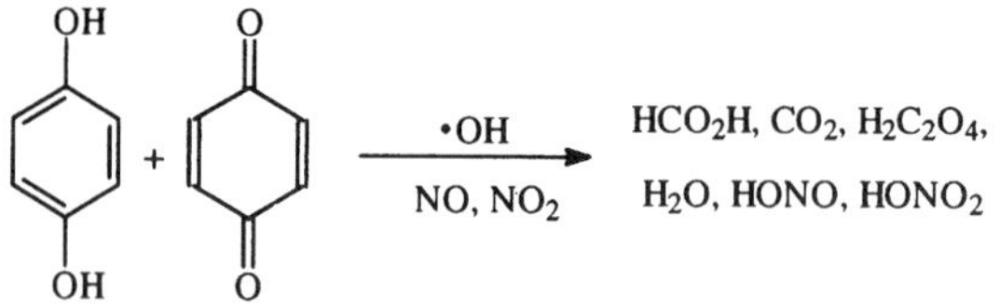

$$\text{HO-C}_6\text{H}_4\text{-OH} + \text{O=C}_6\text{H}_4\text{=O} \xrightarrow[\text{NO, NO}_2]{^\bullet\text{OH}} \text{HCO}_2\text{H, CO}_2, \text{H}_2\text{C}_2\text{O}_4, \text{H}_2\text{O, HONO, HONO}_2 \quad (65)$$

The empirical rate law for the oxidative degradation of PNP consisted of two-terms as follows:

$$-\frac{d[\text{PNP}]}{dt} = k'_0 + k_1[\text{PNP}] \quad (66)$$

where k'_0 is an apparent pseudo zero-order rate constant reflecting degradation initiated by hydroxyl radical attack and k_1 represents degradation initiated by pyrolytic decomposition in the hot interfacial region. Integration of eq. (66) gives the observed concentration of p-nitrophenol as a function of time in the probe-type sonolytic reactor.

$$[\text{PNP}]_t = \left(-\frac{k'_0}{k_1} + [\text{PNP}]_0\right) e^{-k_1 t} - \frac{k'_0}{k_1} \quad (67)$$

Intermediate products resulting from both hydroxyl radical attack and thermal bond cleavage were detected. Hydroxyl radical attack on PNP most likely occurs in a region of the bubble interface with $T < 440$ K, while pyrolysis occurs in a hotter interfacial region with an average temperature of 900 K.

Cost et al. [63] also investigated the kinetics of the sonochemical degradation of p-nitrophenol. However, they investigated the reactions in aqueous solutions containing particulate matter, phosphate, bicarbonate, humic acid and in solutions prepared from lake water. Under conditions that are similar to those of the natural environment, the rate of p-nitrophenol degradation was found to be the same as the decay rate in pure water. Their kinetic results suggested that high-temperature denitration reactions of p-nitrophenol at the interface of cavitating bubbles are not affected by the presence of chemical components of natural waters.

10.3.2 Hydrolysis and Degradation of p-Nitrophenolphenyl Acetate

As described above, Hua et al. [89] have shown that ultrasonic irradiation accelerates the rate of hydrolysis of p-nitrophenyl acetate (PNPA) in aqueous solution by 2 orders of magnitude over the pH range of 3–8. In the presence of ultrasound, the observed first-order rate constant for the hydrolysis of PNPA is found to be independent of pH and ionic strength with $k_{H_2O} = 7.5 \times 10^{-4}\,s^{-1}$ with Kr as the cavitating gas, $k_{H_2O} = 4.6 \times 10^{-4}\,s^{-1}$ with Ar as the cavitating gas, and $k_{H_2O} = 1.2 \times 10^{-4}\,s^{-1}$ with He as the cavitating gas. The apparent activation parameters for sonolytic catalysis were found to be ΔH(sonified) $= 211$ kJ/mol, $\Delta S^{\neq}$(sonified) $= -47$ J/mol K, and $\Delta G^{\neq}$(sonified) $= 248$ kJ/mol. Under ambient conditions, and in the absence of ultrasound, k_{obs} is a strong function of pH where $k_{obs} = k_{H_2O}[H_2O] + k_{OH^-}[OH^-]$ with $k_{H_2O} = 6.0 \times 10^{-7}\,s^{-1}$ and $k_{OH^-} = 11.8\,M^{-1}\,s^{-1}$ at 25°C. The corresponding activation parameters for k_{H_2O} are $\Delta H = 71.5$ kJ/mol, $\Delta S = -107$ J/mol K, and $\Delta G^{\neq} = 155$ kJ/mol. The formation of transient supercritical water SCW appears to be an important factor in the acceleration of chemical reactions in the presence of ultrasound. The apparent activation entropy, $\Delta S^{\neq}$, is decreased substantially during the sonolytic catalysis of PNPA hydrolysis, while $\Delta G^{\neq}$ and $\Delta H^{\neq}$ are increased. The decrease in $\Delta S^{\neq}$ is attributed to differential solvation effects due to the existence of supercritical water (e.g., lower density ρ and lower dielectric constant ε), while the increases in $\Delta G^{\neq}$ and $\Delta H^{\neq}$ are attributed to changes in the heat capacity of the water due to the formation of a transient supercritical state.

10.3.3 Degradation of Carbon Tetrachloride

Hua and Hoffmann [107] have investigated the rapid sonolytic degradation of aqueous CCl_4 at a sound frequency of 20 kHz and at an applied power of 130 W (108 W cm^{-2}). The rate of disappearance of CCl_4 was found to be first-order (Figure 10.3) over a broad range of conditions. The observed first-order degradation rate constant was $3.3 \times 10^{-3}\,s^{-1}$ when $[CCl_4]_i = 195$ μM; k_{obs} was found to increase slightly to $3.9 \times 10^{-3}\,s^{-1}$ when $[CCl_4]_i$ was decreased by a factor of ten (i.e., $[CCl_4]_i = 19.5$ μM). Low concentrations (0.01 to 0.1 μM) of hexachloroethane, tetrachloroethylene and hypochlorous acid (HOCl) were detected as transient intermediates (Figure 10.4), while chloride ion and CO_2 were found to be the stable products.

The highly reactive intermediate, dichlorocarbene, was identified and quantified by means of trapping with 2,3-dimethylbutene. Evidence for involvement of the trichloromethyl radical (Figure 10.5) was also obtained and is indirectly implied by the formation of hexachloroethane. The presence of ozone during sonolysis of CCl_4 did not affect the degradation of carbon tetrachloride but was shown to inhibit the accumulation of hexachloroethane and tetrachloroethylene.

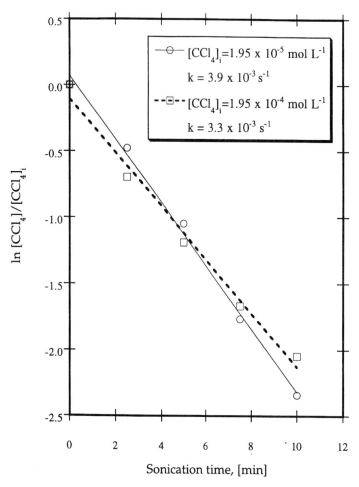

Figure 10.3 Variation of the degradation rate constant with initial carbon tetrachloride concentration.

The following mechanism (eqs. (68)–(75)) is proposed to account for the observed kinetics, reaction intermediates, and final products:

$$CCl_4 \xrightarrow{\Delta} \cdot CCl_3 + Cl\cdot \tag{68}$$

$$\cdot CCl_3 \xrightarrow{\Delta} \cdot\cdot CCl_2 + Cl\cdot \tag{69}$$

10.3. SONOLYTIC REACTIONS IN WATER

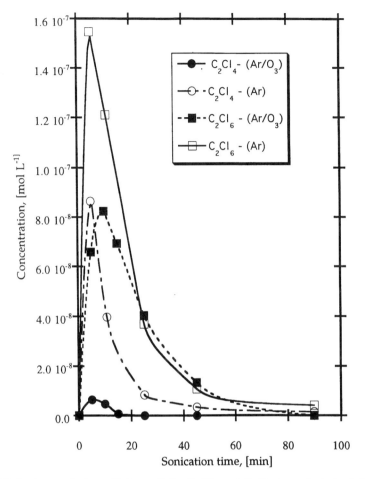

Figure 10.4 Accumulation of intermediates C_2Cl_4 and C_2Cl_6 in sonicated solutions of aqueous CCl_4 in the the presence of Ar and Ar/O_3 mixtures.

$$2 \; {}^{\bullet}CCl_3 \xrightarrow{\Delta} Cl_2C=CCl_2 \tag{70}$$

$$2 \; {}^{\bullet}CCl_3 \xrightarrow{\Delta} Cl_3C-CCl_3 \tag{71}$$

$$ {}^{\bullet}CCl_3 + {}^{\bullet}OH \longrightarrow HOCCl_3 \tag{72}$$

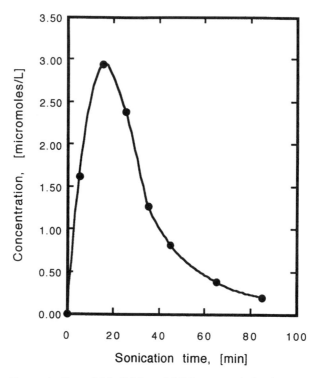

Figure 10.5 Concentration of 1,1-dichloro-2,2,3,3-tetramethylcyclopropane as a function of sonication time.

$$\underset{HO}{\overset{Cl}{\underset{Cl}{C-Cl}}} \longrightarrow \underset{Cl}{\overset{O}{\underset{Cl}{\parallel}}} + H^+ + Cl^- \tag{73}$$

$$\underset{Cl}{\overset{O}{\underset{Cl}{\parallel}}} + H_2O \longrightarrow CO_2 + 2H^+ + 2Cl^- \tag{74}$$

$$Cl^\cdot + Cl^\cdot \longrightarrow Cl_2 \xrightarrow{H_2O} HCl + HOCl \tag{75}$$

The formation of the reactive intermediate, dichlorocarbene, was confirmed by selective trapping of the carbene with 2,3-dimethyl-2-butene as shown in eq. (76):

$$\underset{H_3C}{\overset{H_3C}{>}}C=C\underset{CH_3}{\overset{CH_3}{<}} \xrightarrow{:CCl_2} \text{(cyclopropane product)} \tag{76}$$

10.3. SONOLYTIC REACTIONS IN WATER

to form 1,1-dichloro-2,2,3,3-tetramethylcyclopropane. In a similar fashion, the trichloromethyl radical is trapped by 2,3-dimethyl-2-butene to yield 2-methyl-2-trichloromethyl-1-butene. These trapped intermediates were identified and quantified by GC/MS.

Hua and Hoffmann [107] also noted that the degradation of PNP was accelerated substantially in the presence of CCl_4 (Figure 10.6). In the absence of ultrasonic irradiation, PNP does not react directly with CCl_4. However, addition of CCl_4 during the sonication of PNP in an Ar-saturated, aqueous solution enhanced the rate constant of PNP disappearance by a factor of 4.5 compared to sonication in the absence of CCl_4. In addition, the formation of 4-nitrocatechol as an intermediate was suppressed. The accelerating action appears to be due to the formation of dichlorocarbene which then adds across the double bonds of the PNP ring to enhance its rate of disappearance. Furthermore, the degradation of PNP is enhanced due to its oxidation by HOCl, which is generated as an intermediate of CCl_4 sonolysis.

Other investigators [60, 117–119] have also measured the kinetics of CCl_4 degradation in the presence of ultrasound, but little mechanistic information was provided.

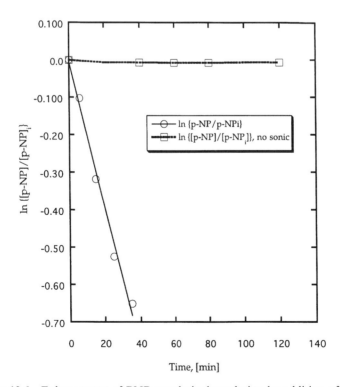

Figure 10.6 Enhancement of PNP sonolytic degradation by addition of CCl_4.

10.3.4 Degradation of TNT

Höchemer and Hoffmann [120] have studied the sonochemical degradation of 2,4,6-trinitrotoluene (TNT) to acetate, formate, glycolate, oxalate, carbon dioxide, nitrite, and nitrate. The overall rate of degradation follows an apparent first-order rate law. 1,3,5-trinitrobenzene appears to be the principal reaction intermediate based on HPLC and Electrospray/MS evidence.

The observed first-order kinetics vary by a factor of three depending on the nature of the dissolved gases and the ultrasonic frequency. At 20 kHz, $k_{obs} = 1.67 \times 10^{-5}$ s^{-1} with O$_2$, $k_{obs} = 4.50 \times 10^{-5}$ s^{-1} with Ar, and $k_{obs} = 5.50 \times 10^{-5}$ s^{-1} with a O$_2$/O$_3$ (99/1 volume-%) mixture as cavitating background gases. The observed rate constants for 500 kHz are 2.00×10^{-5} s^{-1} (O$_2$), 7.17×10^{-5} s^{-1} (Ar), and 8.50×10^{-5} s^{-1} (O$_2$/O$_3$). Ultrasonic irradiation at 500 kHz yielded consistently higher reaction rates than 20 kHz irradiation for all background gases. However, simultaneous irradiation with orthogonally-positioned 20 kHz and 500 kHz ultrasound sources leads to a negative interference as reflected in a diminished reaction rate compared to sonication with 500 kHz ultrasound alone.

The mechanism for TNT degradation is initiated by hydroxyl radical attack on the methyl group of TNT (eq. (77)), yielding a TNT radical, which is stabilized by delocalization of the unpaired electron over the three nitro groups.

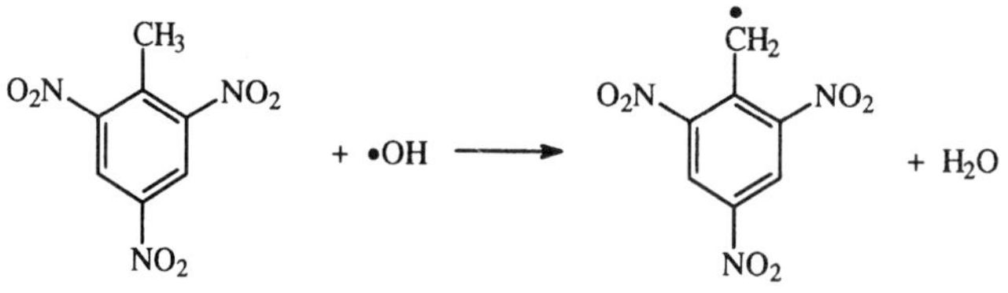

$$\text{TNT} + {}^{\bullet}\text{OH} \longrightarrow \text{TNT}^{\bullet} + \text{H}_2\text{O} \tag{77}$$

$$\text{TNT}^{\bullet} + \text{O}_2 \longrightarrow \text{TNT-O}_2^{\bullet} \tag{78}$$

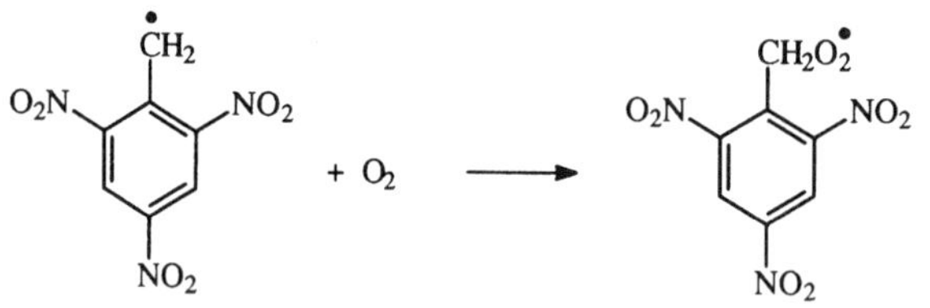

The peroxy radical intermediate (i.e., the oxygen adduct) shown in eq. (78) is further oxidized to 2,4,6-trinitrobenzoic acid, TNBA, by hydroxyl radical as follows:

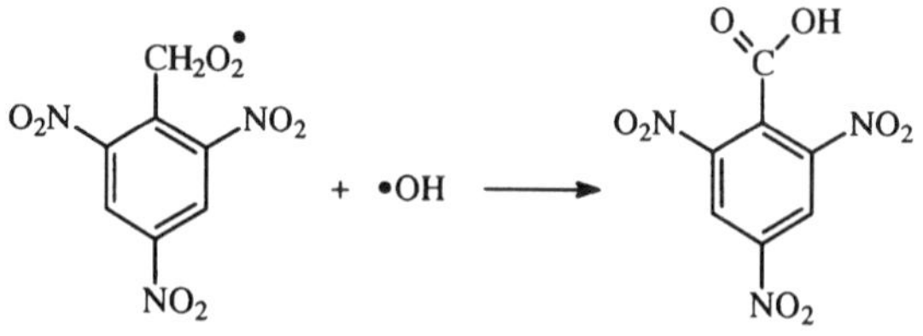

$$\text{TNT-O}_2^{\bullet} + {}^{\bullet}\text{OH} \longrightarrow \text{TNBA} \tag{79}$$

10.3. SONOLYTIC REACTIONS IN WATER

Subsequent pyrolytic decarboxylation of TNBA (eq. (80)) yields trinitrobenzene, which was identified by ES/MS.

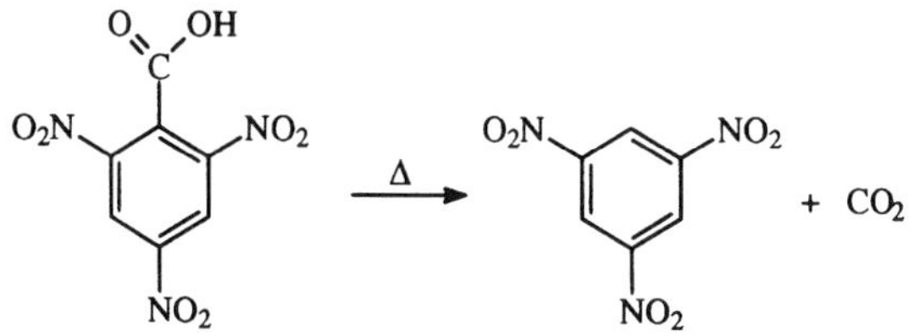

$$\text{TNBA} \xrightarrow{\Delta} \text{trinitrobenzene} + CO_2 \tag{80}$$

Hydroxyl radical addition to the phenyl ring leads to various hydroxylated TNT derivatives. For example, the following addition product (eq. (81)) is a likely intermediate:

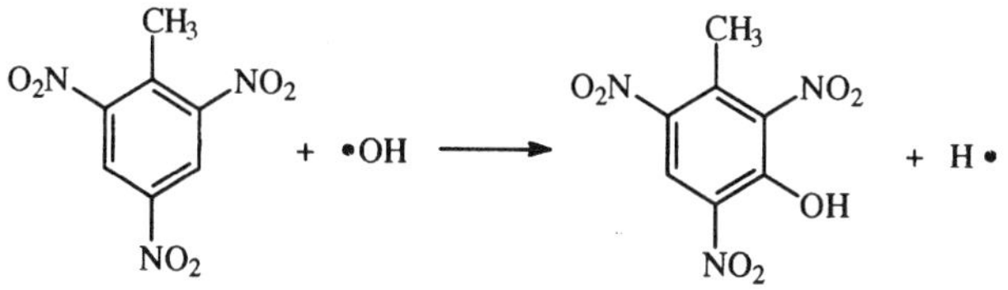

$$\text{TNT} + {}^\bullet OH \longrightarrow \text{hydroxylated intermediate} + H^\bullet \tag{81}$$

In addition to hydroxyl radical addition to the substituted phenyl ring, denitration via pyrolytic cleavage of the C–N bond to release NO (eq. (82)) to solution is also a viable pathway analogous to the denitration of PNP.

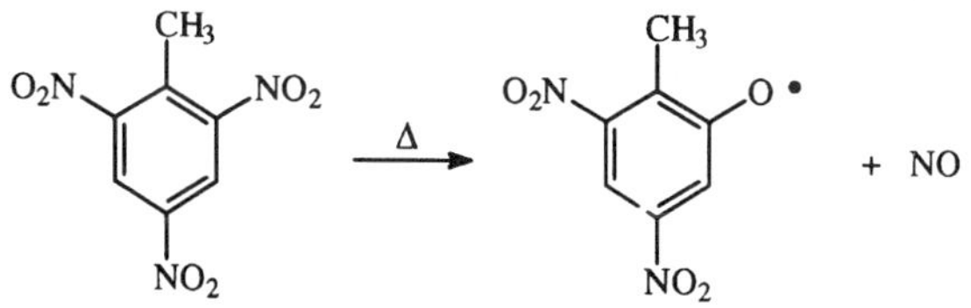

$$\text{TNT} \xrightarrow{\Delta} \text{aryloxy radical} + NO \tag{82}$$

As in the case of PNP denitration, NO is oxidized further to yield nitrite and nitrate. The overall reaction stoichiometry for the autooxidation of TNT is shown in eq. (83):

$$C_7H_5(NO_2)_3 + 9\,O_2 \xrightarrow{))))} 7CO_2 + H_2O + 3H^+ + 3NO_3^- \tag{83}$$

In order to compare the efficiencies for different sonochemical reactor configurations, Höchemer and Hoffmann [120] have introduced the concept of a "G" value for sonochemistry in a fashion similar to those employed in radiation chemistry. The "G" value in this particular case (eq. (84)) is defined as

$$G \equiv \left(\frac{N_A \cdot V}{P}\right) \cdot \frac{d[TNT]}{dt} \qquad (84)$$

where N_A = Avogadro's number, the reaction volume, V, and the power input, P. The derivative represents the initial rate of disappearance of TNT. G values range from $1.25-332 \times 10^{14}$ molecules kJ^{-1} depending on the specific reactor configuration, saturating gas, and frequency of irradiation. The calculations are carried out for P = power taken directly from the power outlet rather than for P that is actually delivered to the reaction solution. Based on these considerations, TNT degradation appears to be more efficient at 500 kHz than at 20 kHz. The apparent frequency dependence of the observed rate of degradation of TNT may be the result of several factors. For example, the power intensity, P_A, of the 500 kHz ortho reactor (17 W cm^{-2}) is substantially less than that of the 20 kHz probe reactor (PA = 83 W cm^{-2}). If a certain intensity threshold is reached, further increases of P_A lead to a decrease in efficiency due to decoupling of the solution from the transmitter. The 20 kHz probe reactor has a relatively high power intensity and correspondingly a lower "G" value caused by an apparent decoupling of the energy transmission between the probe tip and the solution. In addition, the 500 kHz ultrasonic reactor has a higher production rate for radicals, which initiate the degradation of TNT as shown in Figure 10.7. Thus, the sonochemical degradation of TNT dissolved in water at 500 kHz is more efficient than irradiation at 20 kHz by more than one order of magnitude, due to the more effective energy coupling at low power intensities and the higher rate of hydroxyl radical production.

Unlike the parallel-plate reactor utilizing 2 different frequencies (*vide infra*), a synergistic effect between 20 and 500 kHz irradiators positioned orthogonally to one-another is not observed. This effect may be due to negative interference of the sound waves, which will diminish the amount of energy that is available for inducing cavitation. In designing an ultrasonic reactor, care has to be taken in order to avoid a possible diminishing return in chemical effects when the power input is increased.

10.3.5 Sonochemical Autoxidation of Iodide

The reaction of iodide ion with molecular oxygen in aqueous solution as catalyzed by ultrasonic irradiation has been the focus of much research over the years, because it provides a dramatic colorimetric example of the chemical effects of ultrasound and because this reaction serves as a useful dosimeter for the measurement of the flux of hydroxyl emanating from collapsing bubbles into the bulk aqueous solution [47, 51, 77, 87, 118, 121-123].

10.3. SONOLYTIC REACTIONS IN WATER

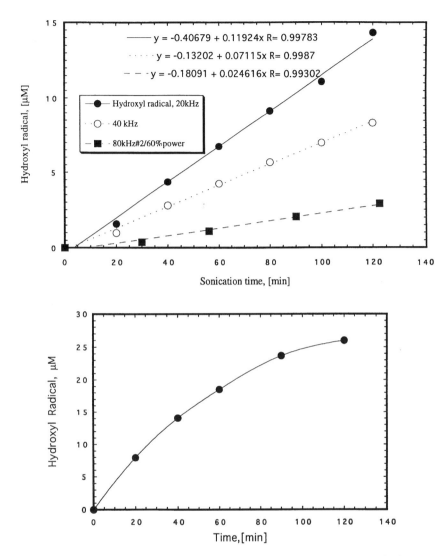

Figure 10.7 Hydroxyl radical production as a function of frequency over the frequency range of 20 to 500 kHz. Hydroxyl radical was detected using the selective trapping reagent, terephthalic acid (benzene-1,4-dicarboxylic acid) to form the fluorescent hydroxy-adduct, 2-hydroxybenzene-1,4-dicarboxylic acid, which is excited at 345 nm and fluoresces at 425 nm. Detection limit 1 µM.

The stoichiometric oxidation of iodide with hydroxyl radical (eq. (85))

$$3I^- + 2 \cdot OH \longrightarrow I_3^- + 2OH^- \tag{85}$$

generated within collapsing cavitation bubbles takes place according to the following mechanism (eqs. (86)–(88)) [51]:

$$I^- + \cdot OH \longrightarrow \cdot I + OH^- \tag{86}$$

$$I^- + \cdot I \longrightarrow I_2^{-\cdot} \tag{87}$$

$$I_2^{-\cdot} + \cdot I \longrightarrow I_3^- \tag{88}$$

However, secondary reactions during sonolyis in water will result in the production of hydrogen peroxide [124] that, in turn, will rapidly oxidize iodide as in eqs. (89)–(95):

$$H_2O \xrightarrow{\Delta} H\cdot + \cdot OH \tag{89}$$

$$H\cdot + O_2 \longrightarrow HO_2 \cdot \tag{90}$$

$$2 HO_2 \cdot \longrightarrow H_2O_2 + O_2 \tag{91}$$

$$2 HO \cdot \longrightarrow H_2O_2 \tag{92}$$

$$I^- + H_2O_2 + H^+ \longrightarrow IOH + H_2O \tag{93}$$

$$I^- + IOH + H^+ \longrightarrow I_2 + H_2O \tag{94}$$

$$I^- + I_2 \underset{}{\overset{K}{\rightleftharpoons}} I_3^- \tag{95}$$

The rate of oxidation of iodide is observed to be pseudo zero-order (as measured by the linear production of triiodide ion) over extended periods of time. The rate of oxidation is fastest in the presence of pure O_2 (due primarily to the trapping of H· and the formation of H_2O_2), next fastest in the presence of argon with the following order $k(O_2) > k(Ar) > k(air) \gg k(N_2)$ [51].

10.3.6 Other Sonochemical Reactions of Interest in Water

Petrier et al. [57] have studied the sonochemical degradation of pentachlorophenol (PCP) at 530 kHz in aqueous solutions saturated with air, oxygen, and argon. Pentachlorophenol, which is commonly used as a wood preservative, it is often found as a groundwater contaminant. They reported that there was a fast sonochemical cleavage of carbon-chlorine bonds leading to Cl^- production and resulting in complete mineralization of PCP to CO_2, when the solution was saturated with air or oxygen. However, when Ar was used as the saturating gas, CO was produced. They were also able to show that the sonochemical treatment of water contaminated with PCP resulted in a decrease in the toxicity of the medium toward the green algae *Scenedesmus subspicatus*.

In a follow-up study, Petrier [56] and co-workers investigated the sonochemical destruction of phenol in water at two different frequencies, 20 and 487 kHz.

Using the same acoustical power (30 W), as determined by a calorimetric method (*vide supra*), they found that the higher frequency resulted in faster zero-order rates of degradation. The initial zero-order rates were found to be dependent on the initial phenol concentration (*i.e.*, a Langmuirian saturation phenomenon related to a limited number of sites on bubble surfaces), reaching saturation values $k(20 \text{ kHz}) = 1.84 \times 10^{-6}$ M min^{-1}, and $k(487 \text{ kHz}) = 11.6 \times 10^{-6}$ M min^{-1}. They argued that the observed intermediates of the reaction (hydroquinone, catechol, benzoquinone) indicate that the hydroxyl radical was the principal oxidant. They also showed that the rate of H$_2$O$_2$ formation was faster at 487 kHz ($k = 4.9 \times 10^{-6}$ M min^{-1}) than at 20 kHz ($k = 0.75 \times 10^{-6}$ M min^{-1}).

Serpone *et al.* [59] studied several different sonochemical reactions in water. They reported that the rate of disappearance of phenols followed zero-order kinetics at initial phenol concentrations in the range of 30 to 70 μM. They observed the formation of catechol (CC), hydroquinone (HQ), and *p*-benzoquinone (BQ) as intermediates at pH 3, while at pH 5.6, they found only catechol and hydroquinone. No intermediate species were detected at pH 12. They also investigated the sonochemical fate of CC, HQ, and BQ at pH 3 and found that BQ is the major species formed from HQ, while HQ is produced during the sonolysis of BQ. In both cases, trace levels of hydroxy-*p*-benzoquinone are formed. Serpone *et al.* argued that their results confirmed the important role of ·OH radicals in sonochemical degradation processes in water. They also noted that sonolysis of phenol resulted in only minor losses of total organic carbon (TOC) after several hours, even though the aromatic substrate and the intermediates had degraded. They presented a mechanism, in which benzoquinone reacted with ·OH and H· radicals at the 'hydrophobic' gas bubble/liquid interface, while the hydrophilic species (phenol, CC, and HQ) reacted, to a greater extent, with the ·OH radicals in the bulk aqueous phase.

In a similar study, Serpone *et al.* [58] examined the sonochemical degradation of a series of chlorophenols, 2-chlorophenol (2-CPOH), 3-chlorophenol (3-CPOH), and 4-chlorophenol (4-CPOH), under pulsed sonolytic conditions (*i.e.*, 20 kHz and 30 W) in air-saturated water solutions. They found that this series of chlorophenols was completely transformed to dechlorinated and hydroxylated intermediate products *via* first-order kinetics after about 10 h of irradiation for 2-CPOH and 3-CPOH and about 15 h of irradiation for 4-CPOH. They reported first-order rate constants of 4.8×10^{-3} min^{-1}, 4.4×10^{-3} min^{-1}, and 3.3×10^{-3} min^{-1}, respectively, when the initial concentrations were 80 μM. They also reported that the sonochemical kinetics showed two distinct regimes as noted by others: a low-concentration regime where the degradation rate was zero order in [CPOH], and a second regime at higher concentrations where the rate displayed saturation-type kinetics. They argued that these results indicated that the reaction occurred primarily in the bulk aqueous phase at low concentrations of chlorophenol, while at the higher concentrations the reaction occurred predominantly at the gas bubble/liquid interface.

Other sonochemical studies in water have been focused on the oxidation of 5-hydroxymethylfurfural [125], the oxidation of homoallylic alcohols [126], the oxidation of naturally-occurring organic matter [55], the oxidation of phenyl acetate [14], the hydrolysis of nitriles [127], the hydrolysis of nitrophenyl esters [128], the sonochemical reaction between water and 3-chloropropionitrile [129], the reduction of 2-nitroalkanols [130], the degradation of methanol [39, 131], the reduction of ferric hexacyanoferrate [132], the inactivation of enzymes [133], and on the oxidative dissolution of $RuO_2 \times H_2O$ by bromate ions [134].

Henglein and co-workers [39] examined the sonolysis of methanol-water mixtures with 1.0 MHz ultrasound under argon and oxygen. They found typical pyrolysis and combustion products of methanol (H_2, CH_2O, CO, CH_4, C_2H_4 and C_2H_6 under argon and CO_2, CO, $HCOOH$, CH_2O, H_2O_2 under oxygen). The relative yields of products changed with the composition of the methanol-water mixture. In a 10 vol-% methanol solution, the chemical yields were much higher than those in pure water. In solutions containing more than 80 vol-% methanol, almost no chemical reactions occurred. Their results were interpreted in terms of the changes in the temperature of adiabatically compressed cavitation bubbles with the changing composition of the methanol-water mixture.

Riesz and co-workers [86, 102, 105, 106, 135–141] have been examining the potential biological effects of ultrasound. In this regard, they investigated the sonochemical oxidation of dimethyl sulfoxide (DMSO) in water [102]. Irradiation at 50 kHz of argon-saturated DMSO-water mixtures was investigated by ESR and by spin trapping with 3,5-dibromo-4-nitrosobenzenesulfonate (DBNBS). At low DMSO concentrations, methyl radicals are formed due to the reaction with hydroxyl radicals. They also observed that SO_2 was produced *via* the pyrolysis of DMSO in the vapor phase. Release of sulfur dioxide led to the production of the sulfite radical, SO_3^-, due to reaction of aquated sulfur dioxide with a hydroxyl radical.

10.4 CHEMICAL REACTIONS INDUCED BY EHD PLASMAS

Pulsed-power discharges into water lead to rapid chemical transformations for a wide variety of compounds [143]. For example, 35% conversion of 4-chlorophenol is obtained with 120 electrohydraulic discharges of 7 kJ (equal to a total power-utilization time of 3 ms). Chloride ion production is stoichiometric. At the end of the experiment 41% of the degraded 4-chlorophenol could be accounted for in the form of *p*-benzoquinone.

These reaction product distributions are consistent with direct photolysis as the primary mechanism of degradation of the 4-chlorophenol (*i.e.*, they are similar to the product distributions reported in the literature for aqueous-phase 4-chlorophenol photolysis [142]). Stoichiometric production of chloride ions is consistent with the photochemical mechanism of 4-chlorophenol which involves cleavage of the C–Cl bond in the first step. Furthermore, the primary photolytic

10.4. CHEMICAL REACTIONS INDUCED BY EHD PLASMAS

degradation product of 4-chlorophenol is p-benzoquinone, with hydroquinone as a secondary product.

It is possible that the shockwaves produced during an EHD discharge could cause cavitational effects similar to those observed in aqueous-phase sonochemistry. Sonolysis of an aqueous solution of 4-chlorophenol is known to oxidize 4-chlorophenol to hydroquinone, 4-chlororesorcinol and 4-chlorocatechol [58]. The sonolytic oxidation of 4-chlorophenol proceeds *via* reactions with hydroxyl radicals rather than by pyrolysis. At a concentration of 200 µM, if electrohydraulic cavitation is occurring, 4-chlorophenol should react with hydroxyl radicals in the hydrophobic bubble interfacial regions. This possibility was tested by running the discharge system in the presence of 20 µM t-butanol, which is a well-known hydroxyl radical trap. The t-butanol showed no significant effect on the rate or extent of 4-chlorophenol degradation. Furthermore, the products of the sonolytic degradation of 4-chlorophenol are not the same as seen during EHD induced degradation of 4-chlorophenol. During sonolysis, the hydroquinone precursor to the quinone was observed as the primary degradation product, and 4-chlororesorcinol and 4-chlorocatechol were also observed [58]. Thus, we conclude that cavitation induced during electrohydraulic discharges does not result in discernible 4-chlorophenol degradation.

The degradation rates of 3,4-dichloroaniline and TNT using the EHD reactor have also been investigated [143]. The initial rates of disappearance of these compounds are also pseudo zero-order. Figure 10.8 shows the experimental results for the 3.5 L experiments on 4-chlorophenol, 3,4-dichloroaniline, and TNT; the values of k_A are plotted against C_i. The determined values of k_0 and k_1 for these compounds are also listed in Table 10.4.

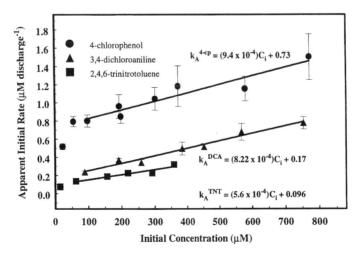

Figure 10.8 Apparent initial rate constants (k_A) *vs.* the initial concentration (C_i) for 4-CP, 3,4-dichloroaniline, and TNT.

Table 10.4 Experimentally determined constants for EHD plasmas.

Experiment	k_0^* (μM)	k_1^* (10^{-4})	V_R (mL)	k (M s^{-1})	G^* (10^{-4})	Φ
4-CP (3.5 L, 7 kJ)	0.73	9.40	3.5	0.022	44.5	0.25
4-CP (1.0 L, 7 kJ)	3.25	24.3	2.4	0.073	51.6	0.25
4-CP (1.0 L, 4 kJ)	2.50	12	1.2	0.054	38	0.25
DCA	0.17	8.2	2.8	0.008	16	0.044
TNT	0.10	5.6	1.9	0.005	10	0.0021
TNT + O_3				0.024	48.3	

These values are reported for a representative 200 μM solution of the substrate with the exception for TNT/Ozone which is at 160 μM. 4-CP = 4-chlorophenol; DCA = 3,4-dichloroaniline; TNT = 2,4,6-trinitrotoluene.

* The units are expressed in terms of per discharge; each discharge has a duration of approximately 40 μs.

The photolysis of ozone produces hydroxyl radicals, which can efficiently oxidize refractory organics which may not be susceptible to direct photolysis. TNT was examined because it is refractory to both direct photolysis and to dark reactions with ozone.

The combined EHD/ozone process was able to completely degrade a 160 μM TNT solution to less than detection limits (<1 μM) over the duration of a 260 discharge experiment (Figure 10.9). The total organic carbon content of the solution was decreased by 34% from 18 ppm to 12 ppm. No organic intermediates were detected by HPLC.

These experimental results show that the EHD reactor operates primarily *via* UV-induced photochemical processes, with a significant secondary contribution from other effects within the plasma channel. The values of k_0 for a compound should therefore scale with its quantum efficiency for direct photolysis. The values for all three compounds are listed in Table 10.4, along with their reported quantum efficiencies for direct photolysis (Φ_r).

As long as all the UV is absorbed by the solution, the total number of moles of substrate converted should depend only on the UV flux emitted from the plasma channel and be independent of the volume of the chamber. Therefore, the value of k_0 should scale inversely with the volume of the chamber. The ratio of $k_0(1.0 \text{ L})/k_0(3.5 \text{ L}) = 3.4$ is in excellent agreement with the simple inverse ratio of the volumes (3.5).

In essence the EHD reactor can be considered to be a high energy analog of a pulsed Xenon flashlamp. Both systems are pulsed sources of radiation and both predominantly emit blackbody spectra. The blackbody temperature, and

10.4. CHEMICAL REACTIONS INDUCED BY EHD PLASMAS

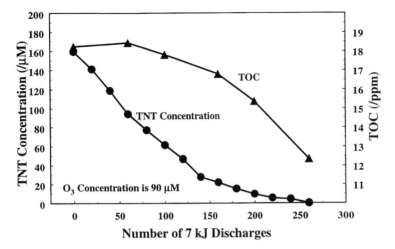

Figure 10.9 TNT concentration and TOC as a function of the number of 7 kJ discharges, where each discharge lasts for approximately 40 µs in the presence of ozone.

the percentage of radiation that is in the 185 to 300 nm region, increase dramatically with the energy of the discharge in a flashlamp. However, the lifetime of flashlamps dramatically decreases with increasing pulse energy; the shockwave generated in the fill gas can damage the quartz envelope of the lamp. Typical operating temperatures for xenon flashlamps are 6000–9000 K.

Since there is no quartz envelope in an EHD reactor, the pulses can be operated at much higher energies and hence reach much higher temperatures. The peak blackbody temperature can be 20,000–50,000 K for the plasma channel of an EHD [113]. The peak temperature and the temperature profile of the pulse are highly dependent on electrical parameters such as initial voltage, spark-gap length (resistance), capacitance, and inductance.

An estimate of the average plasma channel temperature can be made using eq. (96), the observed k_0 values for 4-chlorophenol and 3,4-dichloroaniline, and the average current pulse duration ($\tau = 41$ µs);

$$k_0 = \Phi_r Q(\lambda) = \Phi_r \langle M(\lambda, T) \rangle A \tau \tag{96}$$

$M(\lambda, T)$ is the blackbody emission of the plasma channel (Planck distribution function), and A is the estimated surface area of the plasma channel. The temperature T can be estimated by using this equation to determine the correct $M(\lambda, T)$. In our experiments, the average temperature over the 40 µs period of discharge in the plasma channel is $\approx$ 13,000–15,000 K. The peak temperature of the pulse would actually be at a much higher temperature (Figure 10.10).

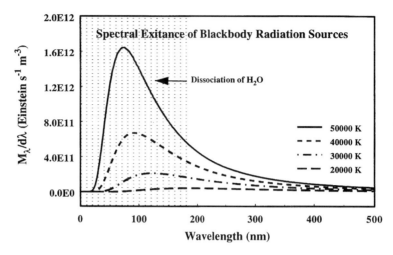

Figure 10.10 Black-body spectra obtained for different plasma channel temperatures. The wavelength cutoff ($\lambda = 185$ nm) for the direct photolysis of water into H· and ·OH is shown.

To determine the energy efficiency of the EHD process the G-values for conversion of all three compounds have been calculated and are listed in Table 10.4. The G-value, in this case, is defined as the number of substrate molecules converted per 100 eV of energy stored in the capacitor bank. If the G-value is defined according to the energy dissipated during the actual EHD discharge then the values would even be higher. The listed G-values have been calculated for representative 200 μM solutions and include both the contributions of the direct photolysis and the plasma channel (k_0 and k_1). It should be reiterated that at present the EHD reactor and power supply have not been optimized for maximum energy efficiency; therefore, the calculated G-values are apparatus specific and are not intrinsic efficiencies.

The N-based rate constants k_A, k_0, and k_1 can be converted to conventional intrinsic rate constants, which are based on real-time. Given that the average current pulse lasts 41 or 51 μs (depending on the reactor configuration), the values of the intrinsic rate constants can be readily computed and are listed in Table 10.4. For example, the zero-order rate constant for TNT/O_3 degradation is 0.024 M s^{-1}. Given these relatively rapid rates of degradation, the overall rate-determining factor for the treatment of a given volume of solution will be the repetition rate of the electronic circuit. The current repetition rate is approximately six discharges per minute. However, rates of 5–10 Hz are easily achievable using our current technology. At a 10 Hz repetition rate, the total treatment time for the degradation of the 3.5 L of 160 μM TNT would be approximately 30 seconds. Current chemical and biochemical technologies require hours to days to achieve the same degree of treatment.

10.5 REACTOR DESIGN CONSIDERATIONS

A wide variety of reactor configurations have been utilized to investigate the chemical effects of ultrasonic irradiation. Most commonly used reactor types are the conventional probe-type reactors and cup-horn reactors [7, 31].

An interesting alternative reactor configuration utilizes a near-field acoustical processor (NAP) (Lewis Corporation, Oxford, CT) in which the vibrational energy is transferred to the reaction mixture *via* two parallel stainless steel plates which are bolted together. Each plate is vibrationally driven by seven magnetostrictive transducers. The transducers convert the electrical energy which is delivered by separate power generators, operating at 16 and 20 kHz, respectively, into the mechanical energy of vibration.

A stainless steel spacer is placed between the two vibrating plates, in order to define the reaction volume, V_S. The seal between spacer and plates is provided by a Gore-tex gasket. Power input into the two frequency generating units can be varied from 0 to 1775 W. Due to energy transformation losses within the system, the acoustical power transferred to the reaction mixture is less than the overall power put into the power generators (*i.e.*, the power drawn from the wall supply).

Since the NAP system does not allow for direct cooling of the plates and transducers, the reaction mixture flowing through the reactor itself is thermostatted externally. This reactor configuration involves the continuous flow of the reaction solution from the reservoir through the NAP and back into the reservoir. The NAP parallel-plate reactor allows for a lower sound intensity but a higher acoustical power per unit volume than conventional probe-type reactors.

Hua *et al.* [54] investigated the sonolytic degradation of *p*-nitrophenol in a NAP, which involved a flow-through reactor and a well-mixed reservoir. They reported that the pseudo first-order rate constant, k, for PNP degradation increased proportionally from 1.00×10^{-4} to 7.94×10^{-4} s^{-1} with increasing power-to-solution-volume density over the range of 0.98 to 7.27 W cm^{-3}. An increase in the power-to-area density (*i.e.*, sound intensity or power-to-vibrating surface area) resulted in an increase in k up to a maximum value of 8.60×10^{-4} s^{-1} at a sound intensity of 1.2 W cm^{-2}.

A mathematical model for a continuous-flow loop (*i.e.*, reservoir) reactor configuration is required in order to extract k from the experimentally observed rate constant, k_{obs}, which is a function of the relative volumes of reactor and reservoir. The nature of the cavitating gas (Ar, O_2) was found to influence the overall degradation rate and the resulting product distribution. The rate constant for PNP degradation in the presence of pure O_2, $k_{O_2} = 5.19 \times 10^{-4}$ s^{-1} was lower than in the presence of pure Ar, $k_{Ar} = 7.94 \times 10^{-4}$ s^{-1}. A 4:1 (v/v) Ar/O_2 mixture yielded the highest degradation rates, $k_{Ar/O_2} = 1.20 \times 10^{-3}$ s^{-1}. Results of these experiments demonstrated the potential application of large-scale, high power ultrasound to the remediation of hazardous compounds present at low concentrations.

An often-used reactor configuration in sonochemistry couples an irradiation chamber with an external reservoir in order to treat larger volumes over extended periods of time. However, when utilizing such a system (*e.g.*, the Harwell Sonochemical Reactor [143–145] or the NAP [54]) care must be taken in interpreting the kinetic results.

We can model either the Harwell Reactor or the NAP as either an ideal continuous-flow stirred-tank reactor (CSTR model) or as an ideal plug-flow reactor (PFR model) and compare the results. The mass balance equation in the CSTR for the reservoir is given in eq. (97)

$$\frac{dC_R(t)}{dt} = \frac{Q}{V_R}[C_S(t) - C_R(t)] \tag{97}$$

where $C_R(t)$ and $C_S(t)$ are the concentrations of PNP in the reservoir and in the reactor, respectively, and Q is the volumetric flow rate of the reaction mixture. Assuming ideal CSTR-behavior, the mass balance equation (eq. (98)) for the reactor is given as

$$\frac{dC_R(t)}{dt} = \frac{Q}{V_R}[C_R(t) - C_S(t)] + vr(k, C_S) \tag{98}$$

where V_S is the sonicated volume in the NAP, $v = -1$ is the stoichiometric coefficient for the reactant PNP, and $r(k, C_S)$ is the first order reaction rate given in eq. (99)

$$r(k, C_S) = k\, C_S(t) \tag{99}$$

where k is the intrinsic first-order rate constant for the rate-limiting step in the degradation reaction in eq. (100):

$$HOC_6H_4NO_2 \xrightarrow{k,\Delta} HOC_6H_4O\cdot + NO \tag{100}$$

The initial conditions for the above equations are given as $C_R(t=0) = C_S(t=0) = C_0$.

The system of equations can be made non-dimensional as follows in eq. (101):

$$\tau \equiv kt, \quad \gamma \equiv \frac{V_S}{V_R}, \quad g \equiv \frac{C_R}{C_0}, \quad f \equiv \frac{C_S}{C_0}, \quad \alpha \equiv \frac{1}{D_a} = \frac{Q}{kV_S} \tag{101}$$

where D_a is the Damkoehler number. D_a represents the ratio of the mean hydrodynamic residence time, V_S/Q, over the characteristic time constant of the reaction, k^{-1}, under the NAP entrance conditions. With these non-dimensional terms eqs. (97) and (98) can be written in dimensionless form as in eqs. (102) and (103):

$$\dot{g}(\tau) = \gamma[f(\tau) - g(\tau)] \tag{102}$$

$$\dot{f}(\tau) = \alpha\, g(\tau) - (1+\alpha)f(\tau) \tag{103}$$

10.5. REACTOR DESIGN CONSIDERATIONS

The dimensionless initial conditions are thereby given as $g(\tau = 0) = f(\tau = 0) = 1$, with the solution for g (eq. (104)):

$$g(\tau) = \frac{1}{\lambda_2 - \lambda_1}[\lambda_2 \exp(\lambda_1) - \lambda_1 \exp(\lambda_2)] \quad (104)$$

and with λ_i given by eqs. (105) and (106):

$$\lambda_1 = -\frac{\tau}{2}[1 + \alpha(1 + \gamma) - \sqrt{\alpha^2(1 + \gamma)^2 + 2\alpha(1 - \gamma) + 1}] \quad (105)$$

$$\lambda_2 = -\frac{\tau}{2}[1 + \alpha(1 + \gamma) - \sqrt{\alpha^2(1 + \gamma)^2 + 2\alpha(1 - \gamma) + 1}] \quad (106)$$

The actual operation conditions for volumetric flow rate, reaction rate constants and volume of NAP and reservoir allow us to simplify eq. (104). For $\tau \geq 10/\alpha$ and $\alpha \gg 1$ (i.e., $D_a \ll 1$), g (i.e., the dimensionless PNP concentration in the reservoir) can be written in terms of γ (eq. (107)):

$$g(\tau) = \exp\left\{-\left(\frac{\gamma}{1 + \gamma}\right)\tau\right\} \quad (107)$$

In dimensional form eq. (107) is given as eq. (108)

$$\frac{C_R(t)}{C_0} = \exp\left[-\left(\frac{V_S}{V_S + V_R}\right)kt\right] \quad (108)$$

where the observed first-order rate constant for PNP degradation, k_{obs}, is given by eq. (109):

$$k_{obs} = \frac{V_S}{V_S + V_R}k \quad (109)$$

Eq. (109) relates the observed rate constant, k_{obs}, to the intrinsic rate constant in for eq. (100), k.

The other extreme condition is to treat the reactor-reservoir system as an ideal plug-flow reactor. The corresponding mass balance for the reactor is given by the following partial differential equation (eq. (110))

$$\left(\frac{\partial C_S(V, t)}{\partial t}\right)_{V=\text{const.}} = -Q\left(\frac{\partial C_S(V, t)}{\partial t}\right)_{t=\text{const.}} + vr(V, t) \quad (110)$$

and the mass balance equation for the reservoir is given by eq. (97).

Using the same dimensionless quantities and the definitions in eq. (111)

$$f_\tau(x, \tau) \equiv \left(\frac{\partial f(x, \tau)}{\partial \tau}\right)_{x = \text{const.}} \quad \text{and} \quad f_x(x, \tau) \equiv \left(\frac{\partial f(x, \tau)}{\partial x}\right)_{\tau = \text{const.}} \quad (111)$$

the mass balances for the reservoir and the NAP, can be written in dimensionless form as

$$\dot{g}(\tau) = \gamma[f(\tau) - g(\tau)] \quad (102)$$

$$f_x(x, \tau) + D_a f_\tau(x, \tau) = -D_a f(x, \tau) \quad (112)$$

with initial conditions $g(\tau = 0) = f(\tau = 0) = 1$, and the boundary condition of $f(x, \tau) = g(\tau)$. Employing the method of characteristics, the differential equations can be solved analytically to obtain an expression for g,

$$g(\tau) = e^{-\lambda \tau} \quad (113)$$

where

$$\lambda = 1 + \frac{1}{D_a} \ln\left(1 + D_a \frac{\lambda}{\gamma}\right) \quad (114)$$

For $D_a(\lambda/\gamma) \ll 1$, eq. (114) can be expanded in a MacLaurin series yielding a series expression for λ

$$\lambda = \frac{\gamma}{1 + \gamma} + O(D_a) \quad (115)$$

Eq. (113) can be written in dimensional form (eq. (116)) by substituting eq. (115) for λ,

$$C_R(t) = C_0 \exp\left[-\left(\frac{V_S}{V_S + V_R}\right) k t\right] \quad (116)$$

The relation between the intrinsic rate constant, k, and the observed rate constant, k_{obs}, is found to be given as (eq. (117)):

$$k_{obs} = \frac{V_S}{V_S + V_R} k \quad (117)$$

which is identical to the solution obtained when assuming that the system was an ideal CSTR. Thus, the final mathematical relationship between k_{obs} and k under the given conditions for D_a and γ is independent of the basic assumption as to reactor type. Therefore, the interpretation of kinetic results for these reactor configurations is possible without incorporating a residence time distribution within the reactor system.

10.6. CONCLUSIONS

The above mathematical analysis of the reactor system shows that for small D_a (i.e., small conversion per pass), the residence time distribution of the reactor system has no influence on the observed rate constants. This result justifies the use of a simple volume correction factor in determining k from k_{obs}. Increasing the rate constant and/or the reactor volume, and decreasing the volumetric flow rate would result in an increased Damkoehler number. Under these conditions, the residence time distribution of the NAP has to be known in order to extract k from k_{obs}.

10.6 CONCLUSIONS

In addition to the standard application of ultrasonic irradiation for industrial cleaning and emulsification, the application of electrohydraulic cavitation in its various forms for the pyrolytic and oxidative control of hazardous chemicals in water has the potential to become economically-competitive with existing technologies. In terms of convenience and simplicity of operation, electrohydraulic cavitation could prove to be far superior to many alternative approaches. For example, the relative efficiency of ultrasound in terms of the total power consumed per mole of *p*-nitrophenol degraded per liter of water is far superior to conventional UV-photolytic degradation (*e.g.*, using a CW Xenon lamp).

As noted above, the sonochemical degradation of a variety of chemical contaminants in aqueous solution has been investigated. Substrates such as chlorinated hydrocarbons, pesticides, phenols, and esters are transformed into short-chain organic acids, CO_2, and inorganic ions as the final products. Time scales of treatment in simple batch reactors are reported to range from minutes to hours for complete degradation. For example, the total elimination of H_2S at pH 7 was achieved within 10 minutes while parathion required 120 min to achieve complete degradation. Ultrasonic irradiation appears to be an effective method for the destruction of organic contaminants in water because of localized high concentrations of oxidizing species such as hydroxyl radical and hydrogen peroxide in solution, high localized temperatures and pressures, and the formation of transient supercritical water.

We have demonstrated that the degradation of chemical compounds by electrohydraulic cavitation involves three distinct pathways. The pathways include oxidation by hydroxyl radicals, pyrolytic decomposition and supercritical water oxidation. We have shown that transient supercritical water is obtained during the collapse of cavitation bubbles generated sonolytically.

Optimization of aqueous-phase organic compound degradation rates within acoustical processors can be achieved by adjusting the energy density, the energy intensity, and the nature and properties of the saturating gas in solution. Observed first-order degradation rate constants increase as the energy density and intensity are increased up to a saturation value. Manipulation of these macroscopic parameters should lead to enhancement of the cavitation chemistry as the number of cavitation bubbles and the chemical events at each bubble are varied. The specific nature of the saturating gas influences the relative proportion of pyrolytic or free-radical reaction steps. Simultaneous acceleration of these pathways results in the maximum rate observed in solutions saturated with Ar/O_2 or $Ar/O_2/O_3$ mixtures.

Mathematical analysis of combined Reactor/Reservoir systems, that are frequently used in sonochemical applications, shows that for small D_a (i.e., small conversion per pass in a PFR), the residence time distribution function of the NAP or Harwell Reactors has no influence on the observed rate constants. To operate an acoustical processor as a steady-state continuous-flow reactor, a higher conversion per pass is needed by either decreasing the volumetric flow rate or increasing the reactor volume.

The pulsed-power plasma discharge process is a technology which has the potential for the cost-effective treatment of contaminated wastewaters. This technology is an adaptation of an existing technology which has been used in a variety of other applications such as in the manufacture of beer cans and automobile springs and in the destruction of kidney stones *via* focused high-pressure shock waves. Pulsed-plasma discharge systems have also been used to knock barnacles off marine structures, to form high strength metal alloys and to pressure-test cannon bores.

An electrohydraulic discharge (EHD) reactor represents a novel and promising approach to the treatment of hazardous chemical wastes in water. For example, experimental results have shown more than 80% of 3,4-dichloroaniline can be degraded over only 2 ms of power utilization time. In the case of 4-chlorophenol, 40% degradation is achieved in 2.4 ms with a series of 120 pulses at 5 kJ per discharge; the corresponding G-value (i.e., the number of molecules converted per 100 eV of input energy) based on the wall voltage utilization is 2.5×10^{-3}. This value is very favorable when compared with other electrodynamic treatment technologies (e.g., electron beam reactors) that are currently being evaluated. The EHD system has the advantage of extremely fast degradation rates for hazardous wastes where the macroscopic rate of degradation is controlled primarily by the repetition rate of the electronic circuit. At higher repetition rates, residence times in a reactor optimized for industrial applications could be on the order of seconds.

ACKNOWLEDGEMENTS

Financial support for much of the research described in this chapter was provided by the Advanced Research Projects Agency, ARPA (Grant #NAV 5HFMN N0001492J1901), through the Office of Naval Research, ONR, and by the Electric Power Research Institute, EPRI (Grant #RP 8003-37). This support is gratefully acknowledged.

10.7 REFERENCES

[1] Berlan, J., Mason, T. J., Ultrasonics, 30 (1992) 203.
[2] Lindley, J., Mason, T. J., Chem. Soc. Rev., 16 (1987) 275.
[3] Lindley, J., Mason, T. J., Lorimer, J. P., Ultrasonics, 25 (1987) 45.
[4] Lorimer, J. P., Mason, T. J., Chem. Soc. Rev., 16 (1987) 239.
[5] Lorimer, J. P., Mason, T. J., Fiddy, K., Ultrasonics, 29 (1991) 338.
[6] Mason, T. J., Chem. Brit., 22 (1986) 661.
[7] Mason, T., Lorimer, J., Sonochemistry: Theory, Applications, and Uses of Ultrasound in Chemistry, Ellis Norwood, Ltd., New York, 1988.
[8] Mason, T. J., Lorimer, J. P., Endeavour, 13 (1989) 123.
[9] Mason, T. J., Lorimer, J. P., Paniwnyk, L., Harris, A. R., Wright, P. W., Syn. Comm., 20 (1990) 3411.
[10] Mason, T. J. (ed.), Advances in Sonochemistry, JAI Press Ltd., London, 1990, Vol. 1.
[11] Mason, T. J., Ultrasonics, 30 (1992) 192.
[12] Mason, T. J., Lorimer, J. P., Bates, D. M., Ultrasonics, 30 (1992) 40.
[13] Mason, T. J., Chem. Ind., (1993) 47.
[14] Walton, D. J., Chyla, A., Lorimer, J. P., Mason, T. J., Syn. Commun., 20 (1990) 1843.
[15] Walton, D. J., Phull, S. S., Bates, D. M., Lorimer, J. P., Mason, T. J., Ultrasonics, 30 (1992) 186.
[16] Flint, E. B., Suslick, K. S., J. Am. Chem. Soc., 111 (1989) 6987.
[17] Suslick, K. S., Schubert, P. F., Wang, H. H., Goodale, J. W., (Chisholm, M. A. (ed.)), Inorganic Chemistry: Toward the 21st Century, Am. Chem. Soc., Washington, DC, 1983, p. 550.
[18] Suslick, K. S., Grinstaff, M. W., Kolbeck, K. J., Wong, M., Ultrasonics Sonochemistry, 1 (1994) S65.
[19] Suslick, K. S., Fang, M. M., Hyeon, T., Cichowlas, A. A., (Gonsalves, K. E., Chow, G. M., Xiao, T. O., Cammarata, R. C. (eds.)), Molecularly Designed Nanostructures, Materials Research Society, Pittsburgh, 1994, pp. 443–448.
[20] Suslick, K. S., Schubert, P. F., J. Am. Chem. Soc., 105 (1983) 6042.
[21] Suslick, K. S., Gawienowski, J. J., Schubert, P. F., Wang, H. H., J. Phys. Chem., 87 (1983) 2299.
[22] Suslick, K. S., Goodale, J. W., Schubert, P. F., Wang, H. H., J. Am. Chem. Soc., 105 (1983) 5781.
[23] Suslick, K. S., Schubert, P. F., Wang, H. H., Goodale, J. W., ACS Symposium Series, 211 (1983) 550.
[24] Suslick, K. S., Johnson, R. E., J. Am. Chem. Soc., 106 (1984) 6856.
[25] Suslick, K. S., Hammerton, D. A., Ultrasonics Intl., (1985) 231.
[26] Suslick, K. S., Gawienowski, J. J., Schubert, P. F., Wang, H. H., Ultrasonics, 22 (1984) 33.
[27] Suslick, K. S., Cline Jr., R. E., Hammerton, D. A., J. Am. Chem. Soc., 108 (1986) 5641.

[28] Suslick, K. S., Modern Synthetic Methods, 4 (1986) 1.
[29] Suslick, K. S., Adv. Organometallic Chem., 25 (1986) 73.
[30] Suslick, K. S., ACS Symposium Series, 333 (1987) 191.
[31] Suslick, K. S. (ed.), Ultrasound: Its Chemical, Physical and Biological Effects, VCH Publishers Inc., New York, 1988.
[32] Suslick, K. S., Doktycz, S. J., J. Am. Chem. Soc., 111 (1989) 2342.
[33] Suslick, K. S., Science, 247 (1990) 1439.
[34] Suslick, K. S., Choe, S. B., Cichowlas, A. A., Grinstaff, M. W., Nature, 353 (1991) 414.
[35] Grinstaff, M. W., Cichowlas, A. A., Choe, S. B., Suslick, K. S., Ultrasonics, 30 (1992) 168.
[36] Tuncay, A., Dustman, J. A., Fisher, G., Tuncay, C. I., Suslick, K. S., Tetrahedron Lett., 33 (1992) 7647.
[37] Suslick, K. S., Flint, E. B., Grinstaff, M. W., Kemper, K. A., J. Phys. Chem., 97 (1993) 7216.
[38] Henglein, A., Herburger, D., Gutierrez, M., J. Phys. Chem., 96 (1992) 1126.
[39] Buttner, J., Gutierrez, M., Henglein, A., J. Phys. Chem., 95 (1991) 1528.
[40] Fischer, C. H., Hart, E. J., Henglein, A., J. Phys. Chem., 90 (1986) 3059.
[41] Fischer, C. H., Hart, E. J., Henglein, A., J. Phys. Chem., 90 (1986) 222.
[42] Gutierrez, M., Henglein, A., Fischer, C.-H., Int. J. Radiation Biol., 50 (1986) 313.
[43] Gutierrez, M., Henglein, A., J. Phys. Chem., 94 (1990) 3625.
[44] Gutierrez, M., Henglein, A., Ibanez, F., J. Phys. Chem., 95 (1991) 6044.
[45] Hart, E. J., Henglein, A., J. Phys. Chem., 91 (1987) 3654.
[46] Henglein, A., Ultrasonics, 25 (1987) 6.
[47] Henglein, A., Gutierrez, M., J. Phys. Chem., 94 (1990) 5169.
[48] Richards, W. T., Loomis, A. L., J. Am. Chem. Soc., 49 (1927) 3086.
[49] Weissler, A. C., Snyder, S. J., J. Am. Chem. Soc., 72 (1950) 1769.
[50] Kotronarou, A., Mills, G., Hoffmann, M. R., J. Phys. Chem., 95 (1991) 3630.
[51] Kotronarou, A., Thesis, California Institute of Technology, 1991.
[52] Kotronarou, A., Mills, G., Hoffmann, M. R., Environ. Sci. Technol., 26 (1992) 1460.
[53] Kotronarou, A., Mills, G., Hoffmann, M. R., Environ. Sci. Technol., 26 (1992) 2420.
[54] Hua, I., Höchemer, R., Hoffmann, M. R., Environ. Sci. Technol., 29 (1995) 2790.
[55] Olson, T. M., Barbier, P. F., Water Res., 28 (1994) 1383.
[56] Petrier, C., Lamy, M. F., Francony, A., Benahcene, A., David, B., Renaudin, V., Gondrexon, N., J. Phys. Chem., 98 (1994) 10514.
[57] Petrier, C., Micolle, M., Merlin, G., Luche, J. L., Reverdy, G., Environ. Sci. Technol., 26 (1992) 1639.
[58] Serpone, N., Terzian, R., Hidaka, H., Pelizzetti, E., J. Phys. Chem., 98 (1994) 2634.
[59] Serpone, N., Terzian, R., Colarusso, P., Minero, C., Pelizzetti, E., Hidaka, H., Res. Chem. Intermed., 18 (1992) 183.
[60] Wu, J. M., Huang, H. S., Livengood, C. D., Environ. Prog., 11 (1992) 195.

10.7. REFERENCES

[61] Price, G., Matthias, P., Lenz, E. J., Process Safety Environ. Protection: Trans. Instit. Chem. Eng. Part B, 72 (1994) 27.
[62] Johnston, A. J., Hocking, P., ACS Symposium Series, 518 (1993) 106.
[63] Cost, M., Mills, G., Glisson, P., Lakin, J., Chemosphere, 27 (1993) 1737.
[64] Noltingk, B. E., Nepprias, E. A., Proc. Phys. Soc. B, 63B (1950) 674.
[65] Nepprias, E. A., Noltingk, B. E., Proc. Phys. Soc. B, 63B (1951) 1032.
[66] Sehgal, C., Steer, R. P., Sutherland, R. G., Verrall, R. E., J. Phys. Chem., 70 (1979) 2242.
[67] Flint, E. B., Suslick, K. S., Science, 253 (1991) 1397.
[68] Becker, L., Bada, J. L., Kemper, K., Suslick, K. S., Marine Chem., 40 (1992) 315.
[69] Barber, B. P., Hiller, R., Arisaka, K., Fetterman, H., Putterman, S., J. Acoust. Soc. Amer., 91 (1992) 3061.
[70] Barber, B. P., Weninger, K., Lofstedt, R., Putterman, S., Phys. Rev. Lett., 74 (1995) 5276.
[71] Chendke, P. K., Fogler, H. S., J. Phys. Chem., 87 (1983) 1362.
[72] Crum, L. A., Roy, R. A., Science, 266 (1994) 233.
[73] Crum, L. A., Phys. Today, 47 (1994) 22.
[74] Crum, L. A., J. Acoust. Soc. Amer., 95 (1994) 559.
[75] Didenko, Y. T., Pugach, S. P., J. Phys. Chem., 98 (1994) 9742.
[76] Flint, E. B., Suslick, K. S., J. Phys. Chem., 95 (1991) 1484.
[77] Henglein, A., Gutierrez, M., J. Phys. Chem., 97 (1993) 158.
[78] Kamath, V., Prosperetti, A., Egolfopoulos, F. N., J. Acoust. Soc. Amer., 94 (1993) 248.
[79] Margulis, M. A., Ultrasonics, 23 (1985) 157.
[80] Seghal, C., Steer, R. P., Sutherland, R. G., Verrall, R. E., J. Chem. Phys., 70 (1979) 2242.
[81] Suslick, K. S., Doktycz, S. J., Flint, E. B., Ultrasonics, 28 (1990) 280.
[82] Suslick, K. S., Kemper, K. A., Ultrasonics, 31 (1993) 463.
[83] Suslick, K. S., Flint, E. B., Grinstaff, M. W., Kemper, K. A., J. Phys. Chem., 97 (1993) 3098.
[84] Young, F. R., J. Acoust. Soc. Am., 60 (1976) 100.
[85] Putterman, S., Scientific Am., 272 (1995) 46.
[86] Misik, V., Miyoshi, N., Riesz, P., J. Phys. Chem., 99 (1995) 3605.
[87] Naidu, D., Rajan, R., Kumar, R., Gandhi, K. S., Arakeri, V. H., Chandrasekaran, S., Chem. Eng. Sci., 49 (1994) 877.
[88] Rayleigh, L., Philos. Mag., 34 (1917) 94.
[89] Hua, I., Höchemer, R. H., Hoffmann, M. R., J. Phys. Chem., 99 (1995) 2335.
[90] Shaw, R. W., Brill, T. B., Clifford, A. A., Eckert, C. A., Franck, E. U., Chem. Eng. News, 69 (1991) 26.
[91] Li, L. L., Egiebor, N. O., Energy Fuels, 6 (1992) 34.
[92] Klein, M. T., Mentha, Y. G., Torry, L. A., Ind. Eng. Chem., 31 (1992) 182.
[93] Lee, D. S., Gloyna, E. F., Environ. Sci. Technol., 26 (1992) 1587.
[94] Townsend, S. H., Abraham, M. A., Huppert, G. L., Klein, M. T., Paspek, S. C., Ind. Eng. Chem. Res., 27 (1988) 143.

[95] Yang, H., Eckert, C. A., Ind. Eng. Chem. Res., 27 (1988) 2009.
[96] Thornton, T. D., LaDue, D. E., Savage, P., E. Environ. Sci. Technol., 25 (1991) 1507.
[97] Suslick, K. S., Hammerton, D. A., Cline, R. E., J. Am. Chem. Soc., 108 (1986) 5641.
[98] Katz, J., Acosta, A., Appl. Sci. Res., 38 (1982) 123.
[99] Suslick, K. S., Hammerton, D. A., IEEE Trans. Ultrasonic Ferroelec. Freq. Contr. 1986, UFCC-33, 143.
[100] Suslick, K. S., Hammerton, D. A., IEEE Trans. Ultrasonics, 2 (1986) 143.
[101] Kondo, T., Kodaira, T., Kano, E., Free Rad. Res. Commun., 19 (1993) S193.
[102] Kondo, T., Kirschenbaum, L. J., Kim, H., Riesz, P., J. Phys. Chem., 97 (1993) 522.
[103] Petrier, C., Jeunet, A., Luche, J. L., Reverdy, G., J. Amer. Chem. Soc., 114 (1992) 3148.
[104] Price, G. J., Lenz, E. J., Ultrasonics, 31 (1993) 451.
[105] Riesz, P., Kondo, T., Free Rad. Biol. Med., 13 (1992) 247.
[106] Riesz, P., Kondo, T., Carmichael, A. J., Free Rad. Res. Commun., 19 (1993) S45.
[107] Hua, I., Hoffmann, M. R., Environ. Sci. Technol., 1995, in press.
[108] Kotronarou, A., Hoffmann, M. R., Adv. Chem. Ser., 244 (1995) 233.
[109] Tsang, W., Robaugh, D., Mallard, G. W., J. Phys. Chem., 90 (1986) 5968.
[110] Robinson, J. W., J. Appl. Phys., 44 (1973) 76.
[111] Martin, E. A., J. Appl. Phys., 31 (1958) 255.
[112] Robinson, J. W., J. Appl. Phys., 38 (1967) 210.
[113] Robinson, J. W., Ham, M., Balaster, A. N., J. Appl. Phys., 44 (1973) 72.
[114] Jakob, L., Hashem, T. M., Burki, S., Guidny, N. M., Braun, A. M., Photochem. Photobiol. A: Chem., 7 (1993) 97.
[115] Buntzen, R. R., Exploding Wires, Plenum Press, New York, 1962, p. 195.
[116] Ben'kovskii, V. G., Golubnichii, P. I., Maslennikov, S. I., Phys. Acoust., 20 (1974) 14.
[117] Bhatnagar, A., Cheung, H. M., Environ. Sci. Technol., 28 (1994) 1481.
[118] Alippi, A., Cataldo, F., Galbato, A., Ultrasonics, 30 (1992) 148.
[119] Cheung, H. M., Bhatnagar, A., Jansen, G., Environ. Sci. Technol., 25 (1991) 1510.
[120] Höchemer, R. H., Hoffmann, M. R., Environ. Sci. Technol., 1996, in press.
[121] Cum, G., Galli, G., Gallo, R., Spadaro, A., Ultrasonics, 30 (1992) 267.
[122] Kawabata, K., Umemura, S., Ultrasonics, 31 (1993) 457.
[123] Voglet, N., Mullie, F., Lepoint, T., New J. Chem., 17 (1993) 519.
[124] Henglein, A., Kormann, C., Int. J. Radiat. Biol., 48 (1985) 251.
[125] Cottier, L., Descotes, G., Lewkowski, J., Skowronski, R., Polish J. Chem., 68 (1994) 693.
[126] Moreno, M., Melo, M., Neves, A., Tetrahedron Lett., 32 (1991) 3201.
[127] Elguero, J., Goya, P., Lissavetzky, J., Valdeomillos, A. M., Comp. Rendus Acad. Sci. Series II, 298 (1984) 877.
[128] Kristol, D. S., Klotz, H., Parker, R. C., Tetrahedron Lett., 22 (1981) 907.
[129] Farhat, F., Berchiesi, G., Syn. Commun., 22 (1992) 3137.

10.7. REFERENCES

[130] Fitch, R. W., Luzzio, F. A., Tetrahedron Lett., 35 (1994) 6013.
[131] Rassokhin, D. N., Bugaenko, L. T., Kovalev, G. V., Radiation Phys. Chem., 45 (1995) 251.
[132] Maeda, M., Wajima, N., Lin, R. J., Teratani, S., Kaneko, M., New J. Chem., 17 (1993) 523.
[133] Matthews, J. C., Harder, W. L., Richardson, W. K., Fisher, R. J., Alkarmi, A. M., Crum, L. A., Dinno, M. A., Membrane Biochem., 10 (1993) 213.
[134] Worsley, D., Mills, A., Ultrasonics, 30 (1992) 333.
[135] Kondo, T., Riesz, P., Free Rad. Biol. Med., 7 (1989) 259.
[136] Krishna, C. M., Kondo, T., Riesz, P., Rad. Phys. Chem. Int. J. Rad. Appl. Instrum. Part C, 32 (1988) 121.
[137] Krishna, C. M., Kondo, T., Riesz, P., J. Phys. Chem., 93 (1989) 5166.
[138] Misik, V., Riesz, P., J. Phys. Chem., 98 (1994) 1634.
[139] Misik, V., Kirschenbaum, L. J., Riesz, P., J. Phys. Chem., 99 (1995) 5970.
[140] Riesz, P., Christman, C. L., Federation Proc., 45 (1986) 2485.
[141] Riesz, P., Kondo, T., Krishna, C. M., Ultrasonics, 28 (1990) 295.
[142] Martin, S. T., Morrison, C. L., Hoffmann, M. R., J. Phys. Chem., 98 (1994) 13695.
[143] Willberg, D. M., Lang, P. S., Höchemer, R. H., Kratel, A., Hoffmann, M. R., Degradation of 4-chlorophenol, 3,4-dichloroaniline and 2,4,6-trinitrotoluene in a Electrohydraulic Discharge Reactor, Environ. Sci. Technol., 1996, 30, in press.
[144] Martin, P. D., Chem. Eng. London, (1990) 15.
[145] Martin, P. D., Chem. Industry, (1993) 233.
[146] Martin, P. D., Ward, L. D., Chem. Eng. Res. Design, 70 (1992) 296.

11

MICROWAVE DIELECTRIC HEATING EFFECTS IN CHEMICAL SYNTHESIS

D. Michael P. Mingos[1]* and A. Gavin Whittaker[2]

[1] *Chemistry Department, Imperial College of Science, Technology and Medicine, South Kensington, London SW1 2AY, U.K.*
[2] *Department of Chemistry, The University of Edinburgh, King's Buildings, West Mains Road, Edinburgh EH9 3JJ, U.K.*

11.1 INTRODUCTION
11.2 MICROWAVES AND THEIR INTERACTIONS WITH MATTER
    11.2.1 Dielectric Polarization
11.3 MULTIMODE MICROWAVE OVENS
    11.3.1 The Magnetron
    11.3.2 Modification of the Magnetron Power Circuit
    11.3.3 Ports
11.4 CYLINDRICAL CAVITY
    11.4.1 Reaction Vessels
11.5 TEMPERATURE MEASUREMENT
11.6 HEALTH CONSIDERATIONS
11.7 APPLICATIONS OF MICROWAVE DIELECTRIC HEATING
    11.7.1 Solutions
    11.7.2 Organic Synthesis
    11.7.3 Synthesis of Organometallic and Coordination Compounds
    11.7.4 Polymers
    11.7.5 Chemical Analyses
    11.7.6 Microwave Syntheses Involving Solids
11.8 SUMMARY
11.9 REFERENCES

11.1 INTRODUCTION

In 1775 chemistry was defined by Dr. Johnson as an art whereby sensible bodies contained in vessels . . . are so changed by means of certain instruments, and principally fire, that their several powers and virtues are thereby discovered, with a view to philosophy or medicine. The heating of chemicals in containment vessels still remains the primary means of stimulating chemical reactions. Photochemical, catalytic, sonic, and high pressure techniques [1] are now also used by chemists to accelerate chemical reactions and this book summarizes the important features of these techniques. This chapter describes an alternative to conventional heating for introducing thermal energy into reactions. The microwave dielectric heating effect uses the ability of some liquids and solids to transform electromagnetic energy into heat and thereby drive chemical reactions. This *in situ* mode of energy conversion has many attractions to the chemist, because its magnitude depends on the properties of the molecules. This may lead to reaction and spatial selectivity. Microwave dielectric heating should not be confused with microwave discharges which can create a plasma with a very high temperature and can cause dramatic fragmentation and recombination reactions of molecules in the gas phase.

The magnetron [2] which is a device for generating fixed frequency microwaves was designed by Randall and Booth during the Second World War and used for RADAR. It was recognized early on that microwaves could heat water in a dramatic fashion, and domestic and commercial appliances for heating and cooking foodstuffs began to appear in the United States in the 1950s. Effective Japanese technology transfer and global marketing led to the widespread domestic use of microwave ovens which occurred during the 1970s and 1980s.

11.2 MICROWAVES AND THEIR INTERACTIONS WITH MATTER

The microwave region of the electromagnetic spectrum is associated with wavelengths of 1 cm to 1 m (frequencies of 30 GHz to 300 MHz, respectively, see Figure 11.1) and lies between infrared radiation and radio frequencies. The wavelengths between 1 cm and 25 cm are extensively used for RADAR transmissions and the remaining wavelength range is used for telecommunications. Domestic and industrial microwave heaters are required to operate at either 12.2 cm (2.45 GHz) or 33.3 cm (900 MHz) to prevent interference with these applications. Domestic microwave ovens generally operate at 2.45 GHz.

In microwave spectroscopy molecules are studied in the gas phase, and the microwave spectrum of a molecule shows many sharp bands [3] in the frequency range 3–60 GHz. The sharp bands in the spectrum arise from transitions between quantized rotational states of the molecule defined by the following relationship:

$$\text{Rotational energy}, E_j = J(J+1)h^2/8\pi I^2$$

11.2. MICROWAVES AND THEIR INTERACTIONS WITH MATTER

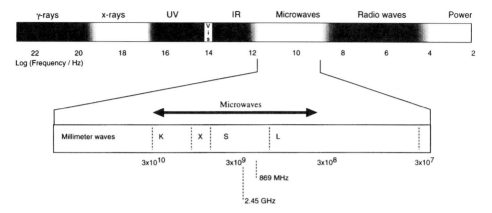

Figure 11.1 Electromagnetic spectrum indicating important microwave frequencies for dielectric heating.

where J is the rotational quantum number, I is the moment of inertia and h is Planck's constant. A pure rotational spectrum is only observed if the molecule possesses a permanent dipole moment [4].

The microwave spectra of small molecules are sharp, and lines less than 1 MHz apart can often be distinguished. Microwave spectroscopy provides an excellent technique for identifying molecules in the gas phase and has, for example, been used to confirm the presence of a wide range of molecules in outer space. Additionally, it can be used for the precise determination of bond lengths and angles.

In the gas phase, at low pressure, the lifetime of the excited state produced by exciting a particular rotational state is long. However, at pressures of around 10^{-1} mm Hg the reduction in the lifetime due to collisions broadens the spectral peaks as a consequence of Heisenberg's uncertainty principle. In liquids, where the molecules are generally not free to rotate independently, the spectra are too broad to be observed. Microwave dielectric loss heating effects are associated with solutions and solids and need to be distinguished from the spectroscopic effects described above. A solid or liquid can be heated by converting electromagnetic energy into thermal energy. The heating effect produced by the electromagnetic waves arises from the ability of an electric field to exert a force on polar molecules and charged particles. If the charged particles present in the substance can move freely through it, then a current may be induced. However, if the charge carriers are bound to certain regions they will move until a counter force balances them and the net result is a dielectric polarization. Both conduction and dielectric polarization are sources of microwave heating. The microwave heating effect depends on the frequency as well as the power applied, but unlike microwave spectroscopy the effect does not result from well spaced discrete quantized energy states, but from a broad band bulk phenomenon which can be treated classically. The theory of microwave heating was developed by Debye [5], Frohlich [6], Daniel [7], Cole [8], Hill [9], and Hasted [10]. Although a detailed analysis is beyond the scope of this review, the important principles are outlined below.

11.2.1 Dielectric Polarization

The complex dielectric constant, ε^*, completely describes the dielectric properties of homogeneous materials and is expressed as the sum of real and imaginary dielectric constants:

$$\varepsilon^* = \varepsilon' + i\varepsilon''$$

The real part of ε^*, ε', represents the ability of a material to be polarized by an external electric field. At very high and very low frequencies, and with static fields, ε' is equal to the total dielectric constant of the material. Where electromagnetic energy is transformed into heat by the material, ε'' is non-zero. ε'' is related to the efficiency with which the electromagnetic energy is converted into thermal energy. The loss angle, δ, is also commonly encountered in the dielectric literature, and is more usually expressed in the form of its tangent. It is related to the complex dielectric constant by the equation:

$$\tan \delta = \varepsilon''/\varepsilon'$$

The angle δ is the phase difference between the electric field and the polarization of the material.

Magnetic polarization may also contribute to the heating effect observed in materials with magnetic properties, and a similar expression for the complex permeability of such materials may be formulated. Although such cases are relatively uncommon, the microwave heating of Fe_3O_4 results from this effect. The total polarization (α_t) of the material arising from the displacement of charges may be expressed as the sum of the following components:

$$\alpha_t = \alpha_e + \alpha_a + \alpha_d + \alpha_i$$

where α_e results from the displacement of electron charges in relation to the nuclei in a material, and α_a from the displacement of nuclei relative to one another in materials with unequal charge distributions. Polarizations of both α_e and α_a operate on time scales which are very much smaller than that required for microwave frequency field oscillations, and therefore follow the microwave frequency field almost exactly. As such they do not result in the conversion of microwave energy into thermal energy. α_d results from the reorientation of polar molecules or other permanent dipoles in the material. The timescale for molecular reorientation is approximately the same as microwave frequencies, therefore this is the most important mechanism for microwave heating, and is discussed further below. The role of the interfacial (Maxwell-Wagner) polarization effect, α_i, which results from interfacial phenomena in inhomogeneous materials is not great at microwave frequencies, and in general its contribution is limited.

Dipolar Polarization, α_d At low frequencies the time taken by the electric field to change direction is longer than the response time of the molecular dipoles, and the dielectric polarization keeps in phase with the electric field. The field provides the energy necessary to make the molecules rotate into alignment. Some of the energy is transferred to random motion each time a dipole is knocked out of alignment and then realigned. The transfer of energy is so small, however, that the temperature hardly rises. If the electric field oscillates rapidly, it changes direction faster than the response time of the dipoles. Since the dipoles do not rotate, no energy is absorbed and the polar liquid does not heat up.

At microwave frequencies the time taken for the field changes is about the same as the response time of the dipoles. They rotate because of the torques they experience, but the resulting polarization lags behind the change of the electric field. When the field reaches its maximum strength, say in the upward direction, polarization may still be low. It keeps rising as the field weakens. The lag indicates that the dipoles absorb energy from the field and the electromagnetic energy is converted into thermal energy.

The time taken for the dipoles in water to become polarized and then depolarized is governed by a relaxation time constant. When this time constant approaches the inverse-excitation frequency the dipolar polarization mechanism becomes important. Since the time constants for electronic and atomic polarization are much faster than 10^{-9} s, these mechanisms do not contribute to microwave-dielectric heating effects [11].

In Figure 11.2 the dielectric properties of water and some alcohols are plotted as a function of frequency at 25 °C [12]. It is apparent that appreciable values of the dielectric loss exist over a wide frequency range. Note that for water ε'' reaches its maximum at around 20 GHz while domestic microwave ovens operate at a much lower frequency, 2.45 GHz. The practical reason for the lower frequency is that it is necessary to heat food efficiently throughout its interior. If the frequency is optimal for a maximum heating rate, the microwaves are absorbed in the outer regions of the food, and penetrate only a short distance. Thus the penetration depth, *i.e.*, the depth into a material where the power falls to one half its value on the surface, is another important parameter in the design of a microwave experiment. An approximate relationship for penetration depth D_p when ε'' is small, is given by:

$$D_p \propto \frac{\lambda_0 \sqrt{\varepsilon''}}{\varepsilon''}$$

where λ_0 is the wavelength of the microwave radiation.

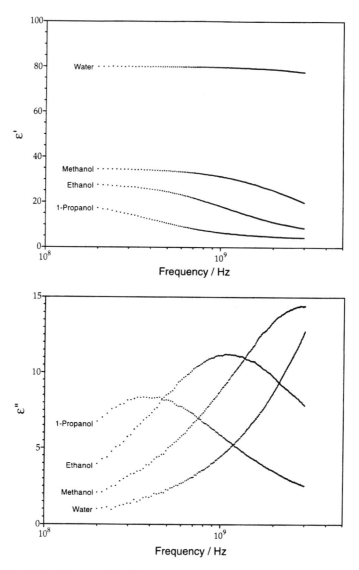

Figure 11.2 Dielectric properties of water and some alcohols as a function of frequency at 25 °C.

The frequency dependences of ε' and ε'' are governed by the Debye equations [5, 12]:

$$\varepsilon' = \varepsilon'_\infty + \frac{(\varepsilon'_0 - \varepsilon''_\infty)}{(1 + \omega^2 \tau^2)}$$

$$\varepsilon'' = \frac{(\varepsilon'_0 - \varepsilon''_\infty)\omega\tau}{(1 + \omega^2 \tau^2)}$$

11.2. MICROWAVES AND THEIR INTERACTIONS WITH MATTER

where ε'_∞ and ε'_0 are defined as the high frequency and static dielectric constants, and ω and τ are the frequency and relaxation times which characterize the rate of build up and decay of polarization. The above results apply to both liquids and solids though different models are used to derive the equations. It is an interesting feature of the Debye equations that the values of ε' and ε'' at the frequency at which the dielectric loss is a maximum (i.e., when $\omega\tau = 1$) is independent of this frequency and of the relaxation time:

$$\varepsilon''_{max} = (\varepsilon'_0 - \varepsilon'_\infty)/2$$

In a liquid the dipoles point randomly and their directions are continually changing due to thermal motions [6]. Debye's interpretation of the relaxation is given in terms of the frictional forces in the medium. He derived the following expression for the relaxation time of spherical dipoles using Stokes' theorem [5].

$$\tau = 4\pi r^3 \eta / kT$$

when η is the viscosity of the medium, r the radius of the dipolar molecule, k is Boltzmann's constant and T is the absolute temperature.

The relaxation times for some common organic solvents are listed in Table 11.1.

For an ideal solid in which opposing dipole positions are separated by a potential barrier U_a, the following relationship between τ and the dielectric constant may be derived from Boltzmann statistics:

$$\tau = \frac{e^{u_a/kT}(\varepsilon'_0 + 2)}{v(\varepsilon'_\infty + 2)}$$

Table 11.1 Relaxation frequencies and loss tangents for some inorganic and organic molecules.

Compound	Relaxation frequency (100 MHz)	Loss tangent @ 2.45 GHz
Water	17	0.137
Methanol	28	0.659
Ethanol	8.5	0.941
1-propanol	4.8	0.777
2-propanol	6.45	0.799
1-butanol	3.0	0.523
2-butanol	2.25	0.580
1-pentanol	2.4	0.412
1-hexanol	1.6	0.355
Benzylalcohol	6.6	0.596

The following form of the Onsager equation [13] has been shown to apply to many liquids and solids.

$$\tau = \frac{4\pi N m^2 \varepsilon_0' + (\varepsilon_0' + \varepsilon_\infty')}{9kT(\varepsilon_0' + \varepsilon_\infty')}$$

where N is the number of molecules and m is the mass of each molecule.

Interfacial Polarization, α_i A suspension of conducting particles in a non-conducting medium in an inhomogeneous material whose dielectric constant is frequency dependent can give rise to dielectric heating. The loss relates to the build up of charges between the interfaces and is known as the Maxwell-Wagner effect. Wagner [14] has shown that for the simplest model, consisting of conducting spheres distributed through a non-conducting medium, the dielectric loss factor, of volume fraction, v, of material, is given by:

$$\varepsilon_i'' = \left(\frac{9v\varepsilon' f_{max}}{1.8 \times 10^{10} \sigma}\right)\left(\frac{\omega\tau}{10 + \omega^2\tau^2}\right)$$

where σ is the conductivity (in S m^{-1}) of the conductive phase and ε' its dielectric constant. The frequency (f) variation of the loss factor is similar to that of dipolar relaxation. An experimental system approximating to Wagner's model has been made by incorporating up to 3% of roughly spherical particles of semiconducting copper phthalocyanine into paraffin wax [15].

Conduction Effects In addition to the dielectric losses described above, many materials may also interact with microwaves through conduction mechanisms. The complex dielectric constant may be expanded to take account of these losses by including a separate conduction term:

$$\varepsilon^* = \varepsilon_\infty' + \frac{(\varepsilon_0' - \varepsilon_\infty'')}{(1 + i\omega\tau)} - \left(\frac{i\sigma}{\omega\varepsilon_s'}\right)$$

This term is relevant to a large number of systems. The addition of dissolved salts in water markedly affects the dielectric properties as the conduction increases and may become important enough to swamp the dielectric dipolar losses. The dielectric losses of the majority of solids arise predominantly from these conduction terms and may be strongly affected by temperature. The conductivity of alumina, for example, increases with temperature as electrons are promoted into the conduction band from the O(2p) valence band leading to increases in the dielectric constants. This is illustrated in Figure 11.3 along with a number of other ceramic materials which exhibit similar behaviour.

The increase in the dielectric constant with increase in temperature is especially important in the microwave heating of solids, as it results in the

11.3. MULTIMODE MICROWAVE OVENS

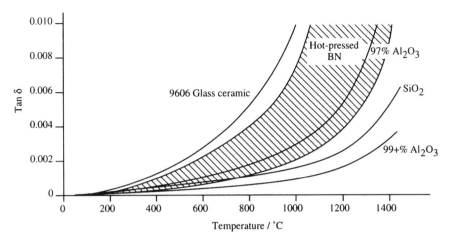

Figure 11.3 Dielectric properties as a function of frequency and temperature for a range of ceramic materials.

phenomenon of thermal runaway. Microwave heating in alumina is poor at room temperature, and dT/dt is therefore small. As the temperature increases so too does the dielectric loss factor and heating becomes more effective [16, 17] and dT/dt increases rapidly, as illustrated for alumina in Figure 11.3. The rapid increase in temperature under these circumstances is described as thermal runaway. Heating in metals and metal powders by microwave radiation depends heavily upon conduction losses.

11.3 MULTIMODE MICROWAVE OVENS

Microwave ovens designed specifically for the chemical laboratory are now available commercially. The main features of these microwave ovens are shown in Figure 11.4. At its most basic, the oven consists of a microwave source, control and power circuits, and a cavity for the sample, into which the microwaves are introduced *via* a wave guide. The cavity has reflective metal walls, which not only act to prevent leakage of radiation from the oven, but also improve its efficiency. The microwave absorbing properties of food or chemical samples which are placed in a domestic oven will seldom, if ever, be matched to the source. The walls act to reflect the microwaves repeatedly through the sample where they are partially absorbed on each pass. This process is particularly important for samples which are either dimensionally very small, or which have low dielectric loss tangents. If small, low loss samples are used, then a large amount of energy may be reflected back to the microwave source which may then be damaged by overheating. For this reason, most commercial ovens are protected by a thermal cut-off which switches off the power circuits if either the source or oven reach a pre-set temperature. In more sophisticated

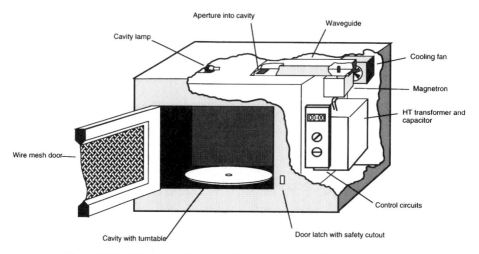

Figure 11.4 The major features of a multi-mode microwave oven.

applications a device known as a circulator may be used, which acts to redirect reflected energy into a dummy load.

Whether or not a sample is present, the microwaves which enter the cavity are reflected from the walls and give rise to complex standing wave patterns. Unless great care is taken in the positioning of samples in the cavity this may lead to unreliable heating behaviour, particularly for small samples, since very high and very low field densities may be present within a few centimetres of each other. In order to compensate for this, oven manufacturers average out the energy which a sample receives either by moving the sample through the field on a turntable, or by using a 'mode stirrer' (a rotating metal paddle) to agitate the field. The turntable is the preferred method in most commercial multimode ovens.

11.3.1 The Magnetron

The microwave source in commercial ovens is almost invariably a magnetron. The multicavity magnetron is one of a large class of microwave tubes which act by the interaction of electrons with magnetic fields. The basic device is shown in Figure 11.5. The magnetron is a thermionic diode and consists of a directly heated cathode and an anode separated under high vacuum by a high voltage (typically ≈ 4 kV), in a strong axial magnetic field. The anode consists of an even number of cavities (typically eight), each of which behaves as a tuned circuit with the open end acting as a capacitance. Each cavity therefore acts as an electrical oscillator which resonates at a specific fundamental frequency. The electrons formed at the cathode interact with the rotating AC field generated by the resonant cavities of the anode. The series of cavities have a number of modes in which they may oscillate, and in order to create the correct phase relationship between them, alternate cavities are linked by metal wires (mode straps). Under

11.3. MULTIMODE MICROWAVE OVENS

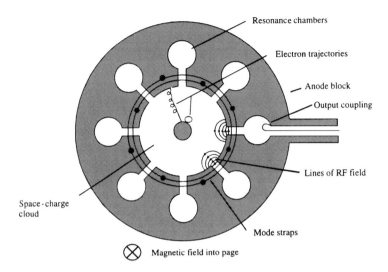

Figure 11.5 Schematic illustration of a magnetron.

the optimum operating conditions, alternate anode cavities have an electrical phase difference of π (the so-called 'π-mode'), and the mode straps act to ensure that this is the operating mode.

A strong axial magnetic field from a permanent magnet bends the path of the electrons as they move towards the anode. Electrons which give up energy to the rotating field as they travel outward in these curved paths are bunched together and form a series of electron cloud 'spokes' which rotate synchronously with the field. The energy which these electrons relinquish to the AC field may then be taken off the anode *via* an antenna. The efficiency of a magnetron is typically of the order of 50% or less, and it is therefore necessary to ensure that the heat which is generated is removed by either water or forced air cooling.

11.3.2 Modification of the Magnetron Power Circuit [18]

Power control in a domestic microwave oven is achieved by switching the magnetron on and off according to a duty cycle. The duty cycle is typically of the order of 20 seconds, and '50% power', for example, is achieved by repeatedly switching the magnetron on and off every 10 seconds. This method is adequate for heating food samples, where it is the total energy absorbed by the sample which is important. In the laboratory it is often necessary to heat samples at a low power level (annealing solid state samples, for example, or where high power levels may damage the sample). Whilst it is possible to achieve this at a basic level by placing a dummy load alongside the sample, it is often preferable to be able to control the output power from the magnetron. Control of the power output of a magnetron is generally achieved by varying either the magnetic field or the anode-cathode voltage. Whilst the former is often used in commercial applications, it requires a bulky electromagnet assembly on the magnetron and is also extremely expensive. In order to obtain relatively cheap microwave power control in the laboratory, the microwave circuit of a commercial oven may be modified to incorporate a variable AC resistor ('Variac') to control the anode-cathode potential. The modified circuit diagram is shown in Figure 11.6. In the unmodified microwave oven, the cathode heating voltage and the anode-cathode voltage are taken from two secondary windings on one transformer, and as a consequence, placing the Variac in series with the primary coil of the transformer does not give desirable power characteristics. It is found that reducing the cathode heating voltage by as little as 0.2 V is sufficient to decrease the number of thermal electrons to a point where the magnetron ceases to operate. In the modified circuit, the cathode heating voltage is made invariant by incorporating a second transformer with a secondary winding potential of 3 volts AC. The relative potential of the heating element is then raised to the required cathode potential by connecting the high voltage output of the original transformer to the heater. The anode-cathode potential is made variable by adding a Variac in series with the primary of the high voltage transformer.

As the scale for the output of the Variac is shown as a percentage of the input voltage, the microwave output power is measured as a function of this percentage. This may be carried out by the standard method used by oven

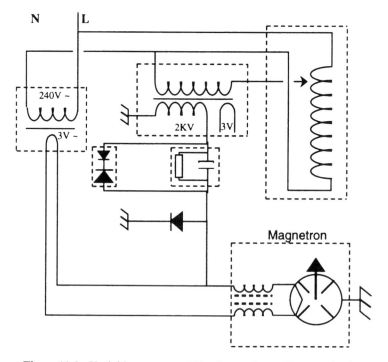

Figure 11.6 Variable power modification to domestic oven circuit.

11.3. MULTIMODE MICROWAVE OVENS

manufacturers. A 500 cm^3 quantity of water is placed in the cavity and heated for 2 minutes. The temperature rise is then measured, and the microwave power output by the magnetron calculated. For the lower powers (78% or less), heating times of 5 minutes are used, so as to obtain measurable temperature rises. A minimum of five measurements is taken at each power setting, and an average of these then taken. The resulting relationship between the applied voltage and the measured power is shown in Figure 11.7.

There is a clear linear relationship between power and the applied voltage down to 80% of the maximum voltage. Above 80%, the magnetron is stable, and the output power is directly proportional to the current which passes from anode to cathode. Once a limiting potential is reached, instabilities in the magnetron operation cause non-linear behavior. This non-linear behaviour seldom presents a significant obstacle to microwave work in the laboratory, although it is clearly necessary to calibrate output from the source in kinetic studies, for example, when precise knowledge of the applied power is important.

11.3.3 Ports

The most common modification to a microwave oven involves the addition of a port to allow the insertion of glass tubes and probes into the cavity. This commonly takes the form of a cylindrical metal tube attached to the cavity wall whose dimensions prevent radiation leakage by attenuation of the microwaves (Figure 11.8). If the tube is considered as a cylindrical wave guide, then the

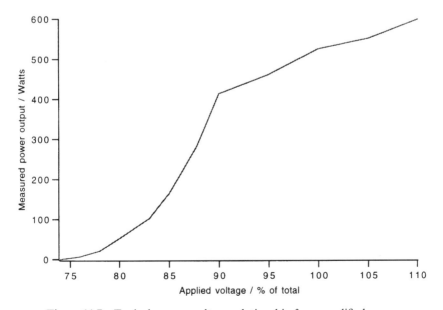

Figure 11.7 Typical power: voltage relationship for a modified oven.

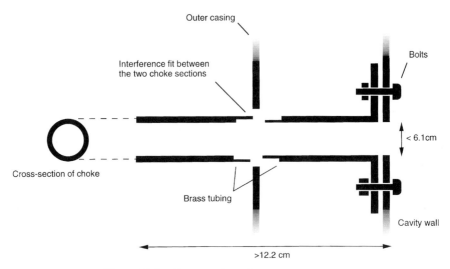

Figure 11.8 Construction of a microwave port.

maximum wavelength which it is capable of sustaining will be transmitted in the TE$_{10}$ mode, whose cut-off wavelength, λ_c, is given by:

$$\lambda_c = 3.412\, a$$

where a is the diameter of the cylindrical wave guide. Thus for an oven operating at 2.45 GHz, a port of approximately 3.5 cm diameter or smaller will not be capable of leaking radiation. Larger diameter wave guides will attenuate waves to varying degrees, and by reference to wave guide tables it is possible to calculate the length of port required for a given diameter and attenuation. The attenuation constant for a cylindrical wave guide may be calculated from the relationship:

$$\alpha = \left(\frac{20}{\ln(10)}\right)\left(\frac{\lambda_0^2}{1-(\lambda_0/\lambda_c)} \cdot \frac{1}{\alpha\lambda_c^2} \cdot \left(\frac{\pi}{\lambda_0\eta\sigma}\right)^{1/2} \cdot \left[1 + \frac{\lambda_c^2}{\lambda_0^2}\left(\frac{n^2}{x^2 - n^2}\right)\right]\right) \text{dB m}^{-1}$$

where λ_c is the wavelength of the microwave in the wave guide, η the susceptibility and σ the conductivity of the wave guide material, and x is a mode dependent variable. For the TE$_{10}$ mode $x = 1.84$, and the resulting relationship between the attenuation constant and the diameter of a copper port for a 2.45 GHz wave is shown in Figure 11.9. It is generally recommended that the port should have a total attenuation of > 60 dB. For most purposes, it is a convenient starting point to note that for a 2.45 GHz microwave source and a 6.1 cm (*i.e.*, $\lambda/2$)

11.3. MULTIMODE MICROWAVE OVENS

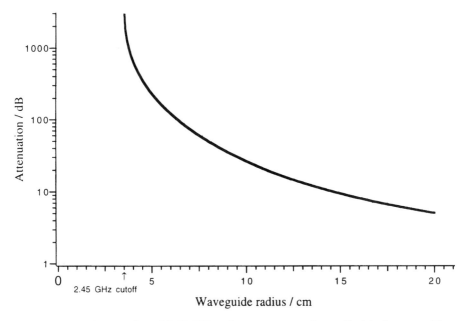

Figure 11.9 Attenuation of 2.45 GHz microwave power in a cylindrical wave guide.

diameter port, a port which is one wavelength (12.2 cm) long will provide complete protection. To enable convenient fitting of the ports through the oven's outer shell, they are constructed in two parts which may be detached when necessary. It is important that there is a good electrical contact between the tube components, and between the tube and the cavity wall. [18]

A port whose dimensions are small enough to prevent leakage may occasionally be quite limiting and allow only a small air condenser to be admitted without danger of microwave leakage. A simple air condenser may be passed through the port, but it does not always provide adequate cooling for reflux procedures, particularly if small reaction volumes or low boiling point solvents are involved. Even in combination with water condensers violent 'bump' boiling of the reactants was occasionally found to be a problem. Where appropriate, a minor modification of the water condenser (Figure 11.10a) allows the vapor to be cooled immediately above the reaction vessel. In an important secondary function, the water introduced into the cavity to provide cooling also behaves as a water load by absorbing excess microwave power. The microwave oven with a modified power circuit gives a variable power output which, in combination with the condenser of Figure 11.10a, yields an effectively lowered output, giving the new power profile shown in Figure 11.10b. The consequence is that the power output is linear over the same range, but at lower, and therefore more useful, power levels.

494   11. MICROWAVE DIELECTRIC HEATING EFEECTS

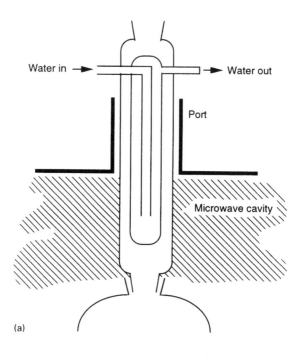

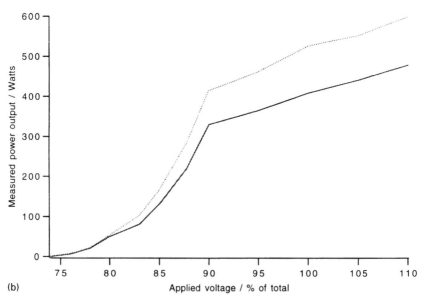

Figure 11.10 (a) Modified condenser and (b) its effect on the available microwave power.

11.4 CYLINDRICAL CAVITY

Figure 11.11 illustrates the application of a cylindrical microwave cavity which is suitable for fluidized bed reactions. In contrast to the other methods, in which the apparatus is placed inside the cavity of a modified microwave oven, heating is provided by a coaxial microwave source coupled to a small tuneable re-entrant cavity. The microwave source is particularly suitable for metal powders which couple very efficiently with the microwaves. Although the fluidized metal is less reflective than the undispersed sample, the sample remains quite opaque to the microwaves, and energy absorption occurs predominantly in the outer few millimetres of the sample. By creating samples of these dimensions, the heating becomes more efficient throughout the samples. The microwave source for this apparatus is slightly more sophisticated than that for a modified

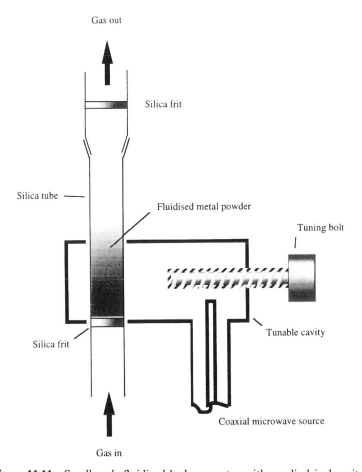

Figure 11.11 Small scale fluidized bed apparatus with a cylindrical cavity.

domestic oven, having a measured output power, and an indicator of the power which is reflected back from the sample. As a result, it is possible to tune the cavity around the sample so as to maximize the microwave power which is absorbed in it [19].

In any common sample of metal powder, there will be a large range of metal particle sizes. It is well known that fluidized beds have the effect of sorting particles into zones of differing particle sizes. Thus, larger metal particles will accumulate towards the base of the bed whilst the smaller particles will dominate the upper layers, with intermediate sized particles being sorted accordingly between the two extremes. For a given frequency of incident radiation, certain dispersions will have superior microwave absorbing properties than others, dependent upon the particle size and degree of dispersion. For this reason it is to be expected that, given a uniform field strength through the sample, these regions will heat up more rapidly than others.

11.4.1 Reaction Vessels

The efficient coupling of microwaves and polar organic solvents can be utilized to heat rapidly reaction solutions and thereby accelerate chemical reactions. If the reactants are enclosed in a closed vessel which is transparent to microwaves the vessel has to be able to sustain the pressures which are generated by rapidly heating the volatile organic solvents. There are a number of plastic materials which are microwave transparent and also sufficiently chemically inert to be used as materials for high pressure reactors. The use of a steel vessel which could effectively sustain the high pressures generated is not possible because metals such as steel are not transparent to microwaves.

The pressures and temperatures generated in an autoclave which is transparent to microwaves depend on the microwave power, the dielectric loss of the solution, the volatility of the solvent, the volume of the container occupied by the solvent and the extent of gas formation during the reaction. It is important to note that for volatile organic solvents such as CH_3OH, CH_3CN and THF it is possible to generate quite large pressures (> than 20 atmospheres) with moderate power inputs, and therefore it is imperative a safety pressure release valve be incorporated into the device if the technique is to be used safely and routinely.

Parr [20] has developed a range of acid digestion vessels which when new can accommodate pressures up to 80 atmospheres and temperatures up to 250 °C. Figure 11.12 illustrates a sectional drawing of such a vessel which, since its reaction vessel is made of Teflon, can be used for a wide range of organic solvents. Teflon has disadvantageous flow and creep properties particularly at temperatures above 150 °C and it is also slightly porous. Therefore, distortions of the vessel occur on repeated use and its pressure holding characteristics deteriorate. The porosity of the Teflon leads to the incorporation of materials such as organic tars and particles of metal which can reduce the transparency of the vessel and can result in the vessel itself reaching high temperatures. The main

11.4. CYLINDRICAL CAVITY

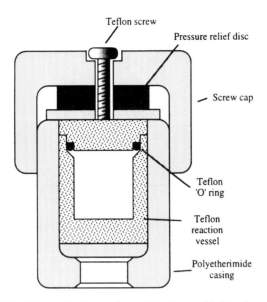

Figure 11.12 Schematic illustration of the Parr acid digestion autoclave.

disadvantage of the Parr vessel is the lack of control over the pressure and temperature within the reactor vessel. Also, the pressure safety release valve is very unsatisfactory and the release of the reactants can be quite violent.

CEM have developed an alternative acid digestion vessel, which has a lower working pressure (14–15 atmospheres) but the temperature and pressure can be continuously monitored within the vessel [21]. A sectional drawing of the vessel is illustrated in Figure 11.13. The reaction vessel is also made of Teflon. A feedback system has been incorporated within the system to allow a constant pressure (or temperature) to be held in the reactor vessel for long periods of time. A safety pressure release valve has also been incorporated.

There are a number of problems associated with the use of Teflon autoclaves, and from the chemist's point of view, the most significant of these is that they are not designed for use with high boiling point or viscous solvents. As a result they are prone to explosive failure under many chemically interesting conditions, despite the incorporation of safety valves. A microwave vessel specifically designed for chemical synthesis which circumvents both of these drawbacks has recently been reported (Figure 11.14) [22]. This apparatus, based upon the familiar Fischer-Porter vessel used in conventional organometallic syntheses, is introduced into the microwave cavity through a suitable port in the oven roof. The glass vessel not only allows convenient control of the reaction atmosphere, but also observation of the reaction as it proceeds. More importantly, it can be used to heat solvents for longer, and to higher temperatures than are possible using PTFE vessels. The apparatus has a specified safe working pressure of

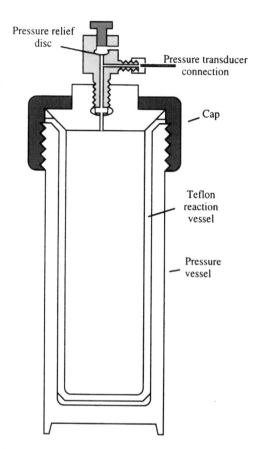

Figure 11.13 Schematic illustration of the CEM digestion vessel.

approximately 7–8 atmospheres, and for this reason a pressure release valve set to 11.5 atmospheres is included in the design. Signals from the pressure sensor are output through control circuitry and may be used to control the vessel pressure by switching the microwave oven on or off while simultaneously running a pressure display.

The types of reaction vessels described above have all been used to shorten the time spans of a wide range of organic and inorganic reactions. In general the reaction rate doubles for each 10 °C temperature rise, and therefore if a reaction in an enclosed microwave cavity can be maintained at 50 °C above the conventional reflux temperature for the reaction, the timescale for the reaction is reduced by a factor of 32, and if maintained at 100 °C above, by a factor of 1024. Furthermore, since the containment materials are transparent to microwaves and the microwave radiation couples directly with the solvents, rate enhancements somewhat greater than these are frequently observed.

11.5. TEMPERATURE MEASUREMENT

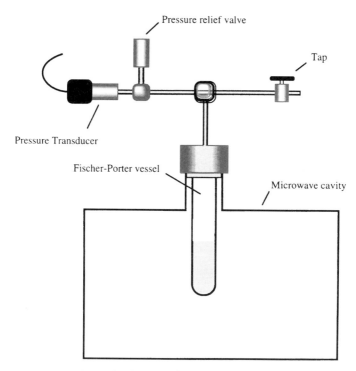

Figure 11.14 A schematic diagram of the Fischer Porter microwave reactor.

11.5 TEMPERATURE MEASUREMENT

Temperature measurement creates some difficulties in the presence of high power microwave fields. Although these are particularly acute when attempting to measure the high temperatures of some solid state reactions, they may be overcome at the lower temperatures of most liquid phase reactions. Reliable and accurate methods remain expensive, however.

Xylene Thermometer This is perhaps the simplest device which has been used to measure temperature in the microwave apparatus. Alcohol and mercury thermometers cannot be reliably used in the microwave cavity since both liquids couple with the microwaves. Alcohol is heated by microwaves and, therefore particularly when used within a liquid medium, which is a poor microwave absorber, the thermometer will tend to give readings above the actual temperature. A mercury-filled thermometer on the other hand may create electrical discharges which not only reduce its reliability, but which may be dangerous in the presence of flammable solvents. The xylene thermometer design simply replaces the alcohol/mercury with xylene. Xylene is inert, non-conducting, and has a very low dielectric loss tangent. Because of these properties, it behaves identically to a conventional thermometer, whilst being compatible with use in a microwave field. This method is of limited practical use since measurements cannot be readily used as the basis of a control system for the magnetron. Measurements taken by this method were found to be accurate to within 0.5 K and were capable of registering superheating effects in boiling solvents.

Gas Pressure Thermometer This thermometer is based on a design published by Whan *et al.* [23]. A capillary tube with a large bulb on one end is fitted with a pressure transducer (0–1 atm) at the opposite end (Figure 11.15). Nitrogen was used as the filler gas, and a T-joint adjacent to the transducer allowed evacuation of the capillary to approximately 25 mm Hg, prior to sealing with a flame. Increasing the temperature of the bulb increases the pressure of the gas, and this is registered by the transducer, which increases its analogue voltage output. This voltage output was directly calibrated against temperature, although it was found to be more convenient to first amplify the signal. The amplified signal may also be used as the basis for a computer controlled microwave heating system.

Fluoroptic Thermometry This technique is particularly useful in microwave heating systems as the probe contains no metallic or high loss materials. The probe is based around a fiber optic cable with a temperature-sensitive phosphor

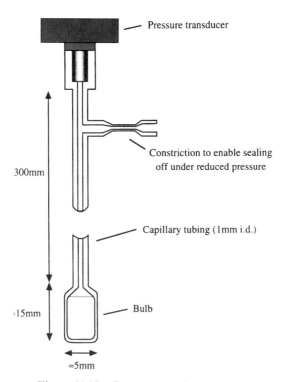

Figure 11.15 Gas pressure thermometer.

11.5. TEMPERATURE MEASUREMENT

painted onto the tip. The phosphor–manganese activated magnesium fluorogerminate–displays a deep red fluorescence on being activated with blue light whose decay time is temperature dependent. In the commercially available apparatus, a brief pulse of blue light from a filtered xenon flashgun is transmitted down the fiber optic line, and the fluorescence thus produced in the phosphor is transmitted back to a sensor. The decay time of the fluorescence is measured, correlated to the detector's internal calibration table, and the temperature of the probe tip displayed (Figure 11.16). The disadvantage of this method is that the equipment is extremely expensive at the present time.

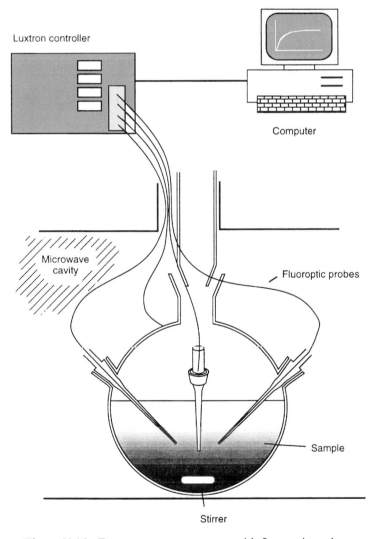

Figure 11.16 Temperature measurement with fluoroptic probes.

11.6 HEALTH CONSIDERATIONS

Once fitted, any modifications to a microwave cavity should be checked for microwave leakage using a hand held detector [24, 25]. The current legislation allows a maximum of 10 mW cm^{-2} at a distance of 5 cm from the source for up to 2 minutes in any 1 hour period, or short-term exposure of up to 25 mW cm^{-2}. The exposure limits are controversial and have been heavily revised during the past decade as medical evidence on the health effects of low frequency radiation has accumulated [26]. In view of the downward trend in recommended exposure levels, caution should be exercized when modifying microwave equipment so as to minimize microwave leakage.

The hazards of microwave radiation are still the subject of intense debate and research, and it is not yet known whether chronic exposure to low levels is harmful. Physiological effects of high levels of microwave radiation result mainly from hyperthermia, and whilst there is little consensus in the literature on the effects of low level radiation [27], it is commonly agreed that the most sensitive organs are the eyes and the testes [28]. Cataracts in the cornea resulting from microwave radiation have long been recognized to arise principally as a consequence of poor heat transfer from this organ. The risk from unmodified domestic ovens is not thought to be significant [29]. It has been suggested that microwaves may be responsible for miscarriages [30], alopecia [31], and hematological changes [32]. However, evidence linking microwaves with the formation of cancer is at present too indefinite for detailed conclusions to be drawn [33].

11.7 APPLICATIONS OF MICROWAVE DIELECTRIC HEATING

We have recently comprehensively reviewed such applications in chemical synthesis [34] and therefore only the important points will be made here.

11.7.1 Solutions [35, 36]

As indicated in the introduction, a wide range of polar organic molecules couple effectively with microwaves at 2.45 GHz and experience large heating rates (1–2 °C/s) even when the power level is 500 W. The high energy input into the solvent, and the fact that the reaction vessels are made of materials which are transparent to microwaves, makes superheating a feature which is commonly observed when organic solvents are dielectrically heated. This phenomenon, in the absence of rapid mixing, leads initially to small regions of superheated liquid which are located in the center of the solution. These enlarge and progressively move away towards the walls of the vessel. As a result a large number of organic solvents show boiling points up to 20 °C higher than their conventional boiling points in a microwave cavity [37]. These results are summarized in Table 11.2. A model has been proposed [38] to account for this behavior based on the

11.7. APPLICATIONS OF MICROWAVE DIELECTRIC HEATING

Table 11.2 Nucleation limited boiling points of some organic solvents in a microwave cavity and their conventional boiling points

Solvent	Nucleation limited boiling point (°C)	Conventional boiling point (°C)
Water	104	100
Methanol	85	64
Ethanol	103	79
2-propanol	100	82
Tetrahydrofuran	81	66
Acetonitrile	107	81
Acetone	81	56
Chlorobenzene	150	132
Ethylacetat	95	78

mechanism of nucleate bubble formation which recognizes that boiling is a kinetic phenomenon requiring bubble formation on the surface of the vessel. Since in microwave dielectric heating the surface is the coolest part, an effective rate of bubble formation requires that the bulk of the liquid is superheated. This superheating is responsible for the rate enhancements of chemical reactions undertaken in a microwave cavity at atmospheric pressures. Of course, these effects are magnified in vessels which can accommodate higher pressures and result in solutions which are maintained at 50–150 °C above their conventional reflux temperatures.

11.7.2 Organic Synthesis

Atmospheric Pressure The exploitation of the acceleration of organic reactions using microwave dielectric heating has become widespread. Microwave dielectric heating is regularly used for a number of multi-step syntheses and has been incorporated within undergraduate laboratory courses. Microwave cavities provide a relatively cheap and clean alternative to oil baths, particularly if they have been modified to incorporate ports for reflux condensers [39]. Alternatively higher boiling and less volatile organic solvents, such as DMF or diglyme, may be used without making the reflux modifications [41]. The Diels-Alder addition of maleic anhydride to anthracene may be carried out in 90% yield in diglyme (b.p. 162 °C) in one minute under microwave conditions, whereas the conventional synthesis in benzene (b.p. 80 °C) requires 90 minutes. Recent applications of this method include amide syntheses [42], hydrogenation of β-lactams [43] and 1,4-dihydropyridones [44].

The question of whether a specific microwave effect is operating rather than a bulk heating effect has been the subject of considerable debate. The majority of early claims for a specific microwave effect have been discounted and reinterpreted in terms of super heating effects [45].

High Pressure Effects The remote nature of the microwave interaction and the transparency to microwaves of a wide range of organic polymers and plastics make microwave heating a very powerful technique for 'temperature enhancement' of reactions. With readily available containers which can sustain 10–40 atmospheres pressure it is possible to generate temperature enhancements of 50–100 °C above the conventional reflux temperatures. No specific microwave effects need to be invoked to account for the majority of rate enhancements observed in such reactions. A wide range of organic reactions has been studied using pressure vessels, including Diels-Alder reactions [46], organic racemizations [47], hydrolyses, esterifications, epoxidations and cyclization reactions [48].

The processing of large quantities of materials by scaling-up small scale microwave-sensitive reactions is not always feasible using conventional microwave equipment. However, high pressure tubing and pumps similar to those used for HPLC may be used to build a flow system which passes through a microwave cavity and which can give a significant throughput of product [49].

A particularly important application of microwave dielectric heating has been developed in positron emission tomography (PET). Radionuclides such as ^{122}I($t_{1/2}$ = 3.6 min), ^{11}C($t_{1/2}$ = 20 min) and ^{18}F($t_{1/2}$ = 110 min) have been successfully incorporated into a range of important drugs using microwave dielectric enhancement reactions [50, 51]. Continuous flow systems have also been incorporated into these microwave reactors [52, 53].

11.7.3 Synthesis of Organometallic and Coordination Compounds

In comparison with the organic syntheses described above, the synthesis of organometallic and coordination compounds has not been exploited to anything like the same extent. A number of post-transition element organometallic compounds have been synthesized using pressure vessels similar to those used for organic syntheses [54]. Reactions were conducted in 100 cm³ Teflon pressure vessels containing relatively small volumes of solvent (12 cm³). Rate enhancements of approximately 40 fold were noted.

A large number of coordination compounds have been synthesized. The technique is most effective for enhancing the rates of reactions leading to compounds with d^3, d^6 (low-spin), and d^8 electronic configurations which are kinetically inert, but thermodynamically labile [55]. The reaction times were reduced to as much as 1/60th using a microwave cavity with a reflux modification, and the yields were comparable to those reported for conventional reactions. Even greater rate enhancements were observed if the reactions were undertaken in sealed pressure reaction vessels.

Improved syntheses of [Fe(η-C$_5$H$_5$)(η-arene)]PF$_6$ and [Fe(η-arene)$_2$]PF$_6$ salts which utilize the ability of metal powders to couple to microwaves, have

11.7. APPLICATIONS OF MICROWAVE DIELECTRIC HEATING

been reported. The apparatus used for these experiments involved a solid CO_2 coolant condenser to prevent the loss of solvent. Yields up to three times those of conventional syntheses were reported based on reaction times of approximately 4 minutes [56].

11.7.4 Polymers

Microwave dielectric heating has been widely applied to polymer chemistry; the ability to control the syntheses by directly heating the reactants has resulted in important cost and energy savings. Epoxy resins have been most widely studied [57] because their dielectric properties are ideal for induced microwave curing. Interestingly the studies have indicated that the rate of curing depends more on the way in which the microwave power is pulsed than the total power input. Investigations into the curing of the diglycidyl ester of bisphenol (DGEBA) with cross linking agents such as 4,4′-diaminophenolsulfane (DNS), dicyanodiamide (DOA), 4,4′-diaminodiphenylether (DDE) and diaminodiphenylethane (DDM) [58, 59] suggest that the pulsing rate may effect the type of bonds formed. The reactions with amines are favored by long pulses, whereas self polymerization is facilitated by short pulses. The optimum pulse frequency for cross linking is dependent on the cross linking agent, and for example with DOM, optimum repetition frequencies of 23.8, 200 and 1000 Hz have been identified [60, 61]. It has been proposed that this behavior is associated with a relaxation process which is derived from the rotations of chain segments in the macromolecular network [62].

The dielectric properties of the reactants leading to the polymer may be influenced by adding secondary materials which couple effectively with microwaves. Conducting materials such as carbon black, carbon fibers, copper and aluminium have been added and have resulted not only in larger temperature rises, but also affected the physical properties of the resultant composite. Conducting organic polymers such as polyaniline *p*-toluene sulfonate and polypyrrole *p*-toluene sulfonate are excellent microwave absorbers and make ideal adhesives for welding plastics [63].

11.7.5 Chemical Analyses

Microwave dielectric heating has had a remarkable influence on analytical procedures. The analyses of foodstuffs and metal containing samples generally require an acid digestion step, and until recently the length of time required for this process influenced the time of the complete analytical procedure. Acceleration of the digestion step using commercially available microwave equipment has therefore had a major impact [64]. There are hundreds of papers dealing with the applications of microwave assisted processes in analytical techniques. References [65–72] give some of the more important reviews and books which have been published in recent years. On line and continuous flow equipment for microwave digestion are now available leading to almost fully automated analytical systems.

11.7.6 Microwave Syntheses Involving Solids

Intercalation Reactions The intercalation of organic molecules within the layers of an inorganic host material is frequently a kinetically redundant process which requires refluxing for days the insoluble inorganic component in an organic solvent containing the guest compound. These heterogeneous reactions may be accelerated using ultrasonic techniques; however, this leads to a degradation of the crystallinity of the crystallites of the host material. It has been shown that the rate of incorporation of substituted pyridines into $VO(PO)_4$ may be enhanced by 10^2-10^3 and the crystallinity of the host maintained by using microwave dielectric heating in sealed Teflon vessels. [73] Moreover, more of the guest compound was incorporated into the host material. In these heterogeneous reactions it is not the $VO(PO)_4$ which couples with the microwaves, but the polar substituted pyridines which behave both as the guest molecules and the solvent.

Intercalation of aryl tin compounds into the synthetic clay laponite has resulted in tin(IV) oxide pillared clays [74]. Although significant rate enhancements were observed, the quality of the sample was inferior to that made by conventional techniques involving mechanical shaking at room temperature.

Ceramic Materials Initial experiments performed in this area centered around the drying of ceramics using microwave dielectric heating. It has been estimated that for ceramic samples containing less than 5% water, microwave dielectric heating is more efficient than conventional heating, and more even drying profiles may also be achieved. Above this level, a hybrid system incorporating both methods at different drying stages is found to be more efficient [75]. However, it was recognized that many dry ceramics have high loss tangents and couple efficiently with microwaves [76]. Specifically oxides [77, 78], sulfides [79] and glass ceramics [80] were shown to have desirable dielectric properties which would be utilized in sintering and processing experiments. The heating characteristics of some typical inorganic solids are summarized in Table 11.3 [81–83].

The sintering of ceramics has been extensively investigated. Ceramics based on aluminium and magnesium oxide, ferrites, boron and silicon carbides and high temperature superconducting ceramics have been particularly widely studied.

The following advantages have been identified for ceramic processing [84]:

1) *Reduced cracking and thermal stress.* In microwave processing, the temperature profile is different from that for conventional heating and the higher temperatures are generated towards the center of the sample. In conventional furnace heating, heat absorbed from the surface may lead to thermal stresses, cracking and related problems. Thus microwave dielectric heating is particularly useful for densifying ceramic components.
2) *Economy.* Since the oven itself is not heated, microwave dielectric heating may be more economical.

11.7. APPLICATIONS OF MICROWAVE DIELECTRIC HEATING

Table 11.3 Heating characteristics of some common inorganic minerals in a microwave cavity

A*			B*		
Material	Temp./°C	Time/min	Material	Temp./°C	Time/min
Al	577	6	CaO	83	30
C	1283	1	CeO_2	99	30
Co_2O_3	1290	3	CuO	701	0.5
CuCl	619	13	Fe_2O_3	88	30
$FeCl_3$	41	4	Fe_3O_4	510	2
$MnCl_2$	53	1.75	La_2O_3	107	30
NaCl	83	7	MnO_2	321	30
Ni	384	1	PbO_2	182	7
NiO	1305	6.25	Pb_3O_4	122	30
$SbCl_3$	224	1.75	SnO	102	30
$SnCl_2$	476	2	TiO_2	122	30
$SnCl_4$	49	8	V_2O_5	701	9
$ZnCl_2$	609	7	WO_3	532	0.5

* A: 25 g samples, 1 kW irradiation with 2 L vented water load; B: 5–6 g samples, 500 W irradiation

3) *Increased strength.* The rapid heating by microwaves reduces the extent of non-isothermal processes such that the segregation of impurities to the grain boundaries is reduced. The ceramic is strengthened by minimizing impurity segregation, decreased grain size, and increased sintered density.
4) *Reduced contamination.* The extent of contamination from the walls of the containment crucible is reduced because for a microwave heated sample the walls are relatively cool. Microwave joining has been achieved for a number of ceramic materials. Palaith *et al.* [85] have butt-welded two mullite rods together at 1300 °C in approximately 10 minutes.

Solid State Reactions If one or more components of a solid state reaction have a high loss tangent, the mixture couples very effectively with the microwaves and the associated temperature rise may be used to drive the chemical reaction to completion. The occurrence of 'thermal runaway' for the reactants leads to particularly fast and dramatic reactions. Indeed, for such reactions melt conditions are rapidly achieved, and the reaction product may be quenched by turning off the microwave power. The remote nature of the microwave source and the absence of effective insulation can lead to particularly efficient quenching.

Table 11.4 summarizes some solid state reactions which have been completed successfully using microwave dielectric heating [86]. The starting materials are intimately mixed and can be pressed into pellets prior to being placed in

Table 11.4 Examples of solid state reactions which have been performed in a microwave cavity

Reactants	Products	Conditions
$V_2O_5 + C$	VO_2	5 min/500 W
$V_2O_5 + 2C$	V_2O_3	20 min/500 W
$U_3O_8 + C$	UO_2	5 min/500 W
$Fe_3O_4 + SrCO_3 + La_2O_3$	$LaSr_2Fe_3O_{8-x}$	36 min/500 W
$Cr + B$	CrB	27 min (3 stages) 500 W
$Zr + 2B$	ZrB_2	5 min/500 W
$Cu + In + 2S$	$CuInS_2$	1 + 3 + 3 min/400 W
$Cu + In + 2Se$	$CuInSe_2$	1 + 3 + 3 min/400 W

alumina, zirconia or silica crucibles. The reaction times are much shorter than those associated with conventional solid state reactions since the reactants reach temperatures of 1000–1500 °C within a matter of minutes. The resultant products show sharp X-ray powder diffraction patterns, and this may be associated with the transitory formation of a melt prior to quenching by switching off the microwave power.

Several groups [87–89] have synthesized samples of superconducting $YBa_2Cu_3O_{7-x}$ utilizing the coupling of CuO in the reactant mixture. The product also has a very high loss tangent which ensures that the reaction proceeds to completion. It has been established that since much shorter reaction times are involved than in conventional synthesis, there is no need to slowly cool the product in an oxygen atmosphere after the reaction is complete in order to obtain the superconducting orthorhombic phase.

Metals were generally regarded as unsuitable reactants for microwave accelerated reactions, because induced voltages in large metal objects often result in electric discharges. We have recently established that metal powders dispersed in a secondary non-conducting medium couple very efficiently with microwaves and this property may be utilized for a wide range of reactions [90]. The microwave accelerated syntheses of several metal chalogenides from the elements were achieved by mixing them in sealed silica ampoules and then placing them in a microwave cavity (500 watt power). The metal particles heat very rapidly on coupling to the microwaves and react directly with the surrounding particles of chalogenides. Therefore, much lower pressures are generated than in conventional reactions because less unreacted chalcogen vapor is formed [91]. Similar techniques have been developed for the synthesis of the semi-conducting ternary sulfides and selenides, $CuInS_2$, $CuInSe_2$ and $CuInSSe$ [92]. Metal powders have also been used in the synthesis of metal borides in a microwave cavity [93].

The coupling of metal powders with microwave radiation has also been used to develop fluidized bed reactions for the synthesis of metal oxides, nitrides and halides. The metal powder is fluidized in a counter stream of gas and the relevant silica container is located in a microwave cavity. When the microwave power is switched on the metal particles couple very efficiently with the microwaves and the resultant temperature rise stimulates the reaction with the gas. The solid product sublimes to the relatively cool walls of the reactor table. [94]

Organic Reactions on Dry Media The possibility of conducting organic reactions without or with a minimal amount of solvent has considerable attractions to industry wishing to limit the environmental impact of chemical processes. In this context some interest has been shown in the possibility of performing microwave assisted reactions on solid inorganic supports. Support materials which have been investigated include: montmorillonite clays, alumina, silica, alkali metal fluorides or acetate doped alumina and bentonite earths. Generally the organic reactants are introduced onto the support as a solution in a small volume of CH_2Cl_2. The reaction is then completed on the dry support by microwave dielectric heating which raises the temperature of the support to 200–300 °C. After the reaction is complete the product is extracted from the support using a small amount of solvent. The vast majority of reactions studied involve condensations and some typical examples are presented in Table 11.5.

Catalysis A number of catalytic procedures have been studied where the energy input is provided by microwave dielectric heating rather than conventional techniques. The catalytic reactions generally appear to give product distributions which are similar to those observed conventionally but at a lower apparent temperature, and there have been some reports of changes in reaction selectivities [101, 102].

11.8 SUMMARY

More than two hundred papers dealing with the applications of microwave dielectric heating in chemistry have been published since the mid-nineteen eighties. Commercially available microwave ovens were used for the initial experiments. Since the early experiments, the mechanism of microwave dielectric heating in chemical systems has been defined more precisely and many advances have been made in the experimental procedures for handling chemical reactions in a microwave cavity in a safe and efficient manner. The following advantages may be associated with dielectric heating:

1) The energy is introduced remotely because of its wave nature.
2) The container materials may be chosen to be transparent to microwaves and therefore the energy is transferred directly to the reacting solution.

Table 11.5 Some examples of organic reactions which have been completed on organic supports

Reaction			Ref.
barbituric acid + R-CHO	Montmorillonite KSF, 350–510 W, 4–5 min	5-arylidene barbiturate	[95]
aniline-PhNH-NH$_2$ + R$_1$COR$_2$	Montmorillonite, 160 W, 5 min	2,3-disubstituted indole	[96]
R-CHO + 3-methyl-1-phenyl-pyrazolone	Montmorillonite KSF, 160 W, 2–5 min	4-arylidene pyrazolone	[97]
MeCOCH$_2$COMe + HNR$_1$R$_2$	Montmorillonite K$_{10}$, 100–250 W, 1–3 min	enaminone	[98]
sugar diol + R-CHO	Montmorillonite, 60 W, 10 min	acetal	[99]
H$_2$C-(CH$_2$)$_n$-NH$_2$ + MeCOCH$_2$CH$_2$CO$_2$Et	Al$_2$O$_3$/Montmorillonite K$_{10}$, 125–440 W, 6–15 min, n = 1, 2, 3	bicyclic pyrrolo product	[100]

3) The coupling of microwave energy is very efficient leading to rapid temperature increases.
4) The method is very convenient for introducing controlled superheating.
5) The rapid cooling of the sample, when the microwave power is switched off, makes it ideal for flash heating experiments.
6) The differences in temperature profiles resulting from (1) to (5) can lead to alternative product distributions for chemical reactions.

ACKNOWLEDGEMENT

The contribution of David Baghurst to our research in the microwave area has been enormous and is gratefully acknowledged.

11.9 REFERENCES

[1] Jolly, W. L., The Synthesis and Characterization of Inorganic Compounds, Prentice Hall, Englewood Cliffs, NJ, 1970.
[2] Harvey, A. F., Microwave Engineering, Academic Press, New York, 1963.
[3] Herzberg, G., Infrared and Raman Spectra, van Nostrand, Reinhold Co., London, 1945.
[4] Banwell, C. N., Fundamentals of Molecular Spectroscopy, 3rd edn., McGraw-Hill, London, 1983.
[5] Debye, P., Polar Molecules, Chemical Catalog, New York, 1929.
[6] Fröhlich, H., Theory of Dielectrics, 2nd edn., Oxford University Press, Oxford, 1958.
[7] Daniel, V., Dielectric Relaxation, Academic Press, London, 1967.
[8] Cole, K. S., Cole, R. H., J. Chem. Phys., 9 (1941) 341.
[9] Hill, N., Vaughan, W. E., Davies, M., Dielectric Properties and Molecular Behaviour, van Nostrand, Reinhold Co., London, 1969.
[10] Hasted, J. B., Aqueous Dielectrics, Chapman and Hall, London, 1973.
[11] Metaxas, A. C., Meredith, R. J., Industrial Microwave Heating, Peter Perigrinus, London, 1983.
[12] Gabriel, S., Mingos, D. M. P., unpublished results.
[13] Böttcher, C. J. F., Theory of Electric Polarisation, Elsevier, Amsterdam, 1952.
[14] Wagner, K. W., Arch. Elektrotech., 2 (1914) 371.
[15] Hamon, B. V., Aust. J. Phys., 6 (1953) 304.
[16] Meakins, R. J., Trans. Faraday Soc., 51 (1955) 953.
[17] Smyth, C. P., Dielectric Behaviour and Structure, McGraw-Hill, New York, 1955.
[18] Whittaker, A. G., D. Phil. Thesis, Oxford University, 1994.
[19] Whittaker, A. G., Mingos, D. M. P., J. Chem. Soc., Dalton Trans., (1993) 2541.
[20] Perry, J. H., (ed.), Chemical Engineers Handbook, 3rd edn., McGraw-Hill, New York, 1950.
[21] CEM Corporation, Matthews, North Carolina 28106, U.S.A..
[22] Baghurst, D. R., Mingos, D. M. P., J. Chem. Soc., Dalton Trans., (1992) 1151.
[23] Bond, G., Moyes, R., Pollington, S., Whan, D., Measurement Science and Technology, 2 (1991) 571.
[24] Apollo XI Microwave Monitor, Apollo Enterprises Ltd., Surrey, England.
[25] Bleaney, B. I., Bleaney, B., Electricity and Magnetism, 3rd edn., Oxford University Press, 1983.
[26] Peterson, R. C., Health Phys., 61 (1991) 59.

[27] Hileman, B., Chem. Eng. News, 171 (1993) 15.
[28] Hayes, R. B., Brown, L. M., Pottern, L. M., Gomez, M., Kardaun, J. W., Hoover, R. N., O'Connell, K. J., Sutzman, R. E., Javadpour, N., Int. J. Epidemol., 19 (1990) 825.
[29] Leitgeb, N., Tropper, K., Biomed. Tech. Berlin, 38 (1993) 17.
[30] Ouellet-Hellstrom, R., Stewart, W. F., Am. J. Epidemol., 138 (1993) 775.
[31] Isa, A. R., Noor, M., Med. J. Malaya, 46 (1991) 235.
[32] Gordoni, J., Health Phys., 58 (1990) 205.
[33] Salvatore, J. R., Weitberg, A. B., R. I. Med. J., 72 (1989) 15.
[34] Whittaker, A. G., Mingos, D. M. P., Journal of Microwave Power and Electromagnetic Energy, 29 (1994) 195.
[35] Gedye, R., Smith, F., Westaway, K., Ali, H., Baldoisa, L., Laberge, L., Tetrahedron Lett., 27 (1986) 279.
[36] Giguere, R. J., Namen, A. M., Lopez, B. O., Arepally, A., Ramos, D. E., Majetich, G., Tetrahedron Lett., 28 (1987) 6553.
[37] Toma, S., Chem. Listy, 87 (1993) 627.
[38] Baghurst, D. R., Mingos, D. M. P., J. Chem. Soc., Chem. Commun., (1992) 674.
[39] Baghurst, D. R., Mingos, D. M. P., Watson, M. J., J. Organomet. Chem., C43 (1989) 368.
[40] Bose, A. K., Manhas, M. S., Ghosh, M., Raju, V. S., Tabei, K., Urbanczyklipkowska, Z., Heterocycles, 30 (1990) 741.
[41] Bose, A. K., Manhas, M. S., Ghosh, M., Shah, M., Raju, V. S., Bari, S., Newaz, S., Banik, B., Chaudhary, A., Barakat, K., J. Org. Chem., 56 (1991) 6968.
[42] Vazquez-Tato, M. P., Synlett, 7 (1993) 506.
[43] Bose, A. K., Banik, B. K., Barakat, K. J., Manhas, M. S., Synlett, 8 (1993) 575.
[44] Alajarin, R., Vaquero, J. J., Navio, J. L. G., Alvarez-Builla, J., Synlett, 4 (1992) 297.
[45] Mingos, D. M. P., Chem. and Ind., (1994) 596.
[46] Stambouli, A., Chastrette, M., Soufiaoui, M., Tetrahedron Lett., 32 (1991) 1723.
[47] Takano, S., Kijima, A., Sugihara, T., Satoh, S., Ogasawara, K., Chemistry Letters, 1 (1989) 87.
[48] Lie Ken Jie, M. S. F., Yan kit, C., Lipids, 23 (1988) 367.
[49] Dema-Khalaf, K., Morales-Rubio, A., de la Guardia, M., Anal. Chim. Acta, 281 (1993) 249.
[50] Hwang, D. R., Moerlein, S. M., Lang, L., Welch, M. J., J. Chem. Soc., Chem. Commun., (1987) 1799.
[51] McCarthy, T. J., Dence, C. S., Welch, M. J., Appl. Radiat. Isot., 44 (1993) 1129.
[52] Peterson, C., New Scientist, 123 (1989) 44.
[53] Chen, S. T., Chiou, S. H., Wang, K. T., J. Chem. Soc., Chem. Commun., (1990) 807.
[54] Ali, M., Bond, S. P., Mbogo, S. A., McWhinnie, W. R., Watts, P. M., J. Organomet. Chem., 371 (1989) 11.
[55] Baghurst, D. R., Mingos, D. M. P., J. Organomet. Chem., 384 (1990) C57.
[56] Dabirmanesh, Q., Roberts, R. M. G., J. Organomet. Chem., 460 (1993) C28.

11.9. REFERENCES

[57] Finzel, M., Hawley, M., Jow, J., Polymer Engineering and Science, 31 (1991) 1240.
[58] Thuillier, F. M., Jullien, H., Makromolekulare Chemie-Macromolecular Symposia, 25 (1989) 63.
[59] Thuillier, F. M., Jullien, H., Grenier-Loustalot, M.-F., Polym. Comm., 27 (1986) 206.
[60] Beldjoudi, N., Bouazizi, A., Doubi, D., Gourdenne, A., Eur. Polym. J., 24 (1988) 49.
[61] Beldjoudi, N., Gourdenne, A., a) Eur. Polym. J., 24 (1988) 53; b) Eur. Polym. J., 24 (1988) 265.
[62] Jullien, H., Valot, H., Polymer, 26 (1985) 506.
[63] Kathirgamanathan, P., Polymer, 43 (1993) 3105.
[64] CEM Corporation, Welcome to Microwave 101, Matthews, N. Carolina 28016, U.S.A..
[65] Matusiewicz, H., Sturgeon, R. E., Prog. Anal. Spectr., 12 (1989) 21.
[66] Rawls, R., Chem. Eng. News, 64 (1986) 31.
[67] Gilman, L. B., Engelhart, W. G., Spectroscopy, 4 (1988) 18.
[68] Paukert, T., Chem. Listy, 86 (1992) 143.
[69] Kokot, S., King, G., Keller, H. R., Massart, D. L., Anal. Chim. Acta, 259 (1992) 267.
[70] Kubrakova, I. V., Su, M. Y., Abuzveida, M., Kuzmin, N. M., J. Anal. Chem. (Engl. Tr.), 47 (1992) 563.
[71] Kuss, H. M., Fresenius J. Anal. Chem., 343 (1992) 788.
[72] Kingston, H. M., Jassie, L. B., (eds.), Introduction to Microwave Sample Preparation, American Chemical Society, Washington D.C., 1988.
[73] Chatakondu, K., Green, M. L. H., Mingos, D. M. P., Reynolds, S. M., J. Chem. Soc., Chem. Commun., (1989) 1515.
[74] Ashcroft, R. C., Bond, S. P., Beevers, M. S., Lawrence, M. A. M., Gelder, A., Mcwhinnie, W. R., Berry, F. J., Polyhedron, 11 (1992) 1001.
[75] Hirai, T., Tari, I., Ohzuku, T., Bull. Chem. Soc. Jpn., 53 (1980) 1477.
[76] Roy, R., Komarneni, S., Yang, L., J. Am. Ceram. Soc., 68 (1985) 392.
[77] Haas, P. A., Am. Ceram. Soc. Bull., 58 (1979) 873.
[78] Holcombe, C. E., Am. Ceram. Soc. Bull., 62 (1983) 1388.
[79] Chen, T. T., Dutrizac, J. E., Haque, K. E., Wyslouzil, W., Kashyap, S., Canadian Metallurgical Quarterly, 23 (1984) 349.
[80] Macdowell, J., Am. Ceram. Soc. Bull., 63 (1984) 282.
[81] McGill, S. L., Walkiewicz, J. W., Myres, G. A., Presented at 91st Annual Meeting Exposition of the American Ceramic Society, Indianapolis, IN, 1989.
[82] Walkiewicz, J. W., Kazonich, G., McGill, S. L., Minerals Metallurgical Processing, 5 (1988) 39.
[83] McGill, S. L., Walkiewicz, J. W., J. Microwave Power Electromagnetic Energy Symp. Summ., 22 (1987) 175.
[84] Dagani, R., Chem. Eng. News, 66 (1988) 7.
[85] Palaith, D., Silberglitt, R., Wu, C. C. M., Kleiner, R., Libelo, E. L., MRS Symp. Proc., 124 (1988) 255.

[86] Baghurst, D. R., Mingos, D. M. P., Br. Ceram. Trans. J., 91 (1992) 124.
[87] Baghurst, D. R., Chippindale, A. M., Mingos, D. M. P., Nature, 332 (1988) 311.
[88] Baghurst, D. R., Mingos, D. M. P., J. Chem. Soc., Chem. Commun., (1988) 829.
[89] Ahmad, I., Chandler, G. T., Clark, D. E., MRS Symp. Proc., (1988) 124.
[90] Whittaker, A. G., Mingos, D. M. P., J. Chem. Soc., Dalton Trans., (1992) 2751.
[91] Whittaker, A. G., Mingos, D. M. P., J. Chem. Soc., Dalton Trans., (1995) 2073.
[92] Barron, A. R., Landry, C. C., Science, 260 (1993) 1653.
[93] Baghurst, D. R., Mingos, D. M. P., Br. Ceram. Trans. J., 91 (1992) 124.
[94] Whittaker, A. G., Mingos, D. M. P., J. Chem. Soc., Dalton Trans., (1993) 2541.
[95] Villemin, D., Labiad, B., Synth. Commun., 20 (1990) 3213.
[96] Villemin, D., Labiad, B., Ouhilal, Y., Chem. and Ind., 18 (1989) 607.
[97] Villemin, D., Labiad, B., Synth. Commun., 20 (1990) 3333.
[98] Rechsteiner, B., Texier-Boullet, F., Hamelin, J., Tetrahedron Lett., 34 (1993) 5071.
[99] Csiba, M., Cleophax, J., Loupy, A., Malthete, J., Gero, S. D., Tetrahedron Lett., 34 (1993) 1787.
[100] Pilard, J. F., Klein, B., Texierboullet, F., Hamelin, J., Synlett, 3 (1992) 219.
[101] Bond, G., Moyes, R. B., Whan, D. A., Catalysis Today, 17 (1993) 427.
[102] Thiebaut, J., Roussy, G., Medjram, M., Garin, F., Seyfried, L., Maire, G., Catalysis Letters, 21 (1993) 133.

12

BIOMOLECULES UNDER EXTREME CONDITIONS

Karel Heremans
Department of Chemistry, Katholieke Universiteit Leuven, B-3001 Leuven, Belgium

12.1 INTRODUCTION
12.2 PRESSURE VERSUS TEMPERATURE EFFECTS
    12.2.1 Physical Transformations
    12.2.2 Chemical Reactions
    12.2.3 Molecular Interpretation of Thermodynamic and Kinetic Parameters
12.3 PHOSPHOLIPIDS
12.4 PROTEINS
    12.4.1 Randomly Coiled Polypeptide Chain: Folded Protein
    12.4.2 Protein Compressibility
    12.4.3 Protein Reactions
12.5 PROTEIN STABILITY
    12.5.1 Pressure versus Temperature Induced Protein Denaturation
    12.5.2 Infrared Spectroscopy Studies with the Diamond Anvil Cell
    12.5.3 High-Pressure NMR Spectroscopy
    12.5.4 Intermediates of Denaturation
    12.5.5 Limits of Extreme Conditions
12.6 MICROORGANISMS
12.7 INDUSTRIAL APPLICATIONS
12.8 PERSPECTIVES FOR THE FUTURE
12.9 REFERENCES

12.1 INTRODUCTION

This chapter has several interwoven themes. Whereas the title suggests that the main topic is *molecules* under extreme conditions, we will also indicate connections with the conditions for *life* under extreme conditions. Although not discussed extensively, *biotechnology* applications will receive some attention. Last but not least, extreme conditions are useful experimental tools. Not only can they be used to obtain thermodynamic and kinetic information for chemical phenomena, but they may also lead to new observations and surprises. The emphasis will be on pressure but, as we hope to show, a combination with temperature leads to new insights into the complexity of natural systems.

About a century ago it was shown that pressures below 100 MPa do not irreversibly affect enzyme processes in deep-sea bacteria. It was also observed that alcoholic fermentation can proceed without living cells despite the fact that pressures of 50 MPa were used to extract the enzymes. On the other hand, it was found that much higher pressures kill bacteria. As a consequence, it was proposed to use pressure as an alternative to heat treatment to prolong the shelf-life of fruit, vegetables and milk [1]. Bridgman [2] showed that a pressure of about 700 MPa is needed during a 30-min treatment of egg-white to obtain an irreversible coagulation of the solution. He also observed that the temperature at which the experiment was performed has a strong influence on the pressure needed to obtain the same stiffening. For lower temperatures, a lower pressure was needed. More recently, a number of papers have appeared on the coagulation of proteins, the inactivation of enzymes, viruses, antigens, antibodies and microorganisms. A review of the older literature may be found in the book by Johnson et al. [3]. The most spectacular effect that has been observed is the conversion of a number of carbon-rich materials, including "peanut-butter", into diamond-like crystals. This and other fascinating stories have been described in detail in a recent book on the breakthrough of high-pressure research in the physical sciences [4].

The history of the effect of other extreme conditions on microorganisms and their enzymes seems to be more recent, but not less exciting. Such organisms are commonly known as extremophiles. Thermophiles are exposed to temperature stresses which may exceed 100 °C or more in the case of hyperthermophiles. Permanent exposure to the cold is a condition for the so-called psychrophiles and exposure to high pressure is the challenge for piezo- or barophiles. The oceans cover nearly 70% of the earth's surface and have an average depth of 3.8 km with an average pressure of 38 MPa. Clearly, pressure plays an important role in the distribution of life in the oceans. Exposure to extreme chemical conditions may involve extremes of pH as in the case of the acidophiles or alkaliphiles, extremely high salt concentrations as in the case of halophiles, or extremely low water activity in other cases. Although the biotechnological possibilities of these organisms and the enzymes that they contain have already been exploited, the perspective towards new applications is very promising. A better understanding of the enzymology, physiology and the genetics of the organisms, in particular the unique *Archaea*, will contribute to possible applications in biotechnology. The physical-chemical basis of protein stability and enzymology poses considerable challenges for the bioscientist [5–11].

If we concentrate on the physical parameters temperature and pressure, it is now becoming clear that the occurrence of life on this planet seems to be restricted to rather narrow ranges of temperature and pressure. Laboratory experiments strongly suggest that these conditions are dictated by the dependence of the stability of proteins on these parameters. Mechanisms of the adaptation to chemical extremes may be quite different.

12.2 PRESSURE VERSUS TEMPERATURE EFFECTS

The physical-chemical reason for using extreme conditions is to explore the effect of temperature and pressure on the conformation, dynamics and reactions of biomolecules. The unique properties of biomolecules are determined by the delicate balance between internal interactions which compete with interactions with the solvent. The primary source of the dynamic behaviour of biomolecules is the free volume of the system. In the absence of strong mechanical constraints, this free volume may be expected to decrease with increasing pressure. As temperature effects act *via* an increased kinetic energy as well as an increase in the free volume, it follows that the study of the combined effect of temperature and pressure is a prerequisite for a full understanding of the dynamic behaviour of biomolecules. Intuitively, pressure effects should be easier to interpret than temperature effects.

12.2.1 Physical Transformations

The effect of pressure on the melting temperature (T_m) of compounds is given by the Clausius-Clapeyron equation

$$\frac{dT_m}{dp} = T_m \frac{\Delta V}{\Delta H}$$

Since the volume (ΔV) and enthalpy change (ΔH) on melting are generally positive, one would predict an increase in the melting temperature with increasing pressure. For many organic compounds dT_m/dp is ca. 15 K/100 MPa. An increase in pressure of 100 MPa would then correspond to an increase in temperature of 15 °C. While this statement is true from a thermodynamic point of view, it is certainly not true at the molecular level. A notable exception to this general rule is water. At room temperature, a pressure of about 1 GPa is needed to obtain the ice phase (VI) which differs from normal ice in its higher density. The amusing history of the discovery of so-called "warm ice" has been exposed recently in the biography of its discoverer, Bridgman [12].

The Clausius-Clapeyron equation as presented before, assumes that ΔV and ΔS in the process are pressure- and temperature-independent. The occurrence of the re-entrant phase in the phase diagram of lipids, proteins and polysaccharides shows that this is not always the case. Removal of these restrictions leads to the following result:

$$\frac{dT_m}{dp} = \frac{\Delta V_0 + \Delta\alpha(T - T_0) - \Delta\beta(P - P_0)}{\Delta S_0 + \Delta C_p(\ln T/T_0) - \Delta\alpha(P - P_0)}$$

with the following definitions: isobaric expansion factor ($\Delta\alpha = (\delta\Delta V/\delta T)_p$), isothermal compressibility factor ($\Delta\beta = (\delta\Delta V/\delta p)_T$), and heat capacity at constant pressure ($\Delta C_p = T(\delta\Delta S/\delta T)_p$). An example of pressure and temperature dependence of dT/dp is given in the phase diagram, Figure 12.1. Whereas the pressure dependence of the transition temperature for the gel and the liquid crystalline phase of the saturated phospholipid DSPC is fairly classical, the interdigitated phase boundary is highly curved.

12.2.2 Chemical Reactions

ΔG, the change in Gibbs free energy of a system as a function of temperature and pressure is given by

$$\Delta G = \Delta E + p\Delta V - T\Delta S$$

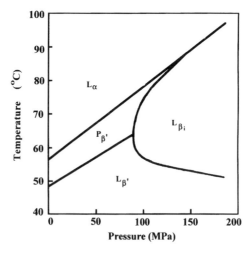

Figure 12.1 Temperature-pressure phase diagram for the phospholipid system DSPC-H_2O in excess water. L_β: Gel phase; P_β: Rippled gel phase; L_α: Liquid-crystalline phase; $L_{\beta i}$: Interdigitated phase. Note the high curvature of the interdigitated phase boundaries. According to [19].

12.2. PRESSURE VERSUS TEMPERATURE EFFECTS

Since the changes in free energy are in general of the order of a few kJ, it is clear that at ambient pressure the term $p\Delta V$ will have a very small contribution to ΔG in condensed phases. For example, for a reaction volume, ΔV, of the order of 82 mL mol^{-1} -quite a large value for a chemical reaction- and $p = 1$ bar, we find $p\Delta V = 8$ J mol^{-1}. Therefore pressures of the order of several 100 MPa are needed to significantly shift the position of chemical equilibria. The relation for the change in equilibrium constant with pressure was first given by Planck:

$$\Delta V = -RT\frac{\delta \ln K}{\delta p}$$

Similar expressions describe the effect of pressure on the reaction rate. From the pressure dependence of the rate constant one obtains $\Delta V^{\neq}$, the activation volume, *i.e.* the partial molar volume change between the transition state and the reactants (see Chapter 2).

It is clear that when thermodynamic or kinetic data are obtained both as a function of temperature and pressure then the activation parameters are both pressure as well as temperature dependent. This may result in *negative* activation energies. This was observed for the first time by Suzuki [13] for the denaturation of ovalbumin and carbonylhemoglobin. Hawley [14] has observed changes in enthalpy for the denaturation of proteins. Zipp and Kauzmann [15] have reported on the effect of pH on the thermodynamics of myoglobin denaturation. Similar effects have also been observed for the inactivation of bacteria by Ludwig *et al.* [16]. Further examples of such effects on the catalytic conversion by enzymes will be given in Section 12.4.3.

12.2.3 Molecular Interpretation of Thermodynamic and Kinetic Parameters

It has been found useful to interpret the observed reaction and activation volumes in terms of intrinsic and solvent effects. Intrinsic effects arise from packing effects, for example the increase in molecular ordering as a consequence of the volume decrease at high pressure, and also from the changes in volume due to the formation or rupture of covalent bonds. The role of the solvent becomes apparent in the non-covalent interactions: electrostatic interactions, hydrogen bonding and hydrophobic interactions.

Because of electrostriction effects, electrostatic interactions become much weaker at elevated pressures. The fact that the dissociation constants of weak acids vary with pressure also has important consequences for the analysis of pressure effects on pH-dependent processes. Closely related is the problem of maintaining a constant pH in high-pressure experiments, since buffers, such as phosphate, have high ionization volumes. Fortunately, amine bases have ionization volumes of almost zero and are used to keep the pH constant over a large pressure range. An interesting example of the effect of pressure on electrostatic interactions in enzymes is the pressure-induced, reversible inactivation of chymotrypsin due to the dissociation of the salt bridge in the active site [17].

Studies on various model systems show that hydrogen bonds are stabilized by high pressure. This results from the smaller interatomic distances in the hydrogen-bonded atoms. The stabilizing effect of pressure on hydrogen bonding may be seen from a comparison of the effect on the intermolecular interactions in hydrogen-bonded versus non-hydrogen-bonded liquid amides [18].

The fact that high pressure stabilizes hydrogen bonds has important consequences for the secondary structures in proteins, such as α-helices and β-structures. The stabilization of hydrogen bonds at high pressure is also the basis for the extreme stability of nucleic acids under pressure. Hydrophobic interactions that play a substantial role in the stabilization of the tertiary structure and in protein-protein interactions, are destabilized under high pressure. On the other hand, stacking interactions between aromatic rings show negative volume changes and are stabilized by pressure.

12.3 PHOSPHOLIPIDS

The primary effect of pressure on phospholipids can be observed in their transition temperatures. Pressure favors the crystalline state, as a result of the Le Chatelier-Braun principle. Whereas the transition temperature of the lipids depends on the length of the hydrocarbon chain, the rate at which the temperature changes with pressure is almost independent of the length, as may be seen from Table 12.1. We also note that cholesterol has no effect. It is of interest that the degree of unsaturation of the hydrocarbon chain lowers the dT/dp values of the lipids. Starting from the phase diagrams of pure phospholipids, one can predict the pressure at which phase transitions occur.

The phase diagram for the unsaturated phospholipid DOPE has been determined by Winter et al. [19]. Two observations are of interest: the pressure dependence of the main transition from the liquid-crysalline (L_β) to the

Table 12.1 Effect of pressure on the main transition temperature (T_m) of some phospholipids. Source: [6].

Phospholipids	T_m (°C)	dT_m/dp (K/100 MPa)
Dilauroyl phosphatidylcholine (C12)	0.5	17
Dimyristoyl phosphatidylcholine (C14)	24	20.5
Dipalmitoyl phosphatidylcholine (C16)	41.5	21.8
Dilauroyl phosphatidylethanolamine (C12)	31	21.5
Dilauroyl phosphatidic acid (C12)	28	20
Dimyristoyl phosphatidyl glycerol (C14)	23	21
Dipalmitoyl phosphatidyl glycerol (C16)	43	22
Dimyristoyl phosphatidylcholine + Cholesterol	23	20

12.3. PHOSPHOLIPIDS

gel-phase (L_α) shows a much flatter slope ($dT/dp = 14\,°C/100$ MPa) than that for the saturated lipids. The transition from L_α to the non-lamellar inverted hexagonal (HII) phase has a much steeper slope ($dT/dp = 40\,°C/100$ MPa). Biological processes in which this phase transition plays a role are thus strongly inhibited at high pressure. Other phases that may develop at high pressure and temperature are the interdigitated phase ($L_{\beta i}$). This phase may be observed in high-pressure neutron diffraction and DTA experiments with DSPC, a saturated phospholipid [20]. The conditions for this phase are given in Figure 12.1. Phase diagrams for lipids may also be determined from high-pressure NMR experiments [11].

High-pressure infrared experiments provide new information on the dynamics of lipid hydrocarbon chains and their interactions with one another. We use the CH_2 bending vibration in DPPC as an example. Under normal conditions the position of the band is a result of the balance between the repulsive and attractive effects. If repulsive effects dominate then a blue shift of the frequency is observed, but if attractive forces dominate, a red shift is observed. This is frequently observed in hydrogen-bonded systems. The situation is more complex when elastic effects occur, such as in conformational changes of the hydrocarbon chains. Wong [21] has shown that as a result of strong intermolecular interactions, splitting of the vibrational bands takes place at high pressures. Such an effect is shown for DPPC in Figure 12.2 [22]. The occurrence of splitting indicates that the conformational disorder and the orientational fluctuations of the chains are dampened. High-pressure infrared spectroscopy provides a rich source of information on the dynamics of molecules.

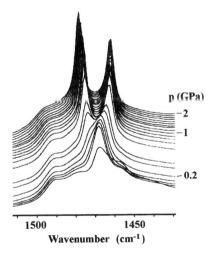

Figure 12.2 Pressure-induced correlation field bandsplitting of the CH_2 bending vibration in DPPC dispersions. The splitting of the bands results from the strong intermolecular interactions. According to [22].

A more complex situation is encountered in biomembranes. It has been found that the physical state of the lipids that surround the membrane proteins plays a crucial role in the activity of membrane-bound enzymes. In addition, the integrity of the membrane of living organisms, such as bacteria, is very sensitive to pressure. This explains the sterilization effect of pressure. It has been observed that the activity of membrane-bound enzymes, such as the Na^+-K^+-ATPase and the Ca^{++}-ATPases, changes in a non-linear fashion with temperature and pressure [23]. There is considerable evidence that the physical state of the lipids controls the activity of the enzyme.

12.4 PROTEINS

The biologically active conformation of proteins and enzymes is the result of the specific folding of a polypeptide chain.

12.4.1 Randomly Coiled Polypeptide Chain: Folded Protein

Formally four levels of organization may be considered. The first level is the sequence of amino acids in the polypeptide chain. There is no report in the literature on the effect of pressure on the covalent bonds, although it is well-known that high temperature may induce chemical reactions in proteins. The second level is that formed by the hydrogen bonds within and between the peptide chains. In general one expects a stabilization of these structures by pressure, but a higher stability is noted for α-helices in some model systems. This may be due to a larger void between the chains in the β-structures. The tertiary level is formed by the specific packing of the secondary structures into a more or less globular shape. This level is stabilized by non-covalent interactions. Pressure is expected to affect these interactions. Several compact structures may assemble to form the quaternary structure. This level is also stabilized by non-covalent interactions. The interactions between the subunits are therefore quite sensitive to pressure [9].

A discussion of the effect of pressure and temperature on proteins and enzymes may be organized within the following simplified scheme of an enzymic reaction:

$$\text{Protein} + \text{Ligand} \rightarrow [\text{P::L}]$$

$$[\text{P::L}] \rightarrow (\text{Active Complex})$$

$$(\text{Active Complex}) \rightarrow \text{Biological effect}$$

Pressure effects are expected to occur at the level of the protein-ligand interaction. If the ligand is a macromolecule, then pressure may affect its conformation making attack by the enzyme easier or more difficult. Pressure may also induce changes in the conformation of the enzyme which may have consequences for

12.4. PROTEINS

the activity. If the enzyme has a quaternary structure, there are also effects on the activity *via* the pressure-induced dissociation of the subunits. Finally, we have to consider the effects on the catalytic process as such. Before we discuss these topics, we will consider the effect of pressure on the compressibility of a protein in solution.

12.4.2 Protein Compressibility

The compressibility of proteins has received considerable attention in recent years in view of its correlation with the volume fluctuations, which, in turn, are related to the conformational dynamics.

$$(\Delta \bar{V})^2 = kTV\beta_T$$

The compressibility is defined as the relative change in the molar volume with pressure. The molar volume of a protein consists of three contributions: the intrinsic volumes of the atoms, the void volume that results from the imperfect packing of the amino acids in the interior of the protein and the volume decrease that results from the hydration of the peptide units and the amino acid residues. Since the molar volume can be calculated from the constitutive atomic volume of the amino acid residues, it is assumed that contributions from the internal cavities, the void volume, and the hydration volume compensate one another.

Secondary structures, such as α-helices and β-sheets, have low compressibilities. The compressibility of protein molecules may reflect the interactions between secondary structure domains. Fluctuations of the protein structure can then be visualized as the movement of secondary structures with respect to one another. The free volume may be considered the source of the conformational dynamics of the proteins. This hypothesis has been tested experimentally by measuring the flexibility of monomeric proteins *via* the tryptophan phosphorescence decay kinetics [24]. A consistent interpretation is given in terms of the opposing effects of the reduction of internal cavities and the increased hydration of the polypeptide.

Sarvazyan [25] has recently reviewed the information available from ultrasonic velocimetric studies performed on biological compounds by his and other groups. The method has been used to follow the change in compressibility of the temperature-induced transition in ribonuclease A [26]. As expected, the high-temperature conformation has a higher compressibility than the low-temperature form.

The compressibility of proteins may also be determined by spectral hole burning spectroscopy. A detailed description of the method is outside the scope of the present chapter. A recent review gives details of the methodology and information on the application to heme proteins, such as myoglobin and horseradish peroxidase [27]. The method is also used to study disordered states in glasses and polymers. Since proteins border between order and disorder, the similarities to and the differences from the glassy state of matter are obvious. The work of Frauenfelder and coworkers on the effect of pressure on protein reactions has given insights into the hierarchical structural disorder in proteins [28]. The advantage of hole burning spectroscopy is that the compressibility of protein molecules can be measured with an optical method. Because of the sensitivity of the method, very small pressures are needed. What is surprising though, is that compressibilities are obtained for proteins that are of the order of those obtained from ultrasonic experiments, even though the hole burning experiments are performed at liquid helium temperatures. This suggests that proteins at room temperature show the behaviour of the glassy state. It is of interest to note that the basic theory which is used to interpret the compressibilities obtained from hole burning spectroscopy is also applicable to absorption and fluorescence spectra at room temperature. The compressibilities that are obtained from the pressure-induced spectral shifts are of a similar order of magnitude [28a].

12.4.3 Protein Reactions

In this section we divide, somewhat artificially, protein reactions into ligand binding, protein-protein interactions, enzyme catalysis and solvent effects. It is important to emphasize the role of non-covalent interactions in these processes.

Ligand-Protein Interactions The binding of carbon monoxide to hemoproteins has been a model system for binding studies for about twenty years. The data of Projahn and van Eldik [29] are a good starting point for an understanding of the work that is still going on in a number of laboratories both on model systems and on heme proteins. It may also be seen from Table 12.2 that the activation volume for the binding of CO is negative, which indicates that the bond formation step is rate-limiting. For oxygen the activation volume is positive, suggesting that the diffusion of the ligand through the protein matrix is rate-limiting.

Whereas these results suggest a simple picture for the reaction between CO, O_2 and myoglobin, the kinetic studies of the binding of azide to sperm whale,

Table 12.2 Activation volumes (mL/mol) for the binding of ligands to different types of myoglobin. Source: [8, 29].

System	$\Delta V^{\neq}$	Rate-limiting Process
Sperm whale + CO	−9	Binding to heme iron
Sperm whale + O_2	+8	Diffusion in protein matrix
Sperm whale + azide	+4	Diffusion in protein matrix
Horse + azide	+11	Diffusion in protein matrix
Dog + azide	+12	Diffusion in protein matrix

12.4. PROTEINS

horse and dog metmyoglobin show a systematic difference between these proteins. The activation volumes are positive indicating rate control from the protein matrix. The difference in activation and reaction volumes for these natural mutants strongly suggests a specific role for the protein. Adachi et al. [30] have studied protein-engineered mutants of human myoglobin. They have concentrated on residues that are part of hydrophobic clusters on the heme distal and proximal sides. They observed a dramatic effect of the Leu(29) mutants on the activation volumes for the binding of CO. Substituting Leu for Ala changes $\Delta V^{\neq}$ from -21 to $+10$ mL/mol.

Lange and coworkers [31] have explored the kinetics of the binding of CO to various reduced hemoproteins by stopped-flow under high pressure (up to 200 MPa). In particular, they studied several varieties of cytochrome P 450, as well as chloroperoxidase and lactoperoxidase, and compared the results with data reported for other hemoproteins. As shown in Figure 12.3, the activation volume is positive for hemoproteins with a sulfur donor as the proximal ligand ($\Delta V^{\neq} = +1$ to $+6$ mL/mol), and negative for those with a nitrogen donor ligand ($\Delta V^{\neq} = -3$ to -36 mL/mol). Furthermore, the transition state volume of the histidine ligand enzymes, but not that of the cysteine ligand enzymes, depends on the solvent composition. This indicates that in the histidine ligand category, protein conformation changes are involved in the reaction.

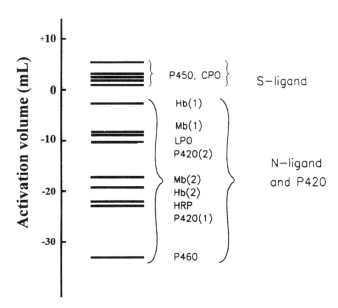

Figure 12.3 Variation of the activation volume for various heme proteins with a sulfur or nitrogen donor ligand at the sixth position of the heme iron, obtained from the kinetics of the CO binding. The negative activation volumes reflect conformational transitions in the proteins. According to [31].

A change in sign of $\Delta V^{\neq}$ between a dissociative and an associative binding mechanism may be predicted from inorganic models for ligand exchange reactions. That is, the rate-limiting step is the dissociation of the sixth ligand or the association, i.e., bond formation, of the incoming ligand. Until now, a dissociative mechanism was not considered for P 450 since the ferrous species was believed to be the pentacoordinated, high-spin form. However, resonance Raman data for several P 450 species show that, at least in the ferric form, the major part of high-spin P 450 is probably hexacoordinated. The sixth rather weakly bound ligand could be a water molecule. It is therefore tempting to speculate that in high-spin, ferrous P 450 the rate-limiting step in the binding of CO is the dissociation of a weakly bound sixth ligand. On the contrary, ferrous, high-spin hemoglobin and myoglobin are pentacoordinated, and the reaction proceeds through an associative mechanism.

Protein-Protein Interactions Protein-protein interactions play a fundamental role in many biological processes and high-pressure studies have contributed substantially to the understanding of the factors that control these interactions [32]. With fluorescence techniques it was shown that in protein dimers the rate of exchange of subunits under pressure approaches the rate of dissociation of the dimers. In tetramers, the situation is different. The rate of subunit exchange is between 5–10 times slower than the dissociation rate. The dissociation of the dimers exhibits the characteristics of classical stochastic chemical equilibria. The dissociation of tetramers is described by the deterministic mechanical equilibria of macroscopic bodies. In extreme cases, the pressure-induced dissociation of these systems is entirely concentration-independent.

Several authors studied various aspects of these processes. Of particular interest is the fact that most of these studies have been performed with optical techniques such as absorption or fluorescence spectroscopy [9]. Quite recently, UV second-derivative spectroscopy of tyrosine side chains located at the interface between the subunits, has been used to follow the pressure-induced dissociation of enolase [33].

Some enzymes show a remarkable resistance to pressure-induced dissociation. By employing high-pressure polyacrylamide gel electrophoresis in capillary gels, Masson et al. [34] have found that the tetrameric form of butyrylcholinesterase does not dissociate up to at least 350 MPa. This unexpected behaviour suggests that the inter-subunit area of this protein is stabilized by either pressure-insensitive interactions, e.g., hydrogen bonds, or pressure-enhanced interactions, such as aromatic-aromatic interactions. The last hypothesis is supported by site-directed mutagenesis experiments.

In appropriate cases, protein-protein interactions may also be studied by observing electron transfer between proteins. Such studies have been performed on the interaction between the quinoprotein methanol dehydrogenase (MDH) and cytochrome c_L (Cyt c_L^{ox}). Both enzymes were isolated from the methylotrophic bacterium *Methylophaga marina 42* [35]. When no conformational changes occur during the reaction, application of Marcus theory predicts quite

12.4. PROTEINS

simple relations between the activation and the reaction volume [36]. The reaction may be represented as follows:

$$\text{MDH}_{\text{sem}} + \text{Cyt } c_L^{ox} \rightleftarrows [\text{MDH}_{\text{sem}} - \text{Cyt } c_L^{ox}] \quad \text{Encounter complex}$$
$$[\text{MDH}_{\text{sem}} - \text{Cyt } c_L^{ox}] \rightleftarrows [\text{MDH}_{ox} - \text{Cyt } c_L^{red}] \quad \text{Electron transfer}$$
$$[\text{MDH}_{ox} - \text{Cyt } c_L^{red}] \rightleftarrows \text{MDH}_{ox} + \text{Cyt } c_L^{red} \quad \text{Successor complex}$$

Surprisingly, a break is observed in the Arrhenius plots at about 15 °C. In addition, the activation volume for the reaction is temperature-dependent as may be seen from Figure 12.4. Since the volume changes for the formation of the encounter complex and the dissociation of the successor complex are expected to cancel each other out, the activation volume may reflect changes in the conformation of the proteins. Independent evidence suggests that MDH may undergo temperature-dependent transitions. In addition, it has been shown recently by high-pressure differential pulse voltammetry and cyclic voltammetry that horse heart cytochrome c exhibits no major volume change during electron transfer [37]. This supports the idea that electron transfer in proteins is a useful tool for studying conformational changes.

Enzyme Reactions The effect of pressure on enzyme reactions has been discussed in detail by Morild [38]. In most cases the enzyme activity is measured either *via* optical signals from the substrate or the enzyme. However, there are

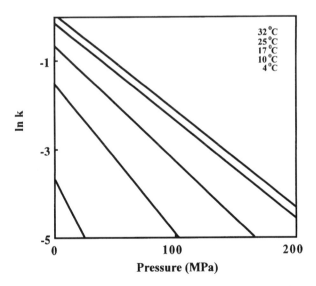

Figure 12.4 Effect of temperature on the pressure dependence of the rate constants for the redox reaction between methanol dehydrogenase and cytochrome c_L from *Methylophaga marina*. The increase in activation volume at low temperature suggests a conformational change in one of the proteins. According to [35].

enzymes which do not show any optical change so that a measure of the activity is not possible without a release of pressure. This approach has been used for the study of aspartate carbamoyltransferase from *Pyrococcus abyssi*, a new deep-sea hyperthermophilic archaeobacterium [39]. This enzyme exhibits a remarkable stability towards high temperature and pressure. It would be of considerable interest to know whether this is a general phenomenon.

We will concentrate further on a number of, at first sight, unusual results. It has been found in a number of instances that the activation energy changes sign under pressure and becomes negative under hydrostatic pressure. One example is the binding of ATP to myosin and the subsequent change in conformation of the protein:

$$\text{Myosin} + \text{ATP} \rightleftarrows \text{Myosin} - \text{ATP} \rightarrow \text{Myosin-ATP*}$$

Experiments were carried out in 40% ethylene glycol to allow experimentation at sub-zero temperatures [40]. The binding of ATP results in perturbations of the tryptophan residues indicating a conformational change. In the temperature range explored, *i.e.*, $+6$ to $-16\,°\text{C}$, the Arrhenius plots are linear at pressures up to 100 MPa, as can be seen from Figure 12.5. The activation enthalpy term is pressure-dependent: $\Delta H^{\neq} = 126$ and -12 kJ/mol at 0.1 and 90 MPa, respectively. At a pressure higher than 80 MPa, a decrease in temperature induces an increase in the velocity of the reaction. Another consequence is that, around $-10\,°\text{C}$, the reaction rate is pressure-independent, at this point $\Delta V^{\neq} = 0$. The

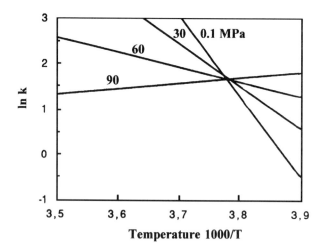

Figure 12.5 Pressure effect on the Arrhenius plots for the reaction of ATP with myosin. It is assumed that temperature and pressure induce different conformational changes in the proteins. According to [40].

12.4. PROTEINS

thermodynamic values obtained fit the Maxwell relationship.

$$\left(\frac{\delta \Delta H^{\neq}}{\delta p}\right)_T = \Delta V^{\neq} - T\left(\frac{\delta \Delta V^{\neq}}{\delta T}\right)_p$$

A similar observation was made for the reduction of hydroxylamine oxidase by hydroxylamine [41]. These findings are consistent with the negative activation energies for the inactivation of enzymes by temperature and pressure [42]. The first indication of such negative activation energies at high pressures was reported for the denaturation of proteins by Suzuki [13]. All these cases are related to the phase diagram for the stability of proteins. Although there is no detailed molecular interpretation, a change in the conformation of the enzyme is assumed. In the case of protein denaturation, it has been established that temperature and pressure induce different conformations.

Solvent Effects In view of the important role of the solvent composition (water activity) in the folding process of proteins, it is more accurate to present the process as

$$Peptide - Solvent + Solvent\text{-}Peptide \rightarrow Folded\ Protein + Solvent$$

Protein conformations are thus the result of a delicate balance between the intramolecular interactions in the polypeptide chain which compete with the solvent interactions.

Timasheff [43] has recently reviewed the stabilizing effect of osmolytes (amino acids, sugars and polyols) on the temperature-induced denaturation of proteins. These compounds are preferentially excluded from the protein surface and thereby affect the binding of water to proteins. Their presence also shifts the conformational equilibrium in macromolecules towards a state with the least amount of bound water. In recent years it has become evident that similar protecting effects take place in a number of other processes, such as protein-nucleic acids interactions [44]. For a number of protein-protein interactions the selection of the cosolvent is determined by its possible secondary effects on the protein [45]. In other cases the effects are quite non-specific, as expected for a purely osmotic effect. Figure 12.6 shows an example of the inactivation of an enzyme by pressure, where various cosolvents protect the enzyme from inactivation.

The possible antagonistic effects of osmotic and hydrostatic pressure have also been studied in the high-spin low-spin equilibrium of cytochrome P 450 [46]. High pressure induces the low-spin form, whereas polyols (not all of those tested, however) induce the reverse reaction. The volume change associated with the spin transition is different in both cases. This suggests that different properties are probed. According to the authors, high pressure probes differences in density, whereas osmotic pressure probes a difference in the number of water molecules associated with the conformation of the protein.

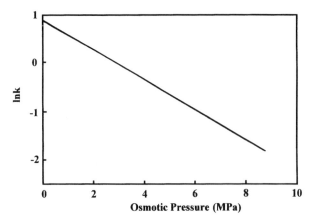

Figure 12.6 Protective effect of the osmotic pressure on the rate of inactivation of the enzyme enolase at 300 MPa. The osmotic pressure was obtained using glycerol, glucose or sucrose. According to [45].

Oliveira et al. [47] have observed that denaturation of the Arc repressor does not occur in the absence of water. Glycerol has been found to stabilize ribonuclease A against pressure denaturation at 0.7 GPa [48].

The solvation of proteins may also be influenced in reversed micelles formed by the anionic surfactant Aerosol OT in octane. The activation volume is dependent on the water content in the system, enabling the enzyme activity to be regulated by pressure. When α-chymotrypsin is solubilized in reversed micelles, high hydrostatic pressure, up to 150 MPa, stabilizes the enzyme against thermal inactivation [49].

12.5 PROTEIN STABILITY

The biologically active structure of proteins is only stable within restricted temperature, pressure and solvent conditions. Outside this range unfolding or denaturation takes place. This may be followed by gel formation. The occurrence of intermediates in the process of unfolding has given impetus to new developments in this field of research.

12.5.1 Pressure versus Temperature Induced Protein Denaturation

At the beginning of this century, the American physicist and Nobel prize winner, Bridgman [2], showed that the appearance of the pressure-induced coagulum of egg white is quite different from that induced by temperature. He also observed that the effect of temperature "seems to be such that the ease (of the pressure-induced coagulation) increases at low temperatures, contrary to what one might expect".

12.5. PROTEIN STABILITY

More detailed investigations by Suzuki [13] and Hawley [14] have shown that these observations can be put together in the phase diagram for the denaturation of proteins. As shown in Figure 12.7, at high temperature pressure stabilizes the protein against temperature denaturation. At room temperature, temperature stabilizes the protein against pressure denaturation. The fact that one can "cook" an egg with pressure is the result of the unique phase diagram of proteins. There are indications that similar phenomena occur in polysaccharides, *e.g.*, the gelatinization of starch [50], certain phospholipids [20] and in more complex structures such as *bacteriophage T4* [51] and *E. coli* cells [16]. We have already indicated that similar diagrams are found for the inactivation of enzymes [42].

The information about the native and the denatured state of the protein that may be obtained, can be derived from the temperature and pressure dependence of the Gibbs free energy difference for the following process:

$$\text{Native (Folded)} \rightleftarrows \text{Denatured (Unfolded)}$$

$$d(\Delta G) = -\Delta S dT + \Delta V dp$$

$$\Delta G = \Delta G_0 + \Delta V_0(p - p_0) - \Delta S_0(T - T_0) + (\Delta \beta/2)(p - p_0)^2$$
$$- (\Delta C_p/2T_0)(T - T_0)^2 + \Delta \alpha(p - p_0)(T - T_0)$$

where the symbols have the same meaning as for the expanded Clausius-Clapeyron equation. The translation of these thermodynamic quantities into molecular terms is one of the main tasks of physical biochemists. Several experimental approaches are available. We will discuss primarily infrared and NMR spectroscopy. Other approaches have been reviewed by Silva and Weber [9].

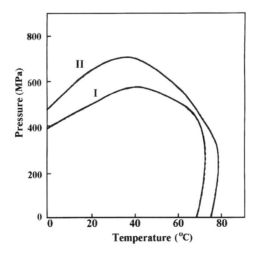

Figure 12.7 Combined effect of temperature and pressure on the rate of denaturation of ovalbumin (I) and carbonylhemoglobin (II). According to [13].

12.5.2 Infrared Spectroscopy Studies with the Diamond Anvil Cell

In the last few years, a high-pressure technique has been developed which allows the observation of biomacromolecules, *in situ*, during the compression and decompression phase. It is a particularly fruitful combination of vibrational spectroscopy and the diamond anvil cell. The application of this technique to the study of biomolecules has been pioneered by Dr. P. T. T. Wong of the NRC, Canada. With this technique pressures of 1 GPa or more can easily be obtained, and it can readily be used with Fourier transform infrared and Raman spectroscopy.

The advantage of Raman spectroscopy is that it can be performed on aqueous samples and no additional preparation of the sample is needed. A disadvantage is that many biological samples show a fluorescence which is much stronger than the Raman effect. In contrast, the infrared technique is an absorption technique. Fluorescence does not interfere, but the strong absorption of water in the region of interest for the protein studies necessitates its replacement by heavy water. We select two cases to demonstrate the power of the infrared technique for the study of protein denaturation. In the case of chymotrypsinogen [52], denaturation starts at about 600 MPa and is complete at 700 MPa. Other proteins are more stable (less pressure sensitive), and it would be of interest to know whether there is a correlation between the stability and the composition in terms of secondary structures. A more detailed analysis of the amide I band of the protein allows a closer look at the molecular events. This is done with a combination of self-deconvolution and band-fitting allows the secondary structures to be followed as a function of pressure. Figure 12.8 gives an example for lipoxygenase [53], an enzyme which plays a role in the biosynthesis of leucotrienes, important bioregulators of inflammation and allergy. The spectrum, whose resolution is enhanced by Fourier self-deconvolution, shows large differences between the temperature- and the pressure-denatured proteins. The bands at around 1620 and 1680 cm^{-1} in the spectrum of the temperature-denatured protein are typical for the intermolecular hydrogen bonds that are formed due to aggregation. However, these bands are absent in the pressure-denatured protein. In all cases, the β-structure plays an important role in the temperature-induced gel formation. There are only very few studies on the effect of pressure-induced gel formation. Pressure favors the formation of a structure which has similarities to the type of gel that is formed by the temperature-induced denaturation. However, different mechanical properties for temperature- and pressure-induced gels have been observed and can be correlated with the infrared data.

12.5.3 High-Pressure NMR Spectroscopy

Major advances in the characterization of structural changes that accompany pressure-induced denaturation in proteins, have been obtained by combining high-resolution NMR techniques with high pressure [11, 54].

12.5. PROTEIN STABILITY

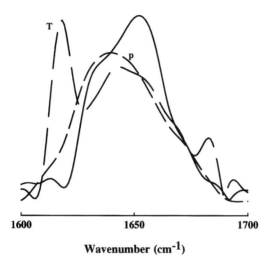

Figure 12.8 Temperature (T) and pressure-induced (p) changes in the infrared spectrum of lipoxygenase. The full line is the spectrum at ambient conditions. The new bands that appear in the temperature-denatured protein are indicative of strong intermolecular hydrogen bonding. Note the absence of these bands in the pressure-denatured protein. According to [53].

By studying the proton resonances of several amino acid residues, the volume changes in different regions of the lysozyme molecule were followed in response to unfolding at high pressure [55]. From the differences in pressure response of these residues, it was established that in lysozyme, unfolding of the β-sheet region is less than the unfolding in α-helices. This is in agreement with the conclusion from X-ray data on the compressibility differences between these regions. Another important finding is that the structural changes which occur during pressure denaturation of lysozyme are distinct from those caused by temperature denaturation of this protein and are smaller in scale. This suggests that the pressure-denatured protein is infiltrated by water. The weakest non-covalent interactions between amino acid residues that support the protein tertiary structure are first destabilized at high pressure and then replaced by protein-water interactions which have a smaller volume. The secondary structure remains essentially unaltered.

Similar conclusions about the structure of pressure-denatured proteins were obtained from ^{1}H-NMR spectra of cold-denatured ribonuclease A at high pressure [11]. By applying pressures of 200 MPa, the freezing point of water is decreased to $-25\,°C$. This approach may be less damaging to proteins than other procedures used to shift the temperature of denaturation experiments to the subzero region, *e.g.*, the application of cryosolvents (like methanol), emulsions of protein solutions in a non-polar solvent or supercooled aqueous solutions.

12.5.4 Intermediates of Denaturation

As suggested in the previous section, several structural features of pressure-denatured proteins, indicate that they resemble the molten globular state. In this state, the proteins retain most of their secondary structure and their hydrodynamic radii are 10–20% greater than those of the native conformations. In addition, patches of hydrophobic residues become solvent-exposed, thus increasing binding of hydrophobic probes and promoting aggregation.

In this section we will discuss some data recently obtained using three different experimental approaches: electrophoresis which monitors the hydrodynamic volume, infrared spectroscopy for examining the secondary structure and fourth-derivative UV spectroscopy which probes the aromatic amino acids in the proteins. Other approaches are: NMR spectroscopy [56]; analysis of adiabatic compressibility data [26] and kinetic data from the pressure-jump relaxation technique [57].

The degree of expansion of the molten globular state can be observed with high-pressure gel capillary electrophoresis [58]. Human butyrylcholinesterase was examined under two conditions. First, the change of the hydrodynamic volume was determined as a function of high pressure by electrophoresis. Figure 12.9 shows the hydrodynamic volume as a function of pressure. It can be seen that at 150 MPa the hydrodynamic volume increases by about 35%. In the presence of 1 M sucrose, the effect is substantially reduced or shifted to a higher pressure. In this case, the occurrence of the molten globular intermediate in pressure-induced denaturation has also been confirmed by the increased binding of the hydrophobic dye ANS to the protein.

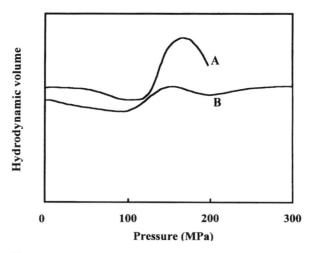

Figure 12.9 Change in the hydrodynamic volume obtained from high-pressure electrophoresis of butyryl cholinesterase, in the absence (A) and in the presence of 1 M sucrose (B). In (A) the hydrodynamic volume increases by 35% at 150 MPa. According to [58].

12.5. PROTEIN STABILITY

In many cases the pressure-induced molten globular state occurs at rather low pressures. In our laboratory, we have obtained evidence that in ribonuclease A an intermediate state with the characteristics of the molten globular state is induced at about 500 MPa [48]. As can be seen in Figure 12.10, the maximum of the amide I band of the infrared spectrum drops to lower frequencies at about 500 MPa before a sudden increase takes place at about 700 MPa. The state between 500 and 600 MPa has the same secondary structure as the native protein. The frequency shift is due to the sudden increase in H/D exchange which is assumed to take place due to swelling of the tertiary structure of the protein. This interpretation was confirmed by repeating the same experiment in 40% glycerol. Under these conditions the pressure-induced molten globular state occurs at about 600 MPa. Since this transition is not seen in the tyrosine side chains and is accompanied by a change in the amide II band, which is absent in the main transition, we consider this good evidence for a pressure-induced intermediate state which has the characteristics of the molten globular state. It should be noted that the pressure (500 MPa) at which the transformation occurs is much higher than for the corresponding transformation which takes place below 250 MPa in the dissociation of the Arc repressor dimer, as observed previously [9]. We also note that the induction of the molten globular state occurs at a higher pressure in the presence of glycerol. This is consistent with the idea that the partial unfolding of the protein is not favored by the presence of the organic cosolvent [43].

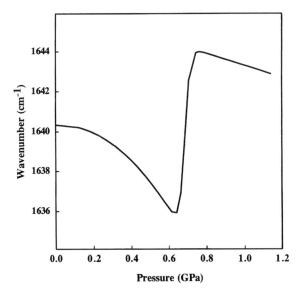

Figure 12.10 Pressure-induced changes of the band maximum of the amide I band of ribonuclease A. The changes below 0.6 GPa are indicative of an intermediate state before the denaturation which takes place at 0.7 GPa. According to [48].

The exposure of the inside of the protein to the solvent may be followed by solvent accessibility to the aromatic side chains of the protein. This may be studied by fourth-derivative UV spectroscopy under high pressure [59]. The selective enhancement of the absorption bands of the aromatic amino acids permits a quantification of the effect of decreasing solvent polarity on the fourth-derivative spectrum. The wavelength of the peak maximum of tyrosine and phenylalanine as well as the amplitude of the tryptophan peak depend linearly on the dielectric constant of the solvent. The only effect of pressure in the 1 to 500 MPa range is a small decrease in the fourth-derivative amplitude. This method appears, therefore, to be a suitable tool for evaluating changes in the dielectric constant in the vicinity of the aromatic amino acids in proteins which undergo pressure-induced structural changes.

This approach also allows one to evaluate the local dielectric constant of ribonuclease and to characterize the thermodynamics of a molten globular-like denaturation intermediate. Figure 12.11 shows the pressure dependence of the fourth-derivative UV spectrum of ribonuclease at low pH. The wavelength shift as well as the change in amplitude suggest that tyrosine residues become accessible to the solvent at high pressure. The combination of the derivative method with the high-pressure technique appears to be particularly useful for characterizing intermediates in the protein unfolding process. This approach also showed that the temperature-induced denaturation of ribonuclease is much more complex than the pressure-induced denaturation.

These examples show convincingly that high-pressure electrophoresis, fourth-derivative UV spectroscopy and infrared spectroscopy are powerful techniques

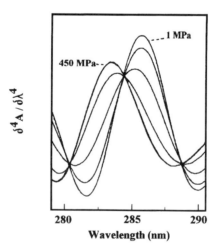

Figure 12.11 Pressure-induced changes in the fourth-derivative UV absorption spectrum of ribonuclease A in the region of the tyrosine side chains. The spectral changes suggest that the side chains become accessible to the solvent. According to [59].

12.5.5 Limits of Extreme Conditions

What is the highest pressure that a protein can resist? This question is not only of fundamental importance, but also has practical implications. There are reports in the literature that some proteins may resist pressures over 1000 MPa while still retaining their biological activity. One of the approaches towards an understanding of this resistance is to study pressure effects on secondary structures in model systems. Studies on poly(L-lysine) reveal that the unordered polypeptide undergoes a reversible pressure-induced change in the α-helix structure at about 900 MPa [60]. The α-helical structure at a pH above 11 and at 4 °C is stable up to 2 GPa. At 50 °C, it forms a β-sheet in the same pH range, which transforms into an α-helix at ca. 200 MPa. In our laboratory we have observed stabilization effects of lipids and detergents on the conformation of hydrophobic peptides, such as gramicidin, incorporated into non-aqueous phases.

Gramicidin is a short polypeptide containing 15 amino acids. It can adopt β-like helices, structures which are not observed in regular proteins. In lipid bilayers, a channel is formed which is assumed to be a dimer of parallel β6.3 helices. This arrangement of the peptide backbone is like a β-sheet folded on a cylindrical surface. In organic solvents, the so-called antiparallel double β5.6 helix is formed. This is a folded, antiparallel β-sheet. The transformation from the double helix to the channel is known to be slow.

Figure 12.12 shows the main bands of gramicidin in DOPC which are tentatively assigned to the double helix and channel structures [61]. The frequency of the main bands decreases with increasing pressure. This suggests that the hydrogen bonds are stabilized by the high pressure. The band at 1624 cm^{-1} shows the largest pressure shift suggesting that this weaker hydrogen bond is more easily compressed. It has been shown by theoretical calculations that the double-helical structure has a small pore so that water cannot enter. The inside represents a considerable free volume, which should be more compressible than the channel form which has a larger pore with easier access for the solvent. Increasing pressure induces a shift from the double-helical form to the channel form. The effect is entirely reversible.

Tortora and coworkers have isolated several enzymes from the extremely thermoacidophilic archaebacterium *Sulfolobus solfataricus*. One class of enzymes, the ribonucleases shows some interesting properties [62]. Ribonuclease P2 (7 kDa) remains active at 80 °C. Complete inactivation occurs at 95 °C. The enzyme has been expressed in *E. coli* and it is now possible to investigate the source of the thermostability by site-directed mutagenesis [63]. Structural studies suggest that an aromatic cluster at the protein center may be responsible for the stability. This is consistent with the extreme pressure stability that has

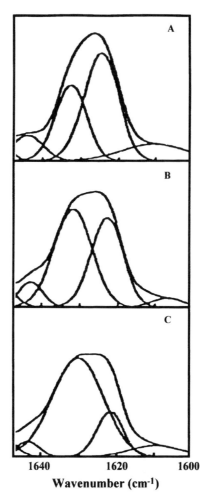

Figure 12.12 Deconvoluted and fitted infrared spectra of gramicidin in DOPC lipids at ambient pressure conditions (A), at 0.9 GPa (B) and at 1.4 GPa (C). The pressure-induced transitions from the double helix to the channel form are reversible. According to [61].

been observed with infrared spectroscopy in the diamond anvil cell as shown in Figure 12.13 [64]. Replacing one of the aromatic residues by an aliphatic side chain reduces the pressure as well as the temperature stability drastically. As far as we are aware, this is the first example of a protein that does not show any pressure-induced changes in solution up to 1.4 GPa. Further research is in progress to determine if there is possibly a correlation between the temperature and the pressure stability of the enzyme.

12.6. MICROORGANISMS

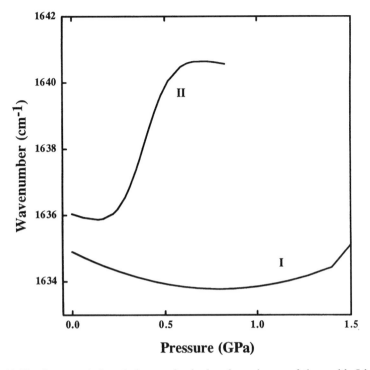

Figure 12.13 Pressure-induced changes in the band maximum of the amide I band of ribonuclease P2 of *Sulfolobus solfactaricus*. No changes take place in the wild-type enzyme (I) up to 1.4 GPa. The decreased stability of a mutant with a single amino acid substitution is shown (II). According to [64].

12.6 MICROORGANISMS

Studies on the adaptation of living organisms to the depths of the oceans have shown that the terms deep and high pressure begin to apply at depths of about 500 m or more. Of particular interest are the studies which concentrate on the characterization of molecules isolated from deep-sea organisms.

It is of particular interest that a correlation has been found between the conditions of stability of proteins and the conditions for the survival of bacterial life. Yayanos [65] has shown that bacteria that thrive in the deep-sea are only capable of surviving under certain conditions of temperature and pressure. A plot of these conditions shows strong similarities with the conditions that are observed for the stability of proteins. More generally, it can be concluded that microorganisms have no difficulties at high temperatures provided that the pressure is not too high (a few hundred bars). As far as high pressure is concerned, life forms seem to prefer these conditions provided the temperature is not too high (below approximately 10 °C). These observations have important implications for the application of high pressure to the food industry. Indeed it has already been observed that far lower pressures are needed to obtain a certain degree of sterilization at low temperatures than at room temperature [16]. The agreement between the trends shown in Figure 12.14, which illustrate the conditions for the inactivation of bacteria, and Figure 12.7, which illustrate the denaturation of proteins, is rather striking.

Using the diamond anvil cell, it is now possible to follow the molecular details of pressure effects on microorganisms with infrared spectroscopy. This approach has found interesting applications in the biomedical sciences [66]. We are using a similar approach in our laboratory to follow, *in situ*, with infrared spectroscopy, the pressure-induced inactivation of *E. coli*. The results correlate remarkably well with the pressure inactivation observed on bacterial cultures [16].

The unusual stability of bacterial spores to pressure treatment is also of interest [67]. The extremely low activity of water may account for this. It is clear that more experimental work, specifically on the effect of solvent conditions on the pressure sensitivity of proteins and organisms, is needed before this conclusion can be accepted as being generally applicabable.

Vibrational spectroscopy is thus a useful tool in the study of cells and cellular components. Molecular details of the resistance of cells and organisms to pressure treatment may therefore be expected from spectroscopic studies.

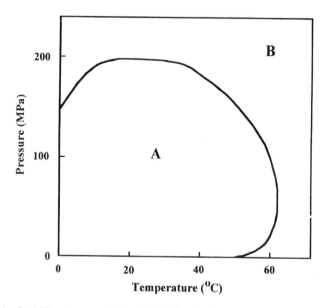

Figure 12.14 Stability diagram of *E. coli*. The line indicates the conditions for a decrease in bacterial count by two orders of magnitude within 5 min. In (A) the cells survive, in (B) they are killed. According to [16].

12.7 INDUSTRIAL APPLICATIONS

High pressure denatures proteins, solidifies lipids and breaks biomembranes, with the consequent killing of bacteria. On the other hand, the observation that pressure treatment does not cause changes in the taste or flavor of food is of special interest to the food industry. Pressure treatment of certain food materials may thus be used as a possible alternative to temperature treatment which has been known since the time of Pasteur. A number of books and reviews have been published and should be consulted for more details [68–71]. Table 12.3 gives a selection of possible applications.

Applications of high pressure technology for the processing of food has a long history. In the United States, Hite and coworkers [1] have made extensive investigations into the possibility of preserving fruit and vegetables by killing microorganisms with high-pressure treatment of about 500 MPa for 30 minutes. The experiments by Bridgman [2] on egg white showed that pressure treatment resulted in coagulation of the white with an appearance like that of a boiled egg. This suggested that microorganisms are killed by the action of pressure on the proteins. The observation that protein denaturation profiles may be correlated with the survival of bacteria strongly suggests that this is the primary mechanism. On the other hand, there remains the distinct possibility that pressure may act upon the colloidal constituents of biological material. This possibility has been suggested previously, but has not been further explored as extensively as the protein denaturation hypothesis. The collodial hypothesis was put forward following the observation that the pressure required to kill bacteria is inversely proportional to their complexity.

Table 12.3 Selected examples of possible applications of high-pressure treatment of food materials. Source: [71].

Microorganisms	Cold "pasteurization"
Chemical reactions	Less than in high-temperature treatment
Proteins	
Egg	Gel formation
Milk	Crystallization of fat/effect on milk coagulation
Fish	Products of novel texture/retained natural aroma
Meat	Novel products when pre-treated with pressure
Enzymes	Selective inactivation
Starch	Gelatinization
Fruit	Pasteurization of citrus juice
Vegetables	Preservation of tomato juice
Cryomethods	High-pressure low-temperature sterilization
	Rapid thawing or freezing

Future developments will determine whether *Bridgmanization*, as the process may be called in honor of the person who established the field of high pressure at the beginning of this century, will be a contribution as vital to biotechnology as the process of *Pasteurization* has been and still is.

12.8 PERSPECTIVES FOR THE FUTURE

The use of pressure as an alternative to temperature treatment, has brought about the need for fundamental studies on the pressure-temperature behavior of macromolecular food constituents such as proteins, lipids and polysaccharides. The mechanisms of protein gelation and the sol-gel behavior of polysaccharides are far from understood. The potential correlation of the results of spectroscopic measurement with those from rheology is certainly a fruitful area of research.

New experimental approaches are needed, however. One new approach that may add considerably to our understanding of temperature and pressure-induced phenomena is the development of molecular dynamics calculations for extreme conditions. Changes in hydration may then be predicted which could be tested experimentally. A recent molecular dynamics simulation on a small protein, bovine pancreatic trypsin inhibitor, has shown that significant changes occur in the protein hydration shell at 1 GPa [72]. No changes in the conformation were detected, however. Recent calculations in our own group have shown that at 1.5 GPa pressure-induced changes in the conformation take place [72a]. No further changes occur at 2 GPa. These results correlate with those from infrared studies performed in our laboratory [72b].

The exploration of the behavior of biomatter under extreme conditions may be compared to mountaineering or exploration of a beach: new horizons unfold before us. The same idea was put in another way by Whitehead: "A fresh instrument serves the same purpose as foreign travel; it shows things in unusual combinations" [73]. And Bridgman has expressed the feeling of a scientist with a new tool superbly: "I am somewhat in the position of a small boy with a new jack-knife who rushes about trying it on every conceivable object" [74].

ACKNOWLEDGEMENTS

Part of the results presented in this review have been obtained in the framework of a COST D6 project (coordinator C. Balny) by R. Ernst and M. Schick (Zürich), J. Frank (Delft), G. Hervé (Paris), G. Hui Bon Hoa (Paris), R. Lange and C. Balny (Montpellier), H. Ludwig (Heidelberg), P. Masson (Grenoble), L. Smeller (Budapest), P. Tortora (Milan) and R. Winter (Dortmund). Research in the laboratory of the author is supported by the Belgian National Science Foundation and the European Union.

12.9 REFERENCES

[1] Hite, B. H., Giddings, N. J., Weakly, C. E., West. Va. Univ. Agr. Expt. Sta. Bull., 146 (1914) 1.
[2] Bridgman, P. W., J. Biol. Chem., 19 (1914) 511.
[3] Johnson, F. H., Eyring, H., Polissar, M. T., The Kinetic Basis of Molecular Biology, John Wiley & Sons, New York 1954.
[4] Hazen, R. M., The New Alchemists. Breaking through the barriers of high pressure, Times Books, New York 1993.
[5] Jaenicke, R., Ann. Rev. Biophys. Bioeng., 10 (1981) 1.
[6] Heremans, K., Ann. Rev. Biophys. Bioeng., 11 (1982) 1.
[7] Weber, G., Drickamer, H. G., Q. Rev. Biophys., 16 (1983) 89.
[8] Heremans, K., (Winter, R., Jonas, J. (eds.)), High Pressure Chemistry, Biochemistry and Material Science, pp 443–469, Kluwer Academic, Dordrecht 1993.
[9] Silva, J. L., Weber, G., Ann. Rev. Phys. Chem., 44 (1993) 89.
[10] Gross, M., Jaenicke, R., Eur. J. Biochem., 221 (1994) 617.
[11] Jonas, J., Jonas, A., Ann. Rev. Biophys. Biomol. Struct., 23 (1994) 287.
[12] Walter, M. L., Science and Cultural Crisis. An Intellectual Biography of Percy Williams Bridgman, Stanford University Press, Stanford 1990.
[13] Suzuki, K., Rev. Phys. Chem. Japan, 29 (1960) 91.
[14] Hawley, S.A., Biochemistry, 10 (1971) 2436.
[15] Zipp, A., Kauzmann, W., Biochemistry, 12 (1973) 4217.
[16] Ludwig, H., Scigalla, W., Sojka, B., (Markley, J. L., Royer, C. A., Northrup, D. (eds.)), Molecular Biophysics and Enzymology, Oxford University Press, 1996, in press.
[17] Heremans, L., Heremans, K., Biochem. Biophys. Acta, 999 (1989) 192.
[18] Goossens, K., Smeller, L., Heremans, K., J. Chem. Phys., 99 (1993) 5736.
[19] Winter, R., Landwehr, A., Brauns, T., Erbes, J., Czeslik, C., Reis, O., (Markley, J. L., Royer, C. A., Northrup, D. (eds.)), Molecular Biophysics and Enzymology, Oxford University Press, 1996, in press.
[20] Winter, R., Pilgrim, W.-C., Ber. Bunseng. Phys. Chem., 93 (1989) 708.
[21] Wong, P. T. T., Biophys. J., 66 (1994) 1505.
[22] Reis, O., Winter, R., Zerda, T. W., Biochim. Biophys. Acta, 1279 (1996) 5.
[23] Heremans, K., Wuytack, F., FEBS Letters, 117 (1980) 161.
[24] Cioni, P., Strambini, G. B., J. Mol. Biol., 242 (1994) 291.
[25] Sarvazyan, A. P., Ann. Rev. Biophys. Biophys. Chem., 20 (1991) 321.
[26] Tamura, Y., Gekko, K., Biochemistry, 34 (1995) 1878.
[27] Friedrich, J., Methods in Enzymology, 246 (1995) 226.
[28] Frauenfelder, H., Alberding, N. A., Anasari, A., Braunstein, D., Cowen, B. R., Hong, M. K., Iben, I. E. T., Johnson, J. B., Luck, S., Marden, M. C., Mourant, J. R., Ormos, P., Reinisch, L., Scholl, R., Schulte, A., Shyamsunder, E., Sorensen, L. B., Steinbach, P. J., Xie, A., Young, R. D., Yue, K. T., J. Phys. Chem., 94 (1990) 1024.
[28a] Jung, C., Hui Bon Hoa, G., Davydov, D., Gill, E., Heremans, K., Eur. J. Biochem. 233 (1995) 600.

[29] Projahn, H.-D., van Eldik, R., Inorg. Chem., 30 (1991) 3288.
[30] Adachi, S., Sunohara, N., Ishimori, K., Morishima, I., J. Biol. Chem., 267 (1992) 12614.
[31] Lange, R., Heiber-Langer, I., Bonfils, C., Fabre, I., Negishi, M., Balny, C., Biophys. J., 66 (1994) 89.
[32] Weber, G., J. Phys. Chem., 99 (1995) 1052.
[33] Kornblatt, J. A., Kornblatt, M. J., Hui Bon Hoa, G., Biochemistry, 34 (1995) 1218.
[34] Masson, P., Clery, C., (Balasubramanian, A. S., Doctor, B. P., Taylor, P., Quinn, D. M. (eds.)), Enzymes of the Cholinesterase Family, Plenum, New York 1996, 113.
[35] Heiber-Langer, I., Clery, C., Frank, J., Masson, P., Balny, C., Eur. J. Biophys., 21 (1992) 241.
[36] Heremans, K., Bormans, M., Snauwaert, J., Vandersypen, H., Faraday Disc. Chem. Soc., 74 (1982) 343.
[37] Sun, J., Wishart, J. F., van Eldik, R., Shalders, R. D., Swaddle, T. W., J. Am. Chem. Soc., 117 (1995) 2600.
[38] Morild, E., Adv. Protein Chem., 34 (1981) 93.
[39] Purcarea, C., Erauso, G., Prieur, D., Hervé, G., Microbiology, 140 (1994) 1967.
[40] Saldana, J.-L., Balny, C., (Balny, C., Hayashi, R., Heremans, K., Masson, P. (eds.)), High Pressure and Biotechnology, pp 529–531, John Libbey/INSERM, Montrouge 1992.
[41] Balny, C., Hooper, A. B., Eur. J. Biochem., 176 (1988) 273.
[42] Heinisch, O., Kowalski, E., Goossens, K., Frank, J., Heremans, K., Ludwig, H., Tauscher, B., Z. Lebensm. Unters. Forsch., 201 (1995) 562.
[43] Timasheff, S. N., Ann. Rev. Biophys. Biomol. Struct., 22 (1993) 67.
[44] Robinson, C. R., Sligar, S. S., Proc. Nat. Acad. Sci. U.S.A., 92 (1995) 3444.
[45] Kornblatt, M. J., Kornblatt, J. A., Hui Bon Hoa, G., Arch. Biochem. Biophys., 306 (1993) 495.
[46] Di Primo, C., Deprez, E., Hui Bon Hoa, G., Douzou, P., Biophys. J., 68 (1995) 2056.
[47] Oliveira, A. C., Gaspar, L. P., Da Poian, A. T., Silva, J. L., J. Mol. Biol., 240 (1994) 184.
[48] Heremans, K., Goossens, K., Smeller, L., (Markley, J. L., Royer, C. A., Northrup, D. (eds.)), Molecular Biophysics and Enzymology, Oxford University Press, New York 1996, in press.
[49] Rariy, R. V., Bec, N., Saldana, J.-L., Nametkin, S. N., Mozhaev, V. V., Klyachko, N. L., Levashov, A. V., Balny, C., FEBS Letters, 364 (1995) 98.
[50] Thevelein, J. M., van Assche, J. A., Heremans, K., Gerlsma, S. Y., Carbohydrate Research, 93 (1981) 304.
[51] Gross, P., Ludwig, H., (Balny, C., Hayashi, R., Heremans, K., Masson, P. (eds.)), High Pressure and Biotechnology, pp 57–59, John Libbey/INSERM, Montrouge 1992.
[52] Wong, P. T. T., Heremans, K., Biochem. Biophys. Acta, 956 (1988) 1.
[53] Goossens, K., Rubens, P., Smeller, L., Frank, J., Heremans, K., (1996) in preparation.

12.9. REFERENCES

[54] Schick, M., Ernst, R. R., (1995), personal communication.
[55] Samarasinghe, S., Campbell, D. M., Jonas, A., Jonas, J., Biochemistry, 31 (1992) 7773.
[56] Peng, X., Jonas, J., Silva, J. L., Biochemistry, 33 (1994) 8323.
[57] Vidugiris, C. A. J., Markley, J. L., Royer, C. A., Biochemistry, 34 (1995) 4909.
[58] Clery, C., Renault, F., Masson, P., FEBS Letters, 370 (1995) 212.
[59] Lange, R., Bec, N., Mozhaev, V. V., Frank, J., Eur. Biophys. J., (1996) in press.
[60] Carrier, D., Mantsch, H. H., Wong, P. T. T., Biopolymers, 29 (1990) 837.
[61] Smeller, L., Goossens, K., Heremans, K., Vibrat. Spectrosc., 8 (1995) 199.
[62] Fusi, P., Tedeschi, G., Aliverti, A., Ronchi, S., Tortora, P., Guerritore, A., Eur. J. Biochem., 211 (1993) 305.
[63] Fusi, P., Grisa, M., Mombelli, E., Consonni, R., Tortora, P., Vanoni, M., Gene, 154 (1995) 99.
[64] Goossens, K., Fusi, P., Tortora, P., Heremans, K., (1996), in preparation.
[65] Yayanos, A. A., Proc. Nat. Acad. Sci. U.S.A., 83 (1986) 9543.
[66] Wong, P. T. T., Lacelle, S., Yadzi, H. M., Applied Spectroscopy, 47 (1993) 1830.
[67] Sojka, B., Ludwig, H., Pharm. Ind., 56 (1994) 660.
[68] Balny, C., Hayashi, R., Heremans, K., Masson, P. (eds.), High Pressure and Biotechnology, John Libbey/INSERM, Montrouge 1992.
[69] Mozhaev, V., Heremans, K., Frank, J., Masson, P., Balny, C., Trends in Biotechnology, 12 (1994) 493.
[70] Ledward, D. A., Johnston, D. E., Earnshaw, R. G., Hasting, A. P. M., High Pressure Processing of Foods, Nottingham University Press, Nottingham 1995.
[71] Tauscher, B., Z. Lebensm. Unters. Forsch., 200 (1995) 3.
[72] Kitchen, D. B., Reed, L. H., Levy, R. M., Biochemistry, 31 (1992) 10083.
[72a] Goossens, K., Smeller, L., Frank, J., Heremans, K., Eur. J. Biochem., 236 (1996) 254.
[72b] Wroblowski, B., Diaz, J. F., Heremans, K., Engelborghs, Y., Proteins, Structure, Function and Genetics (1996) in press.
[73] Whitehead, A. N., Science and the Modern World, Reprinted (1985) by Free Association Books, 1926.
[74] Bridgman, P. W., J. Franklin Inst., 200 (1925) 147.

INDEX

A

ab initio calculations   148 f
ab initio SCF methods   259
acidophiles   516
acoustic cavitation, see cavitation
activation, C–H   26
activation energies, negative   519
activation volume   66–68
   analysis   132
   relationship to ring size   150–155
   solvent dependence   147
active metals   354
addition reactions
   at high pressure   80 f
   sonochemistry   344
aldehydes, activated   164
aldol reactions   178
alkali metals   15 f
   vaporization   17
alkaline earth metals   15 f
alkaliphiles   516
alumina, microwave heating   487
amines, supercritical   211
antisolvents   202
Arc repressor   535
Archaea   517
aspartate carbamoyltransferase   528

B

bacteria, inactivation   519, 540
Barbier reaction   358
barophiles   516
biomass, conversion in sc water   307 f
biomembranes   522
biomolecules
   extreme conditions   517–542
   infrared spectroscopy studies   532
bis(arene) metal complexes   18, 26
Blake threshold   325
bridgmanization   542
bullvalene, Cope rearrangement   143
butyrylcholinesterase   534

C

Ca atoms   18
Ca bulk metal   19
carbon-carbon bond formation, sonochemistry   341
carbon tetrachloride, sonolytic degradation   451–455
carbonyl groups, reduction   356
carbonylhemoglobin   519
carboxy-inversion, in sc fluids   238
catalysis
   chiral   138
   microwave dielectric heating   509
   sonochemically assisted   368–371
catalysts, modified   264
cavitation   325, 384
   electrohydraulic   430–439
   factors affecting   385
cavitation bubble   326
cavitation effects   326–329
ceramic materials, microwave synthesis   506
cheleotropic reactions   122–124, 154
chemical analysis, microwave dielectric heating   505

chemical equilibria
  effect of pressure   106
chemical vapour deposition   201
chemically induced dynamic nuclear polarization   143
chloroperoxidase   525
chlorophenol, sonochemical degradation   461
chloroprene, thermal dimerization   128
chymotrypsin   519
α-chymotrypsin   530
chymotrypsinogen, denaturation   532
Claisen rearrangements   141f
Clausius-Clapeyron equation   517f, 531
cluster compounds   26
Co(II) complexes, reversible binding of dioxygen   89
CO_2
  homogeneous hydrogenation   262
  supercritical properties   275
  phase diagram   221
cocondensation procedure   11, 35
cocondensation reactions   9
co-deposition   42
co-evaporation   42
coke formation   231
colloids, preparation   2
competitive reactions   125–141
composite materials   388
compressibility, of proteins   523
concentration scales   66
condensation polymerization   404f
conducting organic polymers   505
Cope rearrangement   141f, 150
core shell structures   42
corrosion   294
Cottrell equation   241
coupling reactions, sonochemistry   358–360
critical data   190
critical density   190
critical point   221
critical pressure, activation volumes   227
critical temperature   190
cross-over temperature   199
cup-horn reactor   467
cycloaddition reactions   111–124
cycloadditions, 1,3-dipolar   120–122, 154, 177
  effect of solvents   137
  stereoselectivity   134
[2+2] cycloadditions   118, 175
[6+4] cycloadditions   122

[8+2] cycloadditions   122
cyclodimerizations, activation volumes   126–128
cytochrome c, ruthenated   91
cytochrome P450   525, 529

D

Damkoehler number   468, 471
Debye equations   485
decaffeination, supercritical   209
deep-sea organisms   539
deep-well technology   295
depolymerization reactions   237
Dewar benzene, ring-opening   146
diamond anvil cell   532, 538, 540
dichlorocarbene   454
dielectric polarization   482–487
Diels-Alder reactions   111–118, 132
  in supercritical fluids   226–230
  intramolecular   147
  mechanism   125f
  volume data   112–115
  under high pressure   169–178
homo-Diels-Alder reactions   117f
retro-Diels-Alder reactions   116
diffusion coefficients, in sc fluids   234
diffusion restrictions   234
diffusion-controlled processes   110
dilatometry   107
dimerization, in supercritical fluids   238
dimethyl sulfide exchange   79
dimethylformamide, from sc CO_2 262
dipolar polarization   483
diradical intermediates   117
Drude-Nernst equation   109f
dry media   509
dyeing process   278

E

EHD plasmas   462–466
  ozone process   464
  see also electrohydraulic discharge
Eigen-Wilkins mechanism   71
electrochemical etching   210
electrochemical measurements, at elevated pressure   92
electrochemical reactions, in sc fluids   241
electrochemistry, and ultrasound   405
electrohydraulic cavitation   430–439

INDEX 549

electrohydraulic discharge process 441–443
electron beam heating 7, 24
electron paramagnetic resonance spectroscopy (EPR) 253
electron transfer reactions 17, 91, 93
  at high pressure 81–86
electron transfer, in proteins 527
electrostriction 109, 117
electrostriction effects 519
element vapors 6
elements, for high temperature synthesis 2
elimination reactions, at high pressure 80 f
El'yanov equation 111
emulsion polymerization 400
enantioselectivity, pressure-induced increase 138–140
ene reactions 147–150, 177
enolase 526
enzymatic reactions
  effect of pressure 527
  in sc CO_2 277
  in sc fluids 243–254
epoxy resins 505
equilibrium perturbation methods 58
esterification, in sc fluids 238–243
Etard reaction 351
ethylene, supercritical 281
Eyring transition state theory 129

F

Fe-Mg alloy, nano-particles 42
Fischer-Porter vessel 497
Fischer-Tropsch reactions 39
Fischer-Tropsch synthesis, in sc fluids 230–233
  technological 232
flash photolysis 87, 197
flow methods 58
flow-reactors 212
fluorocarbons, solubility in sc CO_2 204
fluoroform, supercritical 251
fluoroptic thermometry 500 f
food, pressure treatment 541
force-field calculations 147
FT-IR-spectroscopy, at high pressure 228
fullerenes
  macroscale synthesis 13, 22
  synthesis 15

furan
  cycloadducts 165
  Diels-Alder reaction 134

G

GaN 68
gas antisolvent crystallization (GAS) 279
gas evaporation method 13 f, 38
gas pressure thermometer 500
GAS process 277
gaseous atoms, heats of formation 4
gases, solubility 205
gel permeation chromatography 383
gramicidin 537
graphitic shells 43
Grignard reagents 18
G-value 466

H

halophiles 516
Harwell reactor 468
hemoproteins, binding of CO 524 f
heterogeneous catalysis 32
  in sc fluids 230–233
heterogeneous isomerization catalysis 231
heterogeneous reactions, sonochemistry 343–353
hexachlorocyclopentadiene, Diels-Alder reactions 117
hexane, supercritical 233
high pressure
  application 164
  generation 104
  in organic synthesis 166
  infrared experiments 521
  instrumentation 55–68
  NMR experiments 59–61, 75, 78, 521
  processing of food 541
  synthesis 55 f
  temperature-jump 78
  see also pressure
high-pressure cell 57, 104
high-pressure chemistry
  basic principles 55–68
  kinetic investigations 56–68
  volume considerations 65
high-pressure experiments 104
high-pressure kinetics 73, 93
high-pressure reactions 68–92
high temperature, see ht

hole burning spectroscopy 524
homogenous reactions, sonochemistry 332
HPNMR, see high pressure NMR
ht chemistry
  development 3
  experimental techniques 6
  group-13–16-elements 19
  transition elements 24–37
ht contact arc process 23
ht species
  applications 14–44
  generation 3–6
  isolation 6
  subvalent 14
ht synthesis
  kinetic control 3
  lanthanide elements 37 f
  main group elements 15–24
  thermodynamic control 3
hydrido sandwich complexes 26
hydroboration 180
hydroboration reaction, sonochemistry 342
hydrocarbons
  extraction 284 f
  supercritical light 280
hydroformylation, thermal 207
hydrogen bonds, at high pressure 520
hydrophobic interactions 520
hyperthermophiles 516

I

impregnation 210
industrial waste 282
inorganic reactions, in supercritical fluids 203
intercalation reactions, microwave synthesis 506
interfacial polarization 486
„inverse Arrhenius" behaviour 395
inverse emulsion polymerization 237
iodide, sonochemical autoxidation 458–460
IR spectroscopy, nanosecond time-resolved 197
isophorone, photodimerization 254
isoprene, Diels-Alder reaction 110 f

K

kinetic analysis 104
kinetics, under high pressure 56–68
Kirkwood equation 117

L

lactoperoxidase 525
lanthanide catalysts 40
lanthanide elements, ht synthesis 37 f
late transition elements 28
Li atoms 18
ligand-protein interactions 524 f
ligand substitution reactions 75
  at high pressure 70
  in square planar complexes 79
lipoxygenase, infrared spectrum 533
liquids, compressed 106

M

macrocycles, high-pressure synthesis 155
macroscale synthesis 2
magnesium activation 357
magnetic polarization 482
magnetron 488 f
magnetron power circuit 489 f
Mannich reaction 184
Marcus-Hush theory 81 f, 85, 93
Maxwell-Wagner-effect 486
melting points, pressure-induced increase 105
Menshutkin reaction 181
metal atom solution 35
metal clusters, ligand stabilized 37
metal hydrides 68
metal-ligand aggregation 32
metal-metal aggregation 32
metal nitrides 68
metal vapor synthesis 28
metal vapors 2
metallic particles, ultrafine 42
metals
  sonochemical activation 354
  vaporization 13
methanol dehydrogenase 526 f
methyl formate, synthesis in sc CO_2 261f
Methylophaga marina 526
Mg atoms 18
micelles, reverse 205
Michael addition 177
microelectrodes 203

microorganisms 539 f
　extreme conditions 516
　pressure effects 540
microwave dielectric heating 481, 502–509
　catalysis 509
　chemical analysis 505
　high pressure effects 504
　organic synthesis 503
　organometallic and coordination compounds 504
　polymer chemistry 505
　solid state reactions 507–509
　solvents 503
microwave fields, high power 499
microwave frequency, field oscillations 482
microwave ovens 487–494
microwave radiation 502
microwave source 495
microwave spectroscopy 480 f
microwave syntheses
　ceramic materials 506
　intercalation reactions 506
microwave vessels 496–499
microwaves
　conducting effects 486
　health considerations 502
　interactions with matter 480–487
　penetration depth 483
　relaxation frequencies 485
Mo oxides 26 f
MODAR process 295
molar volume 107
molecular cluster compounds 31
molecular clusters, macroscale formation 36
molecular rearrangements 141–147
mononuclear complexes 24
monosaccharides, syntheses 164
Monte-Carlo simulations 132 f, 137
myoglobin
　binding of CO 89
　denaturation 519
　mutants 525

N

nanocomposites 44
nanocrystals 44
nanoparticles
　platinum group metal 39
　with special magnetic properties 42
nanopowders 279 f

nanostructured materials 38
nanotubes 23
nitro-aldol reactions 179
nitrogen alkylation, sonochemical 349
p-nitrophenol, degradation 449 f
p-nitrophenolphenyl acetate, sonolytic degradation 451
NMR spectroscopy
　high-pressure 59–61, 75, 78, 532 f
　supercritical 207
　water exchange 71
non-vinyl monomers, sonochemical polymerization 401–412
N_2O, supercritical 201
nucleic acids, under pressure 520
Nylon-6 402

O

Onsager equation 486
organic chemistry, in sc H2O 210
organic reactions, in dry media 509
organic solvents, freezing pressures 168
organic substances
　hydrothermal combustion 297
　partition coefficients 288
　seperation from aqueonus solutions 287–289
organic synthesis
　under high pressure 166
　using microwave dielectric heating 503
organolithium species 399
organometallic reactions, in supercritical fluids 203
organometallic sonochemistry 353
organosamarium chemistry 37
organozinc reagents 362
OSET reactions 82, 93
ovalbumin 519
overpressure, in sonochemistry 331
oxide ceramics, single phase 44
oxygenation reactions 89 f
ozone, photolysis 464

P

packing coefficient 130
Parr vessel 497
pasteurization 542
Peng-Robinson equation 239
pentachlorophenol, sonochemical degradation 460 f
pentane, supercritical 201

pericyclic mechanism 122
pericyclic reactions, activation volumes 129
pericyclic rearrangements 141f
peroxide initiator 398
PGSS process 277, 280
phase transfer catalysis 214, 342
phenol, sonochemical destruction 460f
phenylhalogenocarbenes 124
phospholipid system 518
phospholipids, transition temperatures 520
photochemical investigations, in supercritical fluids 254
photolysis, at high-pressure chemistry 87
photosubstitution 87
piezoelectric materials 323
piezophiles 516
pill box 57
piston-cylinder high pressure apparatus 168
platinum group metal, nanoparticles 39
polarization
  dielectric 482–487
  dipolar 483
  interfacial 486
poly(organosilanes), Wurtz coupling 409–412
polymer chemistry, microwave dielectric heating 505
polymer sonochemistry 385f
polymer surfaces, sonochemical modification 390
polymer synthesis
  sc fluids 280
  ultrasound 387
polymerization reations, in supercritical fluids 233
polymerization
  by ionic mechanisms 399f
  by ring opening 402–404
  electrochemically promoted 405–407
  experimental conditions 396
  metal- atom induced 41
  radical initiated 394
  sonochemical 394–396, 399
polymers
  conducting 382
  fragmention by ultrasound 388f
  tacticity 397f
  ultrasonic degradation 413–419
polyolefins, by Ziegler polymerization 421
polyphenylenes, Grignard coupling 408f

polysaccharides, by sonification 392
ports, microwave oven 491–494
potassium vapor 16
power ultrasound 384, 413
pressure
  and chemical equilibria 104
  effect on free energy of reaction 167
  effect on organic liquids 167
  effect on product ratio 125–141
  effect on reaction equilibrium 167
  effect on stereoselectivity 167
  in macrocycle synthesis 155
  increase of enantioselectivity 138–140
  SI units 55
  see also high pressure
probe-type reactor 467
product ratio, pressure-dependence 125–141
propane, supercritical 233
protein compressibility 523–530
protein conformations 529
protein denaturation 530f
protein folding, solvent effects 529
protein gelation 542
protein-protein interactions 520, 526
protein reactions 524f
protein stability 530–538
proteins 522–530
  ^{1}H-NMR spectra 533
  denaturation 519
  electron transfer 526
  intermediates of denaturation 534–537
  limits of extreme conditions 537
  molar volume 523
  molten globular state 534
  solvation 530
psychrophiles 516
Pt-Sn system 40
pulse-radiolysis techniques 90f
pulsed-plasma discharges 439–449
Pyrococcus abyssi 528

R

radiation-induced methods 58
radiation-induced processes, at high pressure 86–92
radicals, sonochemical initiation 391–394
radiolysis techniques, high-pressure chemistry 87
Raman spectroscopy 532

INDEX 553

Raney nickel 369
rate constants, pressure-dependence 111
Rayleigh-Plesset equation 431
reaction equilibrium, effect of pressure 167
reaction kinetics, effects of SCF 227
reaction suppression system 237
reaction temperature, in sonochemistry 331
reactions
   at elevated pressures 68–92
   cryptocritical 212
   rapid 58
   sonochemical 330–332
Reformatsky reaction 364
residual carbon contents 285
resistive heating 7, 24
RESS process 201, 277, 279
retrograde precipitation 199
rhodium nanoparticles 39
ribonuclease A 533, 536
ribonuclease P2 537 f
ring enlargements 154
ring-closure reactions 87 f
ring-size, relationship to reaction volume 150–155
Roche reactions 212
rotating reactor systems 11
rotations, pressure-independent 150

S

sacrificial metal 42
scaling-up, in sc fluids 212–214
SCF, see supercritical fluids
SCWO conditions 292
SCWO processes
   design and development 298 f
   kinetic data 294
   salt formation 293 f
SCWO reactors 294
SCWO technology
   commercial use 306
   present state 307
secondary orbital interactions 117
self-exchange reactions 82 f
SFC, see supercritical fluid chromatography
SFE, see supercritical fluid extraction
SFE/oxidation process 304
sigmatropic hydrogen shift 147
single bonds, reduction 356
singlet carbene 124
singlet oxygen 149

size-exclusion chromatography 234
SMAD catalsts 39 f
SMAD reactor 12
SMAD technique 41
soil, decontamination in sc water 307
soil extraction 281f
sol-gel behavior 542
solid state reactions, microwave dielectric heating 507–509
solid-fluid extractions 275
solids
   compressed 106
   dielectric losses 486
   microwave heating 486
solvated iron atoms 35, 41
solvated metal atom dispersion, see SMAD
solvated metal atoms 39
solvent density 197
solvent engineering 251
solvent exchange 93
   at high pressure 70
   ligand substitution 80
   on lanthanide ions 79
   organometallic systems 80
   volumes of activation 74
sonication, during catalysis 369
sonochemical polymerization 394–396
sonochemical reactions 330, 332–371
sonochemistry 317–371
   „G" value 458
   addition reactions 344
   alkali metals and magnesium 355
   autoxidation of iodide 458–460
   carbon-carbon bond formation 341
   catalyst preparation 368 f
   coupling reactions 358–360
   equipment 319 f
   formation of anions 348
   formation of bubbles 384
   heterogeneous reactions 343–353
   homogenous reactions 332
   hydroboration reaction 342
   in polymer synthesis 382
   industrial acceptance 420
   laws 329 f
   organometallic 353
   reactions in water 435, 460–462
   reduction and oxidation reactions 350
   solvent 332
   substitution reactions 347
   theoretical aspects 325–333
   zinc 361
   see also ultrasound
sonoisomerization 335

sonolysis 334 f
  in water 449–462
sonoprocessing 387
sound transmission 324
spectral hole burning spectroscopy 523
stacking interactions 520
static reactor systems 8
Stoddart synthesis 171
Stokes' theorem 485
stopped-flow spectrophotometry 58–60, 82
styrene, polymerization in sc fluids 234
substitution reactions
  sonochemistry 347
  thermal 87
subvalent compounds 20
subvalent particles 3
Sulfolobus solfataricus 537
supercritical amines 211
supercritical chemistry
  carboxy-inversion reactions 238–243
  depolymerization reactions 237
  enzymatic reactions 243
  esterifications 238–243
  monitoring 194
  reaction vessels 194
  scale up 212–214
supercritical CO_2
  activation 208–210
  critical data 277
  solubility behavior 283
  solvent power 199
  technical applications 277
  see also CO_2
supercritical conditions 231
supercritical equipment 191 f
supercritical experiments 192 f
supercritical flow-reactor 195
supercritical fluid chromatography 201–203
supercritical fluid extraction
  hydrocarbons 284 f
  in environmental technology 282
supercritical fluid nucleation 279
supercritical fluids
  applications 276
  as reactants 257–263
  as solvent for electrochemistry 203
  definition 190
  density 222
  environmental technology 281
  experimental techniques 192–195
  historical survey 274
  history 191 f
  in food and pharmaceutical industries 277 f
  in textile industry 278
  modifiers 199
  organic chemistry 226–230
  phase diagram 191
  polymer production 280
  properties 223
  solubility of gases 205–208
  thin film formation 210
  tunable solvent power 201
  unusual solubilities 204
  viscosity 210, 222
  voltammetric investigations 244–249
supercritical fluoroform 251
supercritical photochemistry 195
supercritical reactions
  role of the solvent 198
  scale-up 212–214
  see also supercritical fluids
supercritical solubility 199
supercritical solvents 198–208
  environmental concerns 192
  properties 195–197
supercritical synthesis 206
supercritical water 289–309
  as an extraction medium 290
  conversion of biomass and waste 307
  decontamination of soil 307
supercritical water oxidation, see SCWO
suspension polymerization 400

T

teflon autoclaves 497
temperature measurement, microwave vessel 499
tetracyanoethene, [2+2] cycloaddition 119
tetralin, supercritical 212
thermal equilibration, in high-pressure chemistry 62
thermophiles 516
thermoplastics 382
  ultrasound welding 388
thin film formation 201
Timms metal atom reduction technique 18
total organic carbon 303
transducers
  liqid-driven 321 f
  magnetostrictive 322
  piezoelectric 323
transient plasma formation 446–449

INDEX

transition elements, high-temperature chemistry 24–37
transition metal catalysis, in sc fluids 257–263
transition metal complexes, activation volumes for formation 76
transition state theory 93
transition states, in high-pressure chemistry 65
transmetallation 360
trinitrotoluene, sonochemical degradation 456–458
triple point 221
tropone 122
tungsten oxides 26 f

U

Ullmann coupling 366
ultrafine particles 13, 38
ultrasonic energy, transmission 324
ultrasonic frequency 318, 331
ultrasonic irradiation 329
  see also sonochemistry
ultrasonic power 331
ultrasound intensity 385
ultrasound
  as a diagnostic tool 387 f
  biological effects 462
  combination with electrochemistry 405
  condensation reactions 404 f
  degradation of polymers 413–419
  dispersal and encapsulation 388 f
  emulsion polymerization 400 f
  formation of polysaccharides 392
  generation 319–325
  high frequency 384
  irradiation of water 430–439
  modification of polymer surfaces 390
  polymer processing 387
  power 319, 430
  reactor design 467–471
  ring opening reactions 402
  synthesis of vinyl polymers 390–401
  welding of thermoplastics 388
  Ziegler-Natta polymerization 407 f
  see also sonochemistry
UV second-derivative spectroscopy 526
UV/Vis spectroscopy 58

V

van der Waals volumes 116, 130 f
vanadium vapor deposition 26
Vaska's compound 81
vibrational spectroscopy, study of cells 540
vinyl polymers, sonochemical synthesis 390–401
voltammetric investigations, supercritical fluids 244–249
volume contraction 110
volumes of activation 65, 93, 106 f
  solvent exchange 74
vulcanization 401

W

waste
  in thermal processes 286
  SFE 304
waste destruction processes 290
water
  critical parameters 210
  dielectric properties 483 f
  presence in sc fluids 209
  sonolytic reactions 449–462
  supercritical 276, 289–309, 432
  ultrasonic irradiation 430–439
water exchange 72, 78
Wittig reaction 184
Wittig-Horner reaction 341
Wittig-Horner reagent, sonochemical formation 345
Wurtz coupling 18
  poly(organosilanes) 409–412

X

xylene thermometer 499

Z

Ziegler-Natta polymerization 407 f
zinc
  in Barbier type processes 364
  sonochemistry 361